Universitext

Universitext is a series of textbooks that presents material from a wide variety of mathematical disciplines at master's level and beyond. The books, often well class-tested by their author, may have an informal, personal even experimental approach to their subject matter. Some of the most successful and established books in the series have evolved through several editions, always following the evolution of teaching curricula, into very polished texts.

Thus as research topics trickle down into graduate-level teaching, first textbooks written for new, cutting-edge courses may make their way into *Universitext*.

More information about this series at http://www.springer.com/series/223

Antoine Chambert-Loir

(Mostly) Commutative Algebra

 Springer

Antoine Chambert-Loir ⓘD
Institut de mathématiques de Jussieu
Université de Paris
Paris, France

ISSN 0172-5939   ISSN 2191-6675   (electronic)
Universitext
ISBN 978-3-030-61594-9   ISBN 978-3-030-61595-6   (eBook)
https://doi.org/10.1007/978-3-030-61595-6

Mathematics Subject Classification: 13C, 13B

This Springer imprint is published by the registered company Springer Nature Switzerland AG
The registered company address is: Gewerbestrasse 11, 6330 Cham, Switzerland

Mais je ne m'arrête point à expliquer ceci plus en détail, à cause que je vous ôterais le plaisir de l'apprendre par vous-même, et l'utilité de cultiver votre esprit en vous exerçant...

René Descartes (1596–1659)
La géométrie (1638)

Preface

This book stems out of lectures on commutative algebra given to 4th-year university students at two French universities (Paris and Rennes). At that level, students have already followed a basic course in linear algebra and are essentially fluent with the language of vector spaces over fields. The new topics to be introduced were arithmetic of rings, modules, especially principal ideal rings and the classification of modules over such rings, Galois theory, as well as an introduction to more advanced topics such as homological algebra, tensor products, and algebraic concepts involved in algebraic geometry.

Rewriting the text in view of its publication, I have been led to reorganize or expand many sections, for various reasons which range from giving interesting applications of a given notion, adopting a more natural definition which would be valid in a broader context, to calming the anxiety of the author that "this is not yet enough". For example, the "(Mostly)" of the title refers to the fact that basic notions of rings, ideals, modules are equally important in the context of non-commutative rings — important in the sense that this point of view is definitely fruitful in topics such as representation theory — but was initially motivated by the (conceivably naive) desire of being able to say that a structure of an A-module on an abelian group M is a morphism of rings from A to the ring of endomorphisms of M, the latter ring being non-commutative in general.

This makes the present book quite different from classic textbooks on commutative algebra like ATIYAH & MACDONALD (1969), MATSUMURA (1986), JACOBSON (1985), or EISENBUD (1995). I am certainly less terse than the first two, and won't go as far as the last three. Doing so, I believe that this book will already be accessible to younger (undergraduate) students, while giving them an idea of advanced topics which are usually not discussed at that level.

This book has also not been written as an absolute treatise (such as the volumes BOURBAKI (1989a, 2003, 2012) on general algebra and BOURBAKI (1989b) on commutative algebra). While I have taken the time to explain all basic definitions and elementary examples, its reading probably presupposes some habit of abstract algebra. Sometimes, it even assumes some understanding of parts which are discussed in detail only later, but which certainly were the object of former study by the reader. Consequently, and this is especially true for the first chapters, the reader may be willing to skip some sections for a first reading. The 450 pages of the book don't make it fit for a 15-week course anyway. Another hope is that the reader will enjoy more advanced results of each chapter at the occasion of another visit.

The main protagonists of the book are rings, even *commutative rings*, viewed as generalized numbers in which we have an addition and a multiplication. The first chapter introduces them, as well as the language that is needed to understand their properties. Three ways to forge new rings from old ones are introduced: polynomials, quotients and fractions.

In the second chapter, I study *divisibility* in general rings. As was discovered in the twentieth century, classical properties of integers, such as unique factorization, may not hold anymore and this leads to different attitudes. One can either set up new structures — and this leads to the study of prime or maximal ideals — or restrict oneself to rings that satisfy that familiar properties: then, principal ideal domains, euclidean rings, unique factorization domains come in. First applications to algebraic geometry already appear in this chapter: Hilbert's Nullstellensatz and Bézout's theorem on the number of intersection points of plane curves.

Modules are to rings what vector spaces are to fields. This third chapter thus revisits classic notions of linear algebra in this broader context — operations (intersection, direct sums, products, quotients), linear independence, bases, (multi-)linear forms. . . — but also new notions such as Fitting ideals. Conversely, I have found it interesting to retell the story of vector spaces from this general perspective, establishing the existence of bases and the well-definition of dimension. Nakayama's lemma is also discussed in this chapter, as a tool that sometimes allows one to study modules over general rings by reduction to the case of linear algebra over fields.

The fourth chapter presents *field extensions*. While its culminating point is Galois's theory, we however do not discuss classical "number-theoretical" applications of Galois's theory, such as solvability by radicals or geometric constructions with compass and straight-edge, and for which numerous excellent books already exist. The reader will also observe that the first section of this chapter studies *integral dependence* in the context of rings; this is both motivated by the importance of this notion in later chapter, and by the fact that it emphasizes in a deeper way how linear algebra is used to study algebraic/integral dependence.

The description of modules over fields, that is, of vector spaces, is particularly easy: they are classified by a single invariant, their dimension, which in the finitely generated case, is an integer. The next case where such an explicit description is possible is the object of the fifth chapter, namely the classification of *modules over a principal ideal ring*. This classification can be obtained in various ways and we chose the algorithmic approach allowed for by the Hermite forms of matrices; it also has the interest of furnishing information on the structure of the linear group over euclidean domains. Of course, we give the two classical applications of the general theory, to finitely generated abelian groups (when the principal ideal domain is $\mathbf{Z}$) and to the reduction of endomorphisms (when it is the ring $k[T]$ of polynomials in one indeterminate over a field k).

Chapter six introduces various tools to study a module over a general ring. Modules of finite *length*, or possessing the *noetherian*, or the *artinian* properties, are analogues of finite-dimensional vector spaces over a field. I then introduce the support of a module, its associated prime ideals, and establish the existence of a primary decomposition. These sections have a more geometric flavour which can be thought of as a first introduction to the algebraic geometry of schemes.

Homological algebra, one of the important concepts invented in the second half of twentieth century, allows one to systematically quantify the defect of injectivity/surjectivity of a linear map, and is now an unavoidable tool in algebra, geometry, topology,. . . The introduction to this topic I provide here goes in various directions. I define projective and injective modules, and use them as a pretext for elaborating on the language of categories, but fall short of defining resolutions and Tor/Ext functors in general.

The next chapter is devoted to the general study of *tensor products*. It is first used to develop the theory of determinants, via the exterior algebra. The general lack of exactness of this operation leads to the definition of a flat module and, from that point, to the study of *faithfully flat descent* and *Galois descent*. This theory, invented by Alexander Grothendieck, formalizes the following question: imagine you need additional parameters to define an object, how can you get rid of them?

Algebras of finite type over a field are the algebraic building blocks of algebraic geometry, and some of the properties studied in this last chapter, Noether's normalization theorem, dimension, codimension, have an obvious geomet-ric content. Algebraically, these rings are also better behaved than general noetherian rings. A final section is devoted to *Dedekind rings*; an algebraic counterparts of curves, they are also very important in number theory, and I conclude the chapter by proving the finiteness of the class groups of number fields.

In the *appendix*, I first summarize general mathematical conventions. The second section proves two important results in *set theory* that are used all over the text: the Cantor–Bernstein theorem and the Zorn theorem. The main part of the book makes use of the language of categories and functors, in a hopefully progressive and pedagogical way, and I have summarized this language in a final section.

Within the text, 11 one-page notes give historical details about the main contributors of the theory exposed in this book, concentrating on their math-ematical achievements. By their brevity, they do not replace a serious his-torical study of the development of the mathematical ideas, and only aim at shining a dim light on these matters. I hope that these short remarks will incite the readers to read books such as Corry (2004), Fine & Rosenberger (1997), Gray (2018), or Bourbaki (1999), or even read the original texts!

Every chapter comes with a large supply of exercises of all kinds (around 300 in total), ranging from a simple application of concepts to additional results that I couldn't include in the text.

Many important concepts and results in commutative algebra were invented or proved in Germany during the first half of the twentieth century, thanks to David Hilbert, Emmy Noether, Wolfgang Krull... For this reason, some theorems still bear a German name, and I have not failed to conform to this tradition regarding Hilbert's *Nullstellensatz* or Krull's *Hauptidealsatz*.

One tradition I tried to step away from, however, is the use of the German alphabet ("fraktur" or "gothic"), for it is sometimes hard to read, and even harder to write by hand. Consequently, prime and maximal ideals, for example, will be denoted by P or M, rather than $\mathfrak{p}$ or $\mathfrak{m}$. The reader will judge for herself or himself whether this shift from typographic tradition was necessary, or even useful. I have however kept German letters for the symmetric group $\mathfrak{S}_n$ and the set $\mathfrak{P}(S)$ of all subsets of a set S.

Statements of the book are numbered as chapter.section.subsection; in the appendix, I have also made use of subsubsections. Enumerations of mathematical statements are of two typographical kinds: when they consist of independent assertions, I have enumerated them by letters (a), b), c),...) while I have enumerated them by roman figures ((i), (ii), (iii),...) when they consisted of related assertions, such as in lists of equivalent properties.

I tried to make the book essentially self-contained, without any need for exterior references. However, I have included in the bibliography some important reference books (textbooks or research books) on the subject, as well as research articles or notes that helped me describe some results, or prepare some exercises.

The final index will hopefully help the reader to track definitions, properties and theorems in the book. I have usually tried to list them by a noun rather by an adjective, so that, for example, "noetherian" is rather to be found at "ring" and "module" (although I have added an entry for this particularly important case).

François Loeser was teaching this course when I was just a beginning assistant professor, and he suggested that I manage the exercises sessions; I thank him for his trust and friendship all over these years. I would also like to thank Sophie Chemla and Laurent Koelblen, who shared their list of exercises with me.

Many thanks are due to Yuri Tschinkel — not only is he a diligent research collaborator, but he is the one who brought the book to a publisher a long time ago.

I also thank Yves Laszlo, Emmanuel Kowalski, Daniel Ferrand and Javier Fresán for their suggestions or encouragements at various times of the elaboration of the book. I also wish to thank Colas Bardavid, Pierre Bernard, Samit Dasgupta, Brandon Gontmacher, Damien Mégy, Ondine Meyer and Andrey Mokhov for pointing out a few unfortunate typos.

I am also grateful to Tom Artin, Alain Combet, Philippe Douroux, Paulo Ribenboim and to the Department of Mathematics at Princeton University for having graciously allowed me to reproduce some pictures. Although these pictures can be found everywhere on the Internet, I should also thank Elena Griniari, my editor, for having patiently convinced me of the importance of formally requiring these authorizations.

Writing in English is now the daily practice of almost all mathematicians. Nevertheless, this book would have been full of linguistic *bizarreries* without the wonderful work of Barnaby Sheppard, language editor at Springer Nature; I thank him heartily.

I made the final touches to this book during long lockdown months caused by the Covid-19 pandemic, and the following summer. This final work wouldn't have been possible without the availability of an enormous amount of digitized mathematical texts, either behind the closed doors of mathematical publishers (for which my university provides a key), or on public state-funded archives such as the German *Göttinger Digitalisierunszentrum* and the French *Numdam,* as well as on the open websites *Library Genesis* and Alexandra Elbakyan's *Sci-Hub.* Let them all be thanked for their involvement in open science.

Palaiseau, August 2020   *Antoine Chambert-Loir*

Contents

Chapter 1.
Rings

Rings, *the definition of which is the subject of this chapter, are algebraic objects in which one can compute as in classical contexts, integers, real numbers or matrices: one has an addition, a multiplication, two symbols 0 and 1, and the usual computation rules are satisfied.*

The absence of a division, however, gives rise to various subtleties. For example, the product of two elements can be zero, while none of them is zero, or the square of a non-zero element can be zero; the reader has probably already seen such examples in matrix rings. Fields, or division rings, in which every non-zero element is invertible, have more familiar algebraic properties.

Some rings may be given by Nature, but it is the task of a mathematician, be he/she an apprentice mathematician, to construct new rings from old ones. With that aim, this chapter proposes three different tools, each of them having a different goal:

– *Polynomial rings allow the addition to a ring of new "indeterminate" elements, which are not assumed to satisfy any property but be computed with. We introduce them and take the opportunity to review a few facts concerning polynomials in one indeterminate with coefficients in a field and their roots.*

– *Ideals embody in general rings the classical concepts of divisibility and congruences that the reader is certainly familiar with when it comes to integers. Given a (two-sided) ideal I of a ring A, I construct the quotient ring A/I where the elements of I are formally forced to be "equal to zero", while keeping intact the standard computation rules in a ring.*

– *In the opposite direction, fraction rings are defined so as to force given elements to be invertible. A particular case of this construction is the field of fractions of an integral domain.*

1.1. Definitions. First Examples

Definition (1.1.1). — A *ring* is a set A endowed with two binary laws, an *addition* $(a, b) \mapsto a + b$, and a *multiplication* $(a, b) \mapsto ab$, satisfying the following axioms.

© Springer Nature Switzerland AG 2021
A. Chambert-Loir, *(Mostly) Commutative Algebra*, Universitext,
https://doi.org/10.1007/978-3-030-61595-6_1

(i) *Axioms concerning the addition law and stating that* $(A, +)$ *is an abelian group:*

- For every a, b in A, $a + b = b + a$ *(commutativity of addition)*;
- For every a, b, c in A, $(a + b) + c = a + (b + c)$ *(associativity of addition)*;
- There exists an element $0 \in A$ such that $a + 0 = 0 + a = a$ for every a in A *(neutral element for the addition)*;
- For every $a \in A$, there exists an element $b \in A$ such that $a + b = b + a = 0$ *(existence of an additive inverse)*;

(ii) *Axioms concerning the multiplication law and stating that* $(A, \cdot)$ *is a monoid:*

- There exists an element $1 \in A$ such that $1a = a1 = a$ for every $a \in A$ *(neutral element for the multiplication)*;
- For every a, b, and c in A, $(ab)c = a(bc)$ *(associativity of multiplication)*;

(iii) *Axiom relating the addition and the multiplication:*

- For every a, b, and c in A, $a(b + c) = ab + ac$ and $(b + c)a = ba + ca$ *(multiplication distributes over addition)*.

(iv) One says that the ring A is *commutative* if, moreover:

- For every a and b in A, $ab = ba$ *(commutativity)*.

With their usual addition and multiplication, integers, real numbers, and matrices are fundamental examples of rings. In fact, the ring axioms specify exactly the relevant computation rules to which one is accustomed to. We shall give more examples in a moment, but we first state a few computation rules which follow from the stated axioms.

1.1.2. — Let A be a ring.

Endowed with the addition law, A is in particular an abelian group. As a consequence, it admits exactly one *zero element*, which is usually denoted by 0, or by 0_A if it is necessary to specify the ring of which it is the zero element. Moreover, any element $a \in A$ has exactly one additive inverse, and it is denoted by $-a$. Indeed, if b and c are two additive inverses, then $b = b + 0 = b + (a + c) = (b + a) + c = c$.

For every $a \in A$, one has $a \cdot (0 + 0) = a \cdot 0 + a \cdot 0$, hence $a \cdot 0 = 0$; similarly, $0 \cdot a = 0$. Then, for any $a, b \in A$, one has $a(-b) + ab = a(-b + b) = a \cdot 0 = 0$, hence $a(-b) = -(ab)$; similarly, $(-a)b = -(ab)$. Consequently, there is no ambiguity in writing $-ab$ for either $(-a)b$, $-(ab)$ or $a(-b)$.

For any integer n, one defines na by induction, by setting $0a = 0$, and $na = a + (n-1)a$ if $n \geq 1$ and $na = -(-n)a$ if $n \leq -1$. Observe that for any $a, b \in A$ and any integer n, one has $a(nb) = n(ab) = (na)b$. This is proved by induction on n: if $n = 0$, then all three terms are 0; if $n \geq 1$, then $a(nb) = a(b + (n-1)b) = ab + a((n-1)b) = ab + (n-1)(ab) = nab$ and similarly for the other equality; finally, if $n \leq -1$, then $a(nb) = a(-((-n)b)) = -a((-n)b) = (-n)(-ab) = nab$.

Similarly, the multiplicative monoid $(A, \cdot)$ has exactly one neutral element, usually denoted by 1, or by 1_A if it is necessary to specify the ring, and called the *unit element* of A.

If a belongs to a ring A and n is any positive[1] integer $n \geq 0$, one defines a^n by induction by setting $a^0 = 1$, and, if $n \geq 1$, $a^n = a \cdot a^{n-1}$. For any integers m and n, one has $a^{m+n} = a^m a^n$ and $(a^m)^n = a^{mn}$, as can be checked by induction.

However, one should take care that $a^n b^n$ and $(ab)^n$ are generally distinct, unless $ab = ba$, in which case one says that a and b *commute*.

Proposition (1.1.3) (Binomial formula). — *Let A be a ring and let a and b be elements in A such that $ab = ba$. Then, for any positive integer n,*

$$(a + b)^n = \sum_{k=0}^{n} \binom{n}{k} a^k b^{n-k}.$$

Proof. — The proof is the one that is well known when a, b are integers. It runs by induction on n. When $n = 0$, both sides are equal to 1. Assume the formula holds for n; then,

$$(a + b)^{n+1} = (a + b)(a + b)^n = (a + b) \sum_{k=0}^{n} \binom{n}{k} a^k b^{n-k}$$

$$= \sum_{k=0}^{n} \binom{n}{k} a^{k+1} b^{n-k} + \sum_{k=0}^{n} \binom{n}{k} b a^k b^{n-k}$$

$$= \sum_{k=0}^{n} \binom{n}{k} a^{k+1} b^{n-k} + \sum_{k=0}^{n} \binom{n}{k} a^k b^{n+1-k}$$

$$= \sum_{k=1}^{n+1} \binom{n}{k-1} a^k b^{n+1-k} + \sum_{k=0}^{n} \binom{n}{k} a^k b^{n+1-k}$$

$$= \sum_{k=0}^{n+1} \left(\binom{n}{k-1} + \binom{n}{k} \right) a^k b^{n+1-k}$$

$$= \sum_{k=0}^{n+1} \binom{n+1}{k} a^k b^{n+1-k}$$

since $\binom{n}{k-1} + \binom{n}{k} = \binom{n+1}{k}$ for any positive integers n and k. This concludes the proof by induction on n. □

Examples (1.1.4). —   a) As well-known basic examples of commutative rings, let us mention the ring **Z** of integers, the quotient rings **Z**/n**Z** for $n \geq 1$, the fields **Q** of rational numbers, **R** of real numbers, **C** of complex numbers, and the ring K[X] of polynomials in one indeterminate X with coefficients in a field (or a commutative ring) K.

---

[1] Recall that in this book, *positive* means *greater than or equal to 0*, and *negative* means *less than or equal to 0*, see p. 436.

b) Let A be a ring and let S be a set. The set A^S of functions from S to A, with pointwise addition and pointwise multiplication, is a ring. Explicitly, for f and $g \in A^S$, $f + g$ and fg are the functions such that $(f + g)(s) = f(s) + g(s)$ and $(fg)(s) = f(s)g(s)$. The set $\mathscr{C}(X, \mathbf{R})$ of all continuous functions from a topological space X into $\mathbf{R}$ is a ring, as are the sets $\mathscr{C}^k(\Omega, \mathbf{R})$ and $\mathscr{C}^k(\Omega, \mathbf{C})$ of all functions of differentiability class $\mathscr{C}^k$ from an open subset Ω of $\mathbf{R}^n$ into $\mathbf{R}$ or $\mathbf{C}$ (here, $k \in \mathbf{N} \cup \{\infty\}$).

c) Let S be a set and let $(A_s)_{s \in S}$ be a family of rings indexed by s. Let $A = \prod_{s \in S} A_s$ be the product of this family; this is the set of all families $(a_s)_{s \in S}$, where $a_s \in A_s$ for all s. We endow A with termwise addition and termwise multiplication: $(a_s) + (b_s) = (a_s + b_s)$ and $(a_s) \cdot (b_s) = (a_s b_s)$. Then A is a ring; its zero element is the null family (0_{A_s}), its unit element is the family (1_{A_s}).

Let us now give non-commutative examples.

Examples (1.1.5). —   *a*) Let A be a ring, let n be a positive integer and let $M_n(A)$ be the set of $n \times n$ matrices with coefficients in A, endowed with the usual computation rules: the sum of two matrices $P = (p_{i,j})$ and $Q = (q_{i,j})$ is the matrix $R = (r_{i,j})$ such that $r_{i,j} = p_{i,j} + q_{i,j}$ for all i and j in $\{1, \ldots, n\}$; the product of these matrices P and Q is the matrix $S = (s_{i,j})$ given by

$$s_{i,j} = \sum_{k=1}^{n} p_{i,k} q_{k,j}.$$

Endowed with these laws, $M_n(A)$ is a ring; its zero element is the null matrix; its unit element is the matrix $I_n = (\delta_{i,j})$, where $\delta_{i,j} = 1$ if $i = j$, and 0 otherwise.

When $n \geq 2$, or if $n \geq 1$ and A is not commutative, then the ring $M_n(A)$ is not commutative.

b) Let G be an abelian group. When φ and ψ are any two endomorphisms of G, the map $g \mapsto \varphi(g) + \psi(g)$ is again an endomorphism of G, written $\varphi + \psi$; this endows the set End(G) of endomorphisms of G with the structure of an abelian group, whose zero element is the map $g \mapsto 0$. Composition of endomorphisms $(\varphi, \psi) \mapsto \varphi \circ \psi$ is an associative law, and distributes with respect to the addition; the identity map of G, $g \mapsto g$, is the unit element. Consequently, these laws endow the set End(G) of all endomorphisms of the group G with the structure of a ring.

c) Let K be a field (commutative, say). The set $End_K(V)$ of all endomorphisms of a K-vector space V is a ring, non-commutative as soon as $\dim(V) \geq 2$. Here, the addition law is the pointwise addition, $(\varphi + \psi)(v) = \varphi(v) + \psi(v)$ for $\varphi, \psi \in End_K(V)$ and $v \in V$, while the multiplication law is the composition of endomorphisms, given by $\varphi \circ \psi(v) = \varphi(\psi(v))$. In fact, $End_K(V)$ is also a K-vector space and the multiplication is K-linear. One says that $End_K(V)$ is a K-algebra.

If V has finite dimension, say n, we may choose a basis of V and identify V with K^n; then endomorphisms of V identify with matrices of size n, and this identifies the rings $End_K(V)$ and $M_n(K)$.

Here is a possibly less well-known example.

Example (1.1.6). — Let A be a ring and let M be any monoid (a group, for example).

Inside the abelian group A^M of all functions from M to A, let us consider the subgroup $A^{(M)}$ of functions f with finite support, namely such that $f(g) \neq 0$ for only finitely many $g \in M$.

It also possesses a *convolution product*, defined by the formula:

$$(\varphi * \psi)(g) = \sum_{\substack{h,k \in M \\ g=hk}} \varphi(h)\psi(k).$$

This product is well-defined: only finitely many terms in the sum are non-zero, and the convolution of two functions with finite support still has finite support. Moreover, the convolution product is associative: indeed, for $\varphi, \psi, \theta \in A^{(M)}$, one has

$$(\varphi * \psi) * \theta(g) = \sum_{\substack{h,k \in M \\ g=hk}} (\varphi * \psi)(h)\theta(k)$$

$$= \sum_{\substack{h,k \in M \\ g=hk}} \sum_{\substack{i,j \in M \\ h=ij}} \varphi(i)\psi(j)\theta(k)$$

$$= \sum_{\substack{i,j,k \in M \\ g=ijk}} \varphi(i)\psi(j)\theta(k),$$

and a similar computation shows that this is also equal to $\varphi * (\psi * \theta)(g)$.

There is a neutral element for convolution, given by the "Dirac function" δ such that $\delta(g) = 1$ for $g = e$, the neutral element of M, and $\delta(g) = 0$ otherwise. Indeed, for $\varphi \in A^{(M)}$ and $g \in M$, one has

$$\varphi * \delta(g) = \sum_{\substack{h,k \in M \\ g=hk}} \varphi(h)\delta(k) = \varphi(g),$$

and similarly, $\delta * \varphi(g) = \varphi(g)$.

These laws endow $A^{(M)}$ with the structure of a ring, called the monoid ring of M (with coefficients in A). When A is a commutative ring, we also use the expression *monoid algebra*; when M is a group, we use the expressions *group ring* or *group algebra*.

For $k \in M$, let $\delta_k \in A^{(M)}$ be the function such that $\delta_k(g) = 1$ if $g = k$, and $\delta_k(g) = 0$ otherwise. One has $\delta_h * \delta_k = \delta_{hk}$.

Definition (1.1.7). — A *subring* of a ring A is a subgroup B of A for the addition, which contains 1, and is stable under the multiplication, so that the laws of A endow B with the structure of a ring, admitting the same neutral elements 0 and 1.

Examples (1.1.8). —   *a*) Let A be a ring and let $G = A^n$, viewed as an abelian group for addition. With the usual action of matrices on vectors, $M_n(A)$ is a subset of $End(A^n)$, with compatible addition, multiplication and unit element, so that $M_n(A)$ is a subring of $End(A^n)$.

b) Let A be a ring. The set $Z(A)$ of all elements $a \in A$ such that $ax = xa$ for every $x \in A$ is a commutative subring of A, called its *center*.

More generally, let S be a subset of A and let $C_S(A)$ be the set of all elements $a \in A$ such that $ax = xa$ for all $x \in S$. This is a subring of A, called the *centralizer* of S in A. By definition, S is contained in the center of $C_S(A)$, and $C_S(A)$ is the largest subring of A satisfying this property.

c) The intersection of any family of subrings of a ring A is a subring of A.

1.1.9. — Let A be a ring and let S be a subset of A. Let B be the intersection of all subrings of A containing S. It is a subring of A; one calls it the *subring of A generated by* S.

Lemma (1.1.10). — *Let A be a ring and let S be a subset of* A; *assume that any two elements of S commute. Then the subring of A generated by S is commutative.*

Proof. — Let B be this subring. First of all, S is contained in the centralizer C_S of S in A, by assumption, which is a subring of A, hence $B \subset C_S(A)$. Then S is contained in the center of $C_S(A)$, which is a subring of $C_S(A)$, hence of A, so that $B \subset Z(C_S(A))$. Since $Z(C_S(A))$ is a commutative ring, B is commutative.□

Definition (1.1.11). — Let A and B be two rings. A *morphism of rings* $f: A \to B$ is a map that satisfies the following properties:

(i) One has $f(0) = 0$ and $f(1) = 1$;

(ii) For any a and b in A, one has $f(a+b) = f(a)+f(b)$ and $f(ab) = f(a)f(b)$.

Synonyms are *homomorphism of rings* and *ring morphism*. An *endomorphism* of the ring A is a ring morphism from A to A.

If A is a ring, the identity map $id_A: A \to A$, $a \mapsto a$, is a ring morphism. The composition of two ring morphisms is again a ring morphism, and composition is associative.

Consequently, rings and their morphisms form a *category*,[2] which we denote by **Ring**.

As in general category theory, we say that a morphism of rings $f: A \to B$ is an *isomorphism* if there exists a ring morphism $g: B \to A$ such that $f \circ g = id_B$ and $g \circ f = id_A$. In that case, there exists exactly one such morphism g, and we call it the inverse of f; we also say that the rings A and B are isomorphic and write $A \simeq B$. We also write $f: A \xrightarrow{\sim} B$ to say that f is an isomorphism of rings from A to B.

If A is a ring, an *automorphism* of A is an isomorphism from A to A. The set of all automorphisms of a ring A is a group with respect to composition.

---

[2] The theory of categories provides a powerful *language* which enables one to formulate structural properties of mathematics. We have included a basic summary in appendix A.3.

1.1.12. — Let $f: A \to B$ be a morphism of rings. The image $f(A)$ of A by f is a subring of B. The inverse image $f^{-1}(C)$ of a subring C of B is a subring of A.

Let $f, g: A \to B$ be morphisms of rings. The set C of all $a \in A$ such that $f(a) = g(a)$ is a subring of A.

Proposition (1.1.13). — *A ring morphism is an isomorphism if and only if it is bijective.*

Proof. — Let $f: A \to B$ be a morphism of rings.

Assume that f is an isomorphism, and let g be its inverse. The relations $g \circ f = \mathrm{id}_A$ and $f \circ g = \mathrm{id}_B$ respectively imply that f is injective and surjective, so that f is bijective.

Conversely, let us assume that f is bijective and let g be its inverse. We have $g \circ f = \mathrm{id}_A$ and $f \circ g = \mathrm{id}_B$, so we just need to show that g is a ring morphism from B to A. Since $f(0) = 0$ and $f(1) = 1$, one has $g(0) = 0$ and $g(1) = 1$. For any a and $b \in B$,

$$f(g(a + b)) = a + b = f(g(a)) + f(g(b)) = f(g(a) + g(b))$$

and

$$f(g(ab)) = ab = f(g(a))f(g(b)) = f(g(a)g(b)).$$

Since f is a bijection, $g(a + b) = g(a) + g(b)$ and $g(ab) = g(a)g(b)$. □

Example (1.1.14). — *Let A be a commutative ring and let p be a prime number such that $p1_A = 0$. The map $\varphi: A \to A$ defined by $\varphi(x) = x^p$ is a ring homomorphism, called the Frobenius homomorphism.*

Observe that $\varphi(0) = 0$ and $\varphi(1) = 1$. Moreover, since A is commutative, we have $\varphi(xy) = (xy)^p = x^p y^p = \varphi(x)\varphi(y)$ for every $x, y \in A$. It remains to show that φ is additive. Let $x, y \in A$; by the binomial formula, one has

$$\varphi(x + y) = (x + y)^p = \sum_{k=0}^{p} \binom{p}{k} x^k y^{p-k}.$$

But $\binom{p}{k} = p!/k!(p - k)!$, a fraction the numerator of which is a multiple of p. On the other hand, since p is prime, p divides neither $k!$ nor $(p - k)!$, if $1 \le k \le p-1$, hence p does not divide $k!(p - k)!$. Consequently, p divides $\binom{p}{k}$ for $1 \le k \le p-1$. Since $p1_A = 0$, this implies $\binom{p}{k}1_A = 0$ for any $k \in \{1, \ldots, p-1\}$ and all these terms $\binom{p}{k}x^k y^{p-k}$ vanish. Consequently,

$$\varphi(x + y) = x^p + y^p = \varphi(x) + \varphi(y),$$

as was to be shown.

Example (1.1.15). — Let A be a ring and let a, b be elements of A such that $ab = ba = 1$. (We shall soon say that a is invertible and that b is its inverse.) Then, the map $x \mapsto axb$ is an automorphism of A, called an *interior automorphism*.

As an example, one can prove that every $\mathbf{C}$-linear automorphism of $M_n(\mathbf{C})$ is interior (exercise 1.8.12).

Example (1.1.16). — Let $A = \mathbf{C}[X_1, \ldots, X_n]$ be the ring of polynomials in n variables with coefficients in $\mathbf{C}$. (They will be defined in example 1.3.9 below.) Let $Q_1, \ldots, Q_n$ be elements in A. Consider the map φ from A to A which associates to a polynomial P the polynomial $P(Q_1, \ldots, Q_n)$ obtained from P by substituting the polynomial Q_i in place of the indeterminate X_i; it is an endomorphism of A. It is the unique endomorphism of A such that $\varphi(X_i) = Q_i$ for all i, and $\varphi(c) = c$ for all $c \in \mathbf{C}$.

Similarly, for any permutation σ of $\{1, \ldots, n\}$, let Φ_σ be the unique endomorphism of A such that $\Phi_\sigma(X_i) = X_{\sigma(i)}$ for all i and $\Phi_\sigma(c) = c$ for all $c \in \mathbf{C}$. Observe that $\Phi_{\sigma\tau}(X_i) = X_{\sigma(\tau(i))} = \Phi_\sigma(X_{\tau(i)}) = \Phi_\sigma(\Phi_\tau(X_i))$; therefore, the endomorphisms $\Phi_{\sigma\tau}$ and $\Phi_\sigma \circ \Phi_\tau$ are equal. The map $\sigma \mapsto \Phi_\sigma$ is a group morphism from the symmetric group $\mathfrak{S}_n$ to the group $\mathrm{Aut}(\mathbf{C}[X_1, \ldots, X_n])$.

1.2. Nilpotent Elements; Regular and Invertible Elements; Division Rings

Some elements of a ring have nice properties with respect to the multiplication. This justifies a few more definitions.

Definition (1.2.1). — Let A be a ring. One says that a is *nilpotent* if there exists an integer $n \geq 1$ such that $a^n = 0$.

As an example, nilpotent elements of the matrix ring $M_n(\mathbf{C})$ are exactly the nilpotent $n \times n$-matrices.

Definition (1.2.2). — Let A be a ring. We say that an element $a \in A$ is *left regular* if the relation $ab = 0$ in A implies $b = 0$; otherwise, we say that a is a *left zero divisor*. Similarly, we say that a is *right regular* if the relation $ba = 0$ in A only holds for $b = 0$, and that it is a *right zero divisor* otherwise.

An element is said to be *regular* if it is both left and right regular.

We say that a ring A is an *integral domain*, or in short, a *domain*, if it is commutative, non-zero, and if every non-zero element of A is regular.

In a commutative ring, an element which is not regular is simply called a *zero divisor*.

Definition (1.2.3). — Let A be a ring and let a be any element of A.

One says that a is *right invertible* if there exists an element $b \in A$ such that $ab = 1$; we then say that b is a *right inverse* of a. Similarly, we say that a is *left invertible* if there exists an element $b \in A$ such that $ba = 1$; such an element b is called a *left inverse* of a. Finally, we say that a is *invertible*, or a *unit*, if it is both left and right invertible.

Assume that a is right invertible and let b be a right inverse for a, so that $ab = 1$. If $ca = 0$, then $cab = 0$, hence $c = 0$. This shows that a is right regular. Similarly, an element which is left invertible is also left regular.

Assume that a is invertible; let b be a right inverse and c be a left inverse of a. One has $b = 1b = (ca)b = c(ab) = c1 = c$. Consequently, the right and left inverses of a are equal. In particular, a has exactly one left inverse and one right inverse, and they are equal. This element is called the *inverse* of a, and is usually denoted by a^{-1}.

If the ring is commutative, the notions of left and right regular coincide; an element which is left invertible is also right invertible, and conversely.

Let $A^\times$ be the set of invertible elements in a ring A.

Proposition (1.2.4). — *The set of all invertible elements in a ring A is a group with respect to multiplication. It is called the* group of units *of A.*

Any ring morphism $f \colon A \to B$ induces by restriction a morphism of groups from $A^\times$ to $B^\times$.

Proof. — Let a and b be two invertible elements of A, with inverses a^{-1} and b^{-1}. Then, $(ab)(b^{-1}a^{-1}) = a(bb^{-1})a-1 = aa^{-1} = 1$, so that ab is right invertible, with right inverse $b^{-1}a^{-1}$. Similarly, $(b^{-1}a^{-1})(ab) = 1$, so that ab is left invertible too. The multiplication of A induces an associative law on $A^\times$. Moreover, 1 is invertible and is the neutral element for that law. Finally, the inverse of $a \in A^\times$ is nothing but a^{-1}. This shows that $A^\times$ is a group with respect to the multiplication

Let $f \colon A \to B$ be a ring morphism. Let $a \in A^\times$ and let b be its inverse. Since $ab = ba = 1$, one has $1 = f(1) = f(a)f(b) = f(b)f(a)$. This shows that $f(a)$ is invertible, with inverse $f(b)$. Consequently, the map f induces by restriction a map from $A^\times$ to $B^\times$. It maps the product of two invertible elements to the product of their images, hence is a group morphism. □

Let A be a commutative ring. Say that two elements a and $b \in A$ are *associated* if there exists an invertible element $u \in A^\times$ such that $a = bu$. The relation of "being associated" is an equivalence relation.

Definition (1.2.5). — One says that a ring A is a *division ring* (or a *division algebra*) if it is not the zero ring, and if any non-zero element of A is invertible. A *field* is a commutative division ring.[3] A *subfield* of a field is a subring of a field which is still a field.

A few examples of fields are certainly well known to the reader, but it is not obvious that there are any noncommutative division rings at all. Let us begin by quoting the theorem of Wedderburn according to which any finite

---

[3] Terminology is not always consistent among mathematicians, and is sometimes chosen in order to avoid the repetition of adjectives. Some books call fields what we call division rings; others use the awkward expression "skew field" to talk about division rings, but then a skew field is not a field. Since fields, i.e., commutative division rings, will play a larger role in this book than general division rings, it is useful to have a shorter name for them.

On William Hamilton

William Rowan HAMILTON (1805–1865) was an Irish mathematician and physisict. In fact, his official position was that of a professor of astronomy and astronomer; although he is now not remembered as an important astronomer, his introductory lectures were famous and attracted a large audience. As a physicist, he reformulated the classical mechanics of NEWTON and LAGRANGE within the calculus of variations. Featuring a symmetry between position and momentum, the *Hamilton equations* opened the path for quantum mechanics and symplectic geometry. His name is also associated to the notion of a closed path in a graph that visits all vertices of a given graph once and only once.

His main mathematical achievement is the invention of *quaternions*, motivated by "the desire to discover the law of multiplication of triplets": finding a mathematical structure that would be as useful for studying the 3-dimensional space as complex numbers were for the 2-dimensional one. After many failed explorations, the insight came during a walk to his work, on 16th october 1843. In his own words:

> *And here there dawned on me the notion that we must admit, in some sense, a fourth dimension of space for the purpose of calculating with triples.*
>
> W. R. Hamilton,
>
> *Letter to J. Graves, dated 17th october 1843,*
>
> *quoted by* VAN DER WAERDEN *(1976)*

He gave more lively details about this discovery 22 years later, in a letter to his son Archibald that he wrote shortly before his death:

Portrait of Sir William Rowan Hamilton (after 1837)

Etching by John Kirkwood, after Charles Grey.
Source: Wellcome Library, via Wikipedia.
Copyright: CC-BY-4.0.

> But on the 16th day of the same month (October 1843) — which happened to be a Monday and a Council day of the Royal Irish Academy — I was walking in to attend and preside, and your mother was walking with me, along the Royal Canal, to which she had perhaps been driven; and although she talked with me now and then, yet an under-current of thought was going on in my mind, which gave at last a result, whereof it is not too much to say that I felt at once the importance. An electric circuit seemed to close, and a spark flashed forth. (...) I pulled out on the spot a pocket-book, which still exists, and made an entry there and then. Nor could I resist the impulse — unphilosophical as it may have been — to cut with a knife on a stone of Brougham Bridge, as we passed it, the fundamental formula with the symbols i, j, k; $i^2 = j^2 = k^2 = ijk = -1$, which contains the solution of the Problem, but of course as an inscription, has long since mouldered away.

W. R. Hamilton,

quoted by VAN DER WAERDEN *(1976)*

In his study of quaternions, HAMILTON proved the Cayley–Hamilton theorem for certain matrices of size 2 or 4; CAYLEY proved the theorem for matrices of size 2 and 3 in 1858, but the general theorem is due to FROBENIUS (1878).

Despite featuring a fourth dimension, quaternions furnish an efficient representation of 3-dimensional rotations. The algebra of quaternions also allows one to explain why the 3-dimensional sphere admits 3 independent vector fields, or to give a nice description of the Hopf fibration — a partition of the 3-dimensional sphere into circles, parameterized by a 2-dimensional sphere.

Within algebra, their study now belongs to the theory of central simple algebras, a theory with deep connections with number theory, especially class field theory.

division ring is commutative, *i.e.*, is a field; see exercise 4.9.19 for its proof, borrowed from the *Book* AIGNER & ZIEGLER (2014). The ring of *quaternions* is probably the most renowned of all noncommutative division rings; it is in fact the first one to have been discovered, by Hamilton in 1843.

Example (1.2.6). — The underlying abelian group of the *quaternions* is $\mathbf{H} = \mathbf{R}^4$; we write $(1, i, j, k)$ for its canonical basis. Besides the properties of being associative, having 1 as a neutral element, distributing addition, and being $\mathbf{R}$-bilinear (that is, $t(ab) = (ta)b = a(tb)$ for any two quaternions a, b and any real number t), the multiplication $\mathbf{H} \times \mathbf{H} \to \mathbf{H}$ is characterized by the following relations: $i^2 = j^2 = k^2 = -1$ and $ij = k$. Provided these laws give $\mathbf{H}$ the structure of a ring, other relations follow quite easily. Indeed, $i^2 = -1 = k^2 = (ij)k = i(jk)$, hence $i = jk$ after multiplying both sides by $-i$; the equality $j = ki$ is proved similarly; then, $kj = k(ki) = k^2i = -k$, $ik = i(ij) = i^2j = -j$, and $ji = j(jk) = j^2k = -k$.

It is thus a remarkable discovery of W. R. HAMILTON that the following "multiplication table"

	1	i	j	k
1	1	i	j	k
i	i	-1	k	$-j$
j	j	$-k$	-1	i
k	k	j	$-i$	-1

gives rise to a ring structure on $\mathbf{H}$.

Properties of the addition follow from the fact that $\mathbf{H} = \mathbf{R}^4$ is an $\mathbf{R}$-vector space. Only the products of the basic vectors are defined, and distributivity is basically a built-in feature of multiplication. The crucial point is associativity; using distributivity, it is enough to check the relation $a(bc) = (ab)c$ when a, b, c belong to $\{1, i, j, k\}$. This is obvious if $a = 1$ ($1(bc) = bc = (1b)c$), if $b = 1$ ($a(1c) = ac = (a1)c$), or if $c = 1$ ($a(b1) = ab = (ab)1$). We may thus assume that a, b, c belong to $\{i, j, k\}$ but leave the reader to check these twenty-seven remaining cases!

Let $q = a1 + bi + cj + dk$ be any quaternion; set $\overline{q} = a1 - bi - cj - dk$. For any two quaternions q and q', one has $\overline{qq'} = \overline{q'}\,\overline{q}$ and $q\overline{q} = (a^2 + b^2 + c^2 + d^2)1$; in particular, $q\overline{q}$ is a positive real number (multiplied by 1), and vanishes only for $q = 0$.

Any non-zero quaternion q is invertible, with inverse the quaternion $(q\overline{q})^{-1}\overline{q}$.

Let $q = a1 + bi + cj + dk$ be a quaternion. Observe that $i^{-1}qi = -iqi = a1 + bi - cj - dk$, $j^{-1}qj = a1 - bi + cj - dk$ and $k^{-1}qk = a1 - bi - cj + dk$. It follows that the center of $\mathbf{H}$ consists of the elements $a1$, for $a \in \mathbf{R}$; it is a subfield of $\mathbf{H}$, isomorphic to the field of real numbers.

Observe that the set of all quaternions of the form $a + bi$, for $a, b \in \mathbf{R}$, is a subfield of $\mathbf{H}$ isomorphic to $\mathbf{C}$. More generally, for any unit vector $(u_1, u_2, u_3) \in \mathbf{R}^3$, the quaternion $u = u_1 i + u_2 j + u_3 k$ satisfies $u^2 = -1$; moreover, the set of all quaternions of the form $a + bu$, for a and $b \in \mathbf{R}$, is a subfield of $\mathbf{H}$ which is also isomorphic to $\mathbf{C}$.

In particular, we see that the equation $X^2 + 1 = 0$ has infinitely many solutions in **H**, in marked contrast to what happens in a field, where a polynomial equation has at most as many solutions as its degree (see §1.3.20).

Theorem (1.2.7) (Frobenius, 1877). — *Let* A *be a finite dimensional* **R**-*vector space, endowed with an* **R**-*bilinear multiplication law which endows it with the structure of a division algebra. Then,* A *is isomorphic to* **R**, **C**, *or* **H**.

Observe that this statement does not hold if one removes the hypothesis that A is finite-dimensional (as shown by the field **R**(X) of rational functions with real coefficients), or the hypothesis that the multiplication turns A into a division ring (consider the product ring **R** × **R**). It is also false without the assumption that the multiplication is **R**-bilinear, see DESCHAMPS (2001). The proof below follows quite faithfully the paper by PALAIS (1968). It makes use of basic results about polynomials in one variable that will be reviewed in later sections of the book, hence may be skipped for a first reading.

Proof. — Let 1_A be the neutral element of A for the multiplication. Let us identify **R** with the subring of A consisting of the elements $x1_A$, for $x \in$ **R**.

We first make a few observations that will be useful in the proof.

First of all, any subring B of A which is an **R**-vector space is a division algebra. Indeed, for any non-zero $b \in$ B, the map $x \mapsto bx$ from B to itself is **R**-linear and injective; since B is a finite-dimensional vector space, it is also surjective, so that the inverse of b (the preimage of 1 under this map) belongs to B.

Let also α be any element in A $-$ **R**. The ring **R**$[\alpha]$ generated by **R** and α is commutative. It is a vector-subspace of A, hence is a field. The minimal polynomial P of α in **R**$[\alpha]$ is irreducible, hence has degree ≤ 2. Since $\alpha \notin$ **R**, this degree is exactly 2 and there are real numbers u, v such that P $= X^2 + 2uX + v$. Then, $(\alpha + u)^2 = u^2 - v$ so that the element $i = (\alpha + u)/\sqrt{v - u^2}$ of **R**$[\alpha]$ satisfies $i^2 = -1$. Let us observe that **R**$[\alpha] =$ **R**$[i]$ is isomorphic to **C**. In particular, if $\alpha^2 \in$ **R**, then $\alpha^2 < 0$.

Let us now assume that A $\neq$ **R** and let us choose $\alpha \in$ A$-$**R**. As we just saw, there exists an element $i \in$ **R**$[\alpha]$ such that $i^2 = -1$ and **R**$[\alpha] =$ **R**$[i] \simeq$ **C**. We may identify the subfield **R**$[i]$ with the field of complex numbers and view A as a **C**-vector space, complex scalars acting on A by left multiplication.

Let us now assume that A is commutative; we shall show that in this case, A $=$ **C**. More generally, we shall show that there is no element $\beta \in$ A $-$ **C** which commutes with i. By contradiction, let $\beta \in$ A$-$**C** such that $\beta i = i\beta$. The subring **C**$[\beta]$ is commutative, and is a field. Moreover, since $\beta \notin$ **R**, the same argument as above furnishes an element j of the form $(\beta + u')/\sqrt{v' - (u')^2}$ of A such that $j^2 = -1$ and **R**$[j] =$ **R**$[\beta]$. Since $\beta \notin$ **C**, we see that $j \neq \pm i$. It follows that the polynomial $X^2 + 1$ has at least four roots (namely, $i, -i, j$ and $-j$) in the field **C**$[\beta]$, which is absurd.

Let $\varphi:$ A $\rightarrow$ A be the map given by $\varphi(x) = xi$. It is **C**-linear and satisfies $\varphi^2 = -\mathrm{id}_A$. Since the polynomial $X^2 + 1$ is split in **C**, with simple roots, i and $-i$, the space A is the direct sum of the eigenspaces for the eigenvalues i

and $-i$. In other words, A is equal to the direct sum $A_+ \oplus A_-$, where A_+ is the set of all $x \in A$ such that $\varphi(x) = xi = ix$ and A_- is the space of $x \in A$ such that $\varphi(x) = xi = -ix$.

Let us observe that A_+ is stable under multiplication; indeed, if $xi = ix$ and $yi = iy$, then $(xy)i = xiy = ixy$. It also contains the inverse of any of its non-zero elements since, if $x \in A_+$ is not equal to 0, then $xi = ix$, so that $x^{-1}i = ix^{-1}$. Hence A_+ is a subfield of A which contains the field **C**. Since no element of $A - C$ commutes with i, we obtain $A_+ = C$.

Assume finally that $A \neq A_+$ and let us consider any non-zero element $\beta \in A_-$. The map $x \mapsto x\beta$ is **C**-linear, and injective, hence bijective. If $xi = ix$, then $x\beta i = x(-i\beta) = -xi\beta = -ix\beta$, so that $A_+\beta \subset A_-$. Conversely, if $yi = -iy$, let us choose $x \in A$ such that $y = x\beta$; then $xi\beta = -x\beta i = -yi = iy = ix\beta$, hence $xi = ix$ since $\beta \neq 0$, hence the other inclusion and $A_+\beta = A_-$.

The same argument shows that $A_-\beta = A_+$. In particular, $\beta^2 \in C \cap R[\beta]$. These two vector spaces, **C** and $R[\beta]$, are distinct, have dimension 2, and contain **R**. Consequently, their intersection is equal to **R**. Since $\beta^2 \in R$ and $\beta \notin R$, we have $\beta^2 < 0$. Then $j = (-\beta^2)^{-1/2}\beta$ is an element of A_- such that $j^2 = -1$. Moreover, $A_- = A_+\beta = A_+j$ and $A_+ = A_-j$.

Set $k = ij$. One has $A_+ = R1 \oplus Ri$, $A_- = A_+j = Rj \oplus Rk$, and $A = A_+ \oplus A_-$. The **R**-vector space A has dimension 4, and $(1, i, j, k)$ is a basis. One has $k^2 = ijij = i(-ij)j = -i^2j^2 = -1$; more generally, the multiplication table of A coincides with that of **H**. This shows that A and **H** are isomorphic.   □

1.3. Algebras, Polynomials

Definition (1.3.1). — Let k be a commutative ring. A *k-algebra* is a ring A together with a morphism of rings $i: k \to A$ whose image is contained in the center of A.

Formally, a k-algebra is defined as the ordered pair $(A, i: k \to A)$. However, we will mostly say "Let A be a k-algebra", therefore understating the morphism i. For $x \in k$ and $a \in A$, we will also commit the abuse of writing xa for $i(x)a$, even when i is not injective. Observe also that a k-algebra may be non-commutative.

A *subalgebra* of a k-algebra (A, i) is a subring B containing the image of i, so that (B, i) is a k-algebra. The intersection of a family of subalgebras is a subalgebra: it is a subring, since the intersection of a family of subrings is a subring, and it contains the image of i.

Definition (1.3.2). — Let k be a commutative ring and let (A, i) and (B, j) be k-algebras. A *morphism of k-algebras* $f: A \to B$ is a ring morphism f such that $f(i(x)a) = j(x)f(a)$ for every $x \in k$ and every $a \in A$.

A composition of morphisms of k-algebras is a morphism of k-algebras; the inverse of a bijective morphism of k-algebras is a morphism of k-algebras. Consequently, algebras and their morphisms form a category.

Remark (1.3.3). — There is a general notion of k-algebras based on the language of modules, that we introduce later in this book (from chapter 3 on). Namely, a (general) k-algebra is a k-module M endowed with a k-bilinear "multiplication" $\mu_M \colon M \times M \to M$, and a morphism of k-algebras is a k-linear map $f \colon M \to N$ which is compatible with the multiplication: for $a, b \in M$, $f(\mu_M(a, b)) = \mu_N(f(a), f(b))$.

Additional properties of the multiplication can of course be required, giving rise to the notions of unitary k-algebras (if there is a unit for multiplication, and morphisms are also required to map the unit to the unit), commutative k-algebras (if the multiplication is commutative), associative k-algebras (if it is associative), but also Lie k-algebras (if it satisfies the Jacobi triple identity), etc.

If the multiplication is associative and admits a unit element, then we get a k-algebra in our sense; if it is moreover commutative, we get a commutative k-algebra. The terminology we chose reflects the fact that *in this book,* we are essentially interested in unitary and associative algebras.

Examples (1.3.4). —   *a*) Let k be a subring of a commutative ring A. The inclusion $k \hookrightarrow A$ endows A with the structure of a k-algebra.

b) Any ring is, in a unique way, a **Z**-algebra. Indeed, for any ring A, the map defined by $i(n) = n 1_A$ for $n \in \mathbf{Z}$ is the unique morphism of rings $i \colon \mathbf{Z} \to A$.

c) Let K be a division ring and let $i \colon \mathbf{Z} \to K$ the unique morphism of rings.

Assume that i is injective. Then its image is a subring of K which is isomorphic to **Z**. Moreover, the ring K being a division ring, it contains the set K_0 of all fractions a/b, for $a, b \in i(\mathbf{Z})$ and $b \neq 0$. This set K_0 is a subfield of K, isomorphic to the field **Q** of rational numbers; one says that K *has characteristic 0.*

Assume now that i is not injective and let p be the smallest integer such that $p > 0$ and $i(p) = 0$. Let us prove that p is a prime number. Since $i(1) = 1_K \neq 0$, one has $p > 1$. Let m and n be positive integers such that $p = mn$; then $i(m)i(n) = i(p) = 0$, hence $i(m) = 0$ or $i(n) = 0$; by minimality, one has $m \geq p$ or $n \geq p$; this proves that p is a prime number, as claimed. The image of i is then a subfield K_0 of K of cardinality p and one says that K *has characteristic p.*

The field K_0 is sometimes called the *prime field* of K.

d) Let k be a commutative ring. The ring $k[X]$ of polynomials with coefficients in k in one indeterminate X is naturally a k-algebra. We will define below algebras of polynomials in any number of indeterminates.

e) Let k be a commutative ring. The ring $\mathrm{Mat}_n(k)$ of $n \times n$ matrices with coefficients in k is a k-algebra, where the canonical morphism $i \colon k \to \mathrm{Mat}_n(k)$ associates with $a \in k$ the matrix aI_n.

1.3.5. Algebra generated by a set — Let k be a commutative ring, let A be a k-algebra and let S be a subset of A. By definition, the *subalgebra of* A *generated by* S is the smallest subalgebra of A which contains S; it is denoted by $k[S]$. If $S = \{a_1, \ldots, a_n\}$, one writes also $k[a_1, \ldots, a_n]$ for $k[S]$. A subalgebra of A is *finitely generated* if it is generated by a finite subset.

For every $\lambda \in k$ and every $s_1, \ldots, s_n \in S$, one has $\lambda s_1 \ldots s_n \in k[S]$; more generally, every finite sum of elements of this form $\lambda s_1 \ldots s_n$ (where $\lambda \in k$, $n \in \mathbf{N}$, and $s_1, \ldots, s_n \in S$) belongs to $k[S]$. In fact, $k[S]$ is the set B of all such finite sums: by what precedes, $k[S]$ contains B; conversely, one checks that B is a subalgebra of $k[S]$, so that $k[S] \subset B$ by the definition of $k[S]$.

Proposition (1.3.6). — *Let k be a commutative ring, let A be a commutative k-algebra and let B be an A-algebra. Assume that A is generated as a k-algebra by elements $(a_i)_{i \in I}$ and that B is generated as an A-algebra by elements $(b_j)_{j \in J}$. Then B is generated as a k-algebra by the elements $(b_j)_{j \in J}$ and the elements $(a_i)_{i \in I}$.*

Proof. — Let B' be the k-subalgebra of B generated by the a_i and the b_j, and let A' be its inverse image in A. It is a k-subalgebra of A that contains the a_i, hence is equal to A, so that B' contains the image of A in B. Since it also contains the b_j, one has B' = B, as was to be shown.   □

Corollary (1.3.7). — *Let k be a commutative ring. If A is a finitely generated commutative k-algebra and B is a finitely generated A-algebra, then B is finitely generated as a k-algebra*

1.3.8. — Let k be a commutative ring, let A be a commutative k-algebra which is a field, and let S be a subset of A. Then one also considers the subfield of A generated by S over k, denoted by $k(S)$: this is the smallest subfield of A that contains k and S. It is the set of all fractions a/b, where $a, b \in k[S]$, the k-subalgebra of A generated by S, and $b \neq 0$.

Example (1.3.9) (General polynomial rings). — Let A be a ring and let I be any set. The set $\mathbf{N}^{(I)}$ is the set of all multi-indices indexed by I: its elements are families $(n_i)_{i \in I}$ consisting of positive integers, almost all of which are zero. When I is the finite set $\{1, \ldots, d\}$, then $\mathbf{N}^{(I)}$ is naturally identified with the set $\mathbf{N}^d$ of d-tuples of positive integers. The termwise addition of $\mathbf{N}^{(I)}$ endows it with the structure of a monoid.

The ring $\mathscr{P}_I$ is defined as the monoid ring associated with this monoid and coefficients in A. Let us recall from 1.1.6 its definition, specialized to this particular case.

By definition, $\mathscr{P}_I$ is the set $A^{(\mathbf{N}^I)}$ of families $(a_m)_{m \in \mathbf{N}^I}$ of elements of A, indexed by $\mathbf{N}^{(I)}$, with finite support, that is of which all but finitely many terms are zero (we also say that such a family is almost null). Endowed with term-by-term addition, it is an abelian group. Let $P = (p_m)$ and $Q = (q_m)$ be elements of $\mathscr{P}_I$. For any $m \in \mathbf{N}^{(I)}$, one may set

$$r_m = \sum_{m' + m'' = m} p_{m'} q_{m''},$$

because the sum is finite; the family (r_m) is an almost null family of elements of A indexed by $\mathbf{N}^{(I)}$, hence defines an element R of $\mathscr{P}_I$.

One checks that this law $(P, Q) \mapsto R$ is associative: for $P = (p_m)$, $Q = (q_m)$ and $R = (r_m)$, one has $(PQ)R = P(QR) = S$, where $S = (s_m)$ is given by

$$s_m = \sum_{m'+m''+m'''=m} p_{m'} q_{m''} r_{m'''}.$$

The unit element of $\mathscr{P}_I$ is the family (ε_m) such that $\varepsilon_0 = 1$ (for the multi-index $0 = (0, \dots)$) and $\varepsilon_m = 0$ for any $m \in \mathbf{N}^{(I)}$ such that $m \neq 0$.

The map $A \to A[(X_i)]$ given by $a \mapsto a1$ is a morphism of rings.

For any $m \in \mathbf{N}^{(I)}$, write X^m for the element of $\mathscr{P}_I$ whose only non-zero term is at m, and equals 1. One has $X^m X^{m'} = X^{m+m'}$. In particular, the elements of the form X^m pairwise commute.

For $i \in I$, let $\delta_i \in \mathbf{N}^{(I)}$ be the multi-index that equals 1 at i and 0 otherwise; one writes $X_i = X^{\delta_i}$. For any multi-index m, one has $m = \sum_{i \in I} m_i \delta_i$ (a finite sum, since all but finitely many coefficients m_i are zero), so that $X^m = \prod_{i \in I} X_i^{m_i}$. Consequently, for any $P = (p_m) \in \mathscr{P}_I$, one has

$$P = \sum_{m \in \mathbf{N}^{(I)}} p_m X^m = \sum_{m \in \mathbf{N}^{(I)}} p_m \prod_{i \in I} X_i^{m_i}.$$

The ring $\mathscr{P}_I$ is called *the ring of polynomials with coefficients in A in the family of indeterminates* $(X_i)_{i \in I}$; it is denoted by $A[(X_i)_{i \in I}]$.

The map $a \mapsto aX^0$ is a ring morphism from A to $A[(X_i)_{i \in I}]$; the elements of its image are called *constant polynomials*.

The choice of the letter X for denoting the indeterminates is totally arbitrary; x_i, Y_i, T_i, are other common choices. When $I = \{1, \dots, n\}$, one rather writes $A[X_1, \dots, X_n]$, or $A[x_1, \dots, x_n]$, or $A[Y_1, \dots, Y_n]$, etc. When the set I has only one element, the indeterminate is often denoted by x, T, X, etc., so that this ring is denoted by $A[x]$, $A[T]$, $A[X]$, accordingly. When I has few elements, the indeterminates are also denoted by distinct letters, as in $A[X, T]$, or $A[X, Y, Z]$, etc.

When A is a commutative ring, the ring of polynomials is a commutative ring, hence an A-algebra via the morphism $A \to A[(X_i)]$ given by $a \mapsto a1$.

Remark (1.3.10). — Let A be a ring, let I be a set, let J be a subset of I and let $K = I - J$. Extending families $(a_j)_{j \in J}$ by zero, let us identify $\mathbf{N}^{(J)}$ with a subset of $\mathbf{N}^{(I)}$, and similarly for $\mathbf{N}^{(K)}$. Then every element $m \in \mathbf{N}^{(I)}$ can be uniquely written as a sum $m = m' + m''$, where $m' \in \mathbf{N}^{(J)}$ and $m'' \in \mathbf{N}^{(K)}$; in fact, $m'_j = m_j$ for $j \in J$, and $m''_k = m_k$ for $k \in K$.

Then we can write any polynomial $P = \sum_{m \in \mathbf{N}^{(I)}} p_m X^m$ in $\mathscr{P}_I$ in the form

$$P = \sum_{n \in \mathbf{N}^{(K)}} \left(\sum_{m \in \mathbf{N}^{(J)}} p_{m+n} X^m \right) X^n = \sum_{n \in \mathbf{N}^{(K)}} P_n X^n,$$

where $P_n = \sum_{m \in \mathbf{N}^{(I)}} p_{m+n} X^m \in \mathscr{P}_J$. This furnishes a map $P \mapsto (P_n)_n$ from $\mathscr{P}_I$ to the ring $(A[(X_j)_{j \in J}])[(X_k)_{k \in K}]$. One checks readily that this map is bijective and is a ring morphism; it is thus an isomorphism of rings.

1.3.11. Monomials and degrees — For any $m \in \mathbf{N}^{(I)}$, the element $X^m = \prod_{i \in I} X_i^{m_i}$ is called the *monomial* of *exponent* m. One defines its degree with respect to X_i to be m_i, and its total degree as $\sum_{i \in I} m_i$.

Let P be a polynomial in $A[(X_i)]$. Write $P = \sum a_m X^m$, the monomials of P are the polynomials $a_m X^m$ for $a_m \neq 0$. For any $i \in I$, the *degree of P with respect to* X_i, denoted by $\deg_{X_i}(P)$, is the least upper bound of all degrees with respect to X_i of all monomials of P. Similarly, the *total degree* $\deg(P)$ of P is the least upper bound of the total degrees of all monomials of P. These least upper bounds are computed in $\mathbf{N} \cup \{-\infty\}$, so that the degrees of the zero polynomial are equal to $-\infty$; otherwise, they are positive.

The constant term of a polynomial P is the coefficient of the monomial of exponent $0 \in \mathbf{N}^{(I)}$. A polynomial P is *constant* if it is reduced to this constant term, equivalently if its total degree is 0, equivalently if its degree with respect to every indeterminate is 0.

When there is only one indeterminate T, the degree $\deg_T$ and the total degree deg of a polynomial $P \in A[T]$ coincide. Moreover, if $\deg(P) = n$, then T^n is the unique monomial of degree n, and its coefficient is called the *leading coefficient* of P. The polynomial P is said to be monic if its leading coefficient is equal to 1.

Let $d \in \mathbf{N}$. One says that a polynomial $P \in \mathscr{P}_I$ is *homogeneous* of degree d if all of its monomials have total degree d. One can write any polynomial $P \in \mathscr{P}_I$ uniquely as a sum $P = \sum_{k=0}^{d} P_d$, where $P_d \in \mathscr{P}_I$ is homogeneous of degree d. In fact, if $P = \sum_{m \in \mathbf{N}^{(I)}} p_m X^m$, then $P_d = \sum_{\deg(X^m)=d} p_m X^m$.

Let $i \in I$. For all polynomials P and Q, one has

$$\deg_{X_i}(P + Q) \leq \sup(\deg_{X_i}(P), \deg_{X_i}(Q)),$$

with equality if $\deg_{X_i}(P) \neq \deg_{X_i}(Q)$. Moreover,

$$\deg(P + Q) \leq \sup(\deg(P), \deg(Q)).$$

Regarding multiplication, we have

$$\deg_{X_i}(PQ) \leq \deg_{X_i}(P) + \deg_{X_i}(Q),$$
$$\deg(PQ) \leq \deg(P) + \deg(Q),$$

and we shall see below that these inequalities are equalities if A is a domain.

Proposition (1.3.12). — *Let* P *and* $Q \in A[T]$ *be non-zero polynomials in* one *indeterminate* T. *Assume that the leading coefficient of* P *is left regular, or that the leading coefficient of* Q *is right regular. Then,* $\deg(PQ) = \deg(P) + \deg(Q)$. *In particular,* $PQ \neq 0$.

Proof. — Write $P = p_0 + p_1 T + \cdots + p_m T^m$ and $Q = q_0 + \cdots + q_n T^n$, with $m = \deg P$ and $n = \deg Q$, so that $p_m \neq 0$ and $q_n \neq 0$. Then,

$$PQ = p_0 q_0 + (p_0 q_1 + p_1 q_0)T + \cdots + (p_{m-1} q_n + p_m q_{n-1})T^{m+n-1} + p_m q_n T^{m+n}.$$

By assumption, $p_m q_n \neq 0$ and the degree of PQ is equal to $m + n$, as claimed.□

Corollary (1.3.13). — *Let* A *be a domain, and let* I *be a set. Then, for any polynomials* $P, Q \in A[(X_i)_{i \in I}]$, *and any* $i \in I$, *one has*

$$\deg_{X_i}(PQ) = \deg_{X_i}(P) + \deg_{X_i}(Q) \quad and \quad \deg(PQ) = \deg(P) + \deg(Q).$$

In particular, the polynomial ring $A[(X_i)]$ *is a domain.*

Proof. — We first prove that the product of two non-zero polynomials is not equal to 0, and that its degree with respect to the indeterminate X_i is the sum of their degrees.

Since P and Q have only finitely many monomials, we may assume that there are only finitely many indeterminates. We can then argue by induction on the number of indeterminates that appear in P and Q. If this number is 0, then there $P = p$ and $Q = q$, for $p, q \in A$, and $\deg_{X_i}(P) = \deg_{X_i}(q) = 0$. Since A is a domain, one has $pq \neq 0$, hence $PQ = pq \neq 0$, and $\deg_{X_i}(PQ) = 0$.

Let $m = \deg_{X_i}(P)$ and $n = \deg_{X_i}(Q)$. First assume that the indeterminate X_i appears in P or Q, so that $m > 0$ or $n > 0$. Let $J = I - \{i\}$. There are polynomials (P_m) and (Q_n) in the ring $\mathscr{P}_J$ of polynomials with coefficients in A in the indeterminates X_j, for $j \in J$, such that $P = \sum_{k=0}^{m} P_k X_i^k$ and $Q = \sum_{k=0}^{n} Q_k X_i^k$. Then one can write $PQ = \sum_{k=0}^{m+n} R_k X_i^k$ for some polynomials $R_k \in \mathscr{P}_J$, and one has $R_{m+n} = P_m Q_n$. Since X_i appears in P or Q, the number of indeterminates that appear in P_m or Q_n has decreased; by induction, one has $P_m Q_n \neq 0$. This proves that $\deg_{X_i}(PQ) = m + n$; in particular, $PQ \neq 0$.

If the indeterminate X_i appears neither in P, nor in Q, we can redo this argument with some other indeterminate j that appears in P or in Q; this shows that $PQ \neq 0$. Since the indeterminate X_i does not appear in PQ, this shows that $\deg_{X_i}(PQ) = 0$.

To prove the assertion about the total degrees, let $m = \deg(P), n = \deg(Q)$, and let us write $P = \sum_{i=0}^{m} P_i$ and $Q = \sum_{j=0}^{n} P_j$ as the sums of their homogeneous components. Then $PQ = \sum_{k=0}^{m+n} R_k$, where $R_k = \sum_{i+j=k} P_i Q_j$. Expanding the products $P_i Q_j$, we observe that for each k, the polynomial R_k is a sum of monomials of degree k, hence R_k is homogeneous of degree k. By assumption, $P_m \neq 0$ and $Q_n \neq 0$, so that $R_{m+n} = P_m Q_n \neq 0$. This proves that $\deg(PQ) = m + n$, as was to be shown.   □

Corollary (1.3.14). — *Let* A *be a domain and let* I *be a set. The units of the polynomial ring* $A[(X_i)_{i \in I}]$ *are the constant polynomials equal to a unit of* A.

(For the case of a general commutative ring A, see exercise 1.8.21.)

Proof. — Since the map $a \mapsto a1$ is a ring morphism from A to $A[(X_i)]$, it maps units to units, and the constant polynomials equal to a unit of A are units in $A[(X_i)]$.

Conversely, let P be a unit in $A[(X_i)]$ and let $Q \in A[(X_i)]$ be such that $PQ = 1$. Then $\deg(PQ) = \deg(P) + \deg(Q) = \deg(1) = 0$, hence $\deg(P) = \deg(Q) = 0$. This proves that P and Q are constant polynomials. Let $a, b \in A$ be such that $P = a$ and $Q = b$; then $ab = 1$, so that a, b are units in A. In particular, P is constant polynomial equal to a unit of A.   □

Theorem (1.3.15). — *Let A be a ring and let P, Q be two polynomials with coefficients in A in one indeterminate X. One assumes that $Q \neq 0$ and that the coefficient of its monomial of highest degree is invertible. Then, there exists a unique pair* (R, S) *of polynomials in A[X] such that $P = RQ + S$ and $\deg(S) < \deg(Q)$.*

Proof. — Let us begin with uniqueness. If $P = RQ + S = R'Q + S'$ for polynomials R, S, R', S' such that $\deg(S) < \deg(Q)$ and $\deg(S') < \deg(Q)$, then the degree of $(R' - R)Q = S - S'$ is at most $\sup(\deg(S), \deg(S')) < \deg Q$. Assume $R \neq R'$, that is, $R' - R \neq 0$, then $\deg((R'-R)Q) = \deg(R'-R) + \deg(Q) \geq \deg(Q)$, a contradiction. Consequently, $R = R'$ and $S = P - RQ = P - R'Q = S'$.

Let us now prove the existence of a pair (R, S) as in the statement of the theorem. Let $d = \deg(Q)$ and let u be the leading coefficient of Q; by assumption, it is a unit. We argue by induction on $n = \deg(P)$. If $n < d$, it suffices to set $R = 0$ and $S = P$. Otherwise, let aX^n be the monomial of highest degree of P. Then, $P' = P - au^{-1}X^{n-d}Q$ is a polynomial of degree at most n. However, by construction, the coefficient of X^n is equal to $a - au^{-1}u = 0$, so that $\deg(P') < n = \deg P$. By induction, there exist polynomials R' and S' in A[X] such that

$$P' = R'Q + S' \quad \text{and} \quad \deg(S') < \deg(Q).$$

Then,

$$P = P' + au^{-1}X^{n-d}Q = (R' + au^{-1}X^{n-d})Q + S'.$$

It now suffices to set $R = R' + au^{-1}X^{n-d}$ and $S' = S$. This concludes the proof.□

Remark (1.3.16). — Let us consider the particular case where $P = a_nX^n + \cdots + a_0$ and $Q = X - c$. Since $\deg(Q) = 1$, one has $\deg(S) = 0$ and there exists an element $s \in A$ such that $S = s$. In the case where A is commutative, one can evaluate the formula $P = RQ + S$ at c (see below) and obtain $s = P(c) = a_nc^n + \cdots + a_0$. In a non commutative ring, there is no such evaluation morphism that allows one to make sense of this argument, but let us show that this formula holds nevertheless.

Since the leading coefficient of Q is invertible, one has $\deg(Q) + \deg(R) = \deg(P)$, hence $\deg(R) = n - 1$; write $R = b_{n-1}X^{n-1} + \cdots + b_0$. One then has the following relations: $a_n = b_{n-1}, a_{n-1} = b_{n-2} - b_{n-1}c, \ldots, a_1 = b_0 - b_1c$ and $a_0 = -b_0c + s$. From there we get

$$s = a_0 + b_0c = a_0 + a_1c + b_1c^2 = \ldots$$
$$= a_0 + a_1c + \cdots + a_{n-1}c^{n-1} + b_{n-1}c^n$$
$$= a_0 + a_1c + \cdots + a_nc^n,$$

as claimed.

As a particular case of a monoid algebra, polynomial algebras obey an important *universal property*.

Proposition (1.3.17). — *Let k be a commutative ring. Let A be any k-algebra and let I be a set. For any family $(a_i)_{i \in I}$ of elements of A which commute pairwise, there exists a unique morphism $f : k[(X_i)_{i \in I}] \to A$ of k-algebras such that $f(X_i) = a_i$ for every $i \in I$.*

Proof. — If there is such a morphism f, it must satisfy

$$f(\lambda \prod_{i \in I} X_i^{m_i}) = \lambda \prod_{i \in I} f(X_i)^{m_i} = \lambda \prod_{i \in I} a_i^{m_i}$$

for any multi-index $(m_i) \in \mathbf{N}^{(I)}$ and any $\lambda \in k$. (In the previous formulas, all products are essentially finite, and the order of factors is not relevant, because the a_i commute.) Consequently, for any polynomial $P = \sum p_m X^m$, one must have

$$f(P) = \sum_{m \in \mathbf{N}^{(I)}} p_m \prod_{i \in I} a_i^{m_i},$$

which proves first that there exists at most one such morphism f, and second, that if it exists, it has to be defined by this formula.

Conversely, it is straightforward to prove, using that the a_i commute pairwise, that this formula does indeed define a morphism of k-algebras. □

Especially when $A = k$, this morphism is sometimes called the *evaluation morphism* at the point $(a_i)_{i \in I}$, and the image of the polynomial P is denoted by $P((a_i))$. When $I = \{1, \dots, n\}$, one writes simply $P(a_1, \dots, a_n)$. This gives, for example, a morphism of k-algebras $k[X_1, \dots, X_n] \to \mathscr{F}(k^n, k)$ from the algebra of polynomials in n indeterminates to the k-algebra of functions from k^n to k. The functions which belong to the image of this morphism are naturally called *polynomial functions*.

Here is an important example. Let k be a field, let V be a k-vector space and let A be the k-algebra of endomorphisms of V. One can also take A to be the k-algebra $M_n(k)$ of all $n \times n$ matrices with coefficients in k. Then, for any element $a \in A$ and any polynomial $P \in k[X]$, one may compute $P(a)$. For $P, Q \in k[X]$, one has $P(a) + Q(a) = (P + Q)(a)$, and $P(a)Q(a) = (PQ)(a)$. These formulas just express that the map from $k[X]$ to A given by $P \mapsto P(a)$ is a morphism of k-algebras.

Lemma (1.3.18). — *Let k be a commutative ring, let A be a k-algebra and let $a_1, \dots, a_n$ be elements of A that commute pairwise. Then, the subalgebra of A generated by $a_1, \dots, a_n$, $k[a_1, \dots, a_n]$, is the image of the morphism from $k[X_1, \dots, X_n]$ to A given by evaluation at $(a_1, \dots, a_n)$.*

Proof. — Indeed, let φ be this evaluation morphism. Since $\varphi(X_i) = a_i$, the image $\mathrm{Im}(\varphi)$ of φ is a subalgebra of A which contains $a_1, \dots, a_n$. Therefore, $\mathrm{Im}(\varphi)$ contains $k[a_1, \dots, a_n]$. Conversely, any subalgebra of A which contains $a_1, \dots, a_n$ also contains all elements of A of the form $\lambda a_1^{m_1} \dots a_n^{m_n}$, as well as their sums. This shows that $k[a_1, \dots, a_n]$ contains $\mathrm{Im}(\varphi)$, hence the equality.□

1.3.19. Formal derivatives — Let A be a ring and let $P \in A[X]$ be a polynomial; writing $P = \sum_{m=0}^{n} a_m X^m$, its (formal) *derivative* is the polynomial $P' = \sum_{m=1}^{n} m a_m X^{m-1}$.

We observe the following rules: $(aP)' = aP'$, $(P + Q)' = P' + Q'$ and $(PQ)' = P'Q + PQ'$ (*Leibniz rule*). Leaving to the reader the proofs of the first two, let us prove the third one. By additivity, we may assume that $P = X^m$ and $Q = X^n$; then $PQ = X^{m+n}$, $P' = mX^{m-1}$ and $Q' = nX^{n-1}$, hence

$$(PQ)' = (m + n)X^{m+n-1} = mX^{m+n-1} + nX^{m+n-1} = P'Q + PQ',$$

as claimed.

1.3.20. Roots and their multiplicities — Let K be a field and let $P \in K[X]$ be a non-constant polynomial. A *root* of P in K is an element $a \in K$ such that $P(a) = 0$.

If a is a root of P, then the euclidean division of P by $X - a$ takes the form $P = (X - a)Q + R$, where $R \in K[X]$ is a polynomial of degree < 1, hence a constant, and by evaluation, one has $0 = P(a) = (a - a)Q(a) + R(a) = R$, hence $P = (X - a)Q$ and $X - a$ divides P. Conversely, if $X - a$ divides Q, then $P(a) = 0$.

By successive euclidean divisions, one may write $P = Q \prod_{i=1}^{n}(X - a_i)^{m_i}$ where $Q \in K[X]$ is a polynomial without root in K, $a_1, \ldots, a_n$ are distinct elements of K and $m_1, \ldots, m_n$ are strictly positive integers. (The process must stop since the degree of Q decreases by 1 at each step, and is positive.)

Let us prove that m_i *is the largest integer m such that* $(X - a_i)^m$ *divides* P. This integer is called the *multiplicity* of the root a_i. By definition, $(X - a_i)^{m_i}$ divides P. Conversely, assume by contradiction that there exists an integer $m > m_i$ such that $(X - a_i)^m$ divides P; let us write $P = Q_1 (X - a_i)^m$. From the equality $P = Q_1 (X - a_i)^m = Q \prod_{j=1}^{n}(X - a_j)^{m_j}$, we then get $Q_1 (X - a_i)^{m-m_i} = Q \prod_{j \neq i}(X - a_j)^{m_j}$; evaluating both sides of this equality at $X = a_i$, we obtain $0 = Q(a_i) \prod_{j \neq i}(a_i - a_j)^{m_j}$, a contradiction.

Let $a \in K$ be such that $P(a) = 0$. Then there exists an integer $i \in \{1, \ldots, n\}$ such that $a = a_i$: the elements $a_1, \ldots, a_n$ are the roots of P in K.

One has $\deg(P) = \deg(Q) + \sum_{i=1}^{n} m_i$, hence $\sum_{i=1}^{n} m_i \leq \deg(P)$. *The number of roots of a polynomial, counted with their multiplicities, is at most its degree.*

Let us give an application of this fact.

Proposition (1.3.21). — *If K is a field, then every finite subgroup of the multiplicative group of K is cyclic.*

Proof. — Let G be a finite subgroup of $K^\times$ and let n be its cardinality. For every prime number p that divides n, there exists an element $x_p \in G$ such that $(a_p)^{n/p} \neq 1$, because the polynomial $T^{n/p} - 1$ has at most n/p roots and $n > n/p$. Let m_p be the exponent of p in the decomposition of n as a product of prime factors; set $a_p = x_p^{n/p^{m_p}}$. One has $(a_p)^{p^{m_p}} = 1$ and $(a_p)^{p^{m_p-1}} = x_p^{n/p} \neq 1$, so that the order of a_p in G is equal to p^{m_p}.

Set $a = \prod_p a_p$ and let m be its order; let us prove that the $m = n$. Otherwise, there exists a prime number ℓ that divides n/m. In particular, ℓ divides n, m divides n/ℓ and $a^{n/\ell} = 1$. If p is a prime number dividing n which is distinct from ℓ, then the integer p^{m_p} divides n/ℓ, hence $a_p^{n/\ell} = 1$. Consequently, $a^{n/\ell} = a_\ell^{n/\ell} \neq 1$. On the other hand, $a_\ell^{n/\ell} \neq 1$ because the order of a_ℓ, equal to ℓ^{m_ℓ}, does not divide n/ℓ. We thus have $a^{n/\ell} \neq 1$, a contradiction.

The subgroup G_1 of G generated by a is cyclic, of order n, and its elements are the nth roots of the polynomial $T^n - 1$ in K. If $g \in G$, then $g^n = 1$, by definition of n, hence $g \in G_1$, so that $G = G_1$. $\qquad\square$

The following lemma characterizes the multiplicity of a root using derivatives.

Lemma (1.3.22). — *Let $P \in K[X]$ be a non-constant polynomial and let $a \in K$ be a root of P; let m be its multiplicity.*

a) *For every integer $k \in \{0, \dots, m - 1\}$, a is a root of $P^{(k)}$ of multiplicity at least $m - k$.*

b) *If the characteristic of K is zero, or if it is strictly larger than m, then a is not a root of $P^{(m)}$.*

Proof. — Assume that a is a root of P with multiplicity at least m and let us write $P = (X - a)^m Q$, where $Q \in K[X]$ is a polynomial. Taking formal derivatives, we obtain

$$P' = m(X - a)^{m-1}Q + (X - a)^m Q' = (X - a)^{m-1}(mQ + (X - a)Q').$$

In particular, a is a root of P' with multiplicity $\geq m - 1$. By induction, we conclude that a is a root of $P^{(k)}$ of multiplicity at least $m - k$, for every $k \in \{1, \dots, m - 1\}$.

If the characteristic of K is zero, or if it is a prime number that does not divide m, then the polynomial $mQ + (X - a)Q'$ takes the value $mQ(a)$ at a. In this case, we see in particular that a is a root of multiplicity $m - 1$ of P' if and only if a is a root of multiplicity m of P.

By induction, we also have $P^{(m)}(a) = m!Q(a)$.

Assume that m is the multiplicity of a, so that $Q(a) \neq 0$. If the characteristic of K is zero, or if it is strictly greater than m, then $m! \neq 0$ in K, so that a is not a root of $P^{(m)}$. $\qquad\square$

Finally, recall that a field K is said to be *algebraically closed* if every non-constant polynomial $P \in K[T]$ has a root in K. By what precedes, this is equivalent to saying that every non-zero polynomial is a product of factors of degree 1; such a polynomial is said to be *split*. We will return to this topic in more detail in §4.3.

1.4. Ideals

Definition (1.4.1). — Let A be a ring and let I be a subgroup of A (with respect to the addition). One says that I is a *left ideal* if for any $a \in$ I and any $b \in$ A, one has $ba \in$ I. One says that I is a *right ideal* if for any $a \in$ I and any $b \in$ A, one has $ab \in$ I. One says that I is a *two-sided ideal* if it is both a left and right ideal.

In a commutative ring, it is equivalent for a subgroup to be a left ideal, a right ideal or a two-sided ideal; one then just says that it is an *ideal*. Let us observe that in any ring A, the subsets 0 and A are two-sided ideals. Moreover, an ideal I is equal to A if and only if it contains some unit, if and only if it contains 1

For any $a \in$ A, the set Aa consisting of all elements of A of the form xa, for $x \in$ A, is a left ideal; the set aA consisting of all elements of the form ax, for $x \in$ A, is a right ideal. When A is commutative, this ideal is denoted by (a)

To show that a subset I of A is a left ideal, it suffices to prove the following properties:

(i) $0 \in$ I;

(ii) for any $a \in$ I and any $b \in$ I, $a + b \in$ I;

(iii) for any $a \in$ I and any $b \in$ A, $ba \in$ I.

Indeed, since $-1 \in$ A and $(-1)a = -a$ for any $a \in$ A, these properties imply that I is a subgroup of A; the third one then shows that it is a left ideal.

The similar characterization of right ideals is left to the reader.

Example (1.4.2). — *If* K *is a division ring, the only left ideals (or right ideals) of* K *are* (0) *and* K. Indeed, let I be a left ideal of K such that I $\neq$ 0; let a be any non-zero element of I. Let b be any element of K. Since $a \neq 0$, it is invertible in K and, by definition of a left ideal, $b = (ba^{-1})a \in$ I. This shows that I = K.

Example (1.4.3). — *For any ideal* I *of* **Z**, *there exists a unique integer* $n \geq 0$ *such that* I = (n). *More precisely, if* I $\neq$ (0), *then* n *is the smallest strictly positive element of* I.

If I = (0), then $n = 0$ is the only integer such that I = (n). Assume now that I $\neq$ (0).

If I = (n), with $n > 0$, one observes that the strictly positive elements of I are $\{n; 2n; 3n; \dots\}$, and n is the smallest of them. This implies the uniqueness of such an integer n. So let n be the smallest strictly positive element of I. Since $n \in$ I, $an \in$ I for any $a \in$ **Z**, so that $(n) \subset$ I. Conversely, let a be any element of I. Let $a = qn + r$ be the euclidean division of a by n, with $q \in$ **Z** and $0 \leq r \leq n - 1$. Since a and $qn \in$ I, $r = a - qn$ belongs to I. Since n is the smallest strictly positive element of I and $0 \leq r < n$, we necessarily have $r = 0$ and $a = qn \in (n)$. This shows that I = (n).

Example (1.4.4). — *For any non-zero ideal* I *of* K[X], *there exists a unique monic polynomial* P $\in$ K[X] *such that* I = (P); *moreover,* P *is the unique monic polynomial of minimal degree which belongs to* I.

The proof is analogous to that of the previous example. If $I = (P)$, then the degree of any non-zero polynomial in I is at least $\deg(P)$, and P is the unique monic polynomial in I of minimal degree. This shows the uniqueness part of the assertion.

Let P be a monic polynomial in I, of minimal degree. Any multiple of P belongs to I, so $(P) \subset I$. Conversely, let $A \in I$, and let $A = PQ + R$ be the euclidean division of A by P, with $\deg(R) < \deg(P)$. Since A and P belong to I, $R = A - PQ \in I$. If $R \neq 0$, the quotient of R by its leading coefficient is a monic polynomial of degree $< \deg(P)$, which contradicts the definition of P. Consequently, $R = 0$, $A \in (P)$. Hence $I = (P)$.

There are various useful operations on ideals.

1.4.5. Intersection — Let I and J be left ideals of A; their intersection $I \cap J$ is again a left ideal. More generally, the *intersection of any family of left ideals of A is a left ideal of A.*

Proof. — Let $(I_s)_{s \in S}$ be a family of ideals of A and let $I = \bigcap_s I_s$. (If $S = \emptyset$, then $I = A$.) The intersection of any family of subgroups being a subgroup, I is a subgroup of A. Let now $x \in I$ and $a \in A$ and let us show that $ax \in I$. For any $s \in S$, $x \in I_s$, hence $ax \in I_s$ since I_s is a left ideal. It follows that ax belongs to every ideal I_s, hence $ax \in I$. □

We leave it to the reader to state and prove the analogous statements for right and two-sided ideals.

1.4.6. Ideal generated by a subset — Let S be any subset of A; there exists a smallest left ideal of A containing S and it is called the *left ideal generated by S.* Indeed, this ideal is nothing but the intersection of the family of all left ideals containing S. Equivalently, the left ideal generated by S contains S, and is contained in any left ideal which contains S.

Moreover, it is equal to the set of all linear combinations $\sum_{i=1}^{n} a_i s_i$, for $n \in \mathbf{N}$, $a_1, \ldots, a_n \in A$ and $s_1, \ldots, s_n \in S$. Indexing these elements by the elements of S rather than a set of integers, let us prove that this ideal is the set of all linear combinations $\sum_{s \in S} a_s s$, where $(a_s) \in A^{(S)}$ ranges over all almost null families of elements of A.

Proof. — By the preceding proposition, the intersection of the family of all left ideals containing S is a left ideal of A, hence is the smallest left ideal of A that contains S. Let I be the set of all linear combinations $\sum_{s \in S} a_s s$, for $(a_s) \in A^{(S)}$. Moreover, we see by induction on the number of non-zero terms in this sum that $\sum_{s \in S} a_s s$ belongs to any left ideal of A which contains S, so belongs to the left ideal generated by S. This proves that I is contained in the left ideal generated by S. To conclude, it suffices to show that I is a left ideal containing S. Taking a family (a_s) where all members are zero except for one of them equal to 1, we have $S \subset I$. We then check the axioms for a left ideal: we have $0 = \sum_{s \in S} 0s \in I$; if $\sum a_s s$ and $\sum b_s s$ belong to I, then almost all of the terms of the family $(a_s + b_s)_{s \in S}$ are equal to 0 and $\sum (a_s + b_s)s = \sum a_s s + \sum b_s s$; moreover, if $a = \sum a_s s \in I$ and $b \in A$, then $ba = \sum (b a_s)s$ belongs to I. □

Similar arguments show that there exists a smallest right ideal (*resp.* a smallest two-sided ideal) of A containing S; it is the intersection of the family of all right ideals (*resp.* of all two-sided ideals) of A which contain S. They are respectively equal to the set of all linear combinations $\sum_{s \in S} s a_s$ and $\sum_{s \in S} a_s s b_s$, where (a_s) and (b_s) run along $A^{(S)}$.

Definition (1.4.7). — The *kernel* $\mathrm{Ker}(f)$ of a ring morphism $f : A \to B$ is the set of all elements $a \in A$ such that $f(a) = 0$.

Since a ring morphism is in particular a morphism of the underlying additive groups, a ring morphism f is injective if and only if $\mathrm{Ker}(f) = 0$.

Proposition (1.4.8). — *The kernel of a ring morphism is a two-sided ideal.*

Proof. — Let $f : A \to B$ be a ring morphism. Since a morphism of rings is a morphism of abelian groups, $\mathrm{Ker}(f)$ is a subgroup of A. Moreover, if $x \in \mathrm{Ker}(f)$ and $a \in A$, then $f(ax) = f(a)f(x) = f(a)0 = 0$ so that $ax \in \mathrm{Ker}(f)$. Similarly, if $x \in \mathrm{Ker}(f)$ and $a \in A$, then $f(xa) = f(x)f(a) = 0$, so that $xa \in \mathrm{Ker}(f)$. This shows that $\mathrm{Ker}(f)$ is a two-sided ideal of A. $\square$

1.4.9. Image and inverse image of ideals — Let $f : A \to B$ be a morphism of rings and let J be a left ideal of B; the inverse image of J by f,

$$f^{-1}(J) = \{a \in A \,;\, f(a) \in J\}$$

is a left ideal of A.

Proof. — Since $f(0) = 0 \in J$, we have $0 \in f^{-1}(J)$. For any a and $b \in f^{-1}(J)$, $f(a + b) = f(a) + f(b) \in J$ since $f(a)$ and $f(b) \in J$ and J is an ideal of B. Finally, for any $a \in A$ and $b \in f^{-1}(J)$, we have $f(b) \in J$, hence $f(ab) = f(a)f(b) \in J$. $\square$

Similarly, the inverse image of a right ideal (resp. of a two-sided ideal) by a morphism of rings is a right ideal (resp. a two-sided ideal).

However, the image of a left ideal by a ring morphism is not necessarily a left ideal. Let $f : A \to B$ be a morphism of rings and I be a left ideal of A. We shall write $Bf(I)$, or even BI, for the left ideal of B generated by $f(I)$.

1.4.10. Sum of ideals — Let I and J be left ideals of a ring A. The set $I + J$ of all sums $a + b$, for $a \in I$ and $b \in J$, is a left ideal of A. It is also the left ideal of A generated by the subset $I \cup J$. More generally, for any family $(I_s)_{s \in S}$ of left ideals of A, the set of sums $\sum_s a_s$ of all almost null families $(a_s)_{s \in S}$, where, for any s, $a_s \in I_s$, is a left ideal of A, denoted by $\sum_s I_s$. It is also the left ideal of A generated by $\bigcup_s I_s$.

The similar assertions for right and two-sided ideals hold.

Proof. — Let us prove the result for left ideals. Since $0 = \sum_s 0$ and $0 \in I_s$ for every s, one has $0 \in \sum_s I_s$. Then, if $a = \sum_s a_s$ and $b = \sum_s b_s$ are any two elements of $\sum_s I_s$, then $a + b = \sum_s (a_s + b_s)$, where, for every $s \in S$, $a_s + b_s \in I_s$, almost all terms of this sum being null; hence $a + b \in \sum_s I_s$. Finally, if $a = \sum_s a_s$ belongs to $\sum I_s$ and $b \in A$, then $ba = \sum_s (b a_s)$. For every s, $b a_s \in I_s$, so that $ba \in \sum_s I_s$. We have shown that $\sum_s I_s$ is a left ideal of A.

To prove that $\sum I_s$ is the left ideal of A generated by the subset $\bigcup_s I_s$, we must establish two inclusions. Let I denote the latter ideal. First of all, for any $t \in S$ and any $a \in I_t$, we have $a = \sum_s a_s$, where $a_s = 0$ for $s \neq t$ and $a_t = a$. This shows that $a \in \sum_s I_s$ and the ideal $\sum_s I_s$ contains I_t. By definition of the left ideal I, we thus have $I \subset \sum_s I_s$. On the other hand, let $a = \sum_s a_s$ be any element of $\sum_s I_s$, where $a_s \in I_s$ for all s. All terms of this sum belong to I, by definition of I. By definition of a left ideal, this implies $a \in I$, hence $I \supset \sum_s I_s$. □

1.4.11. Product of two-sided ideals — Let A be a ring and let I, J be two-sided ideals of A. The set of all products ab, for $a \in I$ and $b \in J$, is not necessarily a two-sided ideal of A. We define IJ as the two-sided ideal generated by these products.

Let K be the set of all linear combinations $\sum_{s=1}^{n} a_s b_s$, with $n \in \mathbf{N}$, $a_s \in I$ and $b_s \in J$ for every i. It is contained in IJ. Let us show that K is a two-sided ideal of A. The relation $0 = 0 \cdot 0$ shows that $0 \in IJ$. If $c = \sum_{i=1}^{n} a_i b_i$ and $c' = \sum_{i=1}^{n'} a_i' b_i'$ are elements of IJ, then $c + c' = \sum_{i=1}^{n+n'} a_i b_i$, where, for $i \in \{n+1, \ldots, n+n'\}$, we have set $a_i = a_{i-n}'$ and $b_i = b_{i-n}'$; hence $c + c' \in IJ$. Let moreover $c = \sum_{i=1}^{n} a_i b_i \in K$ and let $a \in A$. One has

$$ac = a\left(\sum a_i b_i\right) = \sum (aa_i)b_i;$$

because I is a left ideal, one has $aa_i \in I$ for every i, hence $ac \in K$. Similarly, the equalities

$$ca = \left(\sum a_i b_i\right)a = \sum a_i(b_i a)$$

show that $ca \in K$ since, J being a right ideal, $b_i a \in J$ for every i. Since K contains all products ab, with $a \in I$ and $b \in J$, we have $IJ \subset K$ and, finally, $IJ = K$.

Since I and J are two-sided ideals, for any $a \in I$ and any $b \in J$, the product ab belongs to I and to J. It follows that $IJ \subset I \cap J$.

1.4.12. Nilradical, radical of an ideal — Let A be a *commutative* ring. The *nilpotent radical* of A, also called its *nilradical*, is the set of its nilpotent elements. As we prove below, it is an ideal of A. One says that the ring A is *reduced* if its nilpotent radical is (0), that is, if 0 is its only nilpotent element.

More generally, the *radical* of an ideal I of the commutative ring A is defined by the formula

$$\sqrt{I} = \{a \in A \,;\, \text{there exists an integer } n \geq 1 \text{ such that } a^n \in I\}.$$

It is an ideal of A which contains I. By definition, the nilpotent radical of A is thus the radical $\sqrt{(0)}$ of the null ideal (0).

Proof. — Since $0^1 = 0 \in I$, one has $0 \in \sqrt{I}$. For $a \in \sqrt{I}$ and $b \in \sqrt{I}$, let us choose integers n and $m \geq 1$ such that $a^n \in I$ and $b^m \in I$. By the binomial formula

$$(a + b)^{n+m-1} = \sum_{k=0}^{n+m-1} \binom{n + m - 1}{k} a^k b^{n+m-1-k}.$$

In this sum, all terms belong to I: this holds for those corresponding to $k \geq n$, since, then, $a^k = a^n a^{n-k}$ and $a^n \in$ I; similarly, if $k < n$, then $n + m - 1 - k \geq m$, hence $b^{n+m-1-k} = b^m b^{n-1-k}$ belongs to I. Therefore, $(a + b)^{n+m-1} \in$ I, and $a + b \in \sqrt{I}$. Finally, for any $a \in \sqrt{I}$ and $b \in$ A, let us choose $n \geq 1$ such that $a^n \in$ I. Then, $(ba)^n = b^n a^n \in$ I, hence $ba \in \sqrt{I}$.   □

1.5. Quotient Rings

Given a ring and an adequate equivalence relation on that ring, the goal of this section is to endow the set of equivalence classes with the structure of a ring. This will allow us to formally make all elements of an ideal equal to zero without modifying the other rules for computation.

1.5.1. Construction — Let $\mathscr{R}$ be binary relation on a set X. Recall that one says that $\mathscr{R}$ is an *equivalence relation* if it is reflexive (for any x, $x \mathrel{\mathscr{R}} x$), symmetric (if $x \mathrel{\mathscr{R}} y$, then $y \mathrel{\mathscr{R}} x$) and transitive (if $x \mathrel{\mathscr{R}} y$ and $y \mathrel{\mathscr{R}} z$, then $x \mathrel{\mathscr{R}} z$). The set of all equivalence classes of X for the relation $\mathscr{R}$ is denoted by $X/\mathscr{R}$; the map $\mathrm{cl}_{\mathscr{R}} \colon X \to X/\mathscr{R}$ (such that, for every $x \in X$, $\mathrm{cl}_{\mathscr{R}}(x)$ is the equivalence class of x) is a surjection; its important property is that any two elements have the same image if and only if they are in relation by $\mathscr{R}$.

Let A be a ring. Let us search for all equivalence relations which are *compatible with the ring structure*, namely

if $x \mathrel{\mathscr{R}} y$ and $x' \mathrel{\mathscr{R}} y'$, then $x + x' \mathrel{\mathscr{R}} y + y'$ and $xx' \mathrel{\mathscr{R}} yy'$.

Let I be the equivalence class of 0. If $x \mathrel{\mathscr{R}} y$, since $(-y) \mathrel{\mathscr{R}} (-y)$, one gets $x - y \mathrel{\mathscr{R}} 0$, hence $x - y \in$ I, and conversely. Therefore, the relation $\mathscr{R}$ can be recovered from I by the property: $x \mathrel{\mathscr{R}} y$ if and only if $x - y \in$ I.

On the other hand, let us show that I is a two-sided ideal of A. Of course, one has $0 \in$ I. Moreover, for any $x \in$ I and $y \in$ I, one has $x \mathrel{\mathscr{R}} 0$ and $y \mathrel{\mathscr{R}} 0$, so that $(x + y) \mathrel{\mathscr{R}} 0$, which shows that $x + y \in$ I. Finally, for $x \in$ I and $a \in$ A, one has $x \mathrel{\mathscr{R}} 0$, hence $ax \mathrel{\mathscr{R}} a0$ and $xa \mathrel{\mathscr{R}} 0a$; since $a0 = 0a = 0$, we get that $ax \in$ I and $xa \in$ I.

In the opposite direction, the preceding computations show that the following theorem holds.

Theorem (1.5.2). — *Let A be a ring and let I be a two-sided ideal of A. The binary relation $\mathscr{R}_I$ on A given by $x \mathrel{\mathscr{R}_I} y$ if and only if $x - y \in$ I is an equivalence relation on A which is compatible with its ring structure. There exists a unique structure of a ring on the quotient set $A/\mathscr{R}_I$ for which the canonical surjection $\mathrm{cl}_{\mathscr{R}_I} \colon A \to A/\mathscr{R}_I$ be a morphism of rings. This morphism is surjective and its kernel is equal to I.*

The *quotient ring* $A/\mathscr{R}_I$ is rather denoted by A/I; the morphism $\mathrm{cl}_{\mathscr{R}_I}$ is called the canonical surjection from A to A/I and is often denoted by cl_I

Let a be any element of the center of A; observe that $\mathrm{cl}_I(a)$ belongs to the center of A/I. Let indeed $x \in \mathrm{A}/\mathrm{I}$; there exists an element $b \in \mathrm{A}$ such that $x = \mathrm{cl}(b)$; then, $\mathrm{cl}(a)x = \mathrm{cl}(a)\,\mathrm{cl}(b) = \mathrm{cl}(ab) = \mathrm{cl}(ba)$ since a is central, hence $\mathrm{cl}(a)x = \mathrm{cl}(b)\,\mathrm{cl}(a) = x\,\mathrm{cl}(a)$. Consequently, for any commutative ring k and any ring morphism $i\colon k \to \mathrm{A}$ whose image is contained in the center of A, so that (A, i) is a k-algebra, the composition $\mathrm{cl}_I \circ i\colon k \to \mathrm{A} \to \mathrm{A}/\mathrm{I}$ endows A/I with the structure of a k-algebra (in fact, the unique one!) for which the canonical surjection cl_I is a morphism of k-algebras.

Quotient rings are most often used through their universal property embodied in the following *factorization theorem.*

Theorem (1.5.3). — *Let* A *and* B *be rings and let* $f\colon \mathrm{A} \to \mathrm{B}$ *be a morphism of rings. For any two-sided ideal* I *of* A *which is contained in* $\mathrm{Ker}(f)$, *there exists a unique ring morphism* $\varphi\colon \mathrm{A}/\mathrm{I} \to \mathrm{B}$ *such that* $f = \varphi \circ \mathrm{cl}_I$.

The morphism φ *is surjective if and only if* f *is surjective; the morphism* φ *is injective if and only if* $\mathrm{Ker}(f) = \mathrm{I}$.

It is useful to understand this last equality with the help of the following diagram:

$$
\begin{array}{ccc}
\mathrm{A} & \xrightarrow{\ f\ } & \mathrm{B} \\
{\scriptstyle \mathrm{cl}_I}\big\downarrow & \nearrow_{\varphi} & \\
\mathrm{A}/\mathrm{I} & &
\end{array}
$$

in which the two paths that go from A to B (either the arrow f, or the composition $\varphi \circ \mathrm{cl}_I$ of the two arrows φ and cl_I) coincide. For that reason, one says that this diagram is *commutative.*

In the context of theorem 1.5.3, the morphisms f and cl_I are given, while this theorem asserts the existence and uniqueness of the morphism φ that completes the diagram and makes it commutative.

Proof. — Necessarily, φ has to satisfy $\varphi(\mathrm{cl}_I(a)) = f(a)$ for every $a \in \mathrm{A}$. Since every element of A/I is of the form $\mathrm{cl}_I(a)$, for some $a \in \mathrm{A}$, this shows that there exists at most one ring morphism $\varphi\colon \mathrm{A}/\mathrm{I} \to \mathrm{B}$ such that $f = \varphi \circ \mathrm{cl}_I$.

Let us now show its existence. Let $x \in \mathrm{A}/\mathrm{I}$ and let $a \in \mathrm{A}$ such that $x = \mathrm{cl}_I(a)$. Let a' be any other element of A such that $x = \mathrm{cl}_I(a')$; by definition, $a' - a \in \mathrm{I}$, hence $f(a' - a) = 0$ since $\mathrm{I} \subset \mathrm{Ker}(f)$; this shows that $f(a) = f(a')$. Thus one can set $\varphi(x) = f(a)$, the result is independent of the chosen element a such that $x = \mathrm{cl}_I(a)$. It remains to show that the map φ just defined is a morphism of rings.

Since $\mathrm{cl}_I(0_A) = 0_{A/I}$ and $\mathrm{cl}_I(1_A) = 1_{A/I}$, we have $f(0_{A/I}) = 0_B$ and $f(1_{A/I}) = 1_B$. Moreover, let x and y be elements of A/I and let a and $b \in \mathrm{A}$ be such that $x = \mathrm{cl}_I(a)$ and $y = \mathrm{cl}_I(b)$. One has $x + y = \mathrm{cl}_I(a + b)$ and

$$
\begin{aligned}
\varphi(x + y) = \varphi(\mathrm{cl}_I(a + b)) &= f(a + b) = f(a) + f(b) \\
&= \varphi(\mathrm{cl}_I(a)) + \varphi(\mathrm{cl}_I(b)) = \varphi(x) + \varphi(y).
\end{aligned}
$$

Similarly,

$$\varphi(xy) = f(ab) = f(a)f(b) = \varphi(x)\varphi(y).$$

Therefore, φ is a morphism of rings, as claimed, and this completes the proof of the existence and uniqueness of the morphism φ.

By the surjectivity of cl_I, the relation $f = \varphi \circ \mathrm{cl}_I$ implies that f is surjective if and only if φ is surjective.

Let us assume that φ is injective. Let $a \in \mathrm{Ker}(f)$; one has $0 = f(a) = \varphi(\mathrm{cl}_I(a))$. Consequently, $\mathrm{cl}_I(a) \in \mathrm{Ker}(\varphi)$, hence $\mathrm{cl}_I(a) = 0$ since φ is injective; this proves that $a \in I$, hence $\mathrm{Ker}(f) \subset I$. On the other hand, $I \subset \mathrm{Ker}(f)$ by assumption, hence $I = \mathrm{Ker}(f)$.

Finally, let us assume that $I = \mathrm{Ker}(f)$. Let $x \in A/I$ be such that $\varphi(x) = 0$; let $a \in A$ be such that $x = \mathrm{cl}_I(a)$; then $f(a) = \varphi(x) = 0$, hence $a \in I$ and $x = 0$. This proves that φ is injective.   □

Let $f: A \to B$ be a morphism of rings. We saw (page 7) that $f(A)$ is a subring of B. Consequently, the factorization theorem allows us to write the morphism f as

$$A \xrightarrow{\mathrm{cl}_{\mathrm{Ker}(f)}} A/\mathrm{Ker}(f) \xrightarrow{\varphi} f(A) \hookrightarrow B$$

that is, as the composition of a surjective morphism, an isomorphism, and an injective morphism of rings.

Let A be a ring and let I be a two-sided ideal of A. We want to describe the ideals of the quotient ring A/I. So let $\mathcal{J}$ be a left ideal of A/I. We know that $\mathrm{cl}^{-1}(\mathcal{J})$ is a left ideal of A. By construction, it contains the ideal I, for $\mathrm{cl}(a) = 0$ belongs to $\mathcal{J}$ for any $a \in I$. The important property is the following proposition.

Proposition (1.5.4). — *Let A be a ring and let I be a two-sided ideal of A. The map* cl^{-1}:

$$\text{left ideals of } A/I \to \text{left ideals of A containing I}$$
$$\mathcal{J} \mapsto \mathrm{cl}^{-1}(\mathcal{J})$$

is a bijection. The same result holds for right ideals and for two-sided ideals.

In other words, for every left ideal J of A containing I, there exists a unique ideal $\mathcal{J}$ of A/I such that $J = \mathrm{cl}^{-1}(\mathcal{J})$. Moreover, $\mathcal{J} = \mathrm{cl}(J)$ (so that, in this case, the image of the ideal J by the canonical surjection cl is still a left ideal).

Proof. — Let us first give the inverse map. Let J be a left ideal of A and let us show that $\mathrm{cl}(J)$ is a left ideal of A/I. Obviously, $0 = \mathrm{cl}(0) \in \mathrm{cl}(J)$. Moreover, let x and y belong to $\mathrm{cl}(J)$ and let a and b be elements of J such that $x = \mathrm{cl}(a)$ and $y = \mathrm{cl}(b)$. Then, $x + y = \mathrm{cl}(a) + \mathrm{cl}(b) = \mathrm{cl}(a + b)$; since J is a left ideal of A, $a + b$ belongs to J and $x + y$ is an element of $\mathrm{cl}(J)$. Finally, let x be an element of $\mathrm{cl}(J)$ and let y be an element of A/I. Choose again $a \in J$ and $b \in A$ such that $x = \mathrm{cl}(a)$ and $y = \mathrm{cl}(b)$. Then $yx = \mathrm{cl}(b)\mathrm{cl}(a) = \mathrm{cl}(ba) \in \mathrm{cl}(J)$ since, J being a left ideal of A, $ba \in J$.

If $\mathcal{J}$ is an ideal of A/I, I claim that

$$\boxed{\mathrm{cl}(\mathrm{cl}^{-1}(\mathcal{J})) = \mathcal{J}.}$$

We show the two inclusions separately. Any element x of $\mathrm{cl}(\mathrm{cl}^{-1}(\mathcal{J}))$ is of the form $x = \mathrm{cl}(a)$ for some $a \in \mathrm{cl}^{-1}(\mathcal{J})$, hence $x \in \mathcal{J}$. Conversely, let $x \in \mathcal{J}$ and let $a \in A$ be such that $x = \mathrm{cl}(a)$. Then $\mathrm{cl}(a) = x \in \mathcal{J}$, hence a belongs to $\mathrm{cl}^{-1}(\mathcal{J})$ and x is an element of $\mathrm{cl}(\mathrm{cl}^{-1}(\mathcal{J}))$, as claimed.

Let us show that for any left ideal of A,

$$\boxed{\mathrm{cl}^{-1}(\mathrm{cl}(J)) = I + J.}$$

Again, we show the two inclusions. Let $x \in I + J$ and let us write $x = a + b$ with $a \in I$ and $b \in J$. Consequently, $\mathrm{cl}(x) = \mathrm{cl}(a) + \mathrm{cl}(b) = \mathrm{cl}(b) \in \mathrm{cl}(J)$, hence $x \in \mathrm{cl}^{-1}(\mathrm{cl}(J))$. In the other way, let $x \in \mathrm{cl}^{-1}(\mathrm{cl}(J))$; by definition, $\mathrm{cl}(x) \in \mathrm{cl}(J)$ and there exists an element $a \in J$ such that $\mathrm{cl}(x) = \mathrm{cl}(a)$. We get $\mathrm{cl}(x - a) = 0$, which means that $x - a \in I$. Finally, $x = (x - a) + a$ belongs to $I + J$, as was to be shown.

If, moreover, J contains I, then $I + J = J$ and the two boxed formulas show that the map cl^{-1} induces an bijection from the set of left ideals of A/I onto the set of left ideals of A which contain I, whose inverse map is given by $\mathrm{cl}.\square$

When J is a left ideal of A which contains I, the ideal $\mathrm{cl}(J)$ of A/I is also denoted by J/I. This is useful when the map cl is omitted from the notation. In particular, when we write "let J/I be an ideal of A/I...", we will always mean that J is an ideal of A containing I.

Proposition (1.5.5). — *Let A be a ring, let I be a two-sided ideal of A and let J be a two-sided ideal of A containing I. The composition of the canonical surjections $A \to A/I \to (A/I)/(J/I)$ has kernel J. This gives a canonical isomorphism*

$$A/J \simeq (A/I)/(J/I).$$

In short, a quotient of a quotient is again a quotient.

Proof. — The composition of two surjective morphisms is still a surjective morphism, hence the morphism $A \to (A/I)/(J/I)$ is surjective. An element $a \in A$ belongs to its kernel if and only if the element $\mathrm{cl}(a) \in A/I$ belongs to the kernel of the morphism $\mathrm{cl}_{J/I} \colon A/I \to (A/I)/(J/I)$, which means $\mathrm{cl}(a) \in (J/I)$. Since $J/I = \mathrm{cl}(J)$ and J contains I, this is equivalent to the condition $a \in \mathrm{cl}^{-1}(\mathrm{cl}(J)) = J$.

The factorization theorem (theorem 1.5.3) then asserts the existence of a unique morphism $\varphi \colon A/J \to (A/I)/(J/I)$ which makes the diagram

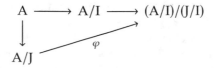

commutative. This map φ is surjective. Let us show that it is injective. Let $x \in A/J$ be such that $\varphi(x) = 0$. Let $a \in A$ be such that $x = \mathrm{cl}_J(a)$. By definition,

$\varphi(x) = \mathrm{cl}_{J/I} \circ \mathrm{cl}_I(a) = 0$, that is, $a \in J$. Consequently, $x = 0$ and the map φ is injective. This proves that φ is an isomorphism.   □

The last part of this proof can be generalized and gives an important complement to the factorization theorem.

Proposition (1.5.6). — *Let $f : A \to B$ be a morphism of rings and let I be a two-sided ideal of A contained in $\mathrm{Ker}(f)$. Let $\varphi : A/I \to B$ be the morphism given by the factorization theorem. Then the kernel of φ is given by $\mathrm{Ker}(f)/I$.*

Proof. — Indeed, let $x \in A/I$ be such that $\varphi(x) = 0$ and let $a \in A$ be any element with $x = \mathrm{cl}_I(a)$. Then $f(a) = 0$, hence $a \in \mathrm{Ker}(f)$ and $x = \mathrm{cl}(a) \in \mathrm{cl}_I(\mathrm{Ker}(f)) = \mathrm{Ker}(f)/I$. Conversely, if $x \in \mathrm{Ker}(f)/I$, there exists an element $a \in \mathrm{Ker}(f)$ such that $x = \mathrm{cl}_I(a)$. It follows that $\varphi(x) = f(a) = 0$ and $x \in \mathrm{Ker}(\varphi)$.   □

1.5.7. Chinese remainder theorem — We say that two two-sided ideals I and J of a ring A are *comaximal* if $I + J = A$. This notion gives rise to the general formulation of the *Chinese remainder theorem*.

Theorem (1.5.8). — *Let A be a ring and let I and J be two-sided ideals of A. Let us assume that I and J are comaximal. Then the canonical morphism $A \to (A/I) \times (A/J)$ given by $a \mapsto (\mathrm{cl}_I(a), \mathrm{cl}_J(a))$ is surjective; its kernel is the two-sided ideal $I \cap J$ of A. Passing to the quotient, we thus have an isomorphism*

$$\boxed{A/(I \cap J) \simeq A/I \times A/J.}$$

Corollary (1.5.9). — *Let I and J be comaximal two-sided ideals of a ring A. For any pair (x, y) of elements of A, there exists an element $a \in A$ such that $a \in x + I$ and $a \in y + J$.*

Proof. — In the diagram of rings

we have to show the existence of a unique morphism φ, drawn as a dashed arrow, so that this diagram is commutative, and that φ is an isomorphism. But the morphism $A \to A/I \times A/J$ maps $a \in A$ to $(\mathrm{cl}_I(a), \mathrm{cl}_J(a))$. Its kernel is thus $I \cap J$. By the factorization theorem, there exists a unique morphism φ that makes the diagram commutative, and this morphism is injective. For any $a \in A$, one has $\varphi(\mathrm{cl}_{I \cap J}(a)) = (\mathrm{cl}_I(a), \mathrm{cl}_J(a))$.

It remains to show that φ is surjective. Since $I + J = A$, there are $x \in I$ and $y \in J$ such that $x + y = 1$. Then, we have $1 = \mathrm{cl}_I(x + y) = \mathrm{cl}_I(y)$ in A/I and $1 = \mathrm{cl}_J(x + y) = \mathrm{cl}_J(x)$ in A/J. Let $\xi = \mathrm{cl}_{I \cap J}(x)$ and $\eta = \mathrm{cl}_{I \cap J}(y)$. For any $a, b \in A$, we get

$$\varphi(b\xi + a\eta) = f(bx + ay) = (\mathrm{cl}_I(a), \mathrm{cl}_J(b)).$$

Since any element of $(A/I) \times (A/J)$ is of the form $(\mathrm{cl}_I(a), \mathrm{cl}_J(b))$, the morphism φ is surjective. □

Remark (1.5.10). — Let I and J be ideals of a commutative ring such that $I + J = A$; then $I \cap J = IJ$.

We had already noticed the inclusion $IJ \subset I \cap J$, which does not require the ring A to be commutative, nor the ideals I and J to be comaximal. Conversely, let $a \in I \cap J$. Since $I + J = A$, there are $x \in I$ and $y \in J$ such that $x + y = 1$. Then $a = (x + y)a = xa + ya$. Since $x \in I$ and $a \in J$, $xa \in IJ$; since $ya = ay$, $y \in J$, and $a \in I$, we get $ya = ay \in IJ$. Finally, $a \in IJ$.

1.6. Fraction Rings of Commutative Rings

In the preceding section devoted to quotients, we somewhat "forced" some elements of a ring to vanish. We now want to perform a quite opposite operation: making *invertible* all the elements of an adequate subset. *Throughout this section, we restrict to the case of commutative rings* (for a preview of the general situation, see remark 1.6.8).

Definition (1.6.1). — Let A be a commutative ring. A subset S of A is said to be *multiplicative* if the following properties hold:

(i) $1 \in S$;

(ii) For any a and b in S, $ab \in S$.

Given a commutative ring A and a multiplicative subset S of A, our goal is to construct a ring $S^{-1}A$ together with a morphism $i\colon A \to S^{-1}A$ such that $i(S)$ consists of invertible elements of $S^{-1}A$, and which is "universal" for this property.

Let us first give a few examples.

Examples (1.6.2). —   *a*) Let $A = \mathbf{Z}$ and $S = \mathbf{Z} - \{0\}$, the ring $S^{-1}A$ will be equal to $\mathbf{Q}$ and $i : \mathbf{Z} \to \mathbf{Q}$ the usual injection. More generally, if A is a domain, then $S = A - \{0\}$ is a multiplicative subset of A and the ring $S^{-1}A$ will be nothing else but the field of fractions of A.

b) If all elements of S are invertible in A, then $S^{-1}A = A$.

c) Let $A = \mathbf{Z}$ and $S = \{1; 10; 100; \dots\}$ be the set of all powers of 10 in $\mathbf{Z}$. Then $S^{-1}A$ is the set of *decimal numbers*, those rational numbers which can be written as $a/10^n$, for some $a \in \mathbf{Z}$ and some positive integer n.

As these examples show, we are simply going to mimic middle school *calculus of fractions*.

1.6.3. Construction — On the set $A \times S$, let us define an equivalence relation $\sim$ by:

$(a, s) \sim (b, t)$ if and only if there exists a $u \in S$ such that $(at - bs)u = 0$.

It is indeed an equivalence relation:

– For any $(a, s) \in A \times S$, $(a, s) \sim (a, s)$, since $1 \in S$ and $(as - as)1 = 0$: the relation is reflexive;

– If $(a, s) \sim (b, t)$, let us choose $u \in S$ such that $(at - bs)u = 0$. Then $(bs - at)u = 0$, hence $(b, t) \sim (a, s)$: the relation is symmetric;

– Finally, if $(a, s) \sim (b, t)$ and $(b, t) \sim (c, u)$, let v and $w \in S$ satisfy $(at - bs)v = (bu - ct)w = 0$. Since

$$(au - cs)t = (at - bs)u + (bu - ct)s,$$

we have $(au - cs)vwt = 0$ (here we use that the ring A is commutative). Since v, w and t belong to S and S is a multiplicative subset, one has $vwt \in S$, hence $(a, s) \sim (c, u)$: the relation is transitive.

We write $S^{-1}A$ for the set of equivalence classes; the class of a pair (a, s) is denoted by a/s; let also $i: A \to S^{-1}A$ be the map that sends $a \in A$ to the class $a/1$. The set $A \times S$ is not a ring, in any reasonable way. However we are going to endow the quotient set $S^{-1}A$ with the structure of a ring in such a way that the map i is a morphism of rings.

The definition comes from the well-known formulas for the sum or product of fractions: we define

$$(a/s) + (b/t) = (at + bs)/st, \quad (a/s) \cdot (b/t) = (ab/st).$$

Let us first check that these formulas make sense: if $(a, s) \sim (a', s')$ and $(b, t) \sim (b', t')$, we have to show that

$$(at + bs, st) \sim (a't' + b's', s't') \quad \text{and} \quad (ab, st) \sim (a'b', s't').$$

Observe that

$$(at + bs)s't' - (a't' + b's')st = (as' - a's)tt' + (bt' - b't)ss'.$$

Let us choose $u, v \in S$ such that $(as' - a's)u = (bt' - b't)v = 0$; we get

$$((at + bs)s't' - (a't' + b's')st)uv = (as' - a's)u(tt'v) + (bt' - b't)v(uss') = 0,$$

hence $(at + bs, st) \sim (a't' + b's', s't')$. Similarly,

$$(abs't' - a'b'st)uv = (as' - a's)u(bt'v) + (bs' - b's)v(ua'sv) = 0,$$

so that $(ab, st) \sim (a'b', s't')$.

Checking that these laws give $S^{-1}A$ the structure of a ring is a bit long, but without surprise, and we shall not do it here. For example, distributivity of addition on multiplication can be proven in the following way: let a/s, b/t and c/u be elements of $S^{-1}A$, then

$$\frac{a}{s}\left(\frac{b}{t} + \frac{c}{u}\right) = \frac{a(bu + ct)}{stu} = \frac{abu}{stu} + \frac{act}{stu} = \frac{ab}{st} + \frac{ac}{su} = \frac{a}{s}\frac{b}{t} + \frac{a}{s}\frac{c}{u}.$$

The unit element of $S^{-1}A$ is $1/1$, the zero element is $0/1$.

The map $i: A \to S^{-1}A$ given, for any $a \in A$, by $i(a) = a/1$ is a morphism of rings. Indeed, $i(0) = 0/1 = 0$, $i(1) = 1/1 = 1$, and, for any a and $b \in A$,

$$i(a + b) = (a + b)/1 = a/1 + b/1 = i(a) + i(b)$$

and

$$i(ab) = (ab)/1 = (a/1)(b/1) = i(a)i(b).$$

Finally, for $s \in S$, $i(s) = s/1$ and $i(s)(1/s) = s/s = 1$, so that $i(s)$ is invertible in $S^{-1}A$ for any $s \in S$.

Remarks (1.6.4). — *a*) A fraction a/s is zero if and only if $(a, s) \sim (0, 1)$. By the definition of the relation $\sim$, this holds if and only if there exists an element $t \in S$ such that $ta = 0$. In particular, *the canonical morphism $i: A \to S^{-1}A$ is injective if and only if every element of S is regular.*

b) In particular, *a ring of fractions $S^{-1}A$ is null if and only if 0 belongs to S.* Indeed, $S^{-1}A = 0$ means that $1/1 = 1 = 0$, hence that there exists an element $t \in S$ such that $t \cdot 1 = t = 0$, in other words, that $0 \in S$. Informally, if you want to divide by 0 and still pretend that all classical rules of algebra apply, then everything becomes 0 (and algebra is not very interesting anymore).

A posteriori, this explains the middle school rule that *thou shalt never divide by zero!* if you could, the rules of calculus of fractions (imposed by algebra) would make any fraction equal to 0!

c) The definition of the equivalence relation used in the construction of the ring of fractions is at first surprising. It is in any case more complicated than the middle school rule that would claim that $a/s = b/t$ if and only if $at = bs$. If the ring A is a domain and $0 \notin S$, more generally when all elements of S are not zero divisors, this simpler definition is sufficient. However, in the general case, the simpler definition does not give rise to an equivalence relation.

The importance of this construction is embodied in its *universal property*, another factorization theorem.

Theorem (1.6.5). — *Let A be a commutative ring and let S be a multiplicative subset of A. Let $i: A \to S^{-1}A$ be the ring morphism that has been constructed above. Then, for any (possibly not commutative) ring B and any morphism $f: A \to B$ such that $f(S) \subset B^{\times}$, there exists a unique ring morphism $f: S^{-1}A \to B$ such that $f = \varphi \circ i$.*

One can sum up this last formula by saying that the diagram

$$
\begin{array}{ccc}
A & \xrightarrow{\;f\;} & B \\
{\scriptstyle i}\downarrow & \nearrow_{\varphi} & \\
S^{-1}A & &
\end{array}
$$

is commutative.

Proof. — If such a morphism φ exists, it has to satisfy

$$\varphi(a/s)f(s) = \varphi(a/s)\varphi(i(s)) = \varphi(a/s)\varphi(s/1) = \varphi(a/1) = \varphi(i(a)) = f(a)$$

hence

$$\varphi(a/s) = f(a)f(s)^{-1},$$

where $f(s)^{-1}$ is the inverse of $f(s)$ in B. Similarly, it also has to satisfy

$$f(s)\varphi(a/s) = \varphi(i(s))\varphi(a/s) = \varphi(s/1)\varphi(a/s) = \varphi(a/1) = f(a),$$

hence $\varphi(a/s) = f(s)^{-1}(f(a))$. This shows that there can exist at most one such morphism φ.

To establish its existence, it now suffices to check that these two formulas are compatible and define a ring morphism $\varphi: S^{-1}A \to B$ such that $\varphi \circ i = f$.

Let us check that the formula is well posed. So let $a, b \in A$, $s, t \in S$ such that $a/s = b/t$ and let $u \in S$ be such that $(at - bs)u = 0$. Then, $f(at)f(u) = f(bs)f(u)$, hence $f(at) = f(bs)$ since $f(u)$ is invertible in B. Since A is commutative, it follows that $f(t)f(a) = f(a)f(t) = f(b)f(s)$, hence $f(a)f(s)^{-1} = f(t)^{-1}f(b)$.

Applying this equality to $(b, t) = (a, s)$, we also obtain the relation $f(b)f(t)^{-1} = f(t)^{-1}f(b)$.

This proves that there is a map $\varphi: S^{-1}A \to B$ such that $f(a/s) = f(a)f(s)^{-1} = f(s)^{-1}f(a)$ for any $a \in A$ and any $s \in S$.

By construction, for any $a \in A$, one has $\varphi \circ i(a) = \varphi(a/1) = f(a)f(1)^{-1} = f(a)$, hence $\varphi \circ i = f$.

Let us now verify that φ is a ring morphism. We have

$$\varphi(0) = f(0/1) = f(1)^{-1}f(0) = 0 \quad \text{and} \quad \varphi(1) = f(1/1) = f(1)^{-1}f(1) = 1.$$

Then,

$$\varphi(a/s) + \varphi(b/t) = f(a)f(s)^{-1} + f(b)f(t)^{-1} = f(st)^{-1}\left(f(at) + f(bs)\right)$$
$$= f(st)^{-1}f(at + bs) = \varphi((at + bs)/st) = \varphi((a/s) + (b/t)).$$

Finally,

$$\varphi(a/s)\varphi(b/t) = f(s)^{-1}f(a)f(t)^{-1}f(b) = f(st)^{-1}f(ab)$$
$$= \varphi(ab/st) = \varphi((a/s)(b/t)).$$

The map φ is thus a ring morphism. That concludes the proof of the theorem.   □

It is also possible to construct the fraction ring as a quotient. I content myself with the particular case of a multiplicative subset consisting of the powers of a single element; the generalization to any multiplicative subset is left as an exercise (exercise 1.8.48).

Proposition (1.6.6). — *Let A be a commutative ring, let s be any element of A and let S = {1; s; s^2; . . . } be the multiplicative subset of A consisting of powers of s. The canonical morphism*

$$f : A[X] \to S^{-1}A, \quad P \mapsto P(1/s)$$

is surjective, its kernel is the ideal $(1 - sX)$. *Passing to the quotient, one deduces an isomorphism*

$$\varphi : A[X]/(1 - sX) \simeq S^{-1}A.$$

Proof. — Any element of $S^{-1}A$ can be written in the form a/s^n for some $a \in A$ and some integer $n \geq 0$. Then, $a/s^n = \varphi(aX^n)$ so that f is surjective. Its kernel contains the polynomial $1 - sX$ since $f(1 - aX) = 1 - a/a = 0$. Therefore, it contains the ideal $(1 - sX)$ generated by this polynomial. By the universal property of quotient rings, we get a well-defined ring morphism $\varphi : A[X]/(1 - sX) \to S^{-1}A$ which maps the class cl(P) modulo $(1 - sX)$ of a polynomial P to $f(P) = P(1/s)$. Let us show that φ is an isomorphism. By proposition 1.5.6, it will then follow that $\mathrm{Ker}(f) = (1 - sX)$.

To show that φ is an isomorphism, we shall construct its inverse. This inverse φ should satisfy $\psi(a/s^n) = \mathrm{cl}(aX^n)$ for all $a \in A$ and all $n \in \mathbf{N}$; we could check this by a direct computation, but let us rather choose a more abstract way. Let $g : A \to A[X]/(1 - sX)$ be the canonical morphism such that $g(a) = \mathrm{cl}(a)$, the class of the constant polynomial a modulo $1 - sX$. In the ring $A[X]/(1 - sX)$, we have $\mathrm{cl}(sX) = 1$, so that $\mathrm{cl}(s)$ is invertible, with inverse $\mathrm{cl}(X)$. By the universal property of fraction rings, there exists a unique morphism $\psi : S^{-1}A \to A[X]/(1 - sX)$ such that $\psi(a/1) = g(a)$ for any $a \in A$. By construction, for any $a \in A$ and any $n \geq 0$, one has $\psi(a/s^n) = a\,\mathrm{cl}(X^n) = \mathrm{cl}(aX^n)$.

Finally, let us show that ψ and φ are inverses of one another. For $P \in A[X]$, one has $\psi(\varphi(\mathrm{cl}(P))) = \psi(P(1/s))$. Consequently, if $P = \sum a_n X^n$, we get

$$\psi(\varphi(\mathrm{cl}(P))) = \psi(f(P)) = \psi(P(1/s))$$
$$= \psi\left(\sum(a_n/s^n)\right) = \sum \psi(a_n/s^n)$$
$$= \sum \mathrm{cl}(a_n X^n) = \mathrm{cl}\left(\sum a_n X^n\right) = \mathrm{cl}(P)$$

and $\psi \circ \varphi = \mathrm{id}$. Finally,

$$\varphi(\psi(a/s^n)) = \varphi(\mathrm{cl}(aX^n)) = f(aX^n) = a/s^n$$

and $\varphi \circ \psi = \mathrm{id}$. The ring morphism φ is therefore an isomorphism, with inverse ψ, as was to be shown.   □

Examples (1.6.7). —   *a)* Let A be a domain. The set $S = A - \{0\}$ is a multiplicative subset of A. The morphism $i : A \to S^{-1}$ is injective and the ring $S^{-1}A$ is a *field*, called the *field of fractions* of A.

Proof. — Since A is a domain, $1 \neq 0$ and $1 \in S$. On the other hand, for any two non-zero elements a and $b \in A$, their product ab is non-zero, for A is a domain. This shows that S is a multiplicative subset of A. Since no element of S is a zero divisor, the morphism $i\colon A \to S^{-1}A$ is injective.

Any element x of $S^{-1}A$ can be written a/s, for some $a \in A$ and some $s \neq 0$. Assume $x = 0$; then, there exists an element $b \in S$ such that $ab = 0$. Then $b \neq 0$ and, since A is a domain, $a = 0$. Reversing the argument, any element a/s of $S^{-1}A$, with $a \neq 0$, is distinct from 0. In particular, $1/1 \neq 0$ and the ring $S^{-1}A$ is non-zero. Let again $x = a/s$ be any non-zero element of $S^{-1}A$. Then $a \neq 0$, so that we can consider the element $y = s/a$ of $S^{-1}A$. Obviously, $xy = (a/s)(s/a) = as/as = 1$, hence x is invertible.

We thus have shown that $S^{-1}A$ is a field.   □

b) Let A be a commutative ring and let $s \in A$ be any element of A which is not nilpotent. Then, the subset $S = \{1; s; s^2; \dots\}$ of A is a multiplicative subset (obviously) which does not contain 0. By remark 1.6.4, b) above, the ring $S^{-1}A$ is non-zero; it is usually denoted by A_s.

c) Let $f\colon A \to B$ be a morphism of commutative rings. For any multiplicative subset S of A, $T = f(S)$ is a multiplicative subset of B. Moreover, there is a unique morphism $F\colon S^{-1}A \to T^{-1}B$ extending f, in the sense that $F(a/1) = f(a)/1$ for any $a \in A$. This follows from the universal property explained below applied to the morphism F_1 from A to $T^{-1}B$ given by $a \mapsto f(a)/1$; one just needs to check that for any $s \in A$, $F_1(s) = s/1 \in T$, which is the very definition of T. This can also be shown by hand, copying the proof of the universal property to this particular case.

If the morphism f is implicit, for example if B is given as an A-algebra, we will commit the abuse of writing $S^{-1}B$ instead of $T^{-1}B$.

d) Let $f\colon A \to B$ be a morphism of commutative rings and let T be a multiplicative subset of B. Then, $S = f^{-1}(T)$ is a multiplicative subset of A such that $f(S) \subset T$. There is a unique morphism $F\colon S^{-1}A \to T^{-1}B$ extending the morphism f.

e) Let I be an ideal of a commutative ring A. The set $S = 1+I$ of elements $a \in A$ such that $a - 1 \in I$ is a multiplicative subset. Indeed, this is the inverse image of the multiplicative subset $\{1\}$ of A/I by the canonical morphism $A \to A/I$.

f) Let A be a commutative ring and let P be an ideal of A such that $P \neq A$ and such that $ab \in P$ implies $a \in P$ or $b \in P$ (we shall say that P is a *prime ideal* of A, see definition 2.2.2). Then $A - P$ is a multiplicative subset of A (proposition 2.2.3) and the fraction ring $(A - P)^{-1}A$ is generally denoted by A_P.

Remark (1.6.8). — Let A be a (not necessarily commutative) ring and let S be a multiplicative subset of A. There are several purely formal ways of proving the existence and uniqueness of a ring A_S, endowed with a ring morphism $i\colon A \to A_S$, that satisfies the universal property: for any ring morphism $f\colon A \to B$ such that $f(S) \subset B^\times$, there exists a unique ring morphism $\tilde{f}\colon A_S \to B$ such that $f = \tilde{f} \circ i$. One of them consists in defining rings of polynomials in noncommuting indeterminates (replacing the commutative monoid $\mathbf{N}^{(I)}$

on a set I by the word monoid I* on I), adding such indeterminates T_s for all $s \in S$, and quotienting this ring by the two-sided ideal generated by the elements $sT_s - 1$ and $T_s s - 1$, for $s \in S$.

However, almost nothing can be said about it. For example, MALCEV (1937) constructed a ring without any non-zero zero divisor that admits no morphism to a field (1 should map to 0!).

For rings satisfying the so-called Ore condition, one can check that the construction using fractions furnishes a ring for which the universal property holds; see exercise 1.8.45.

1.7. Relations Between Quotient Rings and Fraction Rings

Finally, I want to study briefly the ideals of a fraction ring $S^{-1}A$. Here is a first result.

Proposition (1.7.1). — *Let A be a commutative ring and let S be a multiplicative subset of A; let $i\colon A \to S^{-1}A$ be the canonical morphism. For any ideal $\mathcal{I}$ of $S^{-1}A$, there exists an ideal I of A such that $\mathcal{I} = i(\mathrm{I})(S^{-1}A)$. In fact, the ideal $\mathrm{I} = i^{-1}(\mathcal{I})$ is the largest such ideal.*

Proof. — If $\mathcal{I} = i(\mathrm{I})(S^{-1}A)$, we have in particular $i(\mathrm{I}) \subset \mathcal{I}$, that is $\mathrm{I} \subset i^{-1}(\mathcal{I})$. Provided that the ideal $i^{-1}(\mathcal{I})$ satisfies the given assertion, this shows that it is the largest such ideal. We thus need to show that

$$\boxed{\mathcal{I} = i(i^{-1}(\mathcal{I}))(S^{-1}A).}$$

Since $i(i^{-1}(\mathcal{I})) \subset \mathcal{I}$, the ideal generated by $i(i^{-1}(\mathcal{I}))$ is contained in $\mathcal{I}$, hence the inclusion

$$i(i^{-1}(\mathcal{I}))(S^{-1}A) \subset \mathcal{I}.$$

Conversely, let $x \in \mathcal{I}$, let us choose $a \in A$ and $s \in S$ such that $x = a/s$. Then, $sx = i(a) \in \mathcal{I}$ so that $a \in i^{-1}(\mathcal{I})$. It follows that $sx \in i(i^{-1}(\mathcal{I}))$, hence $x = (sx)(1/s)$ belongs to $i(i^{-1}(\mathcal{I}))(S^{-1}A)$, the other required inclusion.   □

To shorten the notation, we can omit the morphism i and write $\mathrm{I}S^{-1}A$ instead of $i(\mathrm{I})S^{-1}A$. In fact, we shall rather denote this ideal by $S^{-1}\mathrm{I}$, this last notation being the one used in the general context of fraction modules (§3.6).

Proposition (1.7.2). — *Let A be a commutative ring, let S be a multiplicative subset of A and let I be an ideal of A. Let $T = \mathrm{cl}_\mathrm{I}(S) \subset A/\mathrm{I}$ be the image of S by the canonical surjection $\mathrm{cl}_\mathrm{I}\colon A \to A/\mathrm{I}$. There exists a unique ring morphism*

$$\varphi\colon S^{-1}A/S^{-1}\mathrm{I} \to T^{-1}(A/\mathrm{I})$$

such that for any $a \in A$, $\varphi(\mathrm{cl}_\mathrm{I}(a/1)) = \mathrm{cl}_\mathrm{I}(a)/1$. Moreover, φ is an isomorphism.

In a more abstract way, the two rings $S^{-1}A/S^{-1}I$ and $T^{-1}(A/I)$ are A-algebras (a quotient, or a fraction ring, of an A-algebra is an A-algebra). The proposition states that there exists a unique morphism of A-algebras between these two rings, and that this morphism is an isomorphism.

Proof. — It is possible to give an explicit proof, but it is more elegant (say, more abstract and less computational) to rely on the universal properties of quotients and fraction rings. Let us consider the composition

$$f: A \to A/I \to T^{-1}(A/I), \quad a \mapsto \mathrm{cl}(a)/1.$$

By this morphism, an element $s \in S$ is sent to $\mathrm{cl}(s)/1$, an invertible element of $T^{-1}(A/I)$ whose inverse is $1/\mathrm{cl}(s)$. By the universal property of fraction rings, there exists a unique ring morphism

$$\varphi_1: S^{-1}A \to T^{-1}(A/I)$$

such that $\varphi_1(a/1) = \mathrm{cl}(a)/1$.

If, moreover, $a \in I$, then

$$\varphi_1(a/1) = \varphi_1(a)/1 = \mathrm{cl}(a)/1 = 0$$

since $\mathrm{cl}(a) = 0$ in A/I. It follows that $\mathrm{Ker}(\varphi_i)$ contains the image of I in $S^{-1}A$; it thus contains the ideal $S^{-1}I$ generated by I in $S^{-1}A$. By the universal property of quotient rings, there exists a unique ring morphism

$$\varphi: S^{-1}A/S^{-1}I \to T^{-1}(A/I)$$

such that $\varphi(\mathrm{cl}(a/s)) = \mathrm{cl}(a)/\mathrm{cl}(s)$ for any $a/s \in S^{-1}A$.

We have shown that there exists a morphism $\varphi: S^{-1}A/S^{-1}I \to T^{-1}(A/I)$ of A-algebras. We can sum up the construction by the commutative diagram

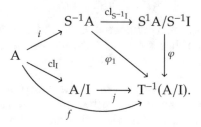

We also see that φ is the only possible morphism of A-algebras. Indeed, by the universal property of quotient rings, the morphism φ is characterized by its restriction to $S^{-1}A$ which, by the universal property of fraction rings, is itself characterized by its restriction to A.

It remains to prove that φ is an isomorphism. For that, let us now consider this diagram in the other direction. The kernel of the morphism $g = \mathrm{cl}_{S^{-1}I} \circ i$ contains I, hence there is a unique morphism of A-algebras

$$\psi_1: A/I \to S^{-1}A/S^{-1}I$$

satisfying $\psi_1(\mathrm{cl}(a)) = \mathrm{cl}(a/1)$ for any $a \in A$. For every $s \in S$, $\psi_1(\mathrm{cl}(s)) = \mathrm{cl}(s/1)$ is invertible, with inverse $\mathrm{cl}(1/s)$. Therefore, the image of T by ψ_1 consists of invertible elements in $S^{-1}A/S^{-1}I$. There thus exists a unique morphism of A-algebras,

$$\psi : T^{-1}(A/I) \to S^{-1}A/S^{-1}I$$

(that is, such that $\psi(\mathrm{cl}(a)/1) = \mathrm{cl}(a/1)$ for any $a \in A$). Again, these constructions are summarized by the commutative diagram:

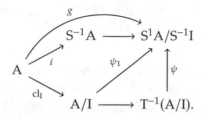

Finally, for $a \in A$ and $s \in S$, we have $\varphi(\mathrm{cl}(a/s)) = \mathrm{cl}(a)/\mathrm{cl}(s)$ in $T^{-1}(A/I)$ and $\psi(\mathrm{cl}(a)/\mathrm{cl}(s)) = \mathrm{cl}(a/s)$ in $S^{-1}A/S^{-1}I$. Necessarily, $\varphi \circ \psi$ and $\psi \circ \varphi$ are the identity maps, so that φ is an isomorphism.   □

This property will appear later under the fancy denomination of *exactness of localization.*

1.8. Exercises

Exercise (1.8.1). — *a*) Let A be a ring. Let us write A° for the abelian group A endowed with the multiplication • defined by $a \bullet b = ba$. Then A° is a ring, called the *opposite ring* of A.

b) Let A be the ring of $n \times n$ matrices with coefficients in **C**. Show that the transposition map, $M \mapsto M^t$, is an isomorphism from A to A°.

c) Show that the conjugation $q \mapsto \bar{q}$ of quaternions induces an isomorphism from **H** to H°.

Exercise (1.8.2). — *a*) Let A be a ring and let (B_i) be a family of subrings of A. Show that the intersection of all B_i is a subring of A.

b) Let A and B be rings, let $f, g : A \to B$ be morphisms of rings, and let C be the set of all $a \in A$ such that $f(a) = g(a)$. Prove that C is a subring of A.

c) Let A be a ring, let B be a subring of A and let I be a two-sided ideal of A. Let R be the set of all sums $a + b$, for $a \in B$ and $b \in I$. Show that R is a subring of A.

Exercise (1.8.3). — Let A be a ring and let S be a subset of A. In the following two cases, determine explicitly the centralizer $C_S(A)$.

a) When $A = \operatorname{End}_K(V)$ is the ring of endomorphisms of a finite dimensional K-vector space V and $S = \{u\}$ consists of a diagonalizable endomorphism u;

b) When $A = M_n(\mathbf{C})$ is the ring of complex $n \times n$ matrices and $S = \{M\}$ consists of a matrix with minimal polynomial X^n.

Exercise (1.8.4). — Let K be a field and let V be a finite-dimensional K-vector space. Show that the center of the ring $\operatorname{End}_K(V)$ consists of the homotheties $x \mapsto ax$, for $a \in K$.

Exercise (1.8.5). — Let A be a ring and let Z be the center of A. Let M be a monoid and let $A^{(M)}$ be the monoid ring of M with coefficients in A.

a) For $m \in M$, let δ_m be the function from M to A with value 1 at m, and 0 elsewhere. Compute the convolution product $\delta_m * \delta_{m'}$ in the monoid ring $A^{(M)}$.

b) For $m \in M$, compute the centralizer of the element δ_m in $A^{(M)}$.

c) Assume that M is a group G. Show that the center of the ring $A^{(G)}$ consists of all functions $f \colon G \to Z$ with finite support which are constant on every conjugacy class in G.

Exercise (1.8.6). — Let k be a commutative ring and let M be a monoid with unit element e. Let $k^{(M)}$ be the monoid algebra of M with coefficients in k; for $m \in M$, let $\delta_m \in A^{(M)}$ be the function from M to k with value 1 at m and 0 elsewhere.

Let A be a k-algebra.

a) Let $f \colon k^{(M)} \to A$ be a morphism of k-algebras; for $m \in M$, let $a_m = f(\delta_m)$. Prove that one has $a_e = 1_A$ and $a_m \cdot a_{m'} = a_{mm'}$ for $m, m' \in M$.

b) Conversely, let $(a_m)_{m \in M}$ be a family of elements of A such that $a_e = 1_A$ and $a_m \cdot a_{m'} = a_{mm'}$ for $m, m' \in M$. Prove that there exists a unique morphism of k-algebras $f \colon k^{(M)} \to A$ such that $f(\delta_m) = a_m$ for every $m \in M$. (*Universal property of a monoid algebra*)

Exercise (1.8.7). — a) Let $\partial \colon \mathbf{C}[X] \to \mathbf{C}[X]$ be the map given by $\partial(P) = P'$, for $P \in \mathbf{C}[X]$. A differential operator on $\mathbf{C}[X]$ is a $\mathbf{C}$-linear map of $\mathbf{C}[X]$ to itself of the form $P \mapsto \sum_{i=0}^n a_i(X)\partial^i(P)$, where the a_i are polynomials.

b) Show that the set of all differential operators on $\mathbf{C}[X]$, endowed with addition and composition, is a ring.

c) Determine its center.

Exercise (1.8.8). — Let $f \colon A \to B$ be a ring morphism.

a) Let R be the set of all ordered pairs $(a, b) \in A \times A$ such that $f(a) = f(b)$. Show that R, with termwise addition and multiplication, is a ring.

b) One says that f is a *monomorphism* if $g = g'$ for any ring C and any pair (g, g') of ring morphisms from C to A such that $f \circ g = f \circ g'$.

Show that a ring morphism is a monomorphism if and only if it is injective.

c) One says that f is an *epimorphism* if $g = g'$ for any ring C and any pair (g, g') of ring morphisms from B to C such that $g \circ f = g' \circ f$.

Show that a surjective ring morphism is an epimorphism. Show also that the inclusion morphism from **Z** into **Q** is an epimorphism.

Exercise (1.8.9). — Let $\mathbf{Z}[\sqrt{2}]$ and $\mathbf{Z}[\sqrt{3}]$ be the subrings of **C** generated by **Z**, and by $\sqrt{2}$ and $\sqrt{3}$ respectively.

a) Show that $\mathbf{Z}[\sqrt{2}] = \{a + b\sqrt{2}\, ; a, b \in \mathbf{Z}\}$ and $\mathbf{Z}[\sqrt{3}] = \{a + b\sqrt{3}\, ; a, b \in \mathbf{Z}\}$.

b) Show that besides the identity, there is only one automorphism of $\mathbf{Z}[\sqrt{2}]$; it maps $a + b\sqrt{2}$ to $a - b\sqrt{2}$ for any $a, b \in \mathbf{Z}$.

c) Show that there does not exist a ring morphism from $\mathbf{Z}[\sqrt{2}]$ to $\mathbf{Z}[\sqrt{3}]$.

d) More generally, what are the automorphisms of $\mathbf{Z}[i]$? of $\mathbf{Z}[\sqrt[3]{2}]$?

Exercise (1.8.10). — Let α be a complex number. Assume that there exists a monic polynomial P with integral coefficients, say $P = X^d + a_{d-1}X^{d-1} + \cdots + a_0$, such that $P(\alpha) = 0$. Show that the set A of all complex numbers of the form $c_0 + c_1\alpha + \cdots + c_{d-1}\alpha^{d-1}$, for $c_0, \ldots, c_{d-1} \in \mathbf{Z}$, is a subring of **C**. Give an example that shows that the result does not hold when P is not monic.

Exercise (1.8.11). — Let K be a field and V be a K-vector space.

a) Let (V_i) be any family of subspaces of V such that $V = \bigoplus V_i$. For $x = \sum x_i$, with $x_i \in V_i$, set $p_j(x) = x_j$. Show that for any j, p_j is a projector in V (namely, $p_j \circ p_j = p_j$) with image V_j and kernel $\bigoplus_{i \neq j} V_i$. Show that $p_j \circ p_i = 0$ for $i \neq j$, and that $\mathrm{id}_V = \sum p_i$.

b) Conversely, let (p_i) be a family of projectors in V such that $p_i \circ p_j = 0$ for $i \neq j$, and such that $\mathrm{id}_V = \sum p_i$. Let V_i be the image of p_i. Show that V is the direct sum of the subspaces V_i, and that p_i is the projector onto V_i whose kernel is the sum of all V_j, for $j \neq i$.

Exercise (1.8.12). — Let $A = \mathrm{Mat}_n(\mathbf{C})$ be the ring of $n \times n$ matrices with complex coefficients and let φ be any automorphism of A. Let Z be the center of A; it consists of all scalar matrices aI_n, for $a \in \mathbf{C}$.

a) Show that φ induces, by restriction, an automorphism of Z.

In the sequel, we assume that $\varphi|_Z = \mathrm{id}_Z$. For $1 \leq i, j \leq n$, let $E_{i,j}$ be the elementary matrix whose only non-zero coefficient is a 1 at position (i, j); let $B_{i,j} = \varphi(E_{i,j})$.

b) Show that $B_{i,i}$ is the matrix of a projector p_i of $\mathbf{C}^n$. Show that $p_i \circ p_j = 0$ for $i \neq j$, and that $\mathrm{id}_{\mathbf{C}^n} = \sum p_i$.

c) Using exercise 1.8.11, show that there exists a basis $(f_1, \ldots, f_n)$ of $\mathbf{C}^n$ such that for any i, p_i is the projector onto $\mathbf{C}f_i$ with kernel the sum $\sum_{j \neq i} \mathbf{C}f_j$.

d) Show that there exist elements $\lambda_i \in \mathbf{C}^*$ such that, denoting $e_i = \lambda_i f_i$, one has $B_{ij}(e_k) = 0$ for $k \neq j$, and $B_{ij}(e_j) = e_i$. Deduce the existence of a matrix $B \in \mathrm{GL}_n(\mathbf{C})$ such that $\varphi(M) = BMB^{-1}$ for any matrix M in $\mathrm{Mat}_n(\mathbf{C})$.

e) What happens when one no longer assumes that φ restricts to the identity on Z?

Exercise (1.8.13). — Let n be an integer such that $n \geq 1$.

a) What are the invertible elements of $\mathbf{Z}/n\mathbf{Z}$? For which integers n is that ring a domain? a field?

b) Let m be an integer such that $m \geq 1$. Show that the canonical map from $\mathbf{Z}/nm\mathbf{Z}$ to $\mathbf{Z}/n\mathbf{Z}$ is a ring morphism. Show that it induces a surjection from $(\mathbf{Z}/mn\mathbf{Z})^\times$ onto $(\mathbf{Z}/n\mathbf{Z})^\times$.

c) Determine the nilpotent elements of the ring $\mathbf{Z}/n\mathbf{Z}$.

Exercise (1.8.14). — Give an example of a ring morphism $f: A \to B$ which is surjective but such that the associated group morphism from $A^\times$ to $B^\times$ is not surjective.

Exercise (1.8.15). — *a*) Let K be a field, let V be a K-vector space and let $A = \mathrm{End}_K(V)$ be the ring of endomorphisms of V.

Show that the elements of A which are right invertible are the surjective endomorphisms, and those which are left invertible are the injective endomorphisms.

b) Give an example of a noncommutative ring and of an element which possesses infinitely many right inverses.

Exercise (1.8.16). — Let A and B be two rings. The set $A \times B$ is endowed with the addition defined by $(a, b) + (a', b') = (a + a', b + b')$ and the multiplication defined by $(a, b) \cdot (a', b') = (aa', bb')$, for a and $a' \in A$, b and $b' \in B$.

a) Show that $A \times B$ is a ring. What is the neutral element for multiplication?

b) Determine the elements of $A \times B$ which are respectively right regular, left regular, right invertible, left invertible, nilpotent.

c) Determine the center of $A \times B$.

d) Show that the elements $e = (1, 0)$ and $f = (0, 1)$ of $A \times B$ satisfy $e^2 = e$ and $f^2 = f$. One says that they are *idempotents*.

e) Prove that the two projections $p: A \times B \to A$ and $q: A \times B \to Q$ are morphisms of rings. Let R be a ring and let $f: R \to A$ and $g: R \to C$ be morphisms of rings. Prove that there exists a unique morphism $h: C \to A \times B$ such that $p \circ h = f$ and $q \circ h = g$. Thus, $(A \times B, p, q)$ is a product of A and B *in the category of rings.*

Exercise (1.8.17). — Let A be a ring such that $A \neq 0$.

a) Let a be an element of A which has exactly one right inverse. Show that a is invertible. (First prove that a is regular.)

b) If every non-zero element of A is left invertible, then A is a division ring.

c) One assumes that A is finite. Show that any left regular element is right invertible. If any element of A is left regular, then A is a division ring. When A is commutative, any prime ideal of A is a maximal ideal.

d) Consider the same question when K is a field and A is a K-algebra which is finite-dimensional as a K-vector space. (In particular, we impose that the multiplication of A is K-bilinear.)

e) Assume that any non-zero element of A is left regular and that the set of all right ideals of A is finite. Show that A is a division ring. (*To show that a non-zero element x is right invertible, introduce the right ideals $x^n A$, for $n \geq 1$.*)

Exercise (1.8.18). — Let A be a ring and let $e \in A$ be any idempotent element of A (this means, we recall, that $e^2 = e$).

a) Show that $1 - e$ is an idempotent element of A.

b) Show that $eAe = \{eae \,; a \in A\}$ is an abelian subgroup of A, and that the multiplication of A endows it with the structure of a ring. (However, it is not a subring of A, in general.)

c) Describe the particular case where $A = \mathrm{Mat}_n(k)$, for some field k, the element e being a diagonal matrix of rank r.

Exercise (1.8.19). — Let A be a ring.

a) Let $a \in A$ be a nilpotent element. Let $n \geq 0$ be such that $a^{n+1} = 0$; compute $(1+a)(1-a+a^2-\cdots+(-1)^n a^n)$. Deduce that $1+a$ is invertible in A.

b) Let x, y be two elements of A; one assumes that x is invertible, y is nilpotent, and $xy = yx$. Show that $x + y$ is invertible.

c) Let x and y be nilpotent elements of A which commute. Show that $x + y$ is nilpotent. (Let n and m be two integers ≥ 1 such that $x^n = y^m = 0$; use the binomial formula to compute $(x + y)^{n+m-1}$.)

Exercise (1.8.20). — Let A be a ring, let a and b be elements of A such that $1 - ab$ is invertible in A.

a) Show that $1 - ba$ is invertible in A and give an explicit formula for its inverse. (One may begin by assuming that ab is nilpotent and give a formula for the inverse of $1 - ba$ in terms of $(1 - ab)^{-1}$, a and b. Then show that this formula works in the general case.)

b) Assume that k is a field and $A = \mathrm{Mat}_n(k)$. Show that ab and ba have the same characteristic polynomial. What is the relation with the preceding question?

Exercise (1.8.21). — Let A be a commutative ring, $a_0, \ldots, a_n \in A$ and let f be the polynomial $f = a_0 + a_1 X + \cdots + a_n X^n \in A[X]$.

a) Show that f is nilpotent if and only if all a_i are nilpotent.

b) Show that f is invertible in A[X] if and only if a_0 is invertible in A and $a_1, \ldots, a_n$ are nilpotent. (*If $g = f^{-1} = b_0 + b_1 X + \cdots + b_m X^m$, show by induction on k that $a_n^{k+1} b_{m-k} = 0$.*)

c) Show that f is a zero-divisor if and only if there exists a non-zero element $a \in A$ such that $af = 0$. (*Let g be a non-zero polynomial of minimal degree such that fg = 0; show that $a_k g = 0$ for every k.*)

Exercise (1.8.22). — Let A be a commutative ring.

a) Use the universal property of polynomial rings to show that there exists a unique morphism of A-algebras $\varphi \colon A[X, Y] \to (A[X])[Y]$ such that $\varphi(X) = X$ and $\varphi(Y) = Y$. Prove that φ is an isomorphism.

b) More generally, let I be a set, let J be a subset of I and let $K = I - J$. Show that there exists a unique morphism of A-algebras $\varphi \colon A[(X_i)_{i \in I}] \to (A[(X_j)_{j \in J}])[(X_k)_{k \in K}]$, and that this morphism is an isomorphism (see remark 1.3.10).

Exercise (1.8.23). — Let A be a commutative ring, let P and Q be polynomials with coefficients in A in one indeterminate X. Let $m = \deg(P)$ and $n = \deg(Q)$, let a be the leading coefficient of Q and $\mu = \sup(1 + m - n, 0)$. Show that there exists a pair (R, S) of polynomials such that $a^\mu P = QR + S$ and $\deg S < n$. Show that this pair is unique if A is a domain, or if a is regular.

Exercise (1.8.24). — Let k be a field and let $A = k[X_1, \dots, X_n]$ be the ring of polynomials with coefficients in k in n indeterminates. One says that an ideal of A is monomial if it is generated by monomials.

a) Let $(M_\alpha)_{\alpha \in E}$ be a family of monomials and let I be the ideal they generate. Show that a monomial M belongs to I if and only if it is a multiple of one of the monomials M_α.

b) Let I be an ideal of A. Show that I is a monomial ideal if and only if, for any polynomial $P \in I$, each monomial of P belongs to I.

c) Let I and J be monomial ideals of A. Show that the ideals $I + J$, IJ, $I \cap J$, $I : J = \{a \in A \,; \, aJ \subset I\}$ and $\sqrt{I}$ are again monomial ideals. Given monomials which generate I and J, give explicit monomials which generate those ideals.

Exercise (1.8.25). — Let k be a field, let $(M_\alpha)_{\alpha \in E}$ be a family of monomials of $A = k[X_1, \dots, X_n]$, and let I be the monomial ideal that they generate (see exercise 1.8.24). The goal of this exercise is to show that there exists a finite subset $F \subset E$ such that $I = (M_\alpha)_{\alpha \in F}$. The proof runs by induction on the number n of indeterminates.

a) Treat the case $n = 1$.

b) In the sequel, one assumes that $n \geq 2$ and that the property holds if there are strictly less than n indeterminates.

For $i \in \{1, \dots, n\}$, one defines a morphism of rings φ_i by

$$\varphi_i(P) = P(X_1, \dots, X_{i-1}, 1, X_{i+1}, \dots, X_n).$$

Using the induction hypothesis, observe that there exists a finite subset $F_i \subset E$ such that for any $\alpha \in E$ the monomial $\varphi_i(M_\alpha)$ can be written $\varphi_i(M_\alpha) = M'_\alpha \times \varphi_i(M_\beta)$ for some $\beta \in F_i$.

c) Let F_0 be the set of all $\alpha \in E$ such that for all $i \in \{1, \ldots, n\}$, one has

$$\deg_{X_i}(M_\alpha) < \sup\{\deg_{X_i}(M_\beta) \, ; \, \beta \in F_i\}\}.$$

Set $F = \bigcup_{i=0}^{n} F_i$. Show that $I = (M_\alpha)_{\alpha \in F}$.

Exercise (1.8.26). — Let A be a ring, let I be a set and let M be the set $\mathbf{N}^{(I)}$ of all multi-indices on I. Let $\mathscr{F}_I = A^M$ be the set of all families of elements of A indexed by M. It is an abelian group with respect to termwise addition.

a) Show that the formulae that define the multiplication of polynomials make sense on $\mathscr{F}_I$ and endow it with the structure of a ring, of which the ring of polynomials $\mathscr{P}_I$ is a subring.

Let X_i be the indeterminate with index i. Any element of $\mathscr{F}_I$ is formally written as an infinite sum $\sum_m a_m X_1^{m_1} \ldots X_n^{m_n}$. The ring $\mathscr{F}_I$ is called the *ring of power series* in the indeterminates $(X_i)_{i \in I}$. When $I = \{1, \ldots, n\}$, it is denoted by $A[[X_1, \ldots, X_n]]$, and by $A[[X]]$ when the set I has a single element, the indeterminate being written X.

Let k be a commutative ring.

b) For any k-algebra A and any family $(a_1, \ldots, a_n)$ of nilpotent elements which commute pairwise, show that there is a unique morphism of k-algebras $\varphi \colon k[[X_1, \ldots, X_n]] \to A$ such that $\varphi(X_i) = a_i$ for all i.

c) Show that an element $\sum a_n X^n$ of $k[[X]]$ is invertible if and only if a_0 is invertible in k.

Exercise (1.8.27). — Let k be a field. Let Φ be the morphism from $k[X_1, \ldots, X_n]$ to $\mathscr{F}(k^n, k)$ that maps a polynomial to the corresponding polynomial function.

a) Let $A_1, \ldots, A_n$ be subsets of k. Let $P \in k[X_1, \ldots, X_n]$ be a polynomial in n indeterminates such that $\deg_{X_i}(P) < \mathrm{Card}(A_i)$ for every i. One assumes that $P(a_1, \ldots, a_n) = 0$ for every $(a_1, \ldots, a_n) \in A_1 \times \cdots \times A_n$. Show that $P = 0$.

b) If k is an infinite field, show that Φ is injective but not surjective.

c) Assume that k is a finite field. Show that Φ is surjective. For any function $f \in \mathscr{F}(k^n; k)$, give an explicit a polynomial P such that $\Phi(P) = f$ (think of the Lagrange interpolation polynomials). Show that Φ is not injective. More precisely, if $q = \mathrm{Card}(k)$, show that $\mathrm{Ker}(\Phi)$ is generated by the polynomials $X_i^q - X_i$, for $1 \le i \le n$.

Exercise (1.8.28). — One says that a ring A admits a right euclidean division if there exists a map $\varphi \colon A \setminus \{0\} \to \mathbf{N}$ such that for any pair (a, b) of elements of A, with $b \ne 0$, there exists a pair (q, r) of elements of A such that $a = qb + r$ and such that either $r = 0$ or $\varphi(r) < \varphi(b)$.

a) Assume that A admits a right euclidean division. Show that any left ideal of A is of the form Aa, for some $a \in A$.

b) Let K be a division ring. Show that the polynomial ring $K[X]$ admits a right euclidean division.

Exercise (1.8.29). — Let A be a ring and let I be a right ideal of A.

a) Show that the left ideal generated by I in A is a two-sided ideal.

b) Show that the set J of all elements $a \in A$ such that $xa = 0$ for any $x \in I$ (the right annihilator of I) is a two-sided ideal of A.

Exercise (1.8.30). — Let A be a commutative ring, and let I, J, L be ideals of A. Show the following properties:

a) $I \cdot J \subset I \cap J$;

b) $(I \cdot J) + (I \cdot L) = I \cdot (J + L)$;

c) $(I \cap J) + (I \cap L) \subset I \cap (J + L)$;

d) If $J \subset I$, then $J + (I \cap L) = I \cap (J + L)$;

e) Let K be a field and assume that $A = K[X, Y]$. Set $I = (X)$, $J = (Y)$ and $L = (X + Y)$. Compute $(I \cap J) + (I \cap L)$ and $I \cap (J + L)$; compare these two ideals.

Exercise (1.8.31). — Let A be a ring.

a) Give an example where the set of all nilpotents elements of A is not an additive subgroup of A. (One may check that $A = \mathrm{Mat}_2(\mathbf{C})$ works.)

b) Let N be the set of all elements $a \in A$ such that ax is nilpotent, for every $x \in A$. Show that N is a two-sided ideal of A, every element of which is nilpotent.

c) Let I be a two-sided ideal of A such that every element of I is nilpotent. Show that $I \subset N$.

Exercise (1.8.32). — Let K be a field, let V be a K-vector space and let A be the ring of endomorphisms of V.

a) For any subspace W of V, show that the set N_W of all endomorphisms whose kernel contains W is a left ideal of A, and that the set I_W of all endomorphisms whose image is contained in W is a right ideal of A.

b) If V has finite dimension, then all left ideals (*resp.* right ideals) are of this form.

c) If V has finite dimension, the only two-sided ideals of A are (0) and A.

d) The set of all endomorphisms of finite rank of V (that is, those whose image has finite dimension) is a two-sided ideal of A. It is distinct from A if V has infinite dimension.

Exercise (1.8.33). — Let A be a commutative ring and let a, b be two elements of A.

a) If there exists a unit u in A such that $a = bu$ (in which case one says that a and b are *associates*), show that the ideals $(a) = aA$ and $(b) = bA$ coincide.

b) Conversely, assuming that A is a domain and that $(a) = (b)$, show that a and b are associates.

c) Assume that A is the quotient of the ring $\mathbf{Z}[a, b, x, y]$ of polynomials in four indeterminates by the ideal generated by $(a - bx, b - ay)$. Prove that the ideals (a) and (b) coincide but that a and b are not associate.

Exercise (1.8.34). — Let A, B be commutative rings and let $f: A \to B$ be a ring morphism. For any ideal I of A, let $f_*(I)$ be the ideal of B generated by $f(I)$; we say it is the *extension* of I to B. For any ideal J of B, we call the ideal $f^{-1}(J)$ the *contraction* of J in A.

Let I be an ideal of A and let J be an ideal of B; show the following properties.

a) $I \subset f^{-1}(f_*(I))$ and $J \supset f_*(f^{-1}(J))$;

b) $f^{-1}(J) = f^{-1}(f_*(f^{-1}(J)))$ and $f_*(I) = f_*(f^{-1}(f_*(I)))$.

Let $\mathscr{C}$ be the set of ideals of A which are contractions of ideals of B, and let $\mathscr{E}$ be the set of ideals of B which are extensions of ideals of A.

c) Show that $\mathscr{C} = \{I; I = f^{-1}(f_*(I))\}$ and $\mathscr{E} = \{J; J = f_*(f^{-1}(J))\}$;

d) The map f_* defines a bijection from $\mathscr{C}$ onto $\mathscr{E}$; what is its inverse?

Let I_1 and I_2 be ideals of A, let J_1 and J_2 be ideals of B. Show the following properties:

e) $f_*(I_1 + I_2) = f_*(I_1) + f_*(I_2)$ and $f^{-1}(J_1 + J_2) \supset f^{-1}(J_1) + f^{-1}(J_2)$;

f) $f_*(I_1 \cap I_2) \subset f_*(I_1) \cap f_*(I_2)$, and $f^{-1}(J_1 \cap J_2) = f^{-1}(J_1) \cap f^{-1}(J_2)$;

g) $f_*(I_1 \cdot I_2) = f_*(I_1) \cdot f_*(I_2)$, and $f^{-1}(J_1 \cdot J_2) \supset f^{-1}(J_1) \cdot f^{-1}(J_2)$;

h) $f_*(\sqrt{I}) \subset \sqrt{f_*(I)}$, and $f^{-1}(\sqrt{J}) = \sqrt{f^{-1}(J)}$.

Exercise (1.8.35). — Let I and J be two ideals of a commutative ring A. Assume that $I + J = A$. Then, show that $I^n + J^n = A$ for any positive integer n.

Exercise (1.8.36). — Let A be a commutative ring, let I be an ideal of A and let S be any subset of A. The conductor of S in I is defined by the formula

$$(I : S) = \{a \in A \, ; \text{ for all } s \in S, \, as \in I\}.$$

Show that it is an ideal of A, more precisely, the largest ideal J of A such that $JS \subset I$.

Exercise (1.8.37). — Let K be a field and let $A = K[X, Y]/(X^2, XY, Y^2)$.

a) Compute the invertible elements of A.

b) Determine all principal ideals of A.

c) Determine all ideals of A.

Exercise (1.8.38). — Let A be a ring and let I be the two-sided ideal of A generated by the elements of the form $xy - yx$, for x and $y \in A$.

a) Show that the ring A/I is commutative.

b) Let J be a two-sided ideal of A such that A/J is a commutative ring. Show that $I \subset J$.

Exercise (1.8.39). — Let A be a ring, let I be a right ideal of A.

a) Show that the set B of all $a \in A$ such that $aI \subset I$ is a subring of A. Show that I is a two-sided ideal of B.

b) Define an isomorphism from the ring $\mathrm{End}_A(A/I)$, where A/I is a right A-module, onto the ring B/I.

c) Let I be a maximal right ideal of A. Show that B/I is a division ring.

Exercise (1.8.40). — Let A be a ring, let I be an ideal of A. We let I[X] be the set of polynomials $P \in A[X]$ all of the coefficients of which belong to I.

a) Show that I[X] is a left ideal of A[X].

b) If I is a two-sided ideal of A, show that I[X] is a two-sided ideal of A[X] and construct an isomorphism from the ring $A[X]/I[X]$ to the ring $(A/I)[X]$.

Exercise (1.8.41). — Let A be a ring, let I be a two-sided ideal of A and let $\mathrm{Mat}_n(I)$ be the set of matrices in $\mathrm{Mat}_n(A)$ all of the coefficients of which belong to I.

a) Show that $\mathrm{Mat}_n(I)$ is a two-sided ideal of $\mathrm{Mat}_n(A)$ and construct an isomorphism of rings from $\mathrm{Mat}_n(A)/\mathrm{Mat}_n(I)$ onto $\mathrm{Mat}_n(A/I)$.

b) Conversely, show that any two-sided ideal of $\mathrm{Mat}_n(A)$ is of the form $\mathrm{Mat}_n(I)$, for some two-sided ideal I of A.

Exercise (1.8.42). — *a*) What are the invertible elements of the ring of decimal numbers?

Let A be a commutative ring and let S be a multiplicative subset of A.

b) Show that an element $a \in A$ is invertible in $S^{-1}A$ if and only if there exists an element $b \in A$ such that $ab \in S$.

c) Let T be a multiplicative subset of A which contains S. Construct a ring morphism from $S^{-1}A$ to $T^{-1}A$.

d) Let $\tilde{S}$ be the set of elements of A whose image in $S^{-1}A$ is invertible. Show that $\tilde{S}$ contains S and that the ring morphism from $S^{-1}A$ to $\tilde{S}^{-1}A$ is an isomorphism.

Give an explicit proof of this fact, as well as a proof relying only on the universal property.

Exercise (1.8.43). — Let A be a commutative ring.

a) Let *s* be an element of A and let $S = \{1; s; s^2; \dots\}$ be the multiplicative subset generated by *s*. We write A_s for the ring of fractions $S^{-1}A$. Show that the following properties are equivalent:

(i) The canonical morphism $i \colon A \to A_s$ is surjective;

(ii) The decreasing sequence of ideals $(s^n A)_n$ is eventually constant;

(iii) For any large enough integer *n*, the ideal $s^n A$ is generated by an idempotent element.

(*To show that* (ii) *implies* (iii), *show by induction on k that a relation of the form* $s^n = s^{n+1}a$ *implies that* $s^n = s^{n+k}a^k$; *then conclude that* $s^n a^n$ *is an idempotent element.*)

b) Let S be a multiplicative subset of A consisting of elements s for which the morphism $A \to A_s$ is surjective. Show that the morphism $A \to S^{-1}A$ is surjective too.

c) Let A be a ring which is either finite, or which is a finite-dimensional vector subspace over a subfield (more generally, an *artinian* ring). Show that condition (ii) holds for any element $s \in A$.

Exercise (1.8.44). — *a)* Let A be a subring of **Q**. Show that there exists a multiplicative subset S of **Z** such that $A = S^{-1}\mathbf{Z}$.

b) Let $A = \mathbf{C}[X, Y]$ be the ring of polynomials with complex coefficients in two indeterminates X and Y. Let $B = A[Y/X]$ be the subring generated by A and Y/X in the field $\mathbf{C}(X, Y)$ of rational functions.

Show that the unique ring morphism from $\mathbf{C}[T, U]$ to B that maps T to X and U to Y/X is an isomorphism. Deduce that $A^{\times} = B^{\times} = \mathbf{C}^{\times}$, hence that B is not the localization of A with respect to some multiplicative subset.

Exercise (1.8.45). — Let A be a (not necessarily commutative) ring and let S be a multiplicative subset of A.

Let A_S be a ring and let $i\colon A \to A_S$ be a morphism of rings. One says that (A_S, i) is a ring of right fractions for S if the following properties hold:

(i) For any $s \in S$, $i(s)$ is invertible in A_S;

(ii) Any element of A_S is of the form $i(a)i(s)^{-1}$ for some $a \in A$ and $s \in S$.

a) Assume that A admits a ring of right fractions for S. Show the following properties (right *Ore conditions*):

(i) For any $a \in A$ and any $s \in S$, there exists $b \in A$ and $t \in S$ such that $at = sb$;

(ii) For any $a \in A$ and any $s \in S$ such that $sa = 0$, there exists an element $t \in S$ such that $at = 0$.

(The second condition is obviously satisfied if every element of S is regular.)

b) Conversely, assume that the right Ore conditions hold. Define an equivalence relation $\sim$ on the set $A \times S$ by "$(a, s) \sim (b, t)$ if and only if there exist $c, d \in A$ and $u \in S$ such that $u = sc = td$ and $ac = bd$." Construct a ring structure on the quotient set A_S such that the map $i\colon A \to A_S$ sending $a \in A$ to the equivalence class of $(a, 1)$ is ring morphism, and such that for any $s \in S$, $i(s)$ is invertible in A_S with inverse the equivalence class of $(1, s)$. Conclude that A_S is a ring of right fractions for S. If every element of S is regular, prove that i is injective.

Exercise (1.8.46). — Let a and b be positive integers.

a) Show that there exist integers m and n such that m is prime to b, each prime divisor of n divides b, and $a = mn$. (Beware, n is *not* the gcd of a and b.)

b) Show that the ring $(\mathbf{Z}/a\mathbf{Z})_b$ is isomorphic to $\mathbf{Z}/n\mathbf{Z}$. With that aim, construct explicitly a ring morphism from $\mathbf{Z}/n\mathbf{Z}$ to $(\mathbf{Z}/a\mathbf{Z})_b$ and show that it is an isomorphism.

Exercise (1.8.47). — *a*) Show that the ring $\mathbf{Z}[i]$ is isomorphic to the quotient ring $\mathbf{Z}[X]/(X^2 + 1)$.

b) Let *a* be an integer. Considering the ring $\mathbf{Z}[i]/(a + i)$ as a quotient of the ring of polynomials $\mathbf{Z}[X]$, define an isomorphism

$$\mathbf{Z}[i]/(a + i) \xrightarrow{\sim} \mathbf{Z}/(a^2 + 1)\mathbf{Z}.$$

c) More generally, let *a* and *b* be two coprime integers. Show that the image of *b* in $\mathbf{Z}[i]/(a + ib)$ is invertible. Write this ring as a quotient of $\mathbf{Z}_b[X]$ and then define an isomorphism

$$\mathbf{Z}[i]/(a + ib) \xrightarrow{\sim} \mathbf{Z}/(a^2 + b^2)\mathbf{Z}.$$

(*Observe that if* $1 = au + bv$, *then* $1 = (a + bi)u + b(v - ui)$.)

Exercise (1.8.48). — Let A be a commutative ring, let S be a multiplicative subset of A.

a) Assume that there are elements *s* and $t \in$ S such that S is the set of all $s^n t^m$, for *n* and *m* in **N**. Show that the morphism $A[X, Y] \to S^{-1}A$, $P(X, Y) \mapsto P(1/s, 1/t)$ is surjective and that its kernel contains the ideal $(1 - sX, 1 - tY)$ generated by $1 - sX$ and $1 - tY$ in $A[X, Y]$. Construct an isomorphism $A[X, Y]/(1 - sX, 1 - tY) \simeq S^{-1}A$.

b) More generally, assume that S is the smallest multiplicative subset of A containing a subset T, and let $\langle 1 - tX_t \rangle_{t \in T}$ be the ideal of the polynomial ring (with a possibly infinite set of indeterminates X_t, for $t \in T$) generated by the polynomials $1 - tX_t$, for $t \in T$. Then show that the canonical morphism

$$A[(X_t)_{t \in T}] \to S^{-1}A, \quad P \mapsto P((1/t)_t)$$

induces an isomorphism

$$A[(X_t)_{t \in T}]/\langle 1 - tX_t \rangle_{t \in T} \simeq S^{-1}A.$$

Exercise (1.8.49). — Let A be a commutative ring and let S be a multiplicative subset of A such that $0 \notin$ S.

a) If A is a domain, show that $S^{-1}A$ is a domain.

b) If A is reduced, show that $S^{-1}A$ is reduced.

c) Let $f \colon A \to S^{-1}A$ be the canonical morphism $a \mapsto a/1$. Show that the nilpotent radical of $S^{-1}A$ is the ideal of $S^{-1}A$ generated by the image of the nilpotent radical of A.

Exercise (1.8.50). — Let $B \subset \mathbf{R}(X)$ be the set of all rational functions with real coefficients of the form $P/(X^2 + 1)^n$, where $P \in \mathbf{R}[X]$ is a polynomial and *n* is an integer. Let A be the subset of B consisting of those fractions $P/(X^2 + 1)^n$ for which $\deg(P) \leq 2n$.

a) Show that A and B are subrings of $\mathbf{R}(X)$.

b) Determine their invertible elements.

c) Show that B is a principal ideal domain. Show that the ideal of A generated by $1/(X^2 + 1)$ and $X/(X^2 + 1)$ is not principal.

Exercise (1.8.51). — *a*) Prove that a surjective morphism of rings is an epimorphism.

b) Let A be a commutative ring and let S be a multiplicative subset of A. Prove that the canonical morphism from A to $S^{-1}A$ is an epimorphism of rings.

Chapter 2.
Ideals and Divisibility

In this second chapter, we study deeper aspects of divisibility in rings. This question appeared historically when mathematicians — notably, in the hope of proving Fermat's Last Theorem! — tried to make use of unique factorization in rings where it didn't hold. Ideals are one device that was then invented to gain a better understanding of divisibility. In this context, there are two natural analogues of prime numbers, namely maximal *and* prime *ideals.*

I introduce maximal ideals in the general framework of possibly noncommutative rings; I then define the Jacobson radical. Matrix rings furnish interesting examples.

However, from that point on, the chapter is essentially focused on commutative rings. I define prime ideals and show their relations with nilpotent elements (via the nilpotent radical), or local rings. The set of all prime ideals of a commutative ring is called its spectrum, *and it is endowed with a natural topology. The detailed study of spectra of rings belongs to the algebraic geometry of schemes, as put forward by Alexander Grothendieck. Although that is a more advanced subject than what is planned for this book, I believe that the early introduction of the spectrum of a ring allows one to add an insightful geometric interpretation to some constructions of commutative algebra.*

Hilbert's "Nullstellensatz" describes the maximal ideals of the ring of polynomials in finitely many indeterminates over an algebraically closed field. After proving it in the particular case of the field of complex numbers (by a short argument which some colleagues find too special), I show how this theorem gives rise to a correspondence between geometry and algebra. (The general case of the theorem will be proved in chapter 9; a few other proofs also appear as exercises in that chapter, as well as in this one.)

Another beautiful geometric example is given by Gelfand's theorem, which describes the maximal ideals of the ring of continuous functions on a compact metric space.

I then go back to the initial goal of understanding uniqueness of a decomposition into prime numbers in general rings, and introduce three classes of commutative rings of increasing generality in which the divisibility relation is particularly well behaved: euclidean rings, principal ideal domains *and* unique factorization

© Springer Nature Switzerland AG 2021   55
A. Chambert-Loir, *(Mostly) Commutative Algebra*, Universitext,
https://doi.org/10.1007/978-3-030-61595-6_2

domains. *In particular, I prove a theorem of Gauss that asserts that polynomial rings are unique factorization domains.*

I conclude the chapter by introducing the resultant *of two polynomials and use it to prove a theorem of Bézout about the number of common roots of two polynomials in two indeterminates — geometrically, this gives an upper bound for the number of intersection points of two plane curves.*

2.1. Maximal Ideals

Definition (2.1.1). — Let A be a ring. One says that a left ideal of A is *maximal* if it is maximal among the left ideals of A which are distinct from A.

In other words, a left ideal I is maximal if one has I $\neq$ A and if the only left ideals of A containing I are A and I. Consequently, to check that a left ideal I $\neq$ A is maximal, it suffices to show that for any $a \in$ A $-$I, the left ideal I + Aa is equal to A.

There is an analogous definition for the right and two-sided ideals.

These notions coincide when the ring A is commutative, but the reader should be cautious in the general case: When the ring A is not commutative, a two-sided ideal can be a maximal ideal as a two-sided ideal but not as a left ideal, or be a maximal ideal as a left ideal but not as a right ideal.

Examples (2.1.2). — a) The ideals of $\mathbf{Z}$ have the form $n\mathbf{Z}$, for some $n \in \mathbf{Z}$ (example 1.4.3). If n divides m, then $m\mathbf{Z} \subset n\mathbf{Z}$, and conversely. It follows that the maximal ideals of $\mathbf{Z}$ are the ideals $p\mathbf{Z}$, where p is a prime number.

Similarly, if K is a field, the maximal ideals of the ring K[X] of polynomials in one indeterminate are the ideals generated by an irreducible polynomial. When K is algebraically closed, the maximal ideals are the ideals $(X - a)$, for some $a \in$ K.

b) Let K be a field and let V be a K-vector space of finite dimension. In exercise 1.8.32, we asked to determine the left, right, and two-sided ideals of $\mathrm{End}_K(V)$.

The left ideals of End(V) are the ideals I_W, where W is a vector subspace of V and I_W is the set of endomorphisms whose kernel contains W. For W $\subset$ W', $I_{W'} \subset I_W$. Consequently, the maximal left ideals of End(V) are those ideals I_W where W is a line in V.

The right ideals of V are the ideals R_W where W is a vector subspace of V and R_W is the set of endomorphisms whose image is contained in W. For W $\subset$ W', $I_W \subset I_{W'}$. Consequently, the maximal right ideals of End(V) are those ideals R_W where W is a hyperplane in V.

The only two-sided ideals of End(V) are (0) and End(V), so that (0) is the only maximal two-sided ideal of End(V).

c) A maximal left ideal of a ring is a maximal right ideal of the opposite ring.

The following general theorem asserts the existence of maximal ideals of a given ring; it makes use of Zorn's lemma (corollary A.2.13).

Theorem (2.1.3) (Krull). — *Let* A *be a ring and let* I *be a left ideal of* A, *distinct from* A. *There exists a maximal left ideal of* A *which contains* I.

In particular, any non-zero ring possesses at least one maximal left ideal (take $I = 0$).

The analogous statements for right ideals and two-sided ideals are true, and proven in the same way.

Proof. — Let $\mathscr{I}$ be the set of left ideals of A which contain I and are distinct from A. Let us endow $\mathscr{I}$ with the ordering given by inclusion.

Let us show that $\mathscr{I}$ is inductive. Let indeed (J_i) be a totally ordered family of left ideals of A such that $I \subset J_i \subsetneq A$ for any i; let us show that it has an upper bound J in $\mathscr{I}$.

If this family is empty, we set $J = I$.

Otherwise, let J be the union of the ideals J_i. Let us show that $J \in \mathscr{I}$. By construction, J contains I, because there is an index i, and $J \supset J_i \supset I$. Since 1 does not belong to J_i, for any index i, we deduce that $1 \notin J$ and $J \neq A$. Finally, we prove that J is a left ideal of A. Let $x, y \in J$; there are indices i and j such that $x \in J_i$ and $y \in J_j$. Since the family (J_i) is totally ordered, we have $J_i \subset J_j$ or $J_j \subset J_i$. In the first case, $x + y \in J_j$, in the second, $x + y \in J_i$; in any case, $x + y \in J$. Finally, if $x \in J$ and $a \in A$, let i be such that $x \in J_i$; since J_i is a left ideal of A, $ax \in J_i$, hence $ax \in J$.

By corollary A.2.13 (Zorn's lemma), the set $\mathscr{I}$ possesses a maximal element J. By definition of the ordering on $\mathscr{I}$, J is a left ideal of A, distinct from A, which contains I, and which is maximal for that property. Consequently, J is a maximal left ideal of A containing I, hence Krull's theorem. □

Corollary (2.1.4). — *Let* A *be a ring. For an element of* A *to be left invertible (resp. right invertible), it is necessary and sufficient that it belongs to no maximal left ideal (resp. to no maximal right ideal) of* A.

Proof. — Let a be an element of A. That a is left invertible means that the left ideal Aa equals A; then no maximal left ideal of A can contain a. Otherwise, if $Aa \neq A$, there exists a maximal left ideal of A containing the left ideal Aa, and that ideal contains a. □

Definition (2.1.5). — Let A be a ring. The *Jacobson radical* of A is the intersection of all maximal left ideals of A.

Lemma (2.1.6). — *Let* A *be a ring and let* J *be its Jacobson radical. Let* $a \in A$. *The following properties are equivalent:*

(i) $a \in J$;

(ii) *For every* $x \in A$, $1 + xa$ *is left invertible;*

(iii) *For every* $x, y \in A$, $1 + xay$ *is invertible.*

Proof. — We may assume that $A \neq 0$, the lemma being trivial otherwise.

(i)$\Rightarrow$(ii). Let $a \in J$, let $x \in A$ and let us show that $1 + xa$ is left invertible. Otherwise, by Krull's theorem 2.1.3, there exists a maximal left ideal I of A such that $1 + xa \in I$. Since $a \in J$, we have $a \in I$ and $xa \in I$; this implies that $1 \in I$, a contradiction.

(ii)$\Rightarrow$(i). Let $a \in A$ be such that $1 + xa$ is left invertible, for every $x \in A$. Let I be a maximal left ideal of A and let us prove that $a \in I$. Otherwise, we have $I + Aa = A$, by the definition of a maximal left ideal, hence there exists $b \in I$ and $x \in A$ such that $b + xa = 1$. Then $b = 1 - xa$ is left invertible, by assumption, contradicting the fact that $b \in I$. This proves that $a \in J$.

(ii)$\Rightarrow$(iii). Let $a \in A$ be such that $1 + xa$ is left invertible, for every $x \in A$.

Let us first prove that $ay \in J$, for every $y \in A$. Let I be a maximal left ideal of A; assume that $ay \notin I$. Then $A = I + Aay$ by the definition of a maximal left ideal. In particular, $y \in I + Aay$, so that there exists $b \in I$ and $x \in A$ such that $y = b + xay$, hence $(1 - xa)y = b \in I$. Since $1 - xa$ is left invertible, this implies that $y \in I$, hence $ay \in I$, a contradiction. In particular, $ay \in I$. This proves that $ay \in J$.

The implication (iii)$\Rightarrow$(ii) being obvious, this concludes the proof of the lemma.   □

Corollary (2.1.7). — *Let A be a ring and let J be its Jacobson radical.*

a) *The ideal J is the intersection of all right maximal ideals of A, and is a two-sided ideal.*

b) *An element $a \in A$ is invertible (resp. left invertible, right invertible) if and only if the same holds for its image in A/J.*

Proof. — *a*) Since characterization (iii) of the Jacobson radical in lemma 2.1.6 is symmetric, it implies that J is equal to the Jacobson radical of the opposite ring A^{o}, that is, the intersection of all maximal right ideals of A. In particular, J is both a left ideal and a right ideal, hence a two-sided ideal.

b) Let $a \in A$. If a is left invertible, then so is its image in A/J. So let us assume that the image of a in A/J is left invertible. Let $b \in A$ be any element whose image modulo J is a left inverse of a. We have $ba \in 1+J$. By lemma 2.1.6, ba is left invertible, so that a is left invertible as well.

By symmetry, a is right invertible if and only if its image in A/J is right invertible. Consequently, a is invertible if and only if its image in A/J is invertible.   □

Definition (2.1.8). — One says that a ring A is *local* if it is non-zero and if, for every $a \in A$, either a or $1 - a$ is invertible.

Proposition (2.1.9). — *Let A be a ring and let J be its Jacobson radical. The following properties are equivalent:*

(i) *The quotient ring A/J is a division ring;*

(ii) *The ring A admits a unique maximal left ideal;*

(ii′) *The ring A admits a unique maximal right ideal;*

(iii) *The ideal J is a maximal left ideal of* A;

(iii') *The ideal J is a maximal right ideal of* A;

(iv) *The set of non-invertible elements of* A *is an additive subgroup of* A;

(v) *The ring* A *is a local ring.*

Proof. — If $A = 0$, then A is not local, A/J is not a division ring, the ring A has no maximal left ideal, no maximal right ideal, and every element of A is invertible; this proves the equivalences of the given properties in this case. In the rest of the proof, we assume that $A \neq 0$.

Assume that A/J is a division ring. Then 0 is the unique maximal left (resp. right) ideal of the division ring A/J, so that J is the unique maximal left (resp. ideal) of A. Moreover, every non-zero element of A/J is invertible, so that $A - J$ is the set of invertible elements of A, by corollary 2.1.7. This proves that (i) implies all other assertions (ii)–(iv). Then A is a local ring since, for every $x \in A$, x and $1 - x$ do not both belong to J.

Let us now assume (ii) that A admits a unique maximal left ideal; then the Jacobson radical of A is equal to this left ideal, so that J is the unique maximal left ideal of A. This shows the equivalence of (ii) and (iii). Similarly, (ii') and (iii') are equivalent.

Let us assume (iii). No element of J is left invertible; conversely, an element of $A - J$ belongs to no left maximal ideal, hence is left invertible. Let $a \in A - J$; let $x \in A$ be such that $xa = 1$. Since J is a right ideal, we have $x \notin J$, hence x is left invertible. Then x is both left and right invertible, hence x is invertible. This implies that a is invertible. We thus have shown that J is the set of non-invertible elements of A. Since $J \neq A$, this implies that A/J is a division ring. We have shown the implication (iii)⟹(i), and the proof of the implication (iii')⟹(i) is analogous.

Let us now assume (iv), i.e., that the set N of non-invertible elements of A is an additive subgroup of A.

We first prove that an element $a \in A$ which is left invertible is invertible. Let $b \in A$ be such that $ba = 1$. The element $1 - ab$ is not left invertible, since the equality $x(1 - ab) = 1$ implies $a = x(1 - ab)a = x(a - aba) = 0$, but $a \neq 0$. Since $1 = ab + (1 - ab)$ is invertible, this proves that ab is invertible; then a is right invertible, hence it is invertible.

Let M be a maximal left ideal of A. No element of M can be invertible, hence $M \subset N$. Conversely, if $a \in A - M$, then a is left invertible, hence is invertible, by what precedes, and $a \notin N$; in other words, $N \subset M$. This proves that N is the unique maximal left ideal of A, hence (ii) holds.

Finally, let us assume that A is a local ring. Let $a \in A - J$. By lemma 2.1.6, there exists an element $x \in A$ such that $1 + xa$ isn't left invertible. Then, by definition of a local ring, xa is invertible, hence a is left invertible. By symmetry, every element of $A - J$ is also right invertible. Consequently, J is exactly the set of all non-invertible elements of A. In particular, (iv) holds. □

2.2. Maximal and Prime Ideals in a Commutative Ring

In this section, we restrict ourselves to the case of a commutative ring.

Proposition (2.2.1). — *Let A be a commutative ring. An ideal I of A is maximal if and only if the ring A/I is field.*

Proof. — Let us assume that A/I is a field. Its ideals are then (0) and A/I. Thus, the ideals of A containing I are I and A, which implies that I is a maximal ideal. Conversely, if I is a maximal ideal, this argument shows that the ring A/I is non-zero and that its only ideals are 0 and A/I itself. Let x be any non-zero element of A/I. Since the ideal (x) generated by x is non-zero, it equals A/I; therefore, there exists an element $y \in$ A/I such that $xy = 1$, so that x is invertible. This shows that A/I is a field. $\square$

Definition (2.2.2). — Let A be a commutative ring. An ideal P of A is said to be *prime* if the quotient ring A/P is an integral domain.

In particular, any maximal ideal of A is a prime ideal. Together with Krull's theorem (theorem 2.1.3), this shows that any non-zero commutative ring possesses at least one prime ideal.

The set of all prime ideals of a ring A is denoted by Spec(A) and called the *spectrum of* A. The set of all maximal ideals of A is denoted by Max(A) and called its *maximal spectrum*.

Proposition (2.2.3). — *Let A be a commutative ring and let P be an ideal of A. The following properties are equivalent:*

(i) *The ideal P is a prime ideal;*

(ii) *The ideal P is distinct from A and the product of any two elements of A which do not belong to P does not belong to P;*

(iii) *The complementary subset A − P is a multiplicative subset of A.*

Proof. — The equivalence of the last two properties is straightforward. Moreover, A/P being an integral domain (property (i), by definition) means that $P \neq 0$ (an integral domain is non-zero) and that the product of any two elements not in P does not belong to P, hence the equivalence with property (ii). $\square$

Examples (2.2.4). — a) The ideal $P = (0)$ is prime if and only if A is an integral domain.

b) In the ring **Z**, the maximal ideals are the ideals (p), for all prime numbers p; the only remaining prime ideal is (0).

These ideals are indeed prime. Conversely, let I be a prime ideal of **Z** and let us show that I is of the given form. We may assume that I $\neq$ (0); let then p be the smallest strictly positive element of I, so that $I = p\mathbf{Z}$. We need to show that p is a prime number. Since **Z** is not a prime ideal of **Z**, one has $p > 1$. If p is not prime, there are integers $m, n > 1$ such that $p = mn$. Then, $1 < m, n < p$, so that neither m nor n belongs to I; but since $p = mn$ belongs to I, this contradicts the definition of a prime ideal.

Proposition (2.2.5). — *Let $f : A \to B$ be a morphism of rings and let Q be a prime ideal of B. Then $P = f^{-1}(Q)$ is a prime ideal of A.*

Proof. — The ideal P is the kernel of the morphism from A to B/Q given by the composition of f and the canonical morphism $B \to B/Q$. Consequently, passing to the quotient, f defines an *injective* morphism from A/P to B/Q. In particular, A/P is an integral domain and the ideal P is prime.

We could also have proved this fact directly. First of all, $P \neq A$ since $f(1) = 1$ and $1 \notin Q$. Let then a and b be elements of A such that $ab \in P$. Then, $f(ab) = f(a)f(b)$ belongs to Q, hence $f(a) \in Q$ or $f(b) \in Q$. In the first case, $a \in P$, in the second, $b \in P$. $\square$

Thus a morphism of rings $f : A \to B$ gives rise to a map $f^* \colon \mathrm{Spec}(B) \to \mathrm{Spec}(A)$, given by $f^*(Q) = f^{-1}(Q)$ for every $Q \in \mathrm{Spec}(B)$.

Remarks (2.2.6). — Let A be a commutative ring.

(i) For every ideal I of A, let $V(I)$ be set of prime ideals P of A which contain I. We have $V(0) = \mathrm{Spec}(A)$ and $V(A) = \emptyset$.

Let (I_j) be a family of ideals of A and let $I = \sum I_j$ be the ideal it generates. A prime ideal P contains I if and only if it contains every I_j; in other words, $\bigcap_j V(I_j) = V(I)$.

Let I and J be ideals of A. Let us show that $V(I) \cup V(J) = V(I \cap J) = V(IJ)$. Indeed, if $P \supset I$ or $P \supset J$, then $P \supset I \cap J \supset IJ$, so that $V(I) \cup V(J) \subset V(I \cap J) \subset V(IJ)$. On the other hand, let us assume that $P \notin V(I) \cup V(J)$. Then, $P \not\supset I$, so that there exists an element $a \in I$ such that $a \notin P$; similarly, there is an element $b \in J$ such that $b \notin P$. Then, $ab \in IJ$ but $ab \notin P$, hence $P \notin V(IJ)$. This proves the claim.

These properties show that the subsets of $\mathrm{Spec}(A)$ of the form $V(I)$ are the closed subsets of a topology on $\mathrm{Spec}(A)$. This topology is called the *Zariski topology*.

(ii) Let I be an ideal of A. If $I = A$, then $V(I) = \emptyset$. Conversely, let us assume that $I \neq A$. By Krull's theorem (theorem 2.1.3), there exists a maximal ideal P of A such that $I \subset P$; then $P \in V(I)$, hence $V(I) \neq \emptyset$.

(iii) Let $f : A \to B$ be a morphism of commutative rings. Let I be an ideal of A; a prime ideal Q of B belongs to the inverse image $(f^*)^{-1}(V(I))$ of $V(I)$ if and only if $f^{-1}(Q)$ contains I; this is equivalent to the fact that Q contains $f(I)$, or, since Q is an ideal, to the inclusion $Q \supset f(I)B$. We thus have shown that $(f^*)^{-1}(V(I)) = V(f(I)B)$. This shows that the inverse image of a closed subset of $\mathrm{Spec}(A)$ is closed in $\mathrm{Spec}(B)$. In other words, the map $f^* \colon \mathrm{Spec}(B) \to \mathrm{Spec}(A)$ induced by f is continuous.

Proposition (2.2.7). — *Let A be a commutative ring, let S be a multiplicative subset of A and let $f : A \to S^{-1}A$ be the canonical morphism. The map $f^* \colon \mathrm{Spec}(S^{-1}A) \to \mathrm{Spec}(A)$ is injective, and its image is the set of prime ideals of A which are disjoint from S.*

Proof. — For every prime ideal Q of $S^{-1}A$, the ideal $P = f^*(Q)$ of A is a prime ideal of A. By proposition 1.7.1, it satisfies $Q = P \cdot S^{-1}A$. Moreover,

$P \cap S = \emptyset$ since, otherwise, one would have $1 \in Q$. This implies that the map $f^* \colon \operatorname{Spec}(S^{-1}A) \to \operatorname{Spec}(A)$ is injective and that its image is contained in the set of prime ideals of A which are disjoint from S.

Conversely, let P be a prime ideal of A which is disjoint from S. The ideal $Q = P \cdot S^{-1}A$ of $S^{-1}A$ is the set of all fractions a/s, with $a \in P$ and $s \in S$.

Let us first show that $f^{-1}(Q) = P$: let $a \in A$ be such that $a/1 \in Q$; then there exists $b \in P$ and $s \in S$ such that $a/1 = b/s$, hence there exists an element $t \in S$ such that $tsa = tb$. In particular $tsa \in P$; one has $ts \notin P$ since $P \cap S = \emptyset$, hence $a \in P$, by definition of a prime ideal.

Let us show that Q is a prime ideal of $S^{-1}A$. Since $f^{-1}(Q) = P \neq A$, we have $Q \neq S^{-1}A$. Let $a, b \in A$ and let $s, t \in S$ be such that $(a/s)(b/t) \in Q$. Then $ab/st \in Q$, hence $ab/1 \in Q$. By what precedes, $ab \in P$. Since P is a prime ideal of A, either a or b belongs to P. This implies that either a/s or b/t belongs to Q.

We thus have proved that f^* is a bijection from $\operatorname{Spec}(S^{-1}A)$ to the subset of $\operatorname{Spec}(A)$ consisting of prime ideals P such that $P \cap S = \emptyset$.   □

Example (2.2.8). — *Let A be a commutative ring and let P be a prime ideal of A. Then the fraction ring* A_P *is a local ring, and* PA_P *is its unique maximal ideal.* Indeed, the prime ideals of the fraction ring A_P are of the form Q_P, where Q is a prime ideal of A such that $Q \cap (A - P) = \emptyset$, that is, $Q \subset P$.

This observation leads to a useful general proof technique in commutative algebra, namely to first prove the desired result in the case for local rings, and to reduce to this case by localization.

Remark (2.2.9). — When $\operatorname{Spec}(A)$ and $\operatorname{Spec}(S^{-1}A)$ are endowed with their Zariski topologies, the map f^* is continuous, by remark 2.2.6, b). Let us prove that it is a homeomorphism from $\operatorname{Spec}(S^{-1}A)$ onto its image in $\operatorname{Spec}(A)$.

It suffices to show that it maps closed sets of $\operatorname{Spec}(S^{-1}A)$ to traces of closed sets of $\operatorname{Spec}(A)$. Thus let J be an ideal of $S^{-1}A$ and let $I = f^{-1}(J)$ be the set of all $a \in A$ such that $a/1 \in J$. Then I is an ideal of A and $I \cdot S^{-1}A = J$. Let us prove that $f^*(V(J)) = V(I) \cap f^*(\operatorname{Spec}(S^A))$.

Let Q be a prime ideal of $S^{-1}A$ and let $P = f^*(Q)$. Assume that Q contains J; let $a \in I$, then $a/1 \in J$, so that $a/1 \in Q$ and finally $a \in P$, so that P contains I. Assume conversely that P contains I; let $a \in P$ and let $s \in S$; then $a/1 \in J$ by definition of I, hence $a/s \in J$, hence Q contains J. This concludes the proof.

Examples (2.2.10). — The next two examples give some geometric content to the terminology "localization" which is used in the context of fraction rings.

a) Let $a \in A$ and let S be the multiplicative subset $\{1; a; a^2; \dots\}$. For a prime ideal P, being disjoint from S is equivalent to not containing a; consequently, f^* identifies $\operatorname{Spec}(S^{-1}A)$ with the complement of the closed subset $V((a))$ of $\operatorname{Spec}(A)$.

b) Let P be a prime ideal of A and let S be the multiplicative subset $A - P$. Being disjoint from S is equivalent to being contained in P. Let us show that f^* identifies $\operatorname{Spec}(S^{-1}A)$ with the intersection of all neighborhoods of P in $\operatorname{Spec}(A)$.

Let Q be a prime ideal of A such that Q $\subset$ P. Let U be an open neighborhood of P in Spec(A) and let I be an ideal of A such that U = Spec(A) $-$ V(I). Then P $\in$ U, hence P $\notin$ V(I), hence P does not contain I. A fortiori, Q does not contain I, so that Q $\notin$ V(I). This proves that the image of Spec(S^{-1}A) is contained in any open neighborhood of P in Spec(A). Conversely, if Q is a prime ideal of A which is not contained in P, then P $\notin$ V(Q), and Spec(A) $-$ V(Q) is an open neighborhood of P that does not contain Q.

Proposition (2.2.11). — *Let A be a commutative ring.*

(i) *An element of A is nilpotent if and only if it belongs to every prime ideal of A. In other words, the intersection of all prime ideals of A is equal to the nilpotent radical of A.*

(ii) *For any ideal I of A, its radical $\sqrt{I}$ is the intersection of all prime ideals of A containing I.*

Proof. — a) Let $a \in$ A and let P be a prime ideal of A. If $a \notin$ P, the definition of a prime ideal implies by induction on n that $a^n \notin$ P for any integer $n \geq 0$. In particular, $a^n \neq 0$ and a is not nilpotent. Nilpotent elements belong to every prime ideal.

Conversely, let $a \in$ A be a non-nilpotent element. The set S = $\{1, a, a^2, \dots\}$ of all powers of a is a multiplicative subset of A which does not contain 0. The localization S^{-1}A is then non-null. Let M be a prime ideal of S^{-1}A (for example, a maximal ideal of this non-zero ring) and let P be the set of all elements $x \in$ A such that $x/1 \in$ M. By definition, P is the inverse image of M by the canonical morphism from A to S^{-1}A. It is therefore a prime ideal of A. Since $a \in$ S, $a/1$ is invertible in S^{-1}A, hence $a/1 \notin$ M and $a \notin$ P. We thus have found a prime ideal of A which does not contain a, and this concludes the proof.

Part b) follows formally. Indeed, let B be the quotient ring A/I. An element $a \in$ A belongs to $\sqrt{I}$ if and only if its class cl$_I(a)$ is nilpotent in A/I. By a), this is equivalent to the fact that cl$_I(a)$ belongs to every prime ideal of A/I. Recall that every ideal of A/I is of the form P/I, for a unique ideal P of A which contains I; moreover, (A/I)/(P/I) is isomorphic to A/P, so that P/I is a prime ideal of A/I if and only if P is a prime ideal of A which contains I. Consequently, a belongs to $\sqrt{I}$ if and only if it belongs to every prime ideal containing I. In other words, $\sqrt{I}$ is the intersection of all prime ideals of A which contain I, as was to be shown. $\square$

Lemma (2.2.12) (Prime avoidance lemma). — *Let A be a commutative ring, let n be an integer ≥ 1 and let $P_1, \dots, P_n$ be prime ideals of A, and let I be an ideal of A. If I $\subset \bigcup_{i=1}^{n} P_i$, then there exists an integer $i \in \{1, \dots, n\}$ such that I $\subset P_i$.*

Proof. — We prove the result by induction on n. It is obvious if $n = 1$. Assume that for every i, I is not contained in P_i. By induction, for every $i \in \{1, \dots, n\}$, the ideal I is not contained in the union of the prime ideals P_j, for $j \neq i$. Consequently, there exists an element $a_i \in$ I such that $a_i \notin P_j$ if $j \neq i$.

Since $I \subset \bigcup_{j=1}^{n} P_j$, we see that $a_i \in P_i$. Let $a = a_1 + a_2 \ldots a_n$; since the elements $a_1, \ldots, a_n$ belong to I, we have $a \in I$. On the other hand, P_1 is prime, hence $a_2 \ldots a_n \notin P_1$; moreover, $a_1 \in P_1$, so that $a \notin P_1$. For $i \in \{2, \ldots, n\}$, $a_2 \ldots a_n \in P_i$, but $a_1 \notin P_i$, hence $a \notin P_i$. Consequently, a does not belong to any of the ideals $P_1, \ldots, P_n$. This contradicts the hypothesis that I is contained in their union.   □

The following lemma is another application of Zorn's lemma, but constructs *minimal* prime ideals.

Lemma (2.2.13). — *Let* A *be a commutative ring, let* I *be an ideal of* A *such that* $I \neq A$. *Then, for any prime ideal* Q *containing* I, *there exists a prime ideal* P *of* A *such that* $I \subset P \subset Q$ *and which is minimal among all prime ideals of* A *containing* I.

In particular, every prime ideal of A contains a prime ideal which is minimal for inclusion.

Proof. — Let us order the set $\mathscr{S}$ of prime ideals of A containing I by reverse inclusion: we say that $P \leq Q$ if P contains Q. The lemma will follow at once from Zorn's lemma (corollary A.2.13) once we show that $\mathscr{S}$ is an inductive set.

With that aim, let $(P_j)_{j \in J}$ be a totally ordered family of prime ideals of A. If this family is empty, the prime ideal Q contains I and is an upper bound. Let us assume that this family is not empty and let P be its intersection. It contains I, because I is contained in each P_j, and is distinct from A because it is contained in $P_j \neq A$, for any $j \in J$. Let us show that P is a prime ideal of A. Let $a, b \in A$ be such that $ab \in P$ but $b \notin P$; let us show that a belongs to every ideal P_j. So let $j \in J$. By definition of P, there exists an element $i \in J$ such that $b \notin P_i$. Since the family (P_j) is totally ordered, we have either $P_i \subset P_j$, or $P_j \subset P_i$. In the first case, the relations $ab \in P_i$ and $b \notin P_i$ imply that $a \in P_i$, hence $a \in P_j$. In the latter case, we have $b \notin P_j$; since $ab \in P_j$ and P_j is prime, this implies $a \in P_j$. Consequently, $P \in \mathscr{S}$, hence $\mathscr{S}$ is an inductive set.

As a consequence, $\mathscr{S}$ has a maximal element P, which is a prime ideal containing I, and minimal among them.   □

Remark (2.2.14). — Let S be a multiplicative subset of A. The map $P \mapsto S^{-1}P$ is a bijection respecting inclusion of the set of prime ideals of A disjoint from S to the set of prime ideals of $S^{-1}A$. In particular, the inverse image of a minimal prime ideal of $S^{-1}A$ by this bijection is a minimal prime ideal of A.

Conversely, if P is a minimal prime ideal of A which is disjoint from S, then $S^{-1}P$ is a minimal prime ideal of $S^{-1}A$.

Lemma (2.2.15). — *Let* A *be a commutative ring and let* P *be a minimal prime ideal of* A. *For every* $a \in P$ *there exists* $b \in A - P$ *and an integer* $n \geq 0$ *such that* $a^n b = 0$.

Proof. — Let $a \in A$ and let S be the set of elements of A of the form $a^n b$, for $b \in A - P$ and $n \geq 0$. It is a multiplicative subset of A that contains $A - P$. Assume $0 \notin S$. Then, $S^{-1}A$ is a non-zero ring, hence contains a prime ideal Q'.

The inverse image of Q in A is a prime ideal P' disjoint from S. In particular, P' ⊂ P. Moreover, $a \notin P'$, because $a \in S$, hence P' ≠ P. This contradicts the hypothesis that P is a minimal prime ideal of A. Consequently, $0 \in S$ and there exists $b \in A - P$ and $n \geq 0$ such that $a^n b = 0$.   □

2.3. Hilbert's Nullstellensatz

The following theorem of Hilbert gives a precise description of maximal ideals of polynomial rings over an algebraically closed field.

Theorem (2.3.1) (Hilbert's Nullstellensatz). — *Let K be an algebraically closed field and let n be a positive integer. The maximal ideals of $K[X_1, \ldots, X_n]$ are the ideals $(X_1 - a_1, \ldots, X_n - a_n)$, for $(a_1, \ldots, a_n) \in K^n$.*

Proof. — Let us first show that for any $(a_1, \ldots, a_n) \in K^n$, the ideal $(X_1 - a_1, \ldots, X_n - a_n)$ of $K[X_1, \ldots, X_n]$ is indeed a maximal ideal. Let us consider the morphism of "evaluation at $(a_1, \ldots, a_n)$", $\varphi \colon K[X_1, \ldots, X_n] \to K$, defined by $\varphi(P) = P(a_1, \ldots, a_n)$. It is surjective and induces an isomorphism $K[X_1, \ldots, X_n]/\mathrm{Ker}(\varphi) \simeq K$; since K is a field, $\mathrm{Ker}(\varphi)$ is a maximal ideal of $K[X_1, \ldots, X_n]$. It thus suffices to show that $\mathrm{Ker}(\varphi)$ coincides with the ideal $(X_1 - a_1, \ldots, X_n - a_n)$. One inclusion is obvious: if $P = (X_1 - a_1)P_1 + \cdots + (X_n - a_n)P_n$ for some polynomials $P_1, \ldots, P_n$, then $\varphi(P) = 0$. Conversely, let $P \in K[X_1, \ldots, X_n]$ be such that $\varphi(P) = 0$. Let us perform the euclidean division of P by $X_1 - a_1$ with respect to the variable X_1; we get a polynomial P_1 and a polynomial $R_1 \in K[X_2, \ldots, X_n]$ such that

$$P = (X_1 - a_1)P_1 + R_1(X_2, \ldots, X_n).$$

Let us repeat the process and divide by $X_2 - a_2$, etc.: we see that there exist polynomials $P_1, \ldots, P_n$, where $P_i \in K[X_i, \ldots, X_n]$ and a constant polynomial R_n, such that

$$P = (X_1 - a_1)P_1 + \cdots + (X_n - a_n)P_n + R_n.$$

Evaluating at $(a_1, \ldots, a_n)$, we get

$$\varphi(P) = P(a_1, \ldots, a_n) = R_n.$$

Since we assumed $\varphi(P) = 0$, we have $R_n = 0$ and P belongs to the ideal $(X_1 - a_1, \ldots, X_n - a_n)$.

We shall only prove the converse assertion under the supplementary hypothesis that the field K is uncountable. (Exercise 2.8.5 explains how to derive the general case from this particular case; see also corollary 9.1.3 for an alternative proof.) Let M be a maximal ideal of $K[X_1, \ldots, X_n]$ and let L be the quotient ring $K[X_1, \ldots, X_n]/M$; it is a field. Let x_i denote the class of X_i in L. The image of K by the canonical homomorphism is a subfield of L which we identify with K.

On David Hilbert

David HILBERT (1862–1943) was a German mathematician whose discoveries and influence encompass a vast number of scientific fields, such as number theory, algebra, mathematical physics, spectral theory, mathematical logic...

In his first mathematical work, he solved GORDAN's problem in invariant theory: if one lets a group G act on a ring of polynomials by linear transformations of the indeterminates, the question is to decide whether the ring of invariant polynomials is a finitely generated algebra (see theorem 6.3.15 for the case of a finite group). In the course of establishing the case $G = SL_n(\mathbf{C})$, HILBERT introduced the finite basis theorem (theorem 6.3.12). By avoiding an explicit search for invariants, on which most of the previous works focused, HILBERT's proof opened a new era in the subject.

On the occasion of his address at the International congress of mathematicians in Paris, 1900, he presented ten mathematical problems (the final paper features thirteen more) which had a strong impact on mathematics during the twentieth century. Remarkably, some of them did not have the answer Hilbert expected, such as the 10th problem (on algorithmic solvability of diophantine equations, answered negatively by MATIYASEVICH in 1970,) or the 14th problem (the general solution to Gordan's problem, solved neg-

Photograph of David Hilbert (1912) for a postcard sold to the students of University of Göttingen

Unknown photographer
Source: Wikipedia
Public domain.

The grave of David Hilbert, at the Göttingen cemetery, featuring the engraved quote "Wir müssen wissen, wir werden wissen."

atively by NAGATA in 1959), others were solved positively (such as the 18th about the densest sphere packing in 3-dimensional space, the 7th about the transcendental nature of a^b, when a, b are algebraic numbers, or the 19th about the analytic nature of solutions of problems in calculus of variations), and some remain open today, such as the Riemann hypothesis (8th problem) or the study of trajectories of polynomial vector fields (16th problem).

HILBERT had proposed an axiomatic framework for geometry in 1899. In the 1920s, he set up a program to formulate mathematics within a complete axiomatic logical system, in the hope that such a system could be found and proved to lead to no contradiction.

> *We must not believe those, who today, with philosophical bearing and deliberative tone, prophesy the fall of culture and accept the* ignorabimus. *For us there is no* ignorabimus, *and in my opinion none whatever in natural science. In opposition to the foolish* ignorabimus *our slogan shall be* Wir müssen wissen — wir werden wissen. *("We must know — we will know.")*
>
> David Hilbert, Address at the Society of German scientists and physicians (8 September 1930)

While GÖDEL's incompleteness theorem (1931) showed that such a program is essentially impossible, HILBERT's work led to the foundation of mathematical logic as an autonomous discipline.

The field L admits the natural structure of a K-vector space. As such, it is generated by the countable family of all $x_1^{i_1} \ldots x_n^{i_n}$, for $i_1, \ldots, i_n \in \mathbf{N}$. Indeed, any element of L is the class of some polynomial, hence of a linear combination of monomials. Moreover, the set $\mathbf{N}^n$ of all possible exponents of monomials is countable.

Let a be any element of L and let $\varphi \colon K[T] \to L$ the ring morphism given by $\varphi(P) = P(a)$. Assume that it is injective. Then, φ extends as a (still injective) morphism of fields, still denoted by φ, from $K(T)$ to L. In particular, the elements $1/(a - c)$ (for $c \in K$) are linearly independent over K since they are the images of the rational functions $1/(T - c)$ which are linearly independent in $K(T)$ in view of the uniqueness of the decomposition of a rational function in simple terms. However, this would contradict lemma 2.3.2 below: any linearly independent family of L must be countable, while K is not. This implies that the morphism φ is not injective.

Let $P \in K[T]$ be a non-zero polynomial such that $P(a) = 0$. The polynomial P is not constant. Since K is algebraically closed, it has the form $c \prod_{i=1}^{m} (T - c_i)$, for some strictly positive integer m and elements $c_1, \ldots, c_m \in K$ and $c \in K^*$. Then, $c \prod_{i=1}^{n} (f - c_i) = 0$ in L. Since L is a field, there exists an element $i \in \{1, \ldots, m\}$ such that $a = c_i$; consequently, $a \in K$ and $L = K$.

In particular, there exists, for any $i \in \{1, \ldots, n\}$, some element $a_i \in K$ such that $x_i = a_i$. This implies the relations $X_i - a_i \in M$, hence the ideal M contains the ideal $(X_1 - a_1, \ldots, X_n - a_n)$. Since the latter is a maximal ideal, we have equality, and this concludes the proof.   □

Lemma (2.3.2). — *Let K be a field, let V be a K-vector space which is generated by a countable family. Then every subset of V which is linearly independent is countable.*

Proof. — This will follow from a general result on the dimension of vector spaces, possibly infinite-dimensional. However, one may give an alternate argument that only makes use of finite-dimensional vector spaces.

Let J be a countable set and $(e_j)_{j \in J}$ be a generating family of V. Let $(v_i)_{i \in I}$ be a linearly independent family in V; we have to prove that I is countable. For every finite subset A of J, the subspace V_A of V generated by the elements e_j, for $j \in A$, is finite-dimensional. In particular, the subset I_A of all $i \in I$ such that $v_i \in V_A$ is finite. Since the set of all finite subsets of J is countable, the union I' of all subsets I_A, when A runs over all finite subsets of J, is countable. For every $i \in I$, there is a finite subset A of J such that v_i is a linear combination of the e_j, for $j \in A$, hence $i \in I'$. Consequently, $I = I'$ and I is countable.   □

The following theorem is a topological analogue.

Theorem (2.3.3) (Gelfand). — *Let X be a topological space; let $\mathscr{C}(X)$ be the ring of real-valued, continuous functions on X.*

For any $x \in X$, the set M_x of all continuous functions on X which vanish at x is a maximal ideal of the ring $\mathscr{C}(X)$.

If X is a compact metric space, the map $x \mapsto M_x$ from X to the set of maximal ideals of $\mathscr{C}(X)$ is a bijection.

Proof. — Let $\varphi_x \colon \mathscr{C}(X) \to \mathbf{R}$ be the ring morphism given by $f \mapsto f(x)$ ("evaluation at the point x"). It is surjective, with kernel M_x. This proves that M_x is a maximal ideal of $\mathscr{C}(X)$.

Let us assume that X is a compact metric space, with distance d. For any point $x \in X$, the function $y \mapsto d(x, y)$ is continuous, vanishes at x but does not vanish at any other point of X. Consequently, it belongs to M_x but not to M_y if $y \neq x$. This shows that $M_x \neq M_y$ for $x \neq y$ and the mapping $x \mapsto M_x$ is injective.

Let I be an ideal of $\mathscr{C}(X)$ which is not contained in any of the maximal ideals M_x, for $x \in X$. For any $x \in X$, there thus exists a continuous function $f_x \in I$ such that $f_x(x) \neq 0$. By continuity, the set U_x of points of X at which f_x does not vanish is an open neighborhood of x. These open sets U_x cover X. Since X is compact, there exists a finite subset $S \subset X$ such that the open sets U_s, for $s \in S$, cover X as well. Let us set $f = \sum_{s \in S}(f_s)^2$. This is a positive continuous function, and it belongs to I. If $x \in X$ is a point such that $f(x) = 0$, then for any $s \in S$, $f_s(x) = 0$, that is to say, $x \notin U_s$. Since the U_s cover X, we have a contradiction and f does not vanish at any point of X. It follows that f is invertible in $\mathscr{C}(X)$ (its inverse is the continuous function $x \mapsto 1/f(x)$). Since $f \in I$, we have shown $I = \mathscr{C}(X)$. Consequently, any ideal of $\mathscr{C}(X)$ distinct from $\mathscr{C}(X)$ is contained in one of the maximal ideals M_x, so that these ideals constitute the whole set of maximal ideals of $\mathscr{C}(X)$.   □

Let K be an algebraically closed field. Hilbert's Nullstellensatz is the basis of an admirable correspondence between algebra (some ideals of the polynomial ring $K[X_1, \ldots, X_n]$) and geometry (some subsets of K^n) which we describe now.

Definition (2.3.4). — An *algebraic set* is a subset of K^n defined by a family of polynomial equations.

Specifically, a subset Z of K^n is an algebraic set if and only if there exists a subset S of $K[X_1, \ldots, X_n]$ such that

$$Z = \mathscr{Z}(S) = \{(a_1, \ldots, a_n) \in K^n \, ; \text{ for every } P \in S, P(a_1, \ldots, a_n) = 0\}.$$

Proposition (2.3.5). —   (i) *If* $S \subset S'$, *then* $\mathscr{Z}(S') \subset \mathscr{Z}(S)$.

(ii) *The empty set and* K^n *are algebraic sets.*

(iii) *If* $\langle S \rangle$ *is the ideal generated by S in* $K[X_1, \ldots, X_n]$, *then* $\mathscr{Z}(\langle S \rangle) = \mathscr{Z}(S)$.

(iv) *The intersection of a family of algebraic sets and the union of two algebraic sets are algebraic sets.*

(v) *If* I *is any ideal of* $K[X_1, \ldots, X_n]$, *then* $\mathscr{Z}(I) = \mathscr{Z}(\sqrt{I})$.

Proof. — a) Let $(a_1, \ldots, a_n) \in \mathscr{Z}(S')$ and let us show that $(a_1, \ldots, a_n) \in \mathscr{Z}(S)$. If $P \in S$, we have to show that $P(a_1, \ldots, a_n) = 0$, which holds since $P \in S'$.

b) We have $\emptyset = \mathscr{Z}(\{1\})$ (the constant polynomial 1 does not vanish at any point of K^n) and $K^n = \mathscr{Z}(\{0\})$ (the zero polynomial vanishes everywhere; we could also write $K^n = \mathscr{Z}(\emptyset)$).

c) Since $S \subset \langle S \rangle$, we have $\mathscr{L}(\langle S \rangle) \subset \mathscr{L}(S)$. Conversely, let $(a_1, \ldots, a_n) \in \mathscr{L}(S)$ and let us show that $(a_1, \ldots, a_n) \in \mathscr{L}(\langle S \rangle)$. Let $P \in \langle S \rangle$; by definition, there are finite families $(P_i)_{i \in I}$ and $(Q_i)_{i \in I}$ of polynomials in $K[X_1, \ldots, X_n]$ such that $P = \sum P_i Q_i$ and $Q_i \in S$ for every $i \in I$. Then,

$$P(a_1, \ldots, a_n) = \sum_{i \in I} P_i(a_1, \ldots, a_n) Q_i(a_1, \ldots, a_n) = 0$$

since $Q_i(a_1, \ldots, a_n) = 0$. Consequently, $(a_1, \ldots, a_n) \in \mathscr{L}(\langle S \rangle)$.

d) Let (Z_j) be a family of algebraic sets, for every j, let S_j be a subset of $K[X_1, \ldots, X_n]$ such that $Z_j = \mathscr{L}(S_j)$. We shall show that

$$\bigcap_j \mathscr{L}(S_j) = \mathscr{L}\left(\bigcup_j S_j\right).$$

Indeed, $(a_1, \ldots, a_n)$ belongs to $\bigcap_j \mathscr{L}(S_j)$ if and only if $P(a_1, \ldots, a_n)$ for every j and every $P \in S_j$, which means exactly that $P(a_1, \ldots, a_n) = 0$ for every $P \in \bigcup_j S_j$, that is, $(a_1, \ldots, a_n) \in \mathscr{L}(\bigcup_j S_j)$.

Let S and S' be two subsets of $K[X_1, \ldots, X_n]$. Let $T = \{PP' ; P \in S, P' \in S'\}$. We are going to show that $\mathscr{L}(S) \cup \mathscr{L}(S') = \mathscr{L}(T)$. Indeed, if $(a_1, \ldots, a_n) \in \mathscr{L}(S)$ and $Q \in T$, we may write $Q = PP'$ with $P \in S$ and $P' \in S'$. Then, $Q(a_1, \ldots, a_n) = P(a_1, \ldots, a_n)P'(a_1, \ldots, a_n) = 0$ since $(a_1, \ldots, a_n) \in \mathscr{L}(S)$. In other words, $\mathscr{L}(S) \subset \mathscr{L}(T)$. Similarly, $\mathscr{L}(S') \subset \mathscr{L}(T)$, hence $\mathscr{L}(S) \cup \mathscr{L}(S') \subset \mathscr{L}(T)$. Conversely, let $(a_1, \ldots, a_n) \in \mathscr{L}(T)$. To show that $(a_1, \ldots, a_n) \in \mathscr{L}(S) \cup \mathscr{L}(S')$, it suffices to prove that if $(a_1, \ldots, a_n) \notin \mathscr{L}(S')$, then $(a_1, \ldots, a_n) \in \mathscr{L}(S)$. By definition, there is a polynomial $P' \in S'$ such that $P'(a_1, \ldots, a_n) \neq 0$. Then, for every $P \in S$, one has $PP' \in T$, hence $(PP')(a_1, \ldots, a_n) = 0 = P(a_1, \ldots, a_n)P'(a_1, \ldots, a_n)$ so that $P(a_1, \ldots, a_n) = 0$, as was to be shown.

e) Since $I \subset \sqrt{I}$, one has $\mathscr{L}(\sqrt{I}) \subset \mathscr{L}(I)$. Conversely, let $(a_1, \ldots, a_n) \in \mathscr{L}(I)$. Let $P \in \sqrt{I}$, let $m \geq 1$ be such that $P^m \in I$. Then $P^m(a_1, \ldots, a_n) = 0$, hence $P(a_1, \ldots, a_n) = 0$ and $(a_1, \ldots, a_n) \in \mathscr{L}(\sqrt{I})$.   □

Remark (2.3.6). — The preceding proposition can be rephrased by saying that there exists a topology on K^n for which the closed sets are the algebraic sets. This topology is called the *Zariski topology.*

We have constructed one of the two directions of the correspondence: with any ideal I of $K[X_1, \ldots, X_n]$, we associate the algebraic set $\mathscr{L}(I)$. From the proof of the previous proposition, we recall the formulas

$$\mathscr{L}(0) = K^n$$
$$\mathscr{L}(K[X_1, \ldots, X_n]) = \emptyset$$
$$\mathscr{L}\left(\sum I_j\right) = \bigcap_j \mathscr{L}(I_j)$$
$$\mathscr{L}(IJ) = \mathscr{L}(I) \cup \mathscr{L}(J).$$

The other direction of the correspondence associates an ideal with any subset of K^n.

Definition (2.3.7). — Let V be a subset of K^n. One defines

$$\mathscr{I}(V) = \{P \in K[X_1, \ldots, X_n] ; P(a_1, \ldots, a_n) = 0 \text{ for every } (a_1, \ldots, a_n) \in V\}.$$

Proposition (2.3.8). — *a) For every $V \subset K^n$, $\mathscr{I}(V)$ is an ideal of $K[X_1, \ldots, X_n]$. Moreover, $\mathscr{I}(V) = \sqrt{\mathscr{I}(V)}$.*
b) If $V \subset V'$, then $\mathscr{I}(V') \subset \mathscr{I}(V)$.
c) For any two subsets V and V' of K^n, one has $\mathscr{I}(V \cup V') = \mathscr{I}(V) \cap \mathscr{I}(V')$.

Proof. — a) For any $(a_1, \ldots, a_n) \in K^n$, the map $P \mapsto P(a_1, \ldots, a_n)$ is a ring morphism from $K[X_1, \ldots, X_n]$ to K, and $\mathscr{I}(V)$ is the intersection of the kernels of those morphisms, for $(a_1, \ldots, a_n) \in V$. Consequently, it is an ideal of $K[X_1, \ldots, X_n]$. Moreover, if $P \in K[X_1, \ldots, X_n]$ and $m \geq 1$ is such that $P^m \in \mathscr{I}(V)$, then $P^m(a_1, \ldots, a_n) = 0$ for every $(a_1, \ldots, a_n) \in V$; this implies that $P(a_1, \ldots, a_n) = 0$, hence $P \in \mathscr{I}(V)$. This shows that $\mathscr{I}(V) = \sqrt{\mathscr{I}(V)}$.

b) Let $P \in \mathscr{I}(V')$. For any $(a_1, \ldots, a_n) \in V$, we have $P(a_1, \ldots, a_n) = 0$, since $V \subset V'$. Consequently, $P \in \mathscr{I}(V)$.

c) By definition, a polynomial P belongs to $\mathscr{I}(V \cup V')$ if and only if it vanishes at any point of V *and* of V'. □

Proposition (2.3.9). — *a) For any ideal I of $K[X_1, \ldots, X_n]$, one has $I \subset \mathscr{I}(\mathscr{Z}(I))$.*
b) For any subset V of K^n, one has $V \subset \mathscr{Z}(\mathscr{I}(V))$.

Proof. — a) Let $P \in I$; let us show that $P \in \mathscr{I}(\mathscr{Z}(I))$. We need to show that P vanishes at every point of $\mathscr{Z}(I)$. Now, for any point $(a_1, \ldots, a_n) \in \mathscr{Z}(I)$, one has $P(a_1, \ldots, a_n) = 0$, since $P \in I$.

b) Let $(a_1, \ldots, a_n) \in V$ and let us show that $(a_1, \ldots, a_n)$ belongs to $\mathscr{Z}(\mathscr{I}(V))$. By definition, we thus need to prove that for any $P \in \mathscr{I}(V)$, one has $P(a_1, \ldots, a_n) = 0$. This assertion is clear, since $(a_1, \ldots, a_n) \in V$. □

We are going to use Hilbert's Nullstellensatz to establish the following theorem.

Theorem (2.3.10). — *For any ideal I of $K[X_1, \ldots, X_n]$, one has*

$$\boxed{\mathscr{I}(\mathscr{Z}(I)) = \sqrt{I}}.$$

Before we pass to the proof, let us show how this gives rise to a bijection between algebraic sets — geometry — and ideals equal to their own radical — algebra.

Corollary (2.3.11). — *The maps $V \mapsto \mathscr{I}(V)$ and $I \mapsto \mathscr{Z}(I)$ induce bijections, each the inverse of the other, between algebraic sets in K^n and ideals I of $K[X_1, \ldots, X_n]$ such that $I = \sqrt{I}$.*

Proof (Proof of the corollary). — Let I be an ideal of $K[X_1, \ldots, X_n]$ such that $I = \sqrt{I}$. By theorem 2.3.10, one has

$$\mathscr{I}(\mathscr{Z}(I)) = \sqrt{I} = I.$$

Conversely, let V be an algebraic set and let S be any subset of $K[X_1, \ldots, X_n]$ such that $V = \mathscr{Z}(S)$. Letting $I = \langle S \rangle$, we have $V = \mathscr{Z}(I)$. By proposition 2.3.5, we even have $V = \mathscr{Z}(\sqrt{I})$. Then,

$$\mathscr{I}(V) = \mathscr{I}(\mathscr{Z}(I)) = \sqrt{I},$$

hence

$$V = \mathscr{Z}(I) = \mathscr{Z}(\sqrt{I}) = \mathscr{Z}(\mathscr{I}(V)).$$

This concludes the proof of the corollary.   □

Proof (Proof of theorem 2.3.10). — The inclusion $\sqrt{I} \subset \mathscr{I}(\mathscr{Z}(I))$ is easy (and has been proved incidentally in the course of the proof of prop. 2.3.5). Indeed, let $P \in \sqrt{I}$, let $m \geq 1$ be such that $P^m \in I$. Then, for any $(a_1, \ldots, a_n) \in \mathscr{Z}(I)$, one has $P^m(a_1, \ldots, a_n) = 0$ hence $P(a_1, \ldots, a_n) = 0$. Consequently, $P \in \mathscr{I}(\mathscr{Z}(I))$.

Conversely, let P be a polynomial belonging to $\mathscr{I}(\mathscr{Z}(I))$. We want to show that there exists an integer $m \geq 1$ such that $P^m \in I$. Let us introduce the ideal J of $K[X_1, \ldots, X_n, T]$ which is generated by I and by the polynomial $1 - TP$. We have $\mathscr{Z}(J) = \emptyset$. Indeed, let $(a_1, \ldots, a_n, t) \in K^{n+1}$ be a point belonging to $\mathscr{Z}(J)$. Since $P \in I$, it follows from the definition of the ideal J that P belongs to J, hence $P(a_1, \ldots, a_n) = 0$. On the other hand, $1 - TP$ belongs to J too, so that we have $tP(a_1, \ldots, a_n) = 1$, a contradiction.

By Hilbert's Nullstellensatz (theorem 2.3.1), the ideal J is contained in no maximal ideal of $K[X_1, \ldots, X_n, T]$, hence $J = K[X_1, \ldots, X_n, T]$. Consequently, there are polynomials $Q_i \in I$, $R_i \in K[X_1, \ldots, X_n, T]$ and $R \in K[X_1, \ldots, X_n, T]$ such that

$$1 = (1 - TP)R + \sum_i Q_i R_i,$$

and the image of the ideal I in the quotient ring $K[X_1, \ldots, X_n, T]/(1 - TP)$ generates the unit ideal.

Now, recall from proposition 1.6.6 that this quotient ring is isomorphic to the localization of $K[X_1, \ldots, X_n]$ by the multiplicative set $S = \{1, P, P^2, \ldots, \}$ generated by P. Consequently, the image of the ideal I in the fraction ring $S^{-1}K[X_1, \ldots, X_n]$ generates the full ideal, which means that the multiplicative subset S meets the ideal I. In other words, there is an integer m such that $P^m \in I$, as was to be shown.   □

Remark (2.3.12). — The introduction of a new variable T in the preceding proof is known as the Rabinowitsch trick. However, in most of the presentations of the proof, it is often followed by the following, more elementary-looking, argument. We start as above until we reach the equality $1 = (1 - TP)R + \sum Q_i R_i$. of polynomials in $K[X_1, \ldots, X_n, T]$. Our goal now is to substitute $T = 1/P$. Formally, we obtain an equality of rational functions in the field $K(X_1, \ldots, X_n)$:

$$1 = \sum_i Q_i(X_1, \ldots, X_n) R_i(X_1, \ldots, X_n, 1/P(X_1, \ldots, X_n)).$$

The denominators are powers of P; chasing them, we shall obtain a relation of the form

$$P^M = \sum_i Q_i(X_1, \ldots, X_n) S_i(X_1, \ldots, X_n)$$

which shows that P^M belongs to I. To make this argument more precise, let M be an integer larger than the T-degrees of all of the polynomials R_i and of the polynomial R. We can then write

$$R(X_1, \ldots, X_n, T) = \sum_{m=0}^{M} S_i(X_1, \ldots, X_n) T^m$$

and, for any i,

$$R_i(X_1, \ldots, X_n, T) = \sum_{m=0}^{M} S_{i,m}(X_1, \ldots, X_n) T^m,$$

for some polynomials S_i and $S_{i,m}$ in $K[X_1, \ldots, X_n]$. Multiplying both sides of the initial equality by P^M, we obtain a relation

$$P^M = (1 - TP) \sum_{m=0}^{M} (PT)^m S_m P^{M-m} + \sum_i \sum_{m=0}^{M} (PT)^m Q_i S_{i,m} P^{M-m}.$$

Collecting the monomials of various T-degrees, we deduce

$$P^M = S_0 P^M + \sum_i Q_i S_{i,0} P^M,$$

$$0 = S_m P^{M-m} - S_{m-1} P^{M-m+1} + \sum_i Q_i S_{i,m} P^{M-m}$$

for $m \in \{1, \ldots, M\}$, and

$$0 = -S_M.$$

Summing all of these relations, we get

$$P^M = \sum_i \sum_{m=0}^{M} Q_i S_{i,m} P^{M-m}.$$

Since the polynomials Q_i belong to I, we conclude that P^M belongs to I, hence $P \in \sqrt{I}$.

2.4. Principal Ideal Domains, Euclidean Rings

2.4.1. — Let A be a commutative ring.

Recall that an ideal of A is said to be *principal* if it is of the form aA, for some $a \in$ A. One also writes (a) for aA.

Let $a, b \in$ A. The ideal aA is contained in the ideal bA if and only if there exists an element $c \in$ A such that $a = bc$, that is, if and only if b divides a.

Assume moreover that A is an integral domain. Let $a, b \in$ A be such that aA = bA. Then, there exist $c, d \in$ A such that $a = bc$ and $b = ad$, hence $a = a(cd)$ and $b = (cd)$. If $a \neq 0$ then $b \neq 0$; simplifying by a, we get $cd = 1$, hence c and d are invertible. In other words, *two non-zero elements a and b of an integral domain A generate the same ideal if and only if there exists a unit $u \in$ A such that $b = au$.*

The units of the ring A = $\mathbf{Z}$ are ± 1; it is thus customary to choose, as a generator of a principal ideal, a positive element. Similarly, the units of the ring K[X] of polynomials in one indeterminate X and with coefficients in a field K are the non-zero constant polynomials and we then often choose a monic polynomial for a generator of a non-zero ideal (see example 1.4.4).

Definition (2.4.2). — One says that a (commutative) ring is a *principal ideal domain* if it is an integral domain and if all of its ideals are principal.

Examples (2.4.3). —   *a)* The ring $\mathbf{Z}$ is a principal ideal domain (example 1.4.3), as well as the ring K[X] of polynomials in one variable with coefficients in a (commutative) field K (example 1.4.4).

b) In the ring K[X, Y] of polynomials in two variables with coefficients in a field K, the ideal (X, Y) is not principal. For if it were generated by a polynomial P, this polynomial would need to divide both X and Y. Necessarily, P has to be a non-zero constant. It follows that there exist Q, R $\in$ K[X, Y] such that $1 = XQ(X, Y) + YR(X, Y)$. This, however, is absurd, since the right-hand term of this equality has no constant term.

2.4.4. Greatest common divisor, least common multiple — Let A be a principal ideal domain. Let (a_i) be a family of elements of A. By the assumption on A, the ideal I generated by the (a_i) is generated by one element, say a. It follows that d divides a_i for any i: d is a common divisor of all of the a_i. Moreover, if d' is a common divisor of the a_i, then $a_i \in (d')$ for every i, hence I $\subset (d')$ and d' divides d. One says that d is a greatest common divisor (gcd) of the a_i. The word "greatest" has to be understood in the sense of divisibility: the common divisors of the a_i are exactly the divisors of their gcd. There is in general no preferred choice of a greatest common divisor, all differ by multiplication by a unit in A.

Let J be the intersection of the ideals (a_i) and let m be a generator of the ideal J. For any i, $m \in (a_i)$, that is, m is a multiple of a_i for every i. Moreover, if $m' \in$ A is a multiple of a_i for every i, then $m' \in (a_i)$ for every i, hence $m' \in (m)$ and m' is a multiple of m. One says that m is a least common multiple (lcm) of the a_i. Again, the word "least" has to be understood in the sense of divisibility. As for the gcd, there is no preferred choice and all least common multiples differ by multiplication by a unit in A.

As explained above, when $A = \mathbf{Z}$ is the ring of integers, one may choose for the gcd and the lcm the unique positive generator of the ideal generated by the a_i, *resp.* of the intersection of the (a_i). Then, except for degenerate cases, d is the greatest common divisor and m is the least common (non-zero) multiple in the naive sense too.

Similarly, when $A = K[X]$ is the ring of polynomials in one indeterminate X, it is customary to choose the gcd and the lcm to be monic polynomials (or the zero polynomial).

Remark (2.4.5). — Let K be a field, let $P, Q \in K[X]$ be polynomials (not both zero) and let $D \in K[X]$ be their gcd, chosen to be monic. Let $P_1, Q_1 \in K[X]$ be defined by $P = DP_1$ and $Q = DQ_1$. Since D is a generator of the ideal (P, Q), there exist polynomials U and V such that $D = UP + VQ$, hence $1 = UP_1 + VQ_1$ and $(P_1, Q_1) = K[X]$.

It follows that for every field L containing K, one has $(P_1, Q_1)_{L[X]} = L[X]$, and D is still a gcd of P and Q in the ring $L[X]$.

Stated in the opposite direction, this result states that the monic gcd of P and Q in $L[X]$ belongs to $K[X]$.

Definition (2.4.6). — Let A be an integral domain. A map $\delta \colon A - \{0\} \to \mathbf{N}$ is called a *euclidean gauge*, or simply a *gauge*[1] on A if it satisfies the following two properties:

 – For any $a, b \in A - \{0\}$, $\delta(ab) \geq \sup(\delta(a), \delta(b))$;

 – For any $a, b \in A$ such that $b \neq 0$, there exists q and $r \in A$ such that $a = bq + r$ and such that either $r = 0$ or $\delta(r) < \delta(b)$.

If there exists a euclidean gauge on A, then one says that A is a *euclidean ring*.

In a euclidean ring, a relation such as $a = bq + r$, where $r = 0$ or $\delta(r) < \delta(b)$ is called a *euclidean division* of a by b; the element q is called its quotient and the element r its remainder.

Examples (2.4.7). — a) The ring of integers and the ring of polynomials in one variable with coefficients in a field are euclidean rings, with gauges given by the usual absolute value and the degree.

b) The ring $\mathbf{Z}[i]$ of Gaussian integers is a euclidean ring, with gauge δ defined by $\delta(z) = z\bar{z} = |z|^2$ (exercise 2.8.18). See also exercises 2.8.19, 2.8.20 and 2.8.17 for other examples of euclidean rings.

Remark (2.4.8). — Property (i) of euclidean gauges implies that $\delta(a) \leq \delta(b)$ when a divides b. Consequently, if u is a unit, then $\delta(a) = \delta(au)$ for any non-zero element of A.

---

[1] According for example to Bourbaki and Wedderburn, the official word is *stathm*.

This property (i) is however not crucial for the definition of a euclidean ring. Indeed, if δ is any map satisfying property (ii), one may modify it in order to get a euclidean gauge; see exercise 2.8.21.

Proposition (2.4.9). — *Any euclidean ring is a principal ideal domain.*

Proof. — Indeed, let A be a euclidean ring with gauge δ, and let I be a non-zero ideal of A. Let a be a non-zero element of I such that $\delta(a)$ is minimal among the values of δ on $I - \{0\}$. Let $x \in I$ and let us consider a euclidean division $x = aq + r$ of x by a; one has $r = x - aq \in I$. If $r \neq 0$, then $\delta(r) < \delta(a)$, which contradicts the choice of a. So $r = 0$ and $x = aq \in aA$, hence $I = (a)$. $\square$

Remark (2.4.10). — There exist principal ideal domains which are not euclidean, for any map δ. One such example is the set of all complex numbers of the form $a + b\frac{1+i\sqrt{19}}{2}$, with a and $b \in \mathbf{Z}$ (see exercises 2.8.23 and 2.8.24).

2.5. Unique Factorization Domains

Definition (2.5.1). — Let A be an integral domain. One says that an element $a \in A$ is *irreducible* if it is not a unit and if the relation $a = bc$, for some b and $c \in A$, implies that b or c is a unit.

Examples (2.5.2). —   *a*) The irreducible elements of $\mathbf{Z}$ are the prime numbers and their opposites.

b) Let k be a field; the irreducible elements of $k[X]$ are the irreducible polynomials, that is, the polynomials of degree ≥ 1 which cannot be written as the product of two polynomials of degree ≥ 1.

c) The element 0 is never irreducible: it can be written as 0×0 and 0 is not a unit (A being an integral domain, one has $1 \neq 0$).

Proposition (2.5.3) (Gauss's lemma). — *Let A be a principal ideal domain. For a non-zero ideal of A to be prime, it is necessary and sufficient that it be generated by an irreducible element; it is then a maximal ideal.*

Proof. — Let I be a prime ideal of A; assume that $I \neq 0$. Since A is a principal ideal domain, there exists an element $a \in A$ such that $I = (a)$. Since $I \neq 0$, we have $a \neq 0$; Let us show that a is irreducible. Since A is not a prime ideal, a is not a unit. Let b and c be elements of A such that $a = bc$. Since I is a prime ideal, b or c belongs to I. Assume that $b \in I$; then, there exists an element $u \in A$ such that $b = au$, hence $a = auc$ and $cu = 1$ after simplifying by a. This shows that c is a unit. Similarly, if $c \in I$ then b is a unit. It follows that a is irreducible, as claimed.

Conversely, let a be an irreducible element of A and let us show that the ideal $I = (a)$ is a maximal ideal of A. Let $x \in A$ be any element which is not a multiple of a and let $J = I + (x)$. Let $b \in A$ be such that $J = (b)$. Since $a \in J$, there exists an element $c \in A$ such that $a = bc$. If c were a unit, one would have $(a) = (b) = I + (x)$, hence $x \in (a)$, contrary to the assumption. Since a is irreducible, it follows that b is a unit, hence $J = A$. This shows that I is a maximal ideal of A. □

The first part of the proof shows, more generally, that in an integral domain, if a non-zero prime ideal is generated by one element, then this element is irreducible. The converse does not hold for general rings and leads to the notion of a unique factorization domain.

Definition (2.5.4). — Let A be a domain. One says that A is a *unique factorization domain* if it satisfies the following two properties:

(i) Every increasing sequence of principal ideals of A is stationary;

(ii) Every ideal generated by an irreducible element is a prime ideal.

The first condition will allow us to write any non-zero element as the product of finitely many irreducible elements; it is automatic if the ring A is noetherian (see definition 6.3.5). The second one is the most important and will guarantee, up to minor tweaks, the uniqueness of such a decomposition into irreducible factors.

Let us rewrite this condition somewhat. Let p be an irreducible element of A. Since p is not a unit, the ideal (p) is a prime ideal if and only if the product of two elements of A cannot belong to (p) unless one of them belongs to p. In other words, ab is divisible by p if and only if either a or b is divisible by p.

Examples (2.5.5). —   *a*) A field is a unique factorization domain.

b) Proposition 2.5.3 states that every irreducible element of a principal ideal domain generates a maximal ideal. Given lemma 2.5.6 below, we conclude that *principal ideal domains are unique factorization domains.*

c) We will prove (corollary 2.6.7) that for any unique factorization domain Λ, the ring $A[X_1, \ldots, X_n]$ of polynomials with coefficients in A is also a unique factorization domain. In particular, polynomial rings with coefficients in a field, or in $\mathbf{Z}$, are unique factorization domains.

Lemma (2.5.6). — *In a principal ideal domain, any increasing sequence of ideals is stationary.*[2]

Proof. — Let A be a principal ideal domain, let (I_n) be an increasing sequence of ideals. Let I be the union of all ideals I_n. Since the sequence is increasing, I is again an ideal of A. Since A is a principal ideal domain, there exists an $a \in I$ such that $I = (a)$. Let then $m \in \mathbf{N}$ be such that $a \in I_m$. For $n \geq m$, we have $I = (a) \subset I_n \subset I$, hence the equality $I_n = I$: the sequence (I_n) is stationary.   □

---

[2] In the terminology that we will introduce later, this says that a principal ideal domain is a noetherian ring (definition 6.3.5).

Theorem (2.5.7). — *Let A be a unique factorization domain and let a be any non-zero element of A.*

(i) *There exist an integer $n \geq 0$, irreducible elements $p_1, \ldots, p_n \in A$, and a unit $u \in A$, such that $a = up_1 \ldots p_n$* (existence of a decomposition into irreducible factors).

(ii) *Let us consider two such decompositions of a, say $a = up_1 \ldots p_n = vq_1 \ldots q_m$. Then $n = m$ and there exists a permutation σ of $\{1, \ldots, n\}$ and units u_i, for $1 \leq i \leq n$, such that $q_i = u_i p_{\sigma(i)}$ for every i* (uniqueness of a decomposition in irreducible factors).

This is often taken as the *definition* of a unique factorization domain; in any case, this explains the chosen denomination!

Let us comment a little bit the use of the word "uniqueness" in the theorem. Strictly speaking, there is no unique decomposition into irreducible factors; indeed, it is always possible to change the order of the factors, or to simultaneously multiply and divide some irreducible factor of the decomposition by the same unit. The content of the uniqueness property is that these are the only two ways in which two decompositions may differ.

Proof. — When a is invertible, property a) is obvious: take $n = 0$ and $u = a$. Otherwise, there exists a maximal ideal containing a, hence an irreducible element $p_1 \in A$ such that p_1 divides a. Let $a_1 \in A$ be such that $a = p_1 a_1$; we then have $(a) \subsetneq (a_1)$. If a_1 is not invertible, we may redo the argument, obtaining irreducible elements $p_1, \ldots, p_n$ of A, and elements $a_1, \ldots, a_n \in A$ such that $a = p_1 \ldots p_n a_n$. If we were to go on forever, we would obtain a strictly increasing sequence of principal ideals $(a) \subsetneq (a_1) \subsetneq (a_2) \subsetneq \ldots$, which contradicts the first axiom of a unique factorization domain. So a_n is a unit for some n, and property a) is proved.

Let now $a = up_1 \ldots p_n = vq_1 \ldots q_m$ be two decompositions of a as the product of a unit and of irreducible elements. Let us prove property b) by induction on m. If $m = 0$, then $a = v$ is a unit; consequently, $up_1 \ldots p_n$ is a unit too, which implies $n = 0$ and $a = u = v$ (indeed, if $n \geq 1$, $up_1 \ldots p_n$ belongs to the maximal ideal (p_1) so cannot be a unit). Assume $m \geq 1$. Since q_m divides $up_1 \ldots p_m$ and the ideal (q_m) is prime, q_m divides one of the factors u, p_1, $\ldots$, p_n. Since u is a unit, there exists an integer $j \in \{1, \ldots, n\}$ such that q_m divides p_j; let $s \in A$ be such that $p_j = sq_m$. Since p_j is irreducible and q_m is not a unit, s is necessarily a unit; we then set $u_m = s^{-1}$ so that $q_m = u_m p_j$. Set $b = a/q_m$. It admits two decompositions as a product of irreducible elements, namely $vq_1 \ldots q_{m-1}$ and $(u/u_m)p_1 \ldots \widehat{p_j} \ldots p_n$, where the hat on p_j indicates that this factor is omitted from the product. By induction, we have $m-1 = n-1$ and there exist a bijection σ from $\{1, \ldots, m-1\}$ to $\{1, \ldots, \widehat{j}, \ldots, n\}$, and units u_i, such that $q_i = u_i p_{\sigma(i)}$ for $1 \leq i \leq m-1$. It follows that $m = n$ and the mapping (still denoted by σ) from $\{1, \ldots, m\}$ to $\{1, \ldots, n\}$ which extends σ and maps m to j is a bijection. For every $i \in \{1, \ldots, n\}$, one has $q_i = u_i p_{\sigma(i)}$. This concludes the proof of the uniqueness property.   □

Remark (2.5.8). — Conversely, let A be a domain which satisfies the conclusion of the theorem.

For any non-zero $a \in A$, let $\omega(a)$ be the number of irreducible factors in a decomposition of a as a product of irreducible elements. It does not depend on the chosen decomposition. Let a, b, c be elements of A such that $a = bc$ and $a \neq 0$. The uniqueness property implies that $\omega(a) = \omega(b) + \omega(c)$.

Let (a_n) be a sequence of elements of A such that the sequence of ideals $(a_0), (a_1), \ldots$ is increasing. For every integer n, there thus exists an element $b_n \in A$ such that $a_n = b_n a_{n+1}$. Consequently, $\omega(a_{n+1}) \leq \omega(a_n)$, so that the sequence $(\omega(a_n))_n$ is a decreasing sequence of positive integers; it is thus stationary. Moreover, if $\omega(a_n) = \omega(a_{n+1})$, then b_n is a unit and the ideals (a_n) and (a_{n+1}) coincide. This implies that the sequence (a_n) of ideals is itself stationary.

Let p be any irreducible element of A. Let us show that it generates a prime ideal of A. Since p is irreducible, p is not a unit and $(p) \neq A$. Let then a, b be elements of A such that p divides ab; let $c \in A$ be given by $ab = pc$. If $c = 0$, then $pc = 0$ and either a or b is zero. Let us assume that $c \neq 0$; then $a \neq 0$ and $b \neq 0$. Pick decompositions of a, b and c as products of irreducible factors, say $a = up_1 \ldots p_n$, $b = vq_1 \ldots q_m$, and $c = wr_1 \ldots r_s$. (Here, u, v, w are units, and n, m, s are positive integers, while the p_i, q_j and r_k are irreducible elements of A.) Then we have two decompositions of ab, namely $uvp_1 \ldots p_n q_1 \ldots q_m$ and $wpr_1 \ldots r_s$. By the uniqueness property, the factor p which appears in the second one has to intervene in the first one; precisely, there exists a unit $\alpha \in A$ and either an integer $i \in \{1, \ldots, n\}$ such that $p = \alpha p_i$, or an integer $j \in \{1, \ldots, m\}$ such that $p = \alpha q_j$. In the first case, p divides a, in the second one, p divides b. This shows that (p) is a prime ideal.

These remarks prove that the ring A is a unique factorization domain.

Remark (2.5.9). — One of the reasons of the non-uniqueness of the decomposition into irreducible factors is that one may multiply those factors by units.

In certain rings, it is possible to distinguish privileged irreducible elements so as to remove this source of ambiguity.

For example, in $\mathbf{Z}$, the irreducible elements are the prime numbers and their negatives, but one may decide to prefer the prime numbers themselves. Up to the order of the factors, any non-zero integer is then uniquely written as the product of ± 1 (the only units in $\mathbf{Z}$) by a product of prime numbers.

Similarly, in the ring K[T] in one indeterminate T over a field K one may prefer the *monic* irreducible polynomials. Still up to the order of factors, any non-zero polynomial can be uniquely written as the product of a non-zero constant (the units in K[T]) by a product of monic irreducible polynomials.

In a general unique factorization domain A, let us show how to normalize the decomposition into irreducible elements so that it can only be modified by the order of the factors.

Let us choose a family (π_i) of irreducible elements of A in such a way that

– For $i \neq j$, π_i and π_j are not associated;

– Every irreducible element of A is associated to one of the π_i.

(To prove the existence of such a family, just choose one element in every equivalence class of irreducible elements for the equivalence relation of being associated.) Then, any non-zero element a of A can be uniquely written in the form $a = u \prod_i \pi_i^{r_i}$ where u is a unit and the r_i are positive integers, all but finitely many of them being equal to zero.

In other words, the map from $A^\times \times \mathbf{N}^{(I)}$ to $A - \{0\}$ which maps $(u, (r_i))$ to the product $u \prod_i \pi_i^{r_i}$ is an isomorphism of monoids.

One interesting aspect of this normalization is that it makes the divisibility relation explicit: an element $a = u \prod_i \pi_i^{r_i}$ divides an element $b = v \prod_i \pi_i^{s_i}$ if and only if $r_i \leq s_i$ for every i. Indeed, it is clear that this condition is sufficient, for it suffices to set $c = (vu^{-1}) \prod_i \pi_i^{s_i - r_i}$ to get $a = bc$. Conversely, if $c \in A$ is such that $b = ac$, decompose c as $w \prod_i \pi_i^{t_i}$; then,

$$b = v \prod_i \pi_i^{s_i} = uw \prod_i \pi_i^{r_i + t_i},$$

hence $s_i = r_i + t_i$ for every i. This implies $s_i \geq r_i$.

2.5.10. Greatest common divisor, least common multiple — Let A be a unique factorization domain.

Let (a_n) be a family of elements of A. We are going to show that it possesses a *greatest common divisor* (gcd) and a *least common multiple* (lcm). As in the case of principal ideal domains, a gcd of the family (a_n) is any element $d \in A$ which divides each of the a_n and such that every such common divisor divides d, and an lcm of this family is an element $m \in A$ which is a multiple of all of the a_n and such that every such common multiple is a multiple of m.

We first treat particular, essentially trivial, cases. Note that every element of A divides 0, but that the only multiple of 0 is 0 itself.

Consequently, if one of the a_n is equal to 0, it has only one common multiple, namely 0, which thus is its lcm. On the other hand, in order to show that the family (a_n) has a gcd, we may remove all the terms equal to 0.

If the family is empty, then every element is a common divisor, so that 0 is a greatest common divisor of the empty family. Similarly, every element is a common multiple, so that 1 is a least common multiple of the empty family.

These remarks allow us to assume that the family (a_n) is non-empty and consists of non-zero elements. To simplify the construction, assume also that we have normalized the decomposition as above into irreducible factors. For every n, let $a_n = u_n \prod_i \pi_i^{r_{n,i}}$ be the decomposition of a_n into irreducible factors. For every i, set $d_i = \inf_n(r_{n,i})$; this is a positive integer, and $d_i = 0$ for all but finitely many i. This allows us to set $d = \prod_i \pi_i^{d_i}$. Let us show that d is a gcd of the family (a_n). Since $d_i \leq r_{n,i}$ for every i, d divides a_n for every n. Let b be a common divisor of the a_n, and let $b = v \prod_i \pi_i^{s_i}$ be its decomposition into irreducible factors. Since b divides a_n, $s_i \leq r_{n,i}$ for every i. It follows that $s_i \leq d_i$ for every i, hence b divides d.

For every i, set $m_i = \sup(r_{n,i})$; this is an element of $\mathbf{N} \cup \{+\infty\}$. If every m_i is an integer, and if all but finitely many of them are 0, we may set $m = \prod_i \pi_i^{m_i}$; otherwise, we set $m = 0$. In each case, m is a common multiple of all of the a_n. Conversely, let b be any non-zero element of A; let $b = v \prod_i \pi_i^{s_i}$ be its decomposition into irreducible elements. For b to be a multiple of a_n, it is necessary and sufficient that $s_i \geq r_{n,i}$ for every i; consequently, b is a common multiple of the family (a_n) if and only if $s_i \geq m_i$ for every i. If m_i is infinite for some i, or if infinitely many terms of the family (m_i) are non-zero, this never holds, so that 0 is the only common multiple of the family (a_n). Otherwise, we see that b is a multiple of m.

Remark (2.5.11). — Unless they are zero, two greatest common divisors (*resp.* two least common multiples) of a family (a_n) differ by multiplication by a unit. As we have seen above, choosing a particular normalization for the decomposition into irreducible factors allows us to get a well-defined representative of the gcd (*resp.* of the lcm).

When we write equalities involving greatest common divisors or least common multiples, we shall always assume that they are properly normalized. In any case, it is always possible to read these equalities up to multiplication by a unit.

Definition (2.5.12). — One says that a family of elements of a unique factorization domain consists of *coprime* elements if this family has 1 for a greatest common divisor.

Remark (2.5.13). — Let (a_n) be a family of elements of A. For every $x \in A$, $\gcd((xa_n)) = x \gcd(a_n)$. This follows easily from the construction above; thanks to the formula $\inf(r_n) + s = \inf(r_n + s)$ for any family (r_n) of integers and any integer s. Conversely, let $d = \gcd((a_n))$. For every n, there is an element $b_n \in A$ such that $a_n = db_n$. Then $d \gcd((b_n)) = \gcd((a_n))$. Assume that at least one of the a_n is non-zero. Then $d \neq 0$ hence $\gcd((b_n)) = 1$, so that the b_n are coprime.

Proposition (2.5.14). — *Let A be a unique factorization domain and let (a_n) be a family of elements of A, with gcd d and lcm m. The ideal dA is the smallest principal ideal containing the ideal $\sum a_n A$, and the ideal mA is the largest principal ideal containing the ideal $\bigcap_n a_n A$.*

Assume in particular that A is a principal ideal domain. Then two elements a and b of A are coprime if and only if the ideals (a) and (b) are comaximal.

Proof. — Since the inclusion of ideals $aA \subset bA$ is equivalent to the fact that b divides a, this is but a reformulation of the discussion above.   □

Corollary (2.5.15) (Bézout's theorem). —*Let A be a principal ideal domain, let (a_n) be a family of elements of A with gcd d. There exist elements $u_n \in A$, all but finitely many of which are zero, such that $d = \sum a_n u_n$.*

In a euclidean ring, there is a simple algorithm to compute the gcd of two elements a and b, as well as a relation $d = au + bv$.

Proposition (2.5.16) (Euclidean algorithm). — *Let A be a euclidean ring; let a, b be elements of A. One defines four sequences (d_n), (u_n), (v_n) and (q_n) by induction on n by setting*

$$d_0 = a \qquad\qquad u_0 = 1 \qquad\qquad v_0 = 0$$
$$d_1 = b \qquad\qquad u_1 = 0 \qquad\qquad v_1 = 1$$

and then, if $d_n \neq 0$, let q_n be the quotient of a euclidean division of d_{n-1} by d_n, and

$$d_{n+1} = d_{n-1} - q_n d_n \qquad u_{n+1} = u_{n-1} - q_n u_n \qquad v_{n+1} = v_{n-1} - q_n v_n.$$

These sequences are finite: there exists a smallest integer n such that $d_{n+1} = 0$; then, $d_n = au_n + bv_n$ is a greatest common divisor of a and b.

Proof. — Let δ be the euclidean gauge of A. If $d_n \neq 0$, then d_{n+1} is the remainder of a euclidean division by d_n, so that either $\delta(d_{n+1}) < \delta(d_n)$, or $d_{n+1} = 0$. Consequently, the sequence of positive integers $(\delta(d_n))$ is strictly decreasing as soon as it is defined. Let us also remark that the pairs $(b, a - bq)$ and (a, b) have the same common divisors, so that $\gcd(b, a - bq) = \gcd(a, b)$. By induction, $\gcd(d_{n-1}, d_n) = \gcd(d_n, d_{n+1})$ for every n. If $d_n \neq 0$ and $d_{n+1} = 0$, we then have $\gcd(a, b) = \gcd(d_0, d_1) = \gcd(d_n, d_{n+1}) = d_n$. By induction, the proof of the relation $d_n = au_n + bv_n$ is immediate. $\qquad\square$

The following result is very useful.

Proposition (2.5.17) (Gauss). — *Let* A *be a unique factorization domain and let* $a, b, c \in$ A.

(i) *If* a *is prime to* b *and* c, *then* a *is prime to* bc.

(ii) *If* a *is prime to* c, *and if* a *divides* bc, *then* a *divides* b.

Proof. — a) Let p be an irreducible element of A that divides a. By assumption, p does not divide b, and p does not divide c. Since A is a unique factorization domain, p does not divide bc. This proves that a is prime to bc.

b) The result is obvious if c is a unit. We then argue by induction on the number of irreducible factors of c. Let p be a prime number that divides c; write $c = pc'$. Then a is prime to c' and a divides $bpc'c$, hence a divides bp by induction. Let $d \in$ A be such that $bp = ad$. Since A is a unique factorization domain, the ideal (p) is prime. One has $a \notin (p)$ by assumption, and $ad \in (p)$; consequently, $d \in (p)$. Let $d' \in$ A be such that $d = pd'$; then $bp = pad'$, hence $b = ad'$: this shows that a divides b. $\qquad\square$

2.6. Polynomial Rings are Unique Factorization Domains

One of the most important and basic results in the theory of unique factorization domains is the theorem of Gauss according to which polynomial rings with coefficients in a unique factorization domain are themselves unique factorization domains.

So let A be a unique factorization domain. We first recall that the units of A[T] are the units of $A^\times$, viewed as constant polynomials (corollary 1.3.14).

Definition (2.6.1). — Let A be a unique factorization domain and let P be any polynomial in A[T]. The *content* of P, denoted by ct(P), is defined as a greatest common divisor of the coefficients of P. One says that a polynomial is *primitive* if its content is a unit, that is to say, if its coefficients are coprime.

As usual for questions of gcd, the content of a non-zero polynomial is only well defined if we have normalized the decomposition into irreducible factors; otherwise, it is defined up to multiplication by a unit. The content of the zero polynomial is 0.

Lemma (2.6.2). — *Let A be a unique factorization domain, let K be its field of fractions.*

(i) *For any polynomial $P \in K[T]$, there exist a primitive polynomial $P_1 \in A[T]$ and an element $a \in K$ such that $P = aP_1$.*

(ii) *Let $P = aP_1$ be such a decomposition. Then, $P \in A[T]$ if and only if $a \in A$; in that case, $a = ct(P)$. In particular, P is a primitive polynomial of A[T] if and only if a is a unit in A.*

Proof. — a) If $P = 0$, we set $a = 0$ and $P_1 = 1$. Assume $P \neq 0$. Let $d \in A - \{0\}$ be a common denominator of all of the coefficients of P, so that $dP \in A[T]$. Let then $b \in A$ be the content of the polynomial dP and set $P_1 = (dP)/b$ and $a = b/d$. The polynomial P_1 belongs to A[T] and is primitive; one has $P = aP_1$.

b) If $a \in A$, it is clear that $P \in A[T]$. Conversely, assume that $P \in A[T]$ and let us show that $a \in A$. Let b and c be elements of A such that $a = b/c$. We write $b = ac$, hence $bP_1 = acP_1 = cP$, from which we deduce that $b = ct(cP) = c \, ct(P)$. It follows that c divides b and $a = ct(P) \in A$.

If a is a unit in A, then P is a primitive polynomial in A[T]. Conversely, assume that P is a primitive polynomial in A[T]. By the preceding paragraph, we have $a \in A$. Then $ct(P) = a \, ct(P_1)$ is a unit, hence a is a unit. □

Proposition (2.6.3). — *Let A be a unique factorization domain and let P, Q be two polynomials in A[T]. Then, $ct(PQ) = ct(P) \, ct(Q)$.*

Proof. — We first treat the particular case where P and Q are primitive. We then need to show that PQ is primitive as well. Let π be any irreducible element of A and let us show that π does not divide all of the coefficients of PQ. Since P is primitive, the reduction cl(P) of P modulo π is a non-zero polynomial with coefficients in the ring $A/(\pi)$. Similarly, cl(Q) is a non-zero polynomial with coefficients in $A/(\pi)$. Since π is irreducible and A a unique factorization domain, the quotient ring $A/(\pi)$ is a domain, hence the polynomial ring $(A/(\pi))[T]$ is again a domain (corollary 1.3.13). It follows that the product cl(P) cl(Q) = cl(PQ) is a non-zero polynomial in $(A/\pi)[T]$. This means exactly that π does not divide all of the coefficients of PQ, as was to be shown.

Carl Friedrich GAUSS (1777–1855) was a German mathematician, astronomer and physicist. In his 1799 dissertation, he gave the first proof of the fundamental theorem of algebra (theorem 4.3.10); taking for granted a result such as the Jordan curve theorem, his beautiful geometric argument is however incomplete. In any case, the necessary analytic or topological foundations for this theorem were only set up later in the nineteenth century.

His *Disquisitiones arithmeticae* (Arithmetical investigations), published in 1801, was the first systematic and rigorous treatment of number theory. Its first chapters cover elementary modular arithmetic, prime numbers and unique factorization, of which he recognized the importance, and linear diophantine equations. He then proves the quadratic reciprocity law (see exercise 4.9.12) that EULER and LEGENDRE had stated; this theorem now admits more than two hundred proofs, eight of them being due to GAUSS himself.

The last chapters of this book concern the arithmetic study of binary quadratic forms. In modern language, this amounts to the study of algebraic number theory in quadratic fields; in particular, he gives an efficient technique to compute the cardinality of the class group for the ring of integers in quadratic fields. GAUSS's *class number problem* states that this cardinality tends to infinity when the discriminant goes to $-\infty$, but that it can be equal to 1 infinitely many times. The first conjecture was solved by HEEGNER, BAKER and STARK in the years 1950–1970 (class numbers of imaginary quadratic fields), but the second one (for real quadratic fields) is still open.

An amazingly gifted calculator, GAUSS was made famous by his computation of the location of the asteroid Ceres, which allowed astronomers to

Carl-Friedrich Gauss at the age of 50 (litho-graph), published in the Astronomische Nachrichten *(1828)*

Artist: Siegried Detlev Bendixen
Source: Wikipedia, public domain

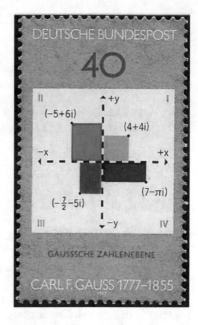

observe it again exactly one year after its first observation. In 1807, he became in fact professor of astronomy and director of the Göttingen observatory. In his work, he made use of a discrete Fourier transform, and of the fast computation method for it — the so-called *Fast Fourier Transform* that Cooley and Tukey would rediscover in 1965.

This work in astronomy led him to study the motion of small planets, when disturbed by larger ones. On this occasion, he introduced the least-square-method, and studied its applicability — the "Gaussian distribution" also originates from this work. With Weber, he worked on magnetism theory; he also formulated the "Gauss law" relating the flux of an electric field through a closed surface with the electric charge within that surface.

After participating in 1818 in a geodetic survey of the kingdom of Hanover, he became interested in the geometry of curves and surfaces. He proved the *theorema egregium* ("remarkable theorem") in 1827 according to which the "gaussian curvature" is unchanged if the surface is moved but not stretched.

In 1831, he proved the "regular" case of Kepler's conjecture, when the spheres are arranged along a lattice. The general case, Hilbert's 18th problem, would only be solved around 2000 by the monumental work of Hales and Ferguson!

Let us now treat the general case. We may assume that P and Q are non-zero, so that ct(P) and ct(Q) are non-zero too. By definition, we can write $P = \text{ct}(P)P_1$ and $Q = \text{ct}(Q)Q_1$, where $P_1, Q_1 \in A[T]$ are primitive polynomials. Then, P_1Q_1 is primitive and the equality $PQ = \text{ct}(P)\text{ct}(Q)P_1Q_1$ shows that, up to a unit, ct(PQ) is thus equal to ct(P) ct(Q).   □

Corollary (2.6.4). — *Let A be a unique factorization domain and let K be its field of fractions. Let $P, Q \in A[T]$ be polynomials.*

(i) *Assume that Q is primitive and that Q divides P in K[T]. Then Q divides P in A[T].*

(ii) *Let $R = \gcd(P, Q)$ be a greatest common divisor of P, Q. Then R is a gcd of P and Q in K[T].*

Proof. — *a*) Let $R \in K[T]$ be such that $P = RQ$. Write $R = aR_1$, where $a \in K$ and $R_1 \in A[T]$ is a primitive polynomial. Then $P = aR_1Q$. Since R_1Q is primitive, we have $a \in A$, hence $R \in A[T]$, which shows that Q divides P in A[T].

b) Let S be a gcd of P and Q in K[T]. Of course, R divides P and Q in K[T], so that R divides S. Conversely, we need to prove that S divides R. Writing $S = aS_1$, with $a \in K$ and $S_1 \in A[T]$ primitive, we assume that $S \in A[T]$ is primitive. By *a*), S divides P and Q in A[T], so that it divides R. This concludes the proof.   □

Thanks to this fundamental proposition, we are now able to determine the irreducible elements of A[T].

Proposition (2.6.5). — *Let A be a unique factorization domain and let K be its field of fractions. The irreducible elements of A[T] are the following:*

– Irreducible elements of A, considered as constant polynomials;

– Primitive polynomials in A[T] which are irreducible as polynomials in K[T].

Proof. — We shall begin by proving that these elements are indeed irreducible in A[T], and then show that there are no others.

Let $a \in A$ be an irreducible element. It is not invertible in A, hence is not a unit of A[T]. Let P, Q be polynomials in A[T] such that $a = PQ$. One has $\deg(P) + \deg(Q) = \deg(PQ) = 0$, hence $\deg(P) = \deg(Q) = 0$. In other words, P and Q are constant polynomials. Since a is irreducible, either P or Q is a unit in A, hence in A[T]. This shows that a is irreducible in A[T].

Let now $P \in A[T]$ be a primitive polynomial which is irreducible in K[T]. Since constant polynomials are units in K[T], P is not constant; in particular, it is not a unit in A[T]. Let Q, R be polynomials in A[T] such that $P = QR$. *A fortiori*, they furnish a decomposition of P in K[T], so that Q or R is a unit in K[T]. In particular, Q or R is constant. To fix the notation, assume that R is the constant a. We thus have $P = aQ$. It follows that the content of P satisfies

$$\text{ct}(P) = \text{ct}(aQ) = a\,\text{ct}(Q).$$

Since P is primitive, a is a unit in A, hence in A[T]. This shows that P is irreducible in A[T].

Conversely, let P be an irreducible element of A[T]. Let P_1 be a primitive polynomial in A[T] such that $P = ct(P)P_1$. Necessarily, $ct(P)$ is a unit in A, or P_1 is a unit in A[T].

Assume first that $ct(P)$ is not a unit. Then, P_1 is a unit in A[T], which means that P_1 is a constant polynomial and a unit in A. In other words, P is an element a of A. Let us show that a is irreducible in A. It is not a unit (otherwise, P would be a unit). And if $a = bc$, for some elements $b, c \in A$, we get $P = bc$. Since P is irreducible in A[T], b or c is a unit in A[T], that is, b or c is a unit in A.

Assume now that P_1 is not a unit in A[T]; then $ct(P)$ is a unit and P is primitive. Let us prove that P is irreducible in K[T]. First of all, $\deg(P) > 0$, for otherwise, P would be a unit in A. In particular, P is a not a unit in K[T]. Let $P = QR$ be a factorization of P as the product of two polynomials in K[T]. By the above lemma 2.6.2, we may write $Q = qQ_1$ and $R = rR_1$, where $q, r \in K$ and Q_1, R_1 are primitive polynomials in A[T]. Then, $P = (qr)Q_1R_1$. Since P is a primitive polynomial in A[T], lemma 2.6.2 implies that qr is a unit in A. Since P is irreducible in A[T], either Q_1 or R_1 is a unit in A[T], hence in K[T]. It follows that either $Q = qQ_1$ or $R = rR_1$ is a unit in K[T].   □

Theorem (2.6.6) (Gauss). — *If A is a unique factorization domain, then so is the ring A[T].*

Proof. — Let us prove that A[T] satisfies the two properties of the definition 2.5.4 of a unique factorization domain.

1) Let $(P_n)_n$ be a sequence of polynomials in A[T] such that the sequence $((P_n))_n$ of principal ideals is increasing; let us prove that it is stationary.

If $P_n \neq 0$, then $P_m \neq 0$ for all $n \geq m$. The case of the constant sequence $((0))$ being obvious, we may assume that $P_n \neq 0$ for all n.

For $n \geq m$, P_n divides P_m, hence $ct(P_n)$ divides $ct(P_m)$. Consequently, the sequence $((ct(P_n)))_n$ of principal ideals of A is increasing too. Since A is a unique factorization domain, it is stationary. Moreover, for $n \geq m$, $\deg(P_n) \leq \deg(P_m)$, so that the sequence $((\deg(P_n))_n$ of integers is decreasing, hence stationary.

Let N be any integer such that $\deg(P_n) = \deg(P_N)$ and $ct(P_n) = ct(P_N)$ for $n \geq N$. Let n be some integer such that $n \geq N$. Since P_n divides P_N, there exists a polynomial Q such that $P_N = QP_n$. Necessarily, $\deg(Q) = 0$ and $ct(Q) = 1$, so that Q is a constant polynomial, with constant term $ct(Q)$, hence is invertible. Consequently, the ideals (P_n) and (P_N) coincide. This shows that the sequence $(P_n)_n$ of principal ideals of A[T] is stationary.

2) Let us now show that the irreducible elements of A[T] generate prime ideals. Since irreducible elements are not units, it suffices to show that if an irreducible element of A[T] divides a product PQ of two polynomials in A[T], then it divides P or Q.

Let first π be an irreducible element of A; assume that π divides PQ. Taking the contents, we see that π divides $ct(PQ) = ct(P) ct(Q)$. Since π is irreducible in A and A is a unique factorization domain, π divides $\mathrm{ct}(P)$ or $\mathrm{ct}(Q)$, hence π divides P or Q.

Let now Π be a primitive polynomial of $A[T]$, irreducible in $K[T]$, such that Π divides PQ. Since $K[T]$ is a principal ideal domain, it is a unique factorization domain and Π divides P or Q in $K[T]$. By corollary 2.6.4, Π divides P or Q in $A[T]$. $\square$

Corollary (2.6.7) (Gauss). — *Let A be a unique factorization domain. For any integer n, $A[X_1, \ldots, X_n]$ is a unique factorization domain. In particular, if K is a field, then $K[X_1, \ldots, X_n]$ is a unique factorization domain.*

Proof. — This is immediate by induction on n, using the isomorphisms (remark 1.3.10):
$$A[X_1, \ldots, X_m] \simeq (A[X_1, \ldots, X_{m-1}])[X_m].$$

2.7. Resultants and Another Theorem of Bézout

Throughout this section, A is a commutative ring. We shall make some use of determinants of matrices.

Definition (2.7.1). — Let m, n be positive integers, let P and Q be two polynomials in $A[X]$ such that $\deg(P) \le n$ and $\deg(Q) \le m$. Write $P = a_n X^n + \cdots + a_0$ and $Q = b_m X^m + \cdots + b_0$, for $a_0, \ldots, a_n, b_0, \ldots, b_m \in A$. The *resultant* (in sizes (n, m)) of (P, Q) is defined as the determinant

$$\mathrm{Res}_{n,m}(P, Q) = \begin{vmatrix} a_0 & & 0 & b_0 & & & 0 \\ a_1 & a_0 & & b_1 & b_0 & & \\ \vdots & & \ddots & \vdots & & \ddots & \\ a_{m-1} & & a_0 & b_{m-1} & & & b_0 \\ \vdots & & \vdots & b_m & & & \ddots \\ \vdots & & \vdots & & \ddots & & b_0 \\ a_n & a_{n-1} & a_{n-m+1} & & b_m & & \vdots \\ & a_n & \vdots & & & \ddots & \vdots \\ & & \ddots & \vdots & & & \ddots & \vdots \\ 0 & & a_n & 0 & & & b_m \end{vmatrix}$$

$$\underbrace{}_{m \text{ columns}} \quad \underbrace{}_{n \text{ columns}}$$

(Precisely, the column vector $(a_0, \ldots, a_n)$ is copied m times, each time shifted by one row, then the column vector $(b_0, \ldots, b_m)$ is copied n times, each time shifted by one row.)

Remark (2.7.2). — Let P and Q $\in$ A[X] be polynomials as above, and let $f: A \rightarrow B$ be a morphism of rings. Let P^f and Q^f be the polynomials in B[X] deduced from P and Q by applying f to their coefficients. One has $\deg(P^f) \leq n$ and $\deg(Q^f) \leq m$. The resultant $\mathrm{Res}_{n,m}(P, Q)$ is the determinant of the matrix of the definition, while the resultant $\mathrm{Res}_{n,m}(P^f, Q^f)$ is the determinant of the matrix obtained by applying f to each entry. Since the determinant is a polynomial expression, we obtain the equality

$$\mathrm{Res}_{n,m}(P^f, Q^f) = f(\mathrm{Res}_{n,m}(P, Q)).$$

Proposition (2.7.3). — *Let* K *be a field. Let* P, Q *be two polynomials in* K[X]*, let* n, m *be strictly positive integers such that* $\deg(P) \leq n$ *and* $\deg(Q) \leq m$. *Then,* $\mathrm{Res}_{n,m}(P, Q) = 0$ *if and only if either*

– P *and* Q *are not coprime; or*

– $a_n = b_m = 0$.

Proof. — If p is any integer ≥ 0, let $K[X]_p$ be the K-vector space of polynomials of degree $\leq p$. The family $(1, X, \ldots, X^p)$ is a basis of $K[X]_p$ hence this space has dimension $p + 1$. Let us observe that $\mathrm{Res}_{n,m}(P, Q)$ is the determinant of the linear map

$$\rho: K[X]_{m-1} \times K[X]_{n-1} \rightarrow K[X]_{m+n-1}, \qquad (U, V) \mapsto UP + VQ$$

in the bases $(1, \ldots, X^{m-1}; 1, \ldots, X^{n-1})$ and $(1, X, \ldots, X^{m+n-1})$ of the vector spaces $K[X]_{m-1} \times K[X]_{n-1}$ and $K[X]_{m+n-1}$. Consequently, $\mathrm{Res}_{n,m}(P, Q) = 0$ if and only if ρ is not invertible. We shall compute the kernel of ρ.

Assume first that $P = 0$ or $Q = 0$. Then some column of this matrix is zero, hence $\mathrm{Res}_{n,m}(P, Q) = 0$.

Let us now assume that P and Q are both non-zero and let D be their gcd; we have $D \neq 0$. We can thus write $P = DP_1$ and $Q = DQ_1$, where P_1 and Q_1 are two coprime polynomials in K[X]. Let $(U, V) \in \mathrm{Ker}(\rho)$. We have $UP + VQ = 0$, hence $UP_1 + VQ_1 = 0$ since $D \neq 0$. Since P_1 and Q_1 are coprime (see §2.5.10), we obtain that Q_1 divides U and P_1 divides V. We can thus write $U = Q_1 S$ and $V = P_1 T$, for two polynomials $S, T \in K[X]$. Then $UP_1 + VQ_1 = P_1 Q_1 (S + T)$, so that $T = -S$, $U = Q_1 S$ and $V = -P_1 S$. Since $U \in K[X]_{m-1}$ and $V \in K[X]_{n-1}$, we have $\deg(S) \leq m - 1 - \deg(Q_1)$ and $\deg(S) \leq n - 1 - \deg(P_1)$. Now,

$$m - 1 - \deg(Q_1) = m - \deg(Q) + \deg(Q) - \deg(Q_1) - 1$$
$$= (m - \deg(Q)) + \deg(D) - 1,$$

and, similarly,

$$n - 1 - \deg(P_1) = (n - \deg(P)) + \deg(D) - 1.$$

Set

$$s = \inf(n - \deg(P), m - \deg(Q))$$

so that $s = 0$ unless $a_n = b_m = 0$, in which case $s \geq 1$. Conversely, every element of $K[X]_{m-1} \times K[X]_{n-1}$ of the form $(Q_1 S, -P_1 S)$, for $S \in K[X]_{s+\deg(D)-1}$, belongs to $\text{Ker}(\rho)$.

This shows that $\text{Res}_{n,m}(P, Q) = 0$ if and only if $s + \deg D > 0$, that is, if and only if either $a_n = b_m = 0$ or $\deg(D) > 0$. $\square$

Corollary (2.7.4). — *Let K be an algebraically closed field, and let $A = K[Y]$; let us identify the ring $A[X]$ with $K[X, Y]$ (remark 1.3.10). Let P, Q be two polynomials in $K[X, Y] = A[X]$. Let m, n be positive integers such that $\deg_X(P) \leq n$ and $\deg_X(Q) \leq m$ and let us write*

$$P = P_n(Y)X^n + \cdots + P_0(Y) \quad and \quad Q = Q_m(Y)X^m + \cdots + Q_0(Y)$$

where $P_0, \ldots, P_n, Q_0, \ldots, Q_m \in K[Y]$. Let $R = \text{Res}_{n,m}(P, Q) \in K[Y]$ be the resultant in sizes (n, m) of the pair (P, Q). Then, an element $y \in K$ is a root of R if and only if either

   – the polynomials $P(X, y)$ and $Q(X, y)$ have a common root in K; or

   – $P_n(y) = Q_m(y) = 0$.

Proof. — By the definition of the resultant (remark 2.7.2), we have

$$R(y) = \left(\text{Res}_{n,m}(P, Q) \right)(y) = \text{Res}_{n,m}(P(X, y), Q(X, y)).$$

It thus suffices to apply the preceding proposition to the polynomials $P(X, y)$ and $Q(X, y)$ of $K[X]$. $\square$

Theorem (2.7.5) (Bézout). — *Let K be an algebraically closed field. Let P, Q be two coprime polynomials in $K[X, Y]$, with total degrees p and q respectively. Then, the set of common roots of P and Q in K^2, that is, the pairs $(x, y) \in K^2$ such that $P(x, y) = Q(x, y) = 0$, has at most pq elements; in particular, it is finite.*

Proof. — Since P and Q are coprime in the ring $K[X, Y]$, they remain coprime in $K(Y)[X]$ (see corollary 2.6.4). By proposition 2.7.3, when viewed as polynomials in $K(Y)[X]$, their resultant R is a non-zero polynomial R_Y of $K[Y]$. Consequently, there are only finitely many possibilities for the ordinates y of the common roots (x, y) of P and Q. Exchanging the roles of X and Y, we prove similarly that there are only finitely many possibilities for the abscissae x of these common roots. It follows that the set Σ of common roots to P and Q is finite.

We now show that $\text{Card}(\Sigma)$ is less than or equal to the product pq of the degrees of P and Q. With that aim, we make a linear change of variables so that any horizontal line contains at most one point of Σ. Since there are only finitely directions to avoid, this may be done. This modifies the polynomials P and Q, but not their degrees p and q.

Let us write

$$P = P_n(Y)X^n + \cdots + P_0(Y) \quad and \quad Q = Q_m(Y)X^m + \cdots + Q_0(Y)$$

where P_n and Q_m are non-zero polynomials. Let $R = \text{Res}_{n,m}(P, Q)$ (resultant with respect to X). We know that for any $y \in K$, $R(y) = 0$ if and only if either $P_n(y) = Q_m(y) = 0$ or y is the ordinate of a point of Σ. It thus suffices to show that $\deg(R) \le pq$.

We first observe that for any integer i, $\deg(P_i) \le p - i$ and $\deg(Q_i) \le q - i$. Let us examine the entry R_{ij} at row i and column j of the matrix whose determinant is R:

– for $1 \le j \le m$, one has $R_{ij} = P_{i-j}$ when $0 \le i - j \le n$, and $R_{ij} = 0$ otherwise;

– for $m + 1 \le j \le m + n$, one has $R_{ij} = Q_{i-j+m}$ when $0 \le i - j + m \le m$, and $R_{ij} = 0$ otherwise.

In particular, $\deg(R_{ij})$ is bounded above by

$$\deg(R_{ij}) \le \begin{cases} p - i + j & \text{if } 1 \le j \le m\,; \\ q - m - i + j & \text{if } m + 1 \le j \le m + n. \end{cases}$$

The determinant R is a sum of products of the form $\prod_{j=1}^{m+n} R_{\sigma(i)i}$, for all permutations σ of $\{1; \ldots; m + n\}$. The degree of such a product is bounded above by

$$\sum_{j=1}^{m+n} \deg(R_{\sigma(j)j}) \le \sum_{j=1}^{m} (p - \sigma(j) + j) + \sum_{j=m+1}^{m+n} (q - m - \sigma(j) + j)$$

$$\le pm + n(q - m) - \sum_{j=1}^{m+n} \sigma(j) + \sum_{j=1}^{m+n} j$$

$$\le pq - (p - n)(q - m) \le pq.$$

Consequently, $\deg(R) \le pq$ and the theorem is proved.   □

Remark (2.7.6). — There is a more precise version of this theorem of Bézout that takes into account the multiplicity of common roots (for example, if the curves with equations P and Q are tangent at some intersection point, this point will have multiplicity ≥ 2), as well as the possible common roots "at infinity". Provided all of this is correctly defined, which belongs to a course in algebraic geometry, the number of common roots then is exactly pq.

Let us now give two additional properties of the resultant.

Proposition (2.7.7). — *Let A be a commutative ring, let P, Q be two polynomials in A[X], and let n, m be integers such that $\deg(P) \le n$ and $\deg(Q) \le m$. Then, the resultant $\text{Res}_{n,m}(P, Q)$ belongs to the ideal $(P, Q)_{A[X]} \cap A$ of A.*

Proof. — The determinant defining the resultant $\text{Res}_{m,n}(P, Q)$ may be computed in any (commutative) overring of A, and we will compute it in A[X]. Let us add to the first row X times the second one, X^2 times the third one, etc.

We obtain that $\mathrm{Res}_{n,m}(P,Q)$ is the determinant of a matrix with coefficients in $A[X]$ whose first row is

$$P \; XP \ldots X^{m-1}P \; Q \; XQ \ldots X^{n-1}Q.$$

If we expand the determinant with respect to this row we see that $\mathrm{Res}_{n,m}(P,Q)$ has the form $UP + VQ$, for two polynomials U and V in $A[X]$. This shows that $\mathrm{Res}_{n,m}(P,Q)$ belongs to the ideal $(P,Q)_{A[X]}$ generated by P and Q in $A[X]$. Since it also belongs to A, this concludes the proof of the proposition.   □

Proposition (2.7.8). — *Let A be a commutative ring and let P,Q be two split polynomials of $A[X]$: $P = a_n \prod_{i=1}^{n}(X - t_i)$ and $Q = b_m \prod_{j=1}^{m}(X - u_j)$. Then,*

$$\mathrm{Res}_{n,m}(P,Q) = (-1)^{mn} a_n^m b_m^n \prod_{i,j}(t_i - u_j)$$

$$= b_m^n \prod_{j=1}^{m} P(u_j) = a_n^m (-1)^{mn} \prod_{i=1}^{n} Q(t_j).$$

Proof. — It is obvious that the three written formulas on the right of the first "=" sign are pairwise equal. We shall prove that they are equal to $\mathrm{Res}_{n,m}(P,Q)$ by induction on n. If $n = 0$, then $P = a_0$ hence $\mathrm{Res}_{0,m}(P,Q) = a_0^m$, so that the formula holds in this case. Let us now show by performing linear combinations on the resultant matrix that

$$\mathrm{Res}_{n+1,m}((X - t)P, Q) = (-1)^m Q(t) \, \mathrm{Res}_{n,m}(P,Q).$$

Let us indeed write $P = a_n X^n + \cdots + a_0$. Then,

$$(X - t)P = a_n X^{n+1} + (a_{n-1} - t a_n)X^n + \cdots + (a_0 - t a_1)X + a_0$$

and $\mathrm{Res}_{n+1,m}((X - t)P, Q)$ equals

$$\begin{vmatrix}
-t a_0 & & & & b_0 & & \\
a_1 - t a_0 & -t a_0 & & & b_1 & \ddots & \\
\vdots & a_0 - t a_1 & -t a_0 & & \vdots & & b_0 \\
& & \ddots & & \vdots & & \vdots \\
& & & -t a_0 & \vdots & & \\
a_{n-1} - t a_n & & & & \vdots & & \\
a_n & \ddots & & & \vdots & & \\
& \ddots & \ddots & & b_{m-1} & & \vdots \\
& & \ddots & & b_m & & \vdots \\
& & \ddots & a_{n-1} - t a_n & & \ddots & b_{m-1} \\
& & & a_n & & & b_m
\end{vmatrix}$$

(There are m "a" columns and $n + 1$ "b" columns.) Beginning from the bottom, let us add to each row t times the next one. This does not change the determinant hence $\mathrm{Res}_{n+1,m}((X - t)P, Q)$ equals

$$\begin{vmatrix} 0 & \ldots & 0 & b_0 + tb_1 + \cdots + t^m b_m & t(b_0 + tb_1 + \ldots) & \ldots & t^n(b_0 + \ldots) \\ a_0 & & & (b_1 + tb_2 + \ldots) & (b_0 + tb_1 + \ldots) & & t^{n-1}(b_0 + \ldots) \\ \vdots & \ddots & a_0 & \vdots & & & \\ \vdots & & \vdots & b_m & & & \vdots \\ a_n & & \vdots & & \ddots & & b_{m-1} + tb_m \\ & & a_n & & & & b_m \end{vmatrix}$$

We observe that $Q(t)$ is a factor of each entry of the first row, so that $\mathrm{Res}_{n+1,m}((X - t)P, Q)$ is equal to

$$Q(t) \begin{vmatrix} 0 & \ldots & 0 & 1 & t & \ldots & t^n \\ a_0 & & & (b_1 + tb_2 + \ldots) & (b_0 + tb_1 + \ldots) & & t^{n-1}(b_0 + \ldots) \\ \vdots & \ddots & a_0 & \vdots & & & \\ \vdots & & \vdots & b_m & & & \vdots \\ a_n & & \vdots & & \ddots & & b_{m-1} + tb_m \\ & & a_n & & & & b_m \end{vmatrix}$$

Beginning from the right, we then may subtract from each "b"-column t times the preceding one; we then get the determinant

$$Q(t) \begin{vmatrix} 0 & \ldots & 0 & 1 & 0 & \ldots & 0 \\ a_0 & & & b_1 + tb_2 + \ldots & b_0 & & \\ \vdots & \ddots & a_0 & \vdots & b_1 & \ddots & \\ \vdots & & \vdots & \vdots & & \ddots & \\ a_n & & \vdots & & & \ddots & b_{m-1} \\ & & a_n & & & & b_m \end{vmatrix}$$

It now suffices to expand the determinant with respect to the first row and we obtain

$$(-1)^m Q(t) \, \mathrm{Res}_{m,n}(P, Q).$$

This concludes the proof of the proposition by induction. $\square$

2.8. Exercises

Exercise (2.8.1). — *a*) Show that the ideal $(2, X)$ of the ring $\mathbf{Z}[X]$ is not principal.

b) Let A be a commutative ring such that the ring $A[X]$ is a principal ideal domain. Show that A is a field.

Exercise (2.8.2). — *a*) Show that the set of continuous functions with compact support, or the set of functions which vanish at any large enough integer, are ideals of the ring $\mathscr{C}(\mathbf{R})$ of continuous functions on the real line $\mathbf{R}$. Prove that they are not contained in any ideal M_x, for $x \in \mathbf{R}$.

b) Let A be the ring of holomorphic functions on a neighborhood of the closed unit disk. Show that any ideal of A is generated by a polynomial $P \in \mathbf{C}[z]$ whose roots have modulus ≤ 1. Prove that the maximal ideals of A are the ideals $(z - a)$, for $a \in \mathbf{C}$ such that $|a| \leq 1$.

c) (Generalization.) Let K be a compact, connected and non-empty subset of $\mathbf{C}$ and let $\mathcal{H}$ be the ring of holomorphic functions on an open neighborhood of K. Show that the ring $\mathcal{H}$ is a principal ideal domain. Show that all maximal ideals of A are the ideals $(z - a)$, for $a \in K$.

Exercise (2.8.3). — Let A be a commutative local ring (definition 2.1.8). Let I and J be two ideals of A, let $a \in A$ be a regular element such that $IJ = (a)$.

a) Show that there exist $x \in I$ and $y \in J$ such that $xy = a$. Check that x and y are regular.

b) Deduce from this that $I = (x)$ and $J = (y)$.

Exercise (2.8.4). — Let A be the product ring of all fields $\mathbf{Z}/p\mathbf{Z}$, where p runs over the set of all prime numbers. Let N be the subset of A consisting of all families (a_p) such that $a_p = 0$ for all but finitely many prime numbers p; let B be the quotient ring A/N.

a) Let M be a maximal ideal of A which does not contain N. Show that thre exists a prime number q such that M is the set of all families (a_p), where $a_q = 0$. What is the quotient ring A/M?

b) Let p be a prime number; show that $pB = B$.

c) Show that the ring B can be endowed in a unique way with the structure of a $\mathbf{Q}$-algebra.

d) Let M be a maximal ideal of A containing N. Show that the field A/M has characteristic 0.

Exercise (2.8.5). — Let k be a field, let A be the ring $k^{\mathbf{N}}$ (with termwise addition and multiplication) and let N be the subset $k^{(\mathbf{N})}$ of all almost null sequences.

a) Show that N is an ideal of A. Explain why there exists a maximal ideal M of A which contains N. Then set $K = A/M$. Show that K is a field extension of k.

b) Let $a = (a_1, \ldots, a_m) \in A^m$ be an element of N^m. Show that the set of integers n such that $a_{i,n} \neq 0$ for some $i \in \{1, \ldots, n\}$ is finite. (*Otherwise, construct $b_1, \ldots, b_m \in A$ such that $\sum_{i=1}^{m} b_i a_i$ maps to 1 in K.*)

c*) If k is infinite, show that the cardinality of K is uncountable.

d) Assuming that k is algebraically closed, show that K is algebraically closed too.

e) Let I be an ideal of $k[X_1, \ldots, X_n]$ and let J be the K-vector subspace of $K[X_1, \ldots, X_n]$ generated by the elements of I. Show that it is an ideal of $K[X_1, \ldots, X_n]$ and that $J \neq (1)$ if $I \neq (1)$.

f) Combine the previous construction with the special case of Hilbert's Nullstellensatz proved in the text to derive the general case. (*You may admit that every ideal of $k[X_1, \ldots, X_n]$ is finitely generated; see corollary 6.3.13.*)

Exercise (2.8.6). — Let A be a commutative ring.

a) Let I and J be ideals of A such that $V(I) \cap V(J) = \emptyset$ in Spec(A). Show that $I + J = A$.

b) Let I and J be ideals of A such that $V(I) \cup V(J) = \text{Spec}(A)$. Show that every element of $I \cap J$ is nilpotent.

c) Let I and J be ideals of A such that $V(I) \cup V(J) = \text{Spec}(A)$ and $V(I) \cap V(J) = \emptyset$. Show that there exist an idempotent $e \in A$ (this means that $e^2 = e$) such that $I = (e)$ and $J = (1 - e)$.

d) Show that Spec(A) is connected if and only if the only idempotents of A are 0 and 1.

Exercise (2.8.7). — Let R be an integral domain which is not a field and let $a \in R - \{0; 1\}$ such that the fraction ring R_a is a field. Prove that $1 - a$ is a unit.

Exercise (2.8.8). — One says that a non-empty topological space T is irreducible if for any closed subsets Z and Z' of T such that $T = Z \cup Z'$, either $Z = T$ or $Z' = T$. Let A be a commutative ring. In this exercise, we study the irreducible closed subsets of Spec(A).

a) Let P be a prime ideal of A. Show that V(P) is irreducible.

b) Show that Spec(A) is irreducible if and only if it has exactly one minimal prime ideal.

c) Let I be an ideal of A such that V(I) is irreducible. Show that there exists a unique minimal prime ideal P containing I and that $V(I) = V(P)$.

Exercise (2.8.9). — An irreducible closed subset of a topological space is called an irreducible component if it is maximal.

a) Let T be a topological space and let A be a closed irreducible subset of T. Show that the closed subsets of T containing A is inductive; conclude that there exists an irreducible component of T that contains A.

b) Deduce from the preceding question that every prime ideal P of A contains a minimal prime ideal of A.

Exercise (2.8.10). — Let K be an algebraically closed field. Let Z be a subset of K^n which is closed in the Zariski topology.

a) Show that the following properties are equivalent:

(i) Z is irreducible;

(ii) There exists a prime ideal P of $K[T_1, \ldots, T_n]$ such that $Z = \mathscr{V}(P)$;

(iii) The ideal $\mathscr{I}(Z)$ is prime.

b) Show that a closed subset Z' of K^n is an irreducible component of Z if and only if $\mathscr{I}(Z')$ is a minimal prime ideal contained in $\mathscr{I}(Z)$.

Exercise (2.8.11). — Let E be a field and let X be an infinite set. Let $A = E^X$ be the ring of functions $X \to E$. For $f \in A$, let $\mathscr{V}(f) = \{x \in X; f(x) = 0\}$.

a) Show that a function $f \in A$ is invertible if and only if it does not vanish.

b) Let I be an ideal of A such that $I \neq A$ and let $\mathscr{F}_I = \{\mathscr{V}(f); f \in I\}$. Show that $\mathscr{F}_I$ is a *filter* of subsets of X: (i) $\emptyset \notin \mathscr{F}$; (ii) If Y_1, Y_2 are subsets of X such that $Y_1 \subset Y_2$ and $Y_1 \in \mathscr{F}$, then $Y_2 \in \mathscr{F}$; (iii) If Y_1, Y_2 are elements of $\mathscr{F}$, then $Y_1 \cap Y_2 \in \mathscr{F}$.

c) Let I, J be ideals of A such that $I \subset J \neq A$; prove that $\mathscr{F}_I \subset \mathscr{F}_J$.

d) Let P be a prime ideal of A. Prove that the filter $\mathscr{F}_P$ is an *ultrafilter*: for every subset Y of X, either Y or $X - Y$ belongs to $\mathscr{F}_P$.

e) Let $\mathscr{F}$ be an ultrafilter on X. Prove that the set of all $f \in A$ such that $\mathscr{V}(f) \in \mathscr{F}$ is a maximal ideal of A.

f) Assume that I is a finitely generated prime ideal. Prove that there exists an element $x \in X$ such that $I = \{f \in A; f(x) = 0\}$, that I is principal, and that $\mathscr{F}_I = \{Y \subset X; x \in Y\}$. (One says that $\mathscr{F}_I$ is a *principal* ultrafilter.)

g) Let I be the set of all $f \in A$ such that $X - \mathscr{V}(f)$ is finite. Prove that I is an ideal of A and $I \neq A$. Prove that if P is a prime ideal that contains I, then P is not principal.

Exercise (2.8.12). — Let $P, Q, R \in C[T_1, \ldots, T_n]$ be polynomials such that P does not divide R.

a) Assume that $P(a) = 0$ for every $a \in C^n$ such that $R(a) \neq 0$. Prove that $P = 0$.

b) Assume that $P(a) = 0$ for every $a \in C^n$ such that $R(a) \neq 0$ and $Q(a) = 0$. Prove that if Q is irreducible, then Q divides P.

c) Under the hypothesis of b), what can be deduced if Q is no longer assumed to be irreducible?

Exercise (2.8.13). — A ring A is called a von Neumann ring if for every $a \in A$, there exists an element $x \in A$ such that $a = axa$.

a) Prove that a division ring is a von Neumann ring.

b) Let $(A_i)_{i \in I}$ be a family of von Neumann rings. Then the product ring $A = \prod_{i \in I} A_i$ is a von Neumann ring.

c) Prove that a local von Neumann ring is a division ring.

d) Let K be a division ring and let V be a K-vector space. Prove that $\mathrm{End_K(V)}$ is a von Neumann ring.

Exercise (2.8.14). — Let A be a commutative von Neumann ring (see exercise 2.8.13).

a) Prove that 0 is the only nilpotent element of A.

b) Let S be a multiplicative subset of A. Prove that the fraction ring $S^{-1}A$ is a von Neumann ring.

c) Prove that all prime ideals of A are maximal.

Exercise (2.8.15). — Let X be a compact metric space and let $A = \mathscr{C}(X)$ be the ring of real-valued, continuous functions on X.

a) If I is an ideal of A, let V(I) be the set of all points $x \in X$ such that $f(x) = 0$ for all $f \in I$. Prove that V(I) is a closed subset of X.

b) Prove that the map $f \mapsto f^{-1}(0)$ gives a bijection between idempotents of A and closed open subsets of X.

c) Let P be a prime ideal of A. Prove that there exists a unique point $x \in X$ such that $V(P) = \{x\}$.

d) Let I be a finitely generated ideal of A such that $I = \sqrt{I}$. Prove that V(I) is open in X. Prove that there exists an idempotent $e \in A$ such that $I = eA$.

Exercise (2.8.16). — Let X be metric space and let $A = \mathscr{C}(X)$ be the ring of real-valued continuous functions on X. Let P be a prime ideal of X.

a) Let $f, g \in A$; assume that $0 \le f \le g$ and $g \in P$. Prove that there exists an element $h \in A$ such that $f^2 = hg$. Conclude that $f \in P$.

b) Prove that the (partial) order on A induces a total order on the quotient ring A/P.

c) Let Q, Q′ be prime ideals of X containing P. Prove that either $Q \subset Q'$ or $Q' \subset Q$. *This is a result of (KOHLS, 1958).*

Exercise (2.8.17). — Let A be the ring $\mathbf{C}[X, Y]/(XY - 1)$. Let x and y be the images of X and Y in A.

a) Show that x is invertible in A. Show that any non-zero element $a \in A$ can be written uniquely in the form $a = x^m P(x)$, for some integer $m \in \mathbf{Z}$ and some polynomial $P \in \mathbf{C}[T]$ whose constant term is non-zero.

b) For a, m and P as above, set $e(a) = \deg(P)$. Show that the map $e \colon A - \{0\} \to \mathbf{N}$ is a gauge on A, hence that A is a euclidean ring.

c) Conclude that A is a principal ideal domain.

Exercise (2.8.18). — Let A be the set of all complex numbers of the form $a + bi$, for a and $b \in \mathbf{Z}$. Let K be the set of all complex numbers of the form $a + bi$, for a and $b \in \mathbf{Q}$.

a) Show that K is a subfield of $\mathbf{C}$ and that A is a subring of K. Show also that any element of A (*resp. of* K) can be written in a unique way in the form $a + bi$ with $a, b \in \mathbf{Z}$ (*resp.* $a, b \in \mathbf{Q}$). (One says that $(1, i)$ is a basis of A as a $\mathbf{Z}$-module, and a basis of K as a $\mathbf{Q}$-vector space.)

b) For $x = a + bi \in \mathbf{C}$, set $\delta(x) = |x|^2 = a^2 + b^2$. Show that $\delta(xy) = \delta(x)\delta(y)$ for any $x, y \in K$.

c) For $x = a + bi \in K$, set $\{x\} = \{a\} + \{b\}i$, where $\{t\}$ denotes the integer which is the closest to a real number t, chosen to be smaller than t if there are two such integers. Show that $\delta(x - \{x\}) \leq \frac{1}{2}$.

d) Show that δ is a gauge on A, hence that A is a euclidean ring.

Exercise (2.8.19). — Let A be the set of all real numbers of the form $a + b\sqrt{2}$, for a and $b \in \mathbf{Z}$. Let K be the set of all real numbers of the form $a + b\sqrt{2}$, for a and $b \in \mathbf{Q}$.

a) Show that K is a subfield of $\mathbf{R}$ and that A is a subring of K. Show also that any element of A (*resp.* of K) can be written in a unique way in the form $a + b\sqrt{2}$ with $a, b \in \mathbf{Z}$ (*resp.* $a, b \in \mathbf{Q}$). — One says that $(1, \sqrt{2})$ is a basis of A as a $\mathbf{Z}$-module, and a basis of K as a $\mathbf{Q}$-vector space.

b) For $x = a + b\sqrt{2} \in K$, set $\delta(x) = |a^2 - 2b^2|$. Show that $\delta(xy) = \delta(x)\delta(y)$ for any $x, y \in K$.

c) For $x = a + b\sqrt{2} \in K$, set $\{x\} = \{a\} + \{b\}\sqrt{2}$, where $\{t\}$ denotes the integer which is the closest to a real number t, chosen to be smaller than t if there are two such integers. Show that $\delta(x - \{x\}) \leq \frac{1}{2}$.

d) Show that δ is a gauge on A, hence that A is a euclidean ring.

Exercise (2.8.20). — Let ω be the complex number given by $\omega = (1 + i\sqrt{3})/2$; let K be the set of all complex numbers of the form $a + b\omega$, with $a, b \in \mathbf{Q}$, and let A be the set of all such elements of K where $a, b \in \mathbf{Z}$.

a) Show that K is a subfield of $\mathbf{C}$ and that A is a subring of K.

b) Show that A is a euclidean ring for the gauge given by $z \mapsto |z|^2$.

Exercise (2.8.21). — Let A be a domain and let $\delta \colon A - \{0\} \to \mathbf{N}$ be a map which satisfies the second property of the definition of a euclidean gauge, namely: for any $a, b \in A$ such that $b \neq 0$, there exists q and $r \in A$ such that $a = bq + r$ and such that, moreover, either $r = 0$, or $\delta(r) < \delta(b)$.

a) For any $a \in A - \{0\}$, set $\delta'(a) = \inf_{b \neq 0} \delta(ab)$.
Show that δ' is a gauge on A, hence that A is a euclidean ring.

b) Let S be a multiplicative subset of A such that $0 \notin S$; prove that $S^{-1}A$ is a euclidean ring. Compare with exercise 2.8.17.

Exercise (2.8.22). — Let A be a euclidean ring, with gauge δ.

a) Let $a \in A$ be a non-zero, and non-unit element, with minimal gauge. Show that for any $x \in A$ which is not a multiple of a, there exists a unit $u \in A$ such that $1 - ux$ is a multiple of a.

b) Let n be the cardinality of the set of units in A. Show that there exists a maximal ideal $M \subset A$ such that the cardinality of the quotient field A/M is smaller than or equal to $n + 1$.

Exercise (2.8.23). — Let A be the subring of $\mathbf{C}$ generated by $\varepsilon = (1 + i\sqrt{19})/2$.

a) Show that $\varepsilon^2 = \varepsilon - 5$. Deduce that any element of A can be written uniquely as $a + \varepsilon b$, for $a, b \in \mathbf{Z}$.

b) Show that for any $a \in A$, $|a|^2$ is an integer. Show that $a \in A$ is a unit if and only if $|a|^2 = 1$. Conclude that $A^{\times} = \{-1, +1\}$.

c) Let M be any maximal ideal of A. Show that there exists a prime number p such that $p \in M$. Show that A/M has cardinality p^2 if $P = X^2 - X + 5$ is irreducible in $\mathbf{Z}/p\mathbf{Z}$, and cardinality p otherwise.

d) Show that the polynomial $X^2 - X + 5$ is irreducible in $\mathbf{Z}/2\mathbf{Z}$ and $\mathbf{Z}/3\mathbf{Z}$. Conclude that the cardinality of A/M is at least 4.

e) Show that A is not a euclidean ring.

Exercise (2.8.24). — This is a continuation of exercise 2.8.23. We will prove that the ring A is a principal ideal domain.

a) Let K be the field of fractions of A. Prove that every element of K can be written uniquely as $a + \varepsilon b$, for $a, b \in \mathbf{Q}$.
Let $x \in K$ and let $a, b \in \mathbf{Q}$ be such that $x = a + \varepsilon b$; let $m, n \in \mathbf{Z}$ be such that $|a - m|$ and $|b - n|$ are smaller than $1/2$.

b) If $|b - n| \leq 1/3$, prove that there exists an element $u \in A$ such that $N(x - u) < 1$.

c) Otherwise, prove that there exists an element $u \in A$ such that $N(2x - u) < 1$.

d) Prove that the ideal (2) is a maximal ideal of A.

e) Prove that A is a principal ideal domain.

Exercise (2.8.25). — Let A be the set of all complex numbers of the form $a + bi\sqrt{5}$, for a and $b \in \mathbf{Z}$.

a) Show that A is a subring of $\mathbf{C}$.

b) Show that the only units of A are 1 and -1.

c) Show that $2, 3, 1 + i\sqrt{5}$ and $1 - i\sqrt{5}$ are irreducible in A.

d) Observing that $2 \cdot 3 = (1 + i\sqrt{5})(1 - i\sqrt{5})$, prove that A is not a unique factorization domain; in particular it is not principal ideal ring.

Exercise (2.8.26). — Let p be a prime number, let n be an integer such that $n \geq 2$ and let P be the polynomial $P = X^n + X + p$.

a) Assume that $p \neq 2$. Show that any complex root z of P satisfies $|z| > 1$.

b) Still assuming $p \neq 2$, show that P is irreducible in $\mathbf{Z}[X]$.

c) Assume now that $p = 2$. If n is even, show that P is irreducible in $\mathbf{Z}[X]$. If n is odd, show that $X + 1$ divides P but that $P/(X + 1)$ is irreducible in $\mathbf{Z}[X]$.

d) More generally, let $P = a_n X^n + \cdots + a_1 X + a_0$ be any polynomial with integer coefficients such that $|a_0|$ is a prime number strictly greater than $|a_1| + \cdots + |a_n|$. Show that P is irreducible in $\mathbf{Z}[X]$.

Exercise (2.8.27). — Let n be any integer ≥ 2 and let S be the polynomial $X^n - X - 1$. The goal of this exercise is to show, following SELMER (1956), that S is irreducible in $\mathbf{Z}[X]$.

a) Show that S has n distinct roots in $\mathbf{C}$.

b) For any polynomial $P \in \mathbf{C}[X]$ such that $P(0) \neq 0$, set

$$\varphi(P) = \sum_{j=1}^{m} \left(z_j - \frac{1}{z_j} \right),$$

where $z_1, \ldots, z_m$ are the complex roots of P, repeated according to their multiplicities.

Compute $\varphi(P)$ in terms of the coefficients of P. In particular, compute $\varphi(S)$.

c) If P and Q are two polynomials in $\mathbf{C}[X]$ such that $P(0)Q(0) \neq 0$, show that $\varphi(PQ) = \varphi(P) + \varphi(Q)$.

d) For any complex root z of S, show that

$$2\Re\left(z - \frac{1}{z}\right) > \frac{1}{|z|^2} - 1.$$

(Set $z = re^{i\theta}$ and estimate $\cos(\theta)$ in terms of r.)

e) For any positive real numbers $x_1, \ldots, x_m$ with $\prod_{j=1}^{m} x_j = 1$, establish the inequality $\sum_{j=1}^{m} x_j \geq m$.

f) Let P and Q be polynomials in $\mathbf{Z}[X]$ of strictly positive degrees such that $S = PQ$. Show that $|P(0)| = 1$ and that $\varphi(P)$ is a strictly positive integer. Deduce a contradiction and conclude that S is irreducible in $\mathbf{Z}[X]$.

Exercise (2.8.28) (Eisenstein's irreducibility criterion). — Let A be an integral domain, let K be its fraction field and let P be a prime ideal of A. Let $f(X) = \sum_{0 \leq k \leq n} a_k X^k$ be a polynomial of degree $n \geq 1$ with coefficients in A. We assume that $a_0, \ldots, a_{n-1} \in P$, $a_n \notin P$ and $a_0 \notin P^2$.

a) Let $g \in A[X]$ be a non-constant polynomial that divides f. Prove that $\deg(g) = n$. (*Consider a factorization $f = gh$ and reduce it modulo P.*)

b) Show that f is irreducible in $K[X]$.

c) Is f necessary irreducible in $A[X]$?

Exercise (2.8.29). — Let $P = X^n + a_{n-1}X^{n-1} + \cdots + a_0$ be a monic polynomial in $\mathbf{Z}[X]$ such that $a_0 \neq 0$ and

$$|a_{n-1}| > 1 + |a_{n-2}| + \cdots + |a_0|.$$

a) Using Rouché's theorem in the theory of holomorphic functions, show that P has exactly one complex root with absolute value > 1, and that all other roots have absolute value < 1.

b) Show that P is irreducible in $\mathbf{Z}[X]$ (*Perron's theorem*).

Exercise (2.8.30). — Let K be a field. The *Newton polytope* N_f of a polynomial $f \in K[T_1, \ldots, T_n]$ in n indeterminates is the convex hull in $\mathbf{R}^n$ of the set of exponents of all monomials that appear in f. If C is a convex subset of $\mathbf{R}^n$, a point $u \in C$ is called a *vertex* if for every $v, w \in C$ such that $u \in [v, w]$ (the line segment with endpoints v and w in $\mathbf{R}^n$), one has $v = w = u$.

a) When $n = 2$, describe all possible Newton polytopes of polynomials of degree ≤ 3.

Let $f, g, h \in K[T_1, \ldots, T_n]$ be polynomials such that $f = gh$.

b) Let u be a vertex of $N_g + N_h$; prove that there exists a unique pair (v, w) where $v \in N_g$ and $w \in N_h$ such that $u = v + w$. Prove that v is a vertex of N_g and that w is a vertex of N_h; conclude that $u \in N_f$.

c) Prove that $N_f = N_g + N_h$.

d) We assume that $n = 2$. Let a, b, u, v be coprime integers. Describe the set of polynomials $f \in K[T_1, T_2]$ whose Newton polytope is a triangle with vertices $(a, 0)$, $(0, b)$ and (u, v). Prove that these polynomials are irreducible.

Exercise (2.8.31). — Consider the **C**-algebra $\mathbf{C}[X, Y]$ of polynomials with complex coefficients in two variables X and Y. Let I be the ring of $\mathbf{C}[X, Y]$ generated by the polynomial $Y^2 - X^3 + X$ and let A be the quotient ring $\mathbf{C}[X, Y]/I$. One writes x and y for the images of X and Y in A.

The goal of the exercise is to show that A is not a unique factorization domain.

a) Show that $Y^2 - X^3 + X$ is an irreducible polynomial and that A is a domain.

b) Show that the canonical morphism from $\mathbf{C}[T]$ to A which sends T to x is injective. Deduce from this that the subring $\mathbf{C}[x]$ of A generated by **C** and x is isomorphic to the ring of polynomials $\mathbf{C}[T]$. We define the degree of an element of $\mathbf{C}[x]$ to be the degree of its unique preimage in $\mathbf{C}[T]$.

c) Let $a \in A$. Show that there exists a unique pair (p, q) of elements of $\mathbf{C}[x]$ such that $a = p + qy$.

d) Show that there exists a unique **C**-linear endomorphism σ of A such that $\sigma(x) = x$ and $\sigma(y) = -y$. Show that σ is an automorphism of A. For $a \in A$, show that $\sigma(a) = a$ if and only if $a \in \mathbf{C}[x]$.

e) For any $a \in A$, set $N(a) = a\sigma(a)$. Show that $N(a) \in \mathbf{C}[x]$, that $N(1) = 1$, and that $\deg(N(a)) \neq 1$. Show also that $N(ab) = N(a)N(b)$ for any $a, b \in A$.

f) Deduce from questions *b)*, *c)* and *e)* that $\mathbf{C} - \{0\}$ is the set of units of A.

g) Show that $x, y, 1 - x$ and $1 + x$ are irreducible in A.

h) Show that y does not divide x, $1 + x$ nor $1 - x$. Conclude that A is not a unique factorization domain.

Exercise (2.8.32). — Let A be a commutative ring.

a) Let P and Q be polynomials in $A[X]$. Assume that the coefficients of P, resp. those of Q, generate the unit ideal (1) of A. Show that the coefficients of PQ generate the unit ideal too. (*Modify the proof of proposition 2.6.3: observe that the hypothesis implies that for any maximal ideal M of A, P and Q are non-zero modulo M.*)

b) Show that there are polynomials

$$W_i \in \mathbf{Z}[P_0, \ldots, P_n, Q_0, \ldots, Q_n, U_0, \ldots, U_n, V_0, \ldots, V_n]$$

(for $0 \le i \le 2n$) such that if $P = \sum p_i X^i$, $Q = \sum q_i X^i$ are polynomials of degree at most n, $1 = \sum p_i u_i$ and $1 = \sum q_i v_i$ are Bézout relations for their coefficients, then, writing $R = PQ = \sum r_i X^i$, the coefficients of R satisfy a Bézout relation

$$1 = \sum_{i=0}^{2n} r_i W_i(p_0, \ldots, p_n, q_0, \ldots, q_n, u_0, \ldots, u_n, v_0, \ldots, v_n).$$

c) For $P \in A[X]$, let ict(P) be the ideal of A generated by the coefficients of P. By the first question, we have $\mathrm{ict}(PQ) = A$ if $\mathrm{ict}(P) = \mathrm{ict}(Q) = A$. If A is a principal ideal domain, ict(P) is the ideal generated by ct(P), hence $\mathrm{ict}(PQ) = \mathrm{ict}(P)\,\mathrm{ict}(Q)$ by Gauss's lemma. Show however that the equality $\mathrm{ict}(PQ) = \mathrm{ict}(P)\,\mathrm{ict}(Q)$ does not hold in general, for example if $A = \mathbf{C}[U, V]$, $P = UX + V$ and $Q = VX + U$.

Exercise (2.8.33). — Let A be an integral domain and let S be a multiplicative subset of A such that $0 \notin S$. This exercise studies the relations between A and $S^{-1}A$ being unique factorization domains (theorems of Nagata).

a) Assume that A is a unique factorization domain; prove that $S^{-1}A$ is a unique factorization domain.

b) Let $p \in A$ be such that the ideal (p) is prime. Assume that $S = \{p^n \; ; \; n \in \mathbf{N}\}$ and that the ring $S^{-1}A$ is a unique factorization domain. Prove that A is a unique factorization domain.

c) Assume that every increasing sequence of principal ideals of A is stationary. Prove that if $S^{-1}A$ is a unique factorization domain, then A is a unique factorization domain.

d) We consider the particular case of a polynomial ring $A = R[T]$, where R is a unique factorization domain. Taking $S = R - \{0\}$, deduce from the preceding question that A is a unique factorization domain.

e) Let k be a field in which -1 is not a square and let

$$A = k[X, Y, Z]/(X^2 + Y^2 + Z^2 - 1).$$

Prove that A is a unique factorization domain. (*Take $S = \{(1 - z)^n, n \in \mathbf{N}\}$, where z is the class of Z in A.*)

f) Let k be an algebraically closed field of characteristic different from 2, let $n \geq 5$ and let $f = X_1^2 + \cdots + X_n^2$. Prove that $k[X_1, \ldots, X_n]/(f)$ is a unique factorization domain. (*Prove that* $X_3^2 + \cdots + X_n^2$ *is irreducible, then take* $S = \{(x_1 + ix_2)^m, m \in \mathbf{N}\}$, *where* $i \in k$ *satisfies* $i^2 = -1$.)

Exercise (2.8.34). — Let K be a field.

a) Let $a, b \in K^\times$ and $d \geq 1$. Prove that every irreducible factor of $aX^d + bY^d$ is homogeneous.

b) Let $a, b, c \in K^\times$ and $d \geq 1$. Prove that $aX^d + bY^d + c$ is irreducible.

c) Let $P_1, P_2, \ldots, P_n$ be non-constant polynomials in $K[T]$, of degrees $d_1, \ldots, d_n$. Let $P = P_1(T_1) + \cdots + P_n(T_n)$. Prove that P is irreducible if $n = 2$ and d_1, d_2 are coprime integers. (*Assign weight* d_1 *to* T_1 *and* d_2 *to* T_2, *hence* $m_1 d_1 + m_2 d_2$ *to a monomial* $T_1^{m_1} T_2^{m_2}$, *and consider the monomials of largest weights in a factorization of* P.)

d) Prove that P is irreducible if $n \geq 3$ (a theorem of TVERBERG (1964)).

Chapter 3.
Modules

Modules *are to rings what vector spaces are to fields and this introductory chapter studies them from the point of view of linear algebra. The first sections have to set a number of definitions up, which are the close companions to similar definitions for vector spaces: modules and their submodules;* morphisms *from a module to another (the analogues of linear maps), their kernel and image, isomorphisms; generating families, independent families,* bases. *A notable difference with the linear algebra of vector spaces is that modules do not possess a basis in general.*

I also present three general constructions of modules: direct sums *and products; the* quotient *of a module by a submodule; and* localization *of a module by a multiplicative subset of the base ring. These last constructions have* universal properties *which I will use systematically in the rest of the book. These universal properties not only characterize the modules they consider; above all, they furnish an explicit way to work with them — in some sense, they have to be understood as mere computation rules. They also furnish an appropriate opportunity to introduce a bit of categorical language.*

As I said in the introduction to the book, I presuppose in this book an acquaintance with basic algebra, such as vector spaces. However, in case the theory of vector spaces *has only been presented to the reader in a restricted case (such as with a base field equal to* **R** *or* **C**), *I decided to prove the main theorems here. One excuse I may present is that I allow for general division rings and not only (commutative) fields.*

Now restricting myself to commutative rings, I then discuss the theory of determinants, *from the point of view of alternating multilinear forms. (I will also revisit this theory in the chapter devoted to tensor products.) As in linear algebra over fields, determinants allow us to characterize injective, resp. surjective endomorphisms of a free module of finite rank, but the proofs are a bit subtler.*

Determinants are also used to define the Fitting ideals *of a finitely generated module. The definition of these ideals will reappear in the classification of modules over a principal ideal domain.*

© Springer Nature Switzerland AG 2021
A. Chambert-Loir, *(Mostly) Commutative Algebra*, Universitext,
https://doi.org/10.1007/978-3-030-61595-6_3

3.1. Definition of a Module

Definition (3.1.1). — Let A be a ring. A *right* A-*module* is a set M endowed with two binary laws: an internal addition, $M \times M \to M$, $(m, n) \mapsto m + n$, and a scalar multiplication $M \times A \to M$, $(m, a) \mapsto ma$, subject to the following properties:

(i) *Axioms concerning the addition law and stating that* (M, +) *is an abelian group:*

  – For every m, n in M, $m + n = n + m$ *(commutativity of addition)*;
  – For every m, n, p in M, $(m + n) + p = m + (n + p)$ *(associativity of addition)*;
  – There exists an element $0 \in M$ such that $m + 0 = 0 + m = m$ for every $m \in M$ *(neutral element for the addition)*;
  – For every $m \in M$, there exists an $n \in M$ such that $m + n = n + m = 0$ *(existence of an additive inverse)*;

(ii) *Axioms concerning the multiplication law:*

  – For any $m \in M$, $m1 = m$ *(the unit of A is neutral for the scalar multiplication)*;
  – For any $a, b \in A$ and any $m \in M$, one has $m(ab) = (ma)b$ *(associativity of the scalar multiplication)*;

(iii) *Axioms relating the ring addition, the module addition and the scalar multiplication:*

  – For any $a, b \in A$ and any $m \in M$, one has $m(a + b) = ma + mb$ *(scalar multiplication distributes over the ring addition)*;
  – For any $a \in A$ and any $m, n \in M$, one has $(m + n)a = ma + na$ *(scalar multiplication distributes over the module addition)*.

Analogously, a *left* A-*module* is a set M endowed with two binary laws: an internal addition, $M \times M \to M$, $(m, n) \mapsto m + n$, and a scalar multiplication $A \times M \to M$, $(a, m) \mapsto am$, subject to the following properties:

(i) *Axioms concerning the addition law and stating that* (M, +) *is an abelian group:*

  – for every m, n in M, $m + n = n + m$ *(commutativity of addition)*;
  – for every m, n, p in M, $(m + n) + p = m + (n + p)$ *(associativity of addition)*;
  – there exists an element $0 \in M$ such that $m + 0 = 0 + m = m$ for every $m \in M$ *(neutral element for the addition)*;
  – for every $m \in M$, there exists $n \in M$ such that $m + n = n + m = 0$ *(existence of an additive inverse)*;

(ii) *Axioms concerning the multiplication law:*

  – for any $m \in M$, $1m = m$ *(the unit of A is neutral for the scalar multiplication)*;

     – for any $a, b \in A$ and any $m \in M$, one has $(ab)m = a(bm)$ *(associativity of the scalar multiplication)*;

(iii) *Axioms relating the ring addition, the module addition and the scalar multiplication:*

     – for any $a, b \in A$ and any $m \in M$, one has $(a + b)m = am + bm$ *(scalar multiplication distributes over the ring addition)*;
     – for any $a \in A$ and any $m, n \in M$, one has $a(m + n) = am + an$ *(scalar multiplication distributes over the module addition)*.

Endowed with its internal addition law, any (left or right) module M is in particular an abelian group. The additive inverse of an element $m \in M$ is written $-m$. The neutral element of M is written 0, as is the neutral element of the ring A for the addition. If needed, one may write 0_M, 0_A to insist on their differences.

Observe that for any right A-module M and any $m \in M$, one has $m0_A = m(0_A + 0_A) = m0_A + m0_A$, hence $m0_A = 0_M$; moreover, $0_M = m0_A = m(1-1) = m1 + m(-1)$, hence $-m = m(-1)$. Similarly, for any left A-module M and any $m \in M$, one has $0_A m = 0_M$ and $(-1)m = -m$.

3.1.2. — Let M be a right A-module, let I be a set, let $(a_i)_{i \in I}$ be a family of elements of A and let $(m_i)_{i \in I}$ be a family of elements of M. Assume that the family (a_i) has finite support. Then the family $(m_i a_i)$ has finite support; its sum $\sum_{i \in I} m_i a_i$ is called a *linear combination* of the elements m_i.

If I', I'' are disjoint subsets of I such that $a_i = 0$ for all $i \in I - (I' \cup I'')$, then one has

$$\sum_{i \in I} m_i a_i = \sum_{i \in I'} m_i a_i + \sum_{i \in I''} m_i a_i.$$

Moreover, for every $b \in A$, one has

$$\left(\sum_{i \in I} m_i a_i \right) b = \sum_{i \in I} m_i (a_i b).$$

Finally, if $(a_i)_{i \in I}$ and $(b_i)_{i \in I}$ are two families with finite support, then the family $(a_i + b_i)_{i \in I}$ has finite support as well, and one has

$$\sum_{i \in I} m_i (a_i + b_i) \sum_{i \in I} m_i a_i + \sum_{i \in I} m_i b_i.$$

Examples (3.1.3). —   a) Let A be a ring; multiplication $A \times A \to A$ endows the abelian group with two structures of modules, one left, denoted A_s (*sinister*, Latin for *left*), and one right, denoted A_d (*dexter*, Latin for *right*). Let I be a left ideal of A; the multiplication of A, $A \times I \to I$, endows the abelian group I with the structure of a left A-module. Similarly, if I is a right ideal of A, the multiplication $I \times A \to I$ endows it with the structure of a right A-module.

b) Let A be a ring and let M be a left A-module. Let A° be the opposite ring to A, with the same addition and multiplication computed in the opposite order. Endowed with the scalar multiplication M×A° → M given by $(m, a) \mapsto am$, M becomes a right A°-module.

This correspondence allows one to deduce from a statement for right A-modules an analogous statement for left A-modules.

When A is commutative, A° is identical to A, hence any left A-module is also a right A-module, with the same addition and the same scalar multiplication, only written in the opposite order.

c) An abelian group G can be converted to a **Z**-module in a unique way. Indeed, for any integer *n* and any element $g \in G$, ng is defined by induction by the rules $0g = 0$, $ng = (n-1)g + g$ if $n \geq 1$ and $ng = -(-n)g$ if $n \leq -1$.

d) When A is a field, in particular commutative, the notion of an A-module coincides with that of an A-vector space, which has presumably already been studied extensively enough. We keep this terminology of a vector space when A is a division ring, thus talking of a left or right A-vector space.

e) Let $f : A \to B$ be a morphism of rings, the scalar multiplication B×A → B given by $(b, a) \mapsto bf(a)$ endows B with the structure of a right A-module.

More generally, if $f : A \to B$ is a morphism of rings and if M is a right B-module, the scalar multiplication $M \times A \to M$ given by $(m, a) \mapsto mf(a)$ endows M with the structure of a right A-module.

We leave it to the reader to write similar statements for left A-modules.

f) Let A and B be rings. An (A, B)-*bimodule* is an abelian group M endowed with the structure of a left A-module and the structure of a right B-module such that $a(mb) = (am)b$ for every $a \in A$, $m \in M$ and $b \in B$. This amounts to the datum of the structure of a left $A \times B°$-module, or to the datum of the structure of a right $A° \times B$-module.

In the sequel, an A-*module* (without further precision) shall be understood as a *right* A-module and most of the stated properties will be only proved for them. I will often leave to the conscientious reader to prove (and sometimes even to find) the analogous statements for left A-modules. This choice may look awkward but will hopefully seem more natural after reading §3.5.10.

Remark (3.1.4). — Let A be a ring and let M be a left A-module. For $a \in A$, let $\mu_a : M \to M$ be the map given by $\mu_a(m) = am$. This is an endomorphism of M as an abelian group. Indeed, $\mu_a(m+n) = a(m+n) = am + an = \mu_a(m) + \mu_a(n)$ for any $m, n \in M$.

Moreover, the map $a \mapsto \mu_a$ is a morphism of rings from A to the ring End(M) of endomorphisms of the abelian group M. Indeed, μ_a is the zero map, $\mu_1 = \mathrm{id}_M$, and for any $a, b \in A$ and any $m \in M$, one has $\mu_{a+b}(m) = (a+b)m = am + bm = \mu_a(m) + \mu_b(m) = (\mu_a + \mu_b)(m)$ and $\mu_{ab}(m) = (ab)m = a(bm) = \mu_a(bm) = \mu_a \circ \mu_b(m)$, so that $\mu_{a+b} = \mu_a + \mu_b$ and $\mu_{ab} = \mu_a \circ \mu_b$.

Conversely, let M be an abelian group and let $\mu : A \to \mathrm{End}(M)$ be a morphism of rings. Let us define a scalar multiplication $A \times M \to M$ by

$(a, m) \mapsto \mu(a)(m)$. One checks that this enriches the abelian group $(M, +)$ with the structure of a left A-module such that for any $a \in A$, $\mu(a)$ is the multiplication by a.

Consequently, given an abelian group M, it is equivalent to endow it with the structure of a left A-module and to give oneself a morphism of rings $A \to \text{End}(M)$. In particular, every abelian group M becomes canonically a left $\text{End}_{Ab}(M)$-module.

A right A-module structure on an abelian group M is equivalent to the datum of a morphism of rings $A \to \text{End}(M)^\circ$.

3.1.5. — Let A be a ring and let M be an A-module.

Let $m \in M$. The set $\text{Ann}_A(m)$ of all $a \in A$ such that $ma = 0$ is a right ideal of A, called the *annihilator* of m. (Indeed, for $a, b \in A$, one has $m0 = 0$, $m(a + b) = ma + mb = 0$ if $ma = mb = 0$, and $m(ab) = (ma)b = 0$ if $ma = 0$.)

The set $\text{Ann}_A(M)$ of all $a \in A$ such that $ma = 0$ for all $m \in M$ is called the *annihilator of* M. Since it is the intersection of all $\text{Ann}_A(m)$, for $m \in M$, it is a right ideal of A. On the other hand, if $a \in \text{Ann}_A(M)$ and $b \in A$, one has $m(ba) = (mb)a = 0$ for all $m \in M$, so that $ba \in \text{Ann}_A(M)$. Consequently, $\text{Ann}_A(M)$ is a two-sided ideal of A.

If $\text{Ann}_A(M) = 0$, one says that the A-module M is *faithful*.

In fact, the ideal $\text{Ann}_A(M)$ is the kernel of the morphism $\mu \colon A \to \text{End}(M)$, $a \mapsto \mu_a$. In this way, M can be viewed as an A/I-module, for every ideal I of A such that $I \subset \text{Ann}_A(M)$. Taking $I = \text{Ann}_A(M)$, we observe that M is a faithful $(A/\text{Ann}_A(M))$-module.

Conversely, if M is an $(A/\text{Ann}_A(M))$-module, the composition $A \to A/\text{Ann}_A(M) \to \text{End}(M)$ endows M with the structure of an A module whose annihilator contains $\text{Ann}_A(M)$.

Definition (3.1.6). — Let A be a ring and let M be a right A-module. A *submodule* of M is a subset N of M satisfying the following properties:

(i) N is an abelian subgroup of M;

(ii) if $a \in A$ and $m \in M$, then $ma \in N$.

There is a similar definition for left A-modules, and for (A, B)-bimodules.

Examples (3.1.7). — a) Let M be a A-module. The subsets $\{0\}$ and M of M are submodules.

b) Let A be a ring. The left ideals of A are exactly the submodules of the left A-module A_s. The right ideals of A are the submodules of the right A-module A_d.

c) Let A be a division ring. The submodules of an A-vector space are its vector subspaces.

Lemma (3.1.8). — *Let A be a ring, let M be a right A-module and let N be a subset of M. To show that N is a submodule of M, it suffices to show that it satisfies the following properties:*

(i) $0 \in N$;

(ii) *If $a \in A$ and $m \in N$, then $ma \in N$;*

(iii) *If $m \in N$ and $n \in N$, then $m + n \in N$.*

Proof. — Indeed, the second property applied to $a = -1$ and $m \in N$ implies that $-m \in N$. Combined with the two other properties, this shows that N is a subgroup of the abelian group M. Then the second property shows that it is a submodule of M   □

Example (3.1.9). — Let A be a commutative ring and let M be an A-module. An element $m \in M$ is said to be *torsion* if there exists a regular element $a \in A$ such that $ma = 0$.

The set $T(M)$ *of all torsion elements of* M *is a submodule of* M. Indeed, one has $0 \in T(M)$, because $0_M \cdot 1_A = 0$. Let $m \in T(M)$ and let $a \in A$; let b be a regular element such that $mb = 0$. Then $(ma)b = m(ab) = (mb)a = 0$, hence $ma \in T(M)$. Finally, let $m, n \in T(M)$, let $a, b \in A$ be regular elements such that $ma = nb = 0$; then ab is a regular element of A and $(m + n)ab = (ma)b + (nb)a = 0$, hence $m + n \in T(M)$.

One says that M *is torsion-free if* $T(M) = 0$, *and that* M *is a torsion A-module if* $M = T(M)$.

3.2. Morphisms of Modules

Definition (3.2.1). — Let A be a ring, and let M and N be two A-modules. A *morphism* from M to N is a map $f: M \rightarrow N$ such that

$$f(ma + nb) = f(m)a + f(n)b$$

for every $a, b \in A$ and every $m, n \in M$. One writes $\mathrm{Hom}_A(M, N)$ for the set of morphisms from M to N.

A morphism from M to itself is called an *endomorphism* of M. The set of all endomorphisms of an A-module M is denoted $\mathrm{End}_A(M)$.

The expressions "A-linear mapping" and "linear mapping" are synonyms for "morphism of A-modules".

The identity map id_M from an A-module M to itself is an endomorphism.

Let A be a ring, and let M, N, P be three A-modules. If $f: M \rightarrow N$ and $g: N \rightarrow P$ are morphisms, then their composition $g \circ f: M \rightarrow P$ is a morphism of A-modules.

We thus can say that A-modules and morphisms of A-modules form a category $\boldsymbol{Mod}_A$. Moreover, if $f: A \rightarrow B$ is a morphism of rings, viewing a B-module as an A-module furnishes a *functor* $f^*: \boldsymbol{Mod}_B \rightarrow \boldsymbol{Mod}_A$.

One says that a morphism of A-modules $f: M \rightarrow N$ is an isomorphism if there exists a morphism $g: N \rightarrow M$ such that $f \circ g = \mathrm{id}_N$ and $g \circ f = \mathrm{id}_M$. Then, there is exactly one such morphism g, called the reciprocal (or inverse) of f. Indeed, if $f \circ g = f \circ h = \mathrm{id}_N$ and $g \circ f = h \circ f = \mathrm{id}_M$, then $h = h \circ \mathrm{id}_N = h \circ (f \circ g) = (h \circ f) \circ g = \mathrm{id}_M \circ g = g$.

One says that a morphism $f \colon M \to N$ is left invertible if there exists a morphism $g \colon N \to M$ such that $g \circ f = \mathrm{id}_M$, and that it is right invertible if there exists a morphism $g \colon N \to M$ such that $f \circ g = \mathrm{id}_N$. A left invertible morphism is injective, a right invertible morphism is surjective, but the converse assertions are not true in general.

Proposition (3.2.2). — *Let A be a ring. For a morphism of A-modules to be an isomorphism, it is necessary and sufficient that it be bijective; its inverse bijection is then a morphism of modules.*

Proof. — If $f \colon M \to N$ is an isomorphism, with reciprocal g, then g is also the inverse bijection of f.

Conversely, let $f \colon M \to N$ be a bijective morphism and let g be its inverse bijection. Then, g is a morphism. Indeed, for $n, n' \in N$ and $a, a' \in A$, we have

$$f\big(g(n)a + g(n')a'\big) = f(g(n))a + f(g(n'))a' = na + n'a',$$

hence $g(n)a + g(n')a' = g(na + n'a')$, so that g is linear. □

Definition (3.2.3). — Let $f \colon M \to N$ be a morphism of A-modules. The *kernel* of f, written $\mathrm{Ker}(f)$, is the set of all $m \in M$ such that $f(m) = 0$.

It is the kernel of f as a morphism of abelian groups.

Proposition (3.2.4). — *Let $f \colon M \to N$ be a morphism of A-modules.*

For any submodule M′ of M, $f(M')$ is a submodule of N. For any submodule N′ of N, $f^{-1}(N')$ is a submodule of M.

In particular, the kernel $\mathrm{Ker}(f)$ and the image $\mathrm{Im}(f) = f(M)$ of f are submodules (respectively of M and N).

Proof. — Let us show that $f(M')$ is a submodule of N. Since $f(0_M) = 0_N$ and $0_M \in M'$, we have $0_N \in f(M')$. On the other hand, for $n, n' \in f(M')$, there exist $m, m' \in M'$ such that $n = f(m)$ and $n' = f(m')$. Then,

$$n + n' = f(m) + f(m') = f(m + m') \in f(M').$$

Finally, if $n = f(m)$ belongs to $f(M')$ and if $a \in A$, then $na = f(m)a = f(ma)$ belongs to $f(M')$ since $ma \in M'$.

Let us show that $f^{-1}(N')$ is a submodule of M. Since $f(0_M) = 0_N \in N'$, we have $0_M \in f^{-1}(N')$. Moreover, for $m, m' \in f^{-1}(N')$ and for $a, b \in A$, we have

$$f(ma + m'b) = f(m)a + f(m')b \in N'$$

since $f(m)$ and $f(m')$ both belong to N′ and N′ is a submodule of N. Hence $ma + m'b$ belongs to $f^{-1}(N')$. □

Example (3.2.5). — Let A be a ring and let M, N be two A-modules. The set $\mathrm{Hom}_{Ab}(M, N)$ of all morphisms of abelian groups from M to N is again an abelian group, the sum of two morphisms f and g being the morphism $f + g$

given by $m \mapsto f(m) + g(m)$. If f and g are A-linear, then $f + g$ is A-linear too, because

$$(f + g)(ma) = f(ma) + g(ma) = f(m)a + g(m)a$$
$$= (f(m)a + g(m)a) = (f + g)(m)a,$$

for $a \in A$ and $m, n \in M$. Similarly, the zero map is a morphism, as well as the map $-f : m \mapsto -f(m)$ for any morphism $f \in \text{Hom}_A(M, N)$. Consequently, $\text{Hom}_A(M, N)$ is a subgroup of $\text{Hom}(M, N)$.

When $M = N$, $\text{End}_A(M)$ is moreover a ring, whose multiplication is given by composition of morphisms; it is a subring of $\text{End}(M)$.

Assume that A is a commutative ring. Let $f \in \text{Hom}_A(M, N)$ and let $a \in A$. Then, the map fa defined by $m \mapsto f(m)a$ is a morphism of abelian groups, but is also A-linear, since

$$(fa)(mb) = f(bm)a = f(m)ba = f(m)ab = (fa)(m)b,$$

for any $m \in M$ and any $b \in A$. This endows the abelian group $\text{Hom}_A(M, N)$ with the structure of an A-module.

However, when the ring A is not commutative, the map $m \mapsto f(m)a$ is not necessarily A-linear, and the abelian group $\text{Hom}_A(M, N)$ may not have any natural A-module structure.

Assume, more generally, that N has the structure of a (B, A)-bimodule. One can then endow $\text{Hom}_A(M, N)$ with the structure of a left B-module by defining a linear map bf, for $f \in \text{Hom}_A(M, N)$ and $b \in B$, by the formula $(bf)(m) = bf(m)$. (It is clearly a morphism of abelian groups; the equalities $bf(ma) = b(f(ma)) = b \cdot f(m) \cdot a = (bf(m))a$ prove that it is A-linear.) Moreover, one has $(bb')f = b(b'f)$ for any $b, b' \in B$ and any $f \in \text{Hom}_A(M, N)$.

Similarly, if M is a (B, A)-bimodule, one can endow $\text{Hom}_A(M, N)$ with the structure of a right B-module by defining, for $f \in \text{Hom}_A(M, N)$ and $b \in B$, a linear map fb by the formula $(fb)(m) = f(bm)$. One has $f(bb') = (fb)b'$ for $b, b' \in B$ and $f \in \text{Hom}_A(M, N)$. Indeed, for any $m \in M$, one has $(fbb')(m) = f(bb'm)$; on the other hand, $(fb)b'$ maps m to $(fb)(b'm) = f(bb'm)$, hence the relation $fbb' = (fb)b'$.

Remark (3.2.6). — We have explained how an abelian group M has the natural structure of an $\text{End}(M)$-module. Similarly, if A is a ring, than any right A-module M is endowed with the canonical structure of a left $\text{End}_A(M)$-module.

Assume that A is commutative, so that $\text{End}_A(M)$ is now naturally an A-module. Let, moreover, $u \in \text{End}_A(M)$. By the universal property of polynomial rings (proposition 1.3.17), there is a unique ring morphism $\mu_u : A[X] \to \text{End}_A(M)$ that coincides with μ on A and maps X to u. It maps a polynomial $P = \sum p_n X^n$ to the endomorphism $P(u) = \sum p_n u^n$. This endows the A-module M with the structure of an A[X]-module.

This construction will be of great use in the case where A is a field, thanks to the fact that A[X] is a principal ideal domain and to the classification of modules over such rings.

Definition (3.2.7). — Let A be a ring, and let M be a right A-module. The *dual* of M, written $M^\vee$, is the abelian group $\mathrm{Hom}_A(M, A_d)$ endowed with its natural structure of a left A-module (for which $(a\varphi)(m) = a\varphi(m)$ for any $a \in A$, $m \in M$ and $\varphi \in M^\vee$). Its elements are called *linear forms* on M.

Similarly, the dual of a left A-module is the abelian group $\mathrm{Hom}_A(M, A_s)$ endowed with its natural structure of a right A-module for which $(\varphi a)(m) = \varphi(m)a$ for any $a \in A$, $m \in M$ and $\varphi \in M^\vee$.

3.3. Operations on Modules

Proposition (3.3.1). — *Let A be a ring, let M be a right A-module and let $(N_s)_{s \in S}$ be a family of submodules of M. Then, its intersection $N = \bigcap_{s \in S} N_s$ is a submodule of M.*

Proof. — Since $0 \in N_s$ for every s, one has $0 \in N$. Let m and n be two elements of N. For any s, m and n belong to N_s, hence so does $m + n$, so that $m + n$ belongs to N. Finally, let $m \in N$ and $a \in A$. For every s, $m \in N_s$, hence $ma \in N_s$ and finally $ma \in N$. Therefore, N is a submodule of M. □

Proposition (3.3.2). — *Let A be a ring, let M be a right A-module and let X be a subset of M. There exists a smallest submodule $\langle X \rangle$ of M that contains X: it is the intersection of the family of all submodules of M which contain X. It is also the set of sums $\sum_{x \in X} x a_x$, where $(a_x)_{x \in X}$ runs among the set of all almost null families of elements of A.*

One says that $\langle X \rangle$ is the *submodule of M generated by X*.

Proof. — The intersection $\langle X \rangle$ of all of the submodules of M that contain X is a submodule of M; it contains X. By construction, $\langle X \rangle$ is contained in every submodule of M which contains X; it is therefore the smallest of them all.

Let $(a_x)_x$ be an almost-null family of elements of A; then, $\sum_{x \in X} x a_x$ is a linear combination of elements in $\langle X \rangle$, hence belongs to $\langle X \rangle$. This shows that the set $\langle X \rangle'$ of all such linear combinations is contained in $\langle X \rangle$.

To obtain the other inclusion, let us first show that $\langle X \rangle'$ is a submodule of M. First of all, $0 = \sum_{x \in X} x0$ belongs to $\langle X \rangle'$. On the other hand, let m and n be two elements of $\langle X \rangle'$, and let $(a_x)_x$ and $(b_x)_x$ be two almost-null families such that $m = \sum_{x \in X} x a_x$ and $n = \sum_{x \in X} x b_x$. Then, the family $(a_x + b_x)_x$ is almost-null and one has

$$m + n = \Big(\sum_{x \in X} x a_x\Big) + \Big(\sum_{x \in X} x b_x\Big) = \sum_{x \in X} x(a_x + b_x)$$

so that $m + n$ belongs to $\langle X \rangle'$. Finally, let $m \in \langle X \rangle'$ and $a \in A$, let $(a_x)_{x \in X}$ be an almost-null family such that $m = \sum_{x \in X} x a_x$. Then, $ma = \sum_{x \in X} x(a_x a)$, so that $ma \in \langle X \rangle'$. This concludes the proof that $\langle X \rangle'$ is a submodule of M. Since it contains X, we have $\langle X \rangle \subset \langle X \rangle'$, the other desired inclusion. $\qquad \square$

Definition (3.3.3). — Let A be a ring, let M be a right A-module and let $(M_s)_{s \in S}$ be a family of submodules of M. Its *sum*, written $\sum_{s \in S} M_s$, is the submodule of M generated by the union $\bigcup_{s \in S} M_s$ of the submodules M_s.

It is also the set of all linear combinations $\sum_s m_s$ where $(m_s)_s$ is an almost-null family of elements of M such that $m_s \in M_s$ for every $s \in S$. Indeed, this set of linear combinations is a submodule of M, it contains $\bigcup_s M_s$, and is contained in every submodule of M which contains all of the M_s.

Definition (3.3.4). — Let A be a ring, and let $(M_s)_{s \in S}$ be a family of right A-modules. Its *direct product*, or simply *product*, is the set $\prod_{s \in S} M_s$ together with the laws

$$(m_s)_s + (n_s)_s = (m_s + n_s)_s, \qquad (m_s)_s a = (m_s a)_s$$

which endow it with the structure of a right A-module.

The *direct sum* of the family $(M_s)_{s \in S}$ is the submodule $\bigoplus_{s \in S} M_s$ of $\prod_{s \in S} M_s$ consisting of families $(m_s)_s$ with finite support, that is, such that $m_s = 0$ for all but finitely many $s \in S$.

Remark (3.3.5). — If all of the M_s are equal to a given module M, one writes $\prod_{s \in S} M_s = M^S$ and $\bigoplus_{s \in S} M_s = M^{(S)}$.

Lemma (3.3.6). — *Let $(M_s)_{s \in S}$ be a family of A-modules. For every $t \in S$, define maps*

$$i_t : M_t \to \bigoplus_{s \in S} M_s, \qquad p_t : \prod_{s \in S} M_s \to M_t$$

by $i_t(m) = (m_s)_{s \in S}$ where $m_t = m$ and $m_s = 0$ for $s \neq t$, and $p_t((m_s)_{s \in S}) = m_t$. These are morphisms of A-modules.

Proof. — Let $t \in S$, let $m, n \in M_t$, and let $a, b \in A$. Then,

$$i_t(ma + nb) = (0, \dots, 0, ma + nb, 0, \dots)$$

(in the right-hand side, the term $ma + nb$ has index t)

$$= (0, \dots, 0, m, 0, \dots)a + (0, \dots, 0, n, 0, \dots, 0)b$$
$$= i_t(m)a + i_t(n)b.$$

Consequently, i_t is a morphism of A-modules. We leave as an exercise the proof that p_t is a morphism. $\qquad \square$

The morphism i_t is injective, and called the *canonical injection* of index t; the morphism p_t is surjective; it is called the *canonical surjection* of index t.

Direct products and direct sums of modules satisfy the following *universal properties*.

Theorem (3.3.7). — *Let A be a ring and let $(M_s)_{s \in S}$ be a family of right A-modules.*

a) *For every right A-module N and any family $(f_s)_{s \in S}$, where $f_s : N \to M_s$ is a morphism, there exists a unique morphism $f : N \to \prod_{s \in S} M_s$ such that $p_s \circ f = f_s$ for every $s \in S$.*

b) *For every right A-module N and any family $(f_s)_{s \in S}$, where $f_s : M_s \to N$ is a morphism, there exists a unique morphism $f : \bigoplus_{s \in S} M_s \to M$ such that $f \circ i_s = f_s$ for every $s \in S$.*

Proof. — a) Assume that $f : N \to \prod_{s \in S} M_s$ satisfies $p_s \circ f = f_s$ for every $s \in S$. Then, if $n \in N$ and $f(n) = (m_s)_s$, we have

$$m_s = p_s((m_s)_s) = p_s(f(n)) = (p_s \circ f)(n) = f_s(n),$$

so that there is at most one morphism f satisfying $p_s \circ f = f_s$ for all s. Conversely, let us define a map $f : N \to \prod_{s \in S} M_s$ by $f(n) = (f_s(n))_{s \in S}$ for $n \in N$ and let us show that it is a morphism. Indeed, for any $a, b \in A$ and any $m, n \in N$,

$$f(ma + nb) = \left(f_s(ma + nb)\right)_s = \left(f_s(m)a + f_s(n)b\right)_s$$
$$= \left(f_s(m)\right)_s a + \left(f_s(n)\right)_s b = f(m)a + f(n)b,$$

so that f is linear.

b) Assume that $f : \bigoplus_{s \in S} M_s \to N$ satisfies $f \circ i_s = f_s$ for every s. Then, the image by f of any element $(0, \ldots, 0, m, 0, \ldots) = i_s(m)$ (where $m \in M_s$ has index s) is necessarily equal to $f_s(m)$. Any element m of $\bigoplus_{s \in S} M_s$ is a family $(m_s)_{s \in S}$, where $m_s \in M_s$ for every s, almost finitely many of them being non-zero. It follows that $m = \sum_{s \in S} i_s(m_s)$ (the sum is finite)

$$f(m) = f(\sum_{s \in S} i_s(m_s)) = \sum_{s \in S} (f \circ i_s)(m_s) = \sum_{s \in S} f_s(m_s),$$

hence the uniqueness of such a map f. Conversely, the map $f : \bigoplus_{s \in S} M_s \to M$ defined for any $(m_s)_s \in \bigoplus_{s \in S} M_s$ by

$$f((m_s)_s) = \sum_{s \in S} f_s(m_s) \qquad \text{(finite sum)}$$

is a morphism of A-modules and satisfies $f \circ i_s = f_s$ for every $s \in S$. Indeed, let $a, b \in A$, and $(m_s)_{s \in S}, (n_s)_{s \in S}$ be two elements of $\bigoplus_s M_s$; then, one has

$$f((m_s)_s a + (n_s)_s b) = f((m_s a + n_s b)_s) = \sum_{s \in S} f_s(m_s a + n_s b)$$

$$= \sum_s (f_s(m_s)a + f_s(n_s)b) = \Big(\sum_{s \in S} f_s(m_s)\Big)a + \Big(\sum_{s \in S} f_s(n_s)\Big)b$$

$$= f((m_s)_s)a + f((n_s)_s)b.$$

Remark (3.3.8). — One can reformulate this theorem as follows: for every right A-module N, the maps

$$\mathrm{Hom}_A\Big(\bigoplus_{s \in S} M_s, M\Big) \to \prod_{s \in S} \mathrm{Hom}_A(M_s, M), \qquad f \mapsto (f \circ i_s)_s$$

and

$$\mathrm{Hom}_A\Big(M, \prod_{s \in S} M_s\Big) \to \prod_{s \in S} \mathrm{Hom}_A(M, M_s), \qquad f \mapsto (p_s \circ f)_s$$

are bijections. In fact, they are isomorphisms of abelian groups (and of A-modules if A is commutative).

In the language of *category theory*, this says that $\prod_s M_s$ and $\bigoplus_s M_s$ (endowed with the morphisms i_s and p_s) are respectively a *product* and a *coproduct* of the family (M_s).

3.3.9. Internal direct sums — Let M be a right A-module and let $(M_s)_{s \in S}$ be a family of submodules of M. Then, there is a morphism of A-modules, $\bigoplus_s M_s \to M$, defined by $(m_s) \mapsto \sum m_s$ for every almost-null family $(m_s)_{s \in S}$, where $m_s \in M_s$ for every s. The image of this morphism is $\sum_s M_s$. One says that the submodules M_s are in *direct sum* if this morphism is an isomorphism; this means that any element of M can be written in a unique way as a sum $\sum_{s \in S} m_s$, with $m_s \in M_s$ for every s, all but finitely many of them being null. In that case, one writes $M = \bigoplus_{s \in S} M_s$ ("internal" direct sum).

When the set S has two elements, say $S = \{1, 2\}$, the kernel of this morphism is the set of all pairs $(m, -m)$, where $m \in M_1 \cap M_2$, and its image is the submodule $M_1 + M_2$. Consequently, M_1 and M_2 are in direct sum if and only if $M_1 \cap M_2 = 0$ and $M_1 + M_2 = M$. The picture is slightly more complicated for families indexed by a set with 3 or more elements; this is already the case for vector spaces, see exercise 3.10.9.

Let M be a right A-module and let N be a submodule of M; a *direct summand* of N is a submodule P of M such that $M = N \oplus P$ (N and P are in direct sum). In contrast to the case of vector spaces, not every submodule has a direct summand. If a submodule N of M has a direct summand, one also says that N *is* a direct summand.

Definition (3.3.10). — Let A be a ring, let M be right A-module and let I be a right ideal of A. One defines the submodule MI of M as the submodule generated by all products ma, for $m \in M$ and $a \in I$.

It is the set of all finite linear combinations $\sum m_i a_i$, where $a_i \in I$ and $m_i \in M$ for every i.

3.4. Quotients of Modules

Let A be a ring and let M be a right A-module. We are interested in the equivalence relations $\sim$ on M which are compatible with its module structure, namely such that for any $m, m', n, n' \in M$, and any $a, b \in A$,

$$\text{if } m \sim m' \text{ and } n \sim n', \text{ then } ma + nb \sim m'a + n'b.$$

Let N be the set of all $m \in M$ such that $m \sim 0$. Since an equivalence relation is reflexive, one has $0 \in N$. If $m, n \in N$, then $m \sim 0$ and $n \sim 0$, hence $ma + nb \sim (a0 + b0) = 0$, hence $ma + nb \in N$. This shows that N is a submodule of M. Moreover, if m and n are elements of M which are equivalent with respect to $\sim$, then the relations $m \sim n$ and $(-n) \sim (-n)$ imply that $m - n \sim 0$, that is, $m - n \in N$. Conversely, if $m, n \in M$ are such that $m - n \in N$, then $m - n \sim 0$, and using that $n \sim n$, we obtain $m \sim n$.

Conversely, let N be a submodule of M and let $\sim$ be the relation on M defined by "$m \sim n$ if and only if $m - n \in N$". This is an equivalence relation on M. Indeed, since $0 \in N$, $m \sim m$ for any $m \in M$; if $m \sim n$, then $m - n \in N$, hence $n - m = -(m - n) \in N$ and $n \sim m$; finally, if $m \sim n$ and $n \sim p$, then $m - n$ and $n - p$ belong to N, so that $m - p = (m - n) + (n - p) \in N$ and $m \sim p$. Moreover, this equivalence relation is compatible with its module structure: if $m \sim m'$, $n \sim n'$ and $a, b \in A$, then

$$(m'a + n'b) - (ma + nb) = (m - m')a + (n - n')b \in N,$$

since $m - m' \in N$ and $n - n' \in N$, hence $m'a + n'b \sim ma + nb$.

Let M/N be the set of all equivalence classes on M for this relation, and let $\mathrm{cl}_N : M \to M/N$ be the canonical projection. (If no confusion can arise, this map shall only be written cl.) From the above calculation, we get the following theorem:

Theorem (3.4.1). — *Let A be a ring, let M be an A-module and let N be a submodule of M. The relation $\sim$ on M given by $m \sim n$ if and only if $m - n \in N$ is an equivalence relation on M which is compatible with its module structure. The quotient set M/N has the unique structure of an A-module for which the map $\mathrm{cl}_N : M \to M/N$ is a morphism of A-modules.*

The map cl_N is surjective; its kernel is N.

We now prove a *factorization theorem*, the universal property of quotient modules.

Theorem (3.4.2). — *Let A be a ring, let M be an A-module and let N be a submodule of M. For any A-module P and any morphism $f : M \to P$ such that $N \subset \mathrm{Ker}(f)$, there exists a unique morphism $\varphi : M/N \to P$ such that $f = \varphi \circ \mathrm{cl}_N$.*

Moreover, $\mathrm{Im}(\varphi) = \mathrm{Im}(f)$ and $\mathrm{Ker}(\varphi) = \mathrm{cl}_N(\mathrm{Ker}(f))$. In particular, φ is injective if and only if $\mathrm{Ker}(f) = N$, and φ is surjective if and only if f is surjective.

One can represent graphically the equality $f = \tilde{f} \circ \mathrm{cl}$ of the theorem by saying that the diagram

$$M \xrightarrow{\;p\;} P$$

$$\text{cl}_N \downarrow \quad \nearrow \varphi$$

$$M/N$$

is commutative. This theorem allows us to factor any morphism $f \colon M \to N$ of A-modules as a composition

$$M \xrightarrow{\text{cl}_N} M/\text{Ker}(f) \xrightarrow{\varphi} \text{Im}(f) \hookrightarrow N$$

of a surjective morphism, an isomorphism, and an injective morphism.

Proof. — Necessarily, $\varphi(\text{cl}(m)) = f(m)$ for any $m \in M$. Since every element of M/N is of the form $\text{cl}(m)$ for a certain $m \in M$, this shows that there is at most one morphism $\varphi \colon M/N \to P$ such that $\varphi \circ \text{cl} = f$.

Let us show its existence. Let $x \in M/N$ and let $m, m' \in M$ be two elements such that $x = \text{cl}(m) = \text{cl}(m')$. Then $m - m' \in N$, hence $f(m - m') = 0$ since $N \subset \text{Ker}(f)$. Consequently, $f(m) = f(m')$ and one may define φ by setting $\varphi(x) = f(m)$, where m is any element of M such that $\text{cl}(m) = x$, the result does not depend on the chosen element m.

It remains to show that φ is a morphism. So let $x, y \in M/N$ and let $a, b \in A$. Let us choose $m, n \in M$ such that $x = \text{cl}(m)$ and $y = \text{cl}(n)$. Then $xa + yb = \text{cl}(m)a + \text{cl}(n)b = \text{cl}(ma + nb)$ hence

$$\varphi(xa + yb) = \varphi(\text{cl}(ma + nb)) = f(ma + nb)$$
$$= f(m)a + f(n)b = \varphi(x)a + \varphi(y)b.$$

Hence, φ is linear.

It is obvious that $\text{Im}(f) \subset \text{Im}(\varphi)$. On the other hand, if $p \in \text{Im}(\varphi)$, let us choose $x \in M/N$ such that $p = \varphi(x)$, and $m \in M$ such that $x = \text{cl}(m)$. Then, $p = \varphi(\text{cl}(m)) = f(m) \in \text{Im}(f)$, so that $\text{Im}(\varphi) \subset \text{Im}(f)$, hence the equality.

Finally, let $x \in M/N$ be such that $\varphi(x) = 0$; write $x = \text{cl}(m)$ for some $m \in M$. Since $f(m) = \varphi \circ \text{cl}(m) = \varphi(x) = 0$, we have $m \in \text{Ker}(f)$. Conversely, if $x = \text{cl}(m)$ for some $m \in \text{Ker}(f)$, then $\varphi(x) = \varphi(\text{cl}(m)) = f(m) = 0$, hence $x \in \text{Ker}\,\varphi$. This shows that $\text{Ker}(\varphi) = \text{cl}(\text{Ker}(f))$.   □

Example (3.4.3). — Let A be a ring and let $f \colon M \to N$ be a morphism of A-modules. The A-module $N/f(M)$ is called the *cokernel* of f and is denoted by $\text{Coker}(f)$.

One has $\text{Coker}(f) = 0$ if and only f is surjective. Consequently, a morphism of A-modules $f \colon M \to N$ is an isomorphism if and only if $\text{Ker}(f) = 0$ and $\text{Coker}(f) = 0$.

The next proposition describes the submodules of a quotient module M/N.

Proposition (3.4.4). — *Let* A *be a ring, let* M *be an* A-*module, and let* N *be a submodule of* M. *The map* cl^{-1} :

$$\text{submodules of } M/N \to \text{submodules of } M \text{ containing } N$$
$$\mathcal{P} \mapsto \text{cl}_N^{-1}(\mathcal{P})$$

is a bijection.

So, for any submodule P of M such that $N \subset P$, there is exactly one submodule $\mathcal{P}$ of M/N such that $P = \text{cl}_N^{-1}(\mathcal{P})$. Moreover, one has $\mathcal{P} = \text{cl}_N(P)$. *The submodule* $\text{cl}_N(P)$ *of* M/N *will be written* P/N. This is a coherent notation. Indeed, the restriction of cl to the submodule P is a morphism $\text{cl}_N|_P : P \to M/N$ with kernel $P \cap N = N$ and image $\text{cl}_N(P)$. By the factorization theorem, $\text{cl}_N|_P$ induces an isomorphism $P/N \to \text{cl}_N(P)$.

Proof. — The proof is an immediate consequence of the two following formulas: if P is a submodule of M, then

$$\boxed{\text{cl}_N^{-1}(\text{cl}_N(P)) = P + N}$$

while if $\mathcal{P}$ is a submodule of M/N,

$$\boxed{\text{cl}_N(\text{cl}_N^{-1}(\mathcal{P})) = \mathcal{P}.}$$

Indeed, if $P \subset N$, $P + N = P$ and these formulas precisely show that the map cl_N^{-1} in the statement of the proposition is bijective, with reciprocal map cl_N.

Let us show the first formula. Let $m \in \text{cl}_N^{-1}(\text{cl}(P))$; we have $\text{cl}_N(m) \in \text{cl}_N(P)$, hence there is a $p \in P$ such that $\text{cl}_N(m) = \text{cl}_N(p)$ and it follows that $\text{cl}_N(m - p) = 0$, hence $n = m - p \in N$; then, $m = p + n$ belongs to $P + N$. Conversely, if $m \in P + N$, write $m = p + n$ for some $p \in P$ and some $n \in N$; then, $\text{cl}_N(m) = \text{cl}_N(p + n) = \text{cl}_N(p)$ belongs to $\text{cl}_N(P)$, hence $m \in \text{cl}_N^{-1}(\text{cl}_N(P))$.

Let us now show the second formula. By definition, one has $\text{cl}_N(\text{cl}_N^{-1}(\mathcal{P})) \subset \mathcal{P}$. Conversely, if $x \in \mathcal{P}$, let $m \in M$ be such $x = \text{cl}_N(m)$. Then, $\text{cl}_N(m) \in \mathcal{P}$ so that $m \in \text{cl}_N^{-1}(\mathcal{P})$, hence $x = \text{cl}_N(m) \in \text{cl}_N(\text{cl}_N^{-1}(\mathcal{P}))$. □

We now compute quotients of quotients.

Proposition (3.4.5). — *Let A be a ring, let M be an A-module and let N, P be two submodules of M such that $N \subset P$. Then, there is a unique map*

$$(M/P) \to (M/N)/(P/N)$$

such that $\text{cl}_P(m)$ *maps to* $\text{cl}_{P/N}(\text{cl}_N(m))$ *for every* $m \in M$, *and this map is an isomorphism.*

Proof. — The uniqueness of such a map follows from the fact that every element of M/P is of the form $\text{cl}_P(m)$, for some $m \in M$. Let f be the morphism given by

$$f : M \to (M/N) \to (M/N)/(P/N), \qquad m \mapsto \text{cl}_{P/N}(\text{cl}_N(m)).$$

It is surjective, as the composition of two surjective morphisms. An element $m \in M$ belongs to $\mathrm{Ker}(f)$ if and only if $\mathrm{cl}_N(m) \in \mathrm{Ker}\,\mathrm{cl}_{P/N} = P/N = \mathrm{cl}_N(P)$; since P contains N, this is equivalent to $m \in P$. Consequently, $\mathrm{Ker}(f) = P$ and the factorization theorem (theorem 3.4.2) asserts that there is a unique morphism $\varphi \colon M/P \to (M/N)/(P/N)$ such that $\varphi(\mathrm{cl}_P(m)) = \mathrm{cl}_{P/N}(\mathrm{cl}_N(m))$ for every $m \in M$. Since $\mathrm{Ker}(f) = P$, the morphism φ is injective; since f is surjective, the morphism φ is surjective. Consequently, φ is an isomorphism.   □

Remark (3.4.6). — *Let M be an A-module and let N be a submodule of M. Assume that N has a direct summand P in M. The morphism f from P to M/N given by the composition of the injection of P into M and of the surjection of M onto M/N is an isomorphism.*

Indeed, let $m \in P$ be such that $f(m) = 0$. Then, $m \in N$ hence $m \in N \cap P = 0$; this shows that $\mathrm{Ker}(f) = 0$ so that f is injective. Observe now that f is surjective: for $m \in M$, let $n \in N$ and $p \in P$ be such that $m = n + p$; then $\mathrm{cl}_N(m) = \mathrm{cl}_N(n) + \mathrm{cl}_N(p) = f(p)$, hence $\mathrm{Im}\, f = \mathrm{cl}_N(M) = M/N$.

3.5. Generating Sets, Free Sets; Bases

Definition (3.5.1). — Let A be a ring and let M be a right A-module.
One says that a family $(m_i)_{i \in I}$ of elements of M is:

– *generating* if the submodule of M generated by the m_i is equal to M;

– *free* if for every almost-null family $(a_i)_{i \in I}$ of elements of A, the relation $\sum_{i \in I} m_i a_i = 0$ implies $a_i = 0$ for every $i \in I$;

– *bonded* if it is not free;

– a *basis* of M if it is free and generating.

On defines similar notions of *generating subset, free subset, bonded subset,* and *basis,* as subsets S of M for which the family $(s)_{s \in S}$ is respectively generating, free, bonded, and a basis.

When the ring is non-null, the members of a free family are pairwise distinct (otherwise, the linear combination $x - y = 0$, for $x = y$, contradicts the definition of a free family since $1 \neq 0$). Consequently, with regards to these properties, the only interesting families $(m_i)_{i \in I}$ are those for which the map $i \mapsto m_i$ is injective and the notions for families and subsets correspond quite faithfully one to another.

A subfamily (subset) of a free family (subset) is again free; a family (set) possessing a generating subfamily (subset) is generating.

Proposition (3.5.2). — *Let A be a ring and let M be a right A-module. Let $(m_i)_{i \in I}$ be a family of elements of M and let φ be the morphism given by*

$$\varphi \colon A_d^{(I)} \to M, \qquad (a_i)_{i \in I} \mapsto \sum_{i \in I} m_i a_i.$$

Then,

- *the morphism φ is injective if and only if the family $(m_i)_{i \in I}$ is free;*
- *the morphism φ is surjective if and only if the family $(m_i)_{i \in I}$ is generating;*
- *the morphism φ is an isomorphism if and only if the family $(m_i)_{i \in I}$ is a basis.*

Proof. — The kernel of φ is the set of all almost-null families (a_i) such that $\sum_{i \in I} m_i a_i = 0$. So $(m_i)_{i \in I}$ is free if and only if $\mathrm{Ker}(\varphi) = 0$, that is, if and only if φ is injective.

The image of φ is the set of all linear combinations of terms of (m_i). It follows that $\mathrm{Im}(\varphi) = \langle \{m_i\} \rangle$, so that φ is surjective if and only if $(m_i)_{i \in I}$ is generating.

By what precedes, (m_i) is a basis if and only if φ is bijective, that is, an isomorphism. □

Example (3.5.3). — Let I be a set and let $M = A_d^{(I)}$. For every $i \in I$, let $e_i \in M$ be the family $(\delta_{i,j})_{j \in I}$ whose only non-zero term is that of index i, equal to 1. For every almost-null family $(a_i)_{i \in I}$, one has $(a_i)_{i \in I} = \sum_{i \in I} e_i a_i$. Consequently, the family $(e_i)_{i \in I}$ is a basis of $A_d^{(I)}$, called its *canonical basis*. The module $A_d^{(I)}$ is often called the *free right A-module on* I.

Definition (3.5.4). — One says that a module is *free* if it has a basis. If a module has a finite generating subset, it is said to be *finitely generated*.

Proposition (3.5.5). — *Let M be an A-module and let N be a submodule of M.*

a) *If M is finitely generated, then M/N is finitely generated;*

b) *If N and M/N are finitely generated, then M is finitely generated;*

c) *If N and M/N are free A-modules, then M is a free A-module.*

However, with this notation, it may happen that M is finitely generated but that N is not, as it may happen that M is free, but not N or M/N.

More precisely, we shall prove the following statements:

a′) *If M has a generating subset of cardinality r, then so does M/N;*

b′) *If N and M/N have generating subsets of cardinalities respectively r and s, then M has a generating subset of cardinality $r + s$;*

c′) *If N and M/N have bases of cardinalities respectively r and s, then M has a basis of cardinality $r + s$.*

Proof. — a′) The images in M/N of a generating subset of M generate M/N, since the canonical morphism from M to M/N is surjective. In particular, M/N is finitely generated if M is.

b′) Let $(n_1, \ldots, n_r)$ be a generating family of N, and let $(m_1, \ldots, m_s)$ be a family of elements of M such that $(\mathrm{cl}_N(m_1), \ldots, \mathrm{cl}_N(m_s))$ generate M/N. Let us show that the family $(m_1, \ldots, m_s, n_1, \ldots, n_r)$ generates M.

Let $m \in M$. By hypothesis, its image $\mathrm{cl}_N(m)$ in M/N is a linear combination of $\mathrm{cl}_N(m_1), \ldots, \mathrm{cl}_N(m_s)$, hence there exist elements $a_i \in A$ such that $\mathrm{cl}_N(m) =$

$\sum_{i=1}^{s} \mathrm{cl_N}(m_i)a_i$. Consequently, $n = m - \sum_{i=1}^{s} m_i a_i$ belongs to N and there exist elements $b_j \in A$ such that $n = \sum_{j=1}^{r} n_j b_j$. Then $m = \sum_{i=1}^{s} m_i a_i + \sum_{j=1}^{r} n_j b_j$ is a linear combination of the m_i and of the n_j.

c') Let us moreover assume that $(n_1, \ldots, n_r)$ is a basis of N and that $(\mathrm{cl_N}(m_1), \ldots, \mathrm{cl_N}(m_s))$ is a basis of M/N and let us show that the compound family $(m_1, \ldots, m_s, n_1, \ldots, n_r)$ is a basis M. Since we already proved that this family generates M, it remains to show that it is free. So let $0 = \sum_{i=1}^{s} m_i a_i + \sum_{j=1}^{r} n_j b_j$ be a linear dependence relation between these elements. Applying $\mathrm{cl_N}$, we get a linear dependence relation $0 = \sum_{i=1}^{s} \mathrm{cl_N}(m_i)a_i$ for the family $\mathrm{cl_N}(m_i)$. Since this family is free, one has $a_i = 0$ for every i. It follows that $0 = \sum_{j=1}^{r} n_j b_j$; since the family $(n_1, \ldots, n_r)$ is free, $b_j = 0$ for every j. The considered linear dependence relation is thus trivial, as was to be shown.   □

Corollary (3.5.6). — *Let M be an A-module and let $M_1, \ldots, M_n$ be submodules of M which are finitely generated. Their sum $\sum_{i=1}^{n} M_i$ is finitely generated.*

In particular, the direct sum of a finite family of finitely generated modules is finitely generated.

Proof. — By induction, it suffices to treat the case of two modules, that is, $n = 2$. The second projection $\mathrm{pr}_2 : M_1 \times M_2 \to M_2$ is a surjective linear map, and its kernel, equal to $M_1 \times \{0\}$, is isomorphic to M_1, hence is finitely generated. Since M_1 and M_2 are finitely generated, their direct product $M_1 \times M_2$ is finitely generated. The morphism $(m_1, m_2) \mapsto m_1 + m_2$ from $M_1 \times M_2$ to M has image $M_1 + M_2$; consequently, $M_1 + M_2$ is finitely generated.   □

Remark (3.5.7). — When the ring A is commutative (and non-zero), or when A is a division ring, we shall prove in the next section that all bases of a finitely generated free module have the same cardinality. This is not true in the general case. However, one can prove that if a module M over a non-zero ring is finitely generated and free, then all bases of M are finite (see exercise 3.10.12).

The case of a zero ring $A = 0$ is pathological: there is no other A-module than the zero module $M = 0$, and every family $(m_i)_{i \in I}$ is a basis...

Proposition (3.5.8). — *Let A be a ring, let M be a free A-module and let $(e_i)_{i \in I}$ be a basis of M. For any A-module N and any family $(n_i)_{i \in I}$ of elements of N, there is a unique morphism $\varphi : M \to N$ such that $\varphi(e_i) = n_i$ for every $i \in I$.*

This is the *universal property* satisfied by free modules (once endowed with a basis); it applies in particular to the free module $A^{(I)}$ on a set I, with its canonical basis.

Proof. — The map $A^{(I)} \to M$ given by $(a_i) \mapsto \sum e_i a_i$ is an isomorphism of modules, so that M inherits the universal property of direct sums of modules. To give a linear map from M to N is equivalent to giving the images of the elements of a basis. The morphism φ such that $\varphi(e_i) = n_i$ for every i is given by $\varphi(\sum e_i a_i) = \sum u(e_i)a_i$.   □

Corollary (3.5.9). — *Let A be a ring, let M be a free A-module and let $(e_i)_{i \in I}$ be a basis of M. There exists, for every $i \in I$, a unique linear form φ_i on M such that $\varphi_i(e_i) = 1$ and $\varphi_i(e_j) = 0$ for $j \neq i$. For any element $m = \sum_{i \in I} e_i a_i$ of M, one has $\varphi_i(m) = a_i$.*

The family $(\varphi_i)_{i \in I}$ is free.

If I is finite, then it is a basis of the dual module $M^{\vee}$.

Proof. — Existence and uniqueness of the linear forms φ_i follow from the proposition.

Let $(a_i)_{i \in I}$ be an almost-null family such that $\sum a_i \varphi_i = 0$. Apply this relation to e_i; one gets $a_i = 0$. This proves that the family (φ_i) is free.

Assume that I is finite and let us show that (φ_i) is a generating family in $M^{\vee}$. Let φ be a linear form on M and let $m = \sum_i e_i a_i$ be an element of M. One has $\varphi(m) = \sum_i \varphi(e_i) a_i$, hence $\varphi(m) = \sum_i \varphi(e_i) \varphi_i(m)$. This proves that $\varphi = \sum_{i \in I} \varphi(e_i) \varphi_i$. $\quad\square$

When I is finite, the family $(\varphi_i)_{i \in I}$ defined above is called the *dual basis* of the basis $(e_i)_{i \in I}$.

3.5.10. Matrices — One of the features of free modules is that they allow for matrix computations. Let $\Phi = (a_{i,j}) \in \mathrm{Mat}_{m,n}(A)$ be a matrix with m rows and n columns and entries in A. One associates to Φ a map $\varphi \colon A^n \to A^m$ given by the formula

$$\varphi(x_1, \ldots, x_n) = \left(\sum_{j=1}^{n} a_{i,j} x_j \right)_{1 \leq i \leq m}.$$

This is a morphism of right A-modules from $(A_d)^n$ to $(A_d)^m$. Indeed, for every $x = (x_1, \ldots, x_n) \in A^n$, every $y = (y_1, \ldots, y_n) \in A^n$, and any $a, b \in A$, one has

$$\varphi(xa + yb) = \varphi((x_1 a + y_1 b, x_2 a + y_2 b, \ldots, x_n a + y_n b))$$
$$= \sum_{j=1}^{n} a_{i,j}(x_j a + y_j b)$$
$$= \left(\sum_{j=1}^{n} a_{i,j} x_j \right) a + + \left(\sum_{j=1}^{n} a_{i,j} y_j \right) b$$
$$= \varphi(x) a + \varphi(y) b.$$

Conversely, any morphism $\varphi \colon A_d^n \to A_d^m$ of right A-modules has this form.

Indeed, let $(e_1, \ldots, e_n)$ be the canonical basis of $(A_d)^n$ (recall that $e_j \in A^n$ has all of its coordinates equal to 0, except for the jth coordinate, which is equal to 1). We represent any element of $(A_d)^n$ as a column matrix, *i.e.*, with one column and n rows. Let $\Phi = (a_{i,j})$ be the matrix with m rows and n columns whose jth column is equal to $\varphi(e_j)$. In other words, we have

$\varphi(e_j) = (a_{1,j}, \ldots, a_{m,j})$ for $1 \leq j \leq n$. For any $x = (x_1, \ldots, x_n) \in A^n$, one has $x = \sum_{j=1}^{n} e_j x_j$, hence

$$\varphi(x) = \sum_{j=1}^{n} \varphi(e_j) x_j = \left(\sum_{j=1}^{n} a_{i,j} x_j \right) = \Phi X,$$

if X is the column matrix $(x_1, \ldots, x_n)$. This shows that φ is given by the $n \times m$ matrix $\Phi = (a_{i,j})$.

The identity matrix I_n is the $n \times n$ matrix representing the identity endomorphism of A^n; its diagonal entries are equal to 1, all the other entries are 0.

Let $\Phi' = (b_{j,k}) \in \mathrm{Mat}_{n,p}$ be a matrix with n rows and p columns, let $\varphi' \colon (A_d)^p \to (A_d)^n$ be the morphism it represents. The product matrix $\Phi'' = \Phi \Phi'$ has m rows and p columns and its coefficient $c_{i,k}$ with index (i, k) is given by the formula

$$c_{i,k} = \sum_{j=1}^{n} a_{i,j} b_{j,k}.$$

It represents the morphism $\varphi'' = \varphi \circ \varphi' \colon (A_d)^p \to (A_d)^m$. Indeed one has by definition

$$\varphi''(x_1, \ldots, x_p) = \left(\sum_k c_{i,k} x_k \right)_{1 \leq i \leq m} = \left(\sum_{j,k} a_{i,j} b_{j,k} x_k \right)_i = \left(\sum_j a_{i,j} y_j \right)_i,$$

where $(y_1, \ldots, y_n) = \varphi'(x_1, \ldots, x_p)$. Consequently,

$$\varphi''(x_1, \ldots, x_p) = \varphi(y_1, \ldots, y_n) = \varphi(\varphi'(x_1, \ldots, x_p)),$$

so that $\varphi'' = \varphi \circ \varphi'$.

This shows in particular that the map $\mathrm{Mat}_n(A) \to \mathrm{End}_A(A_d^n)$ that sends a square $n \times n$ matrix to the corresponding endomorphism of $(A_d)^n$ is an isomorphism of rings.

Here is the very reason why we preferred right A-modules!

3.6. Localization of Modules (Commutative Rings)

Let A be a commutative ring and let S be a multiplicative subset of A. Let M be an A-module. Through the calculus of fractions we constructed a localized ring $S^{-1}A$ as well as a morphism of rings $A \to S^{-1}A$. We are now going to construct by a similar process an $S^{-1}A$-module $S^{-1}M$ together with a *localization morphism* of A-modules $M \to S^{-1}M$.

Define a relation $\sim$ on the set $M \times S$ by the formula

$$(m, s) \sim (n, t) \quad \Leftrightarrow \quad \text{there exists a } u \in S \text{ such that } u(tm - sn) = 0.$$

Let us check that it is an equivalence relation. In fact, the argument runs exactly as for the case of rings (p. 34). The equality $1 \cdot (sn - sn) = 0$ shows that this relation is reflexive; if $u(tm - sn) = 0$, then $u(sn - tm) = 0$, which shows that it is symmetric. Finally, let $m, n, p \in M$ and $s, t, u \in S$ be such that $(m, s) \sim (n, t)$ and $(n, t) \sim (p, u)$; let $v, w \in S$ be such $v(tm - sn) = 0$ and $w(tp - un) = 0$; then

$$tvw(um - sp) = uw\, v(tm - sn) + svw\, (un - tp) = 0$$

so that $(m, s) \sim (p, u)$, since $uw \in S$ and $sv \in S$: this relation is transitive.

Let $S^{-1}M$ be the set of equivalence classes; for $m \in M$ and $s \in S$, write m/s for the class in $S^{-1}M$ of the pair $(m, s) \in S^{-1}M$. We then define two laws on $S^{-1}M$: first, an addition, given by

$$(m/s) + (n/t) = (tm + sn)/(st)$$

for $m, n \in M$ and $s, t \in S$, and then an external multiplication, defined by

$$(a/t)(m/s) = (am)/(ts)$$

for $m \in M$, $a \in A$, s and $t \in S$,

Theorem (3.6.1). — *Endowed with these two laws, $S^{-1}M$ is an $S^{-1}A$-module. If one views $S^{-1}M$ as an A-module through the canonical morphism of rings $A \to S^{-1}A$, then the map $i: M \to S^{-1}M$ given by $i(m) = m/1$ is a morphism of A-modules.*

The proof is left as an *exercise*; anyway, the computations are completely similar to those done for localization of rings.

Proposition (3.6.2). — *Let A be a commutative ring, let S be a multiplicative subset of A and let M be an A-module. Let $i: M \to S^{-1}M$ be the canonical morphism; for $s \in S$, let $\mu_s: M \to M$ be the morphism given by $m \mapsto sm$.*

(i) *An element $m \in M$ belongs to $\mathrm{Ker}(i)$ if and only if there exists an $s \in S$ such that $sm = 0$;*

(ii) *In particular, i is injective if and only if μ_s is injective, for every $s \in S$;*

(iii) *The morphism i is an isomorphism if and only if μ_s is bijective, for every $s \in S$.*

Proof. — Let $m \in M$. If $i(m) = 0$, then $m/1 = 0/1$, hence there exists an $s \in S$ such that $sm = 0$, by the definition of $S^{-1}M$. Conversely, if $sm = 0$, then $m/1 = 0/1$, hence $i(m) = 0$. This proves $a)$, and assertion $b)$ is a direct consequence.

Let us prove $c)$. Let us assume that i is an isomorphism. All morphisms μ_s are then injective by $b)$. Moreover, in the $S^{-1}A$-module $S^{-1}M$, multiplication by an element $s \in S$ is surjective, since $s \cdot (m/st) = m/t$ for every $m \in M$ and $t \in S$. Therefore, μ_s is surjective as well, so that it is an isomorphism. Conversely, let us assume that all morphisms $\mu_s: M \to M$ are isomorphisms. Then i is injective, by $b)$. Let moreover $m \in M$ and $s \in S$; let $m' \in M$ be such that $sm' = m$; then $m/s = m'/1 = i(m')$; this proves that i is surjective. $\square$

Remark (3.6.3). — Let us recall a few examples of multiplicative subsets.

(i) First of all, for any $s \in A$, the set $S = \{1; s; s^2; \dots\}$ is multiplicative. In that case, the localized module $S^{-1}M$ is notated with an index s, so that $M_s = S^{-1}M$ is an A_s-module.

(ii) If P is a prime ideal of A, then $S = A - P$ is also a multiplicative subset of A. Again, the localized module and ring are denoted via an index P; that is, $A_P = (A - P)^{-1}A$ and $M_P = (A - P)^{-1}M$.

Proposition (3.6.4). — *Let A be a commutative ring and let S be a multiplicative subset of A. Let $f : M \to N$ be a morphism of A-modules. There exists a unique morphism of $S^{-1}A$-modules $\varphi : S^{-1}M \to S^{-1}N$ such that $\varphi(m/1) = f(m)/1$ for every $m \in M$. For every $m \in M$, and every $s \in S$, one has $\varphi(m/s) = f(m)/s$.*

In other words, the diagram

$$
\begin{array}{ccc}
M & \xrightarrow{\ f\ } & N \\
\downarrow{\scriptstyle i} & & \downarrow{\scriptstyle i} \\
S^{-1}M & \xrightarrow{\ \varphi\ } & S^{-1}N
\end{array}
$$

is commutative.

Proof. — To check uniqueness, observe that $m/s = (1/s)(m/1)$, so that if such a morphism φ exists, then $\varphi(m/s)$ must be equal to $(1/s)\varphi(m/1) = (1/s)(f(m)/1) = f(m)/s$. To establish the existence of φ, we want to use the formula $\varphi(m/s) = f(m)/s$ as a definition, but we must verify that it makes sense, that is, we must prove that if $m/s = n/t$, for some $m, n \in M$ and $s, t \in S$, then $f(m)/s = f(n)/t$.

So assume that $m/s = n/t$; by the definition of the equivalence relation, there exists an element $u \in S$ such that $u(tm - sn) = 0$. Then,

$$
\frac{f(m)}{s} = \frac{utf(m)}{uts} = \frac{f(utm)}{uts} = \frac{f(usn)}{uts} = \frac{f(n)}{t},
$$

which shows that φ is well defined.

Now, for $m, n \in M$, $s, t \in S$, we have

$$
\varphi\left(\frac{m}{s} + \frac{n}{t}\right) = \varphi\left(\frac{tm + sn}{st}\right) = \frac{f(tm + sn)}{st}
$$
$$
= \frac{tf(m)}{st} + \frac{sf(n)}{st} = \frac{f(m)}{s} + \frac{f(n)}{t} = \varphi\left(\frac{m}{s}\right) + \varphi\left(\frac{n}{t}\right),
$$

hence φ is additive. Finally, for $m \in M$, $a \in A$, s and $t \in S$, we have

$$
\varphi\left(\frac{a}{s}\frac{m}{t}\right) = \varphi\left(\frac{am}{st}\right) = \frac{f(am)}{st} = \frac{af(m)}{st} = \frac{a}{s}\frac{f(m)}{t} = \frac{a}{s}\varphi\left(\frac{m}{t}\right),
$$

which shows that φ is $S^{-1}A$-linear. $\qquad\square$

Localization of modules gives rise to a universal property, analogous to the one we established for localization of rings.

Corollary (3.6.5). — *Let $f: M \to N$ be a morphism of A-modules; assume that for every $s \in S$, the morphism $\mu_s: N \to N$ defined by $n \mapsto sn$ is an isomorphism. Then, there exists a unique morphism of A-modules $\tilde{f}: S^{-1}M \to N$ such that $\tilde{f}(m/1) = f(m)/1$ for every $m \in M$.*

Proof. — In fact, if $\varphi: S^{-1}M \to S^{-1}N$, $m/s \mapsto f(m)/s$, is the morphism constructed in the previous proposition, and $i: N \to S^{-1}N$ is the canonical morphism, then the desired property for φ is equivalent to the equality $i \circ \tilde{f} = \varphi$. By proposition 3.6.2, i is an isomorphism of A-modules; this implies the existence and uniqueness of an A-linear map $\tilde{f}: S^{-1}M \to N$ such that $\tilde{f}(m/1) = f(m)/1$ for every $m \in M$.   □

As we already observed for localization of rings and ideals, localization behaves nicely with respect to submodules; this is the second appearance of *exactness of localization*.

Proposition (3.6.6). — *Let A be a commutative ring, let S be a multiplicative subset of A. Let M be an A-module and let N be a submodule of M.*

Then, the canonical morphism $\tilde{\imath}: S^{-1}N \to S^{-1}M$ deduced from the injection $i: N \to M$ induces an isomorphism from $S^{-1}N$ to a submodule of $S^{-1}M$. Moreover, the canonical morphism $\tilde{p}: S^{-1}M \to S^{-1}(M/N)$ deduced from the surjection $p: M \to M/N$ is surjective and its kernel is $\tilde{\imath}(S^{-1}N)$.

In practice, the morphism $\tilde{\imath}$ is removed from the notation, and we still denote by $S^{-1}N$ the submodule of $S^{-1}M$ image of this morphism. Passing to the quotient, we then obtain from $\tilde{p}$ an isomorphism

$$S^{-1}M/S^{-1}N \xrightarrow{\sim} S^{-1}(M/N).$$

Proof. — Let $n \in N$ and let $s \in S$. The image in $S^{-1}M$ of the element $n/s \in S^{-1}N$ is still n/s, but n is now seen as an element of M. It vanishes if and only if there exists an element $t \in S$ such that $tn = 0$ in M. But then $tn = 0$ in N so that $n/s = 0$ in $S^{-1}N$. This shows that the map $S^{-1}N \to S^{-1}M$ is injective. It is thus an isomorphism from $S^{-1}N$ to its image in $S^{-1}M$.

Since p is surjective, the morphism $\tilde{p}$ is surjective too. Indeed, any element of $S^{-1}(M/N)$ can be written $\mathrm{cl}_N(m)/s$, for some $m \in M$ and some $s \in S$. Then, $\mathrm{cl}_N(m)/s$ is the image of $m/s \in S^{-1}M$. Obviously, $S^{-1}N$ is contained in the kernel of $\tilde{p}$; indeed, $\tilde{p}(\mathrm{cl}_N(n)/s) = \mathrm{cl}_N(p(n))/s = 0$. On the other hand, let $m \in M$ and let $s \in S$ be such that $m/s \in \mathrm{Ker}(\tilde{p})$; then $0 = \mathrm{cl}_N(m)/s$, so that there exists an element $t \in S$ such that $t\,\mathrm{cl}_N(m) = 0$ in M/N. This implies that $tm \in N$. Then, $m/s = tm/ts = \tilde{\imath}(tm/ts)$ belongs to $\tilde{\imath}(S^{-1}N)$. This concludes the proof that $\mathrm{Ker}(\tilde{p}) = \tilde{\imath}(S^{-1}N)$.   □

Proposition (3.6.7). — *Let A be a commutative ring, let S be a multiplicative subset of A and let M be an A-module. Let $i: M \to S^{-1}M$ be the canonical morphism.*

Let $\mathcal{N}$ be an $S^{-1}A$-submodule of $S^{-1}M$. Then $N = i^{-1}(\mathcal{N})$ is an A-submodule of M such that $\mathcal{N} = S^{-1}N$; it is the largest of all such submodules.

Proof. — First of all, $S^{-1}N \subset \mathcal{N}$. Indeed, if $m \in N$, then $m/1 \in \mathcal{N}$ by definition of N; it follows that $m/s \in \mathcal{N}$ for every $s \in S$. Conversely, let $x \in \mathcal{N}$. One may write $x = m/s$ for some $m \in M$ and $s \in S$. This implies that $sx = m/1$ belongs to $\mathcal{N}$, hence $m \in N$ and $x \in S^{-1}N$. We thus have shown that $\mathcal{N} = S^{-1}N$.

Finally, let P be a submodule of M such that $S^{-1}P = \mathcal{N}$. Let $m \in P$; one has $m/s \in \mathcal{N}$, hence $s(m/s) = m/1 \in \mathcal{N}$. Then, $m \in N$, so that $P \subset N$. This shows that N is the largest submodule of M such that $S^{-1}N = \mathcal{N}$.   □

Proposition (3.6.8). — *Let* A *be a commutative ring, let* S *be a multiplicative subset of* A. *Let* M *be an* A-*module and let* (N_i) *be a family of submodules of* M. *Then, one has the following equalities of submodules of* $S^{-1}M$:

$$\sum_i S^{-1}N_i = S^{-1} \sum_i N_i.$$

Proof. — Let $N = \sum N_i$. For every i, $N_i \subset N$, hence an inclusion $S^{-1}N_i \subset S^{-1}N$. It follows that $\sum_i S^{-1}N_i \subset S^{-1}N$. Conversely, let $n/s \in S^{-1}N$. There exists an almost null family (n_i), where $n_i \in N_i$ for every i, such that $n = \sum n_i$. Then, n_i/s belongs to $S^{-1}N_i$ and $n/s = \sum_i (n_i/s)$ belongs to $\sum S^{-1}N_i$, so that $S^{-1}N \subset \sum S^{-1}N_i$.   □

Corollary (3.6.9). — *If* M *is a finitely generated* A-*module, then* $S^{-1}M$ *is a finitely generated* $S^{-1}A$-*module.*

Proof. — Let $(m_1, \ldots, m_n)$ be a generating family. For $i \in \{1, \ldots, n\}$, let $N_i = Am_i$. By construction, $S^{-1}N_i$ is generated by m_i as an $S^{-1}A$-module. By the proposition, $\sum S^{-1}N_i = S^{-1}M$, hence $S^{-1}M$ is finitely generated.   □

3.7. Vector Spaces

Let us recall that a vector space is a module over a division ring.

Proposition (3.7.1). — *Let* K *be a division ring and let* M *be a right* K-*vector space. Let* F, B, G *be subsets of* M *such that* $F \subset B \subset G$. *We assume that* F *is free and that* G *is generating.*

The following assertions are equivalent:

  (i) *The set* B *is a basis;*
 (ii) *The set* B *is maximal among the free subsets of* G *containing* F;
(iii) *The set* B *is minimal among the generating subsets of* G *containing* F.

Proof. — Let us start with a remark and prove that if G, G' are subsets of M such that $G \subsetneq G'$, and G is generating, then G' is not free. Indeed, let $x \in G' - G$; let us write x as a linear combination of elements of G, $x = \sum_{g \in G} g a_g$, where $(a_g)_{g \in G}$ is an almost-null family. Set $a_x = -1$ and $a_y = 0$ for $y \in G' - G$ such that $y \neq x$. The family $(a_g)_{g \in G'}$ is almost null, but not identically null, and one has $\sum_{g \in G'} g a_g = 0$, so that G' is not free.

We now prove the equivalence of the three given assertions.

Assume that B is a basis of M. Then, B is free and generating. By the initial remark, for every subset G of M such that $B \subset G$ and $B \neq G$, G is not free. In other words, B is maximal among all free subsets of M, which proves (i)$\Rightarrow$(ii).

Let B′ be a subset of M such that $F \subset B' \subsetneq B$; by the initial remark, if B′ is generating, then B is not free. Consequently, B is minimal among the generating subsets of G containing F, hence (i)$\Rightarrow$(iii).

We now assume that B is a free subset of G, maximal among those containing F. Let us show that B is generating. This holds if $G = B$. Otherwise, let $m \in G - B$. The subset $B \cup \{m\}$ is not free, so that there are elements a and $(a_b)_{b \in B}$ of K, all but finitely many of them being zero, but not all of them, such that $\sum_{b \in B} ba_b + ma = 0$. If $a = 0$, this is a non-trivial linear dependence relation among the elements of B, contradicting the hypothesis that B is free. So $a \neq 0$ and $m = -\sum_{b \in B} ba_b a^{-1}$ belongs to the subspace M′ of M generated by B. Consequently, M′ contains G. Since G is generating, $M' = M$, and B generates M. It is thus a basis of M.

It remains to show that a generating subset of G which is minimal among those containing F is a basis. If such a subset B were not free, there would exist a non-trivial linear dependence relation $\sum_{b \in B} ba_b = 0$ among the elements of B. Let $\beta \in B$ be such that $a_\beta \neq 0$; we have $\beta = -\sum_{b \neq \beta} ba_b a_\beta^{-1}$. It follows that β belongs to the vector subspace M′ generated by the elements of $B - \{\beta\}$. Consequently, M′ contains B, hence $M' = M$ since B is generating. In particular, $B - \{b\}$ is generating, contradicting the minimality hypothesis on B. $\qquad\square$

Proposition (3.7.2) (Exchange lemma). — *Let K be a division ring and let M be a right K-vector space. Let $(u_i)_{i \in I}$ and $(v_j)_{j \in J}$ be families of elements of M. One assumes that (u_i) is generating and that (v_j) is free. For every $k \in J$, there exists an $i \in I$ such that the family $(w_j)_{j \in J}$ defined by $w_k = u_i$ and $w_j = v_j$ for $j \neq k$ is free.*

In other words, given a free subset and a generating subset of a vector set, one can replace any chosen element of the free subset by some element of the generating set without losing its property of being free.

Proof. — Set $J' = J - \{k\}$ and let M′ be the subspace of M generated by the family $(v_j)_{j \in J'}$. Let $i \in I$ and assume that we cannot replace v_k by u_i. This means that the family $(w_j)_{j \in J}$ defined by $w_k = u_i$ and $w_j = v_j$ for $j \in J'$ is not free; let $\sum_{j \in J} w_j a_j = 0$ be a nontrivial linear dependence relation. Since the family $(v_j)_{j \in J'}$ is free, one has $a_k \neq 0$. Then we can write

$$u_i = w_k = w_k a_k a_k^{-1} = -\sum_{j \in J'} v_j a_j a_k^{-1}$$

so that $w_k \in M'$. Consequently, if the exchange property does not hold, then all u_i belong to M′. Since the family (u_i) is generating, one has $M' = M$. But then, v_k belongs to M′, hence one may write $v_k = \sum_{j \in J'} v_j b_j$, for some almost-null family $(b_j)_{j \in J'}$. Passing v_k to the right-hand side, this gives a linear dependence relation for the family $(v_j)_{j \in J}$, contradicting the hypothesis that it is free.   □

One of the fundamental results in the theory of vector spaces is the following theorem, presumably well known to the reader in the case of (commutative) fields!

Theorem (3.7.3). — *Let K be a division ring and let M be a right K-vector space.*

(i) *M has a basis.*

(ii) *More precisely, if F is free subset of M and G is a generating subset of M such that F ⊂ G, there exists a basis B of M such that F ⊂ B ⊂ G.*

(iii) *All bases of M have the same cardinality.*

The common cardinality of all bases of M is called the *dimension* of M and is denoted $\dim_K(M)$, or $\dim(M)$ if no confusion on K can arise.

Proof. — Assertion *a*) results from *b*), applied to the free subset $L = \emptyset$ and to the generating subset $G = M$.

Let us prove *b*). Let $\mathcal{F}$ be the set of all free subsets of G.

Let F be a subset of G. If F is free, then every subset of F is free, in particular every finite subset. Conversely, assume that every finite subset of F is free and let us show that F is free. Let then $(a_m)_{m \in F}$ be an almost-null family of elements of K such that $0 = \sum_{m \in F} m a_m$. Let F′ be the set of all $m \in F$ such that $a_m \neq 0$; it is finite by definition of an almost-null family, hence free, by assumption. Since $\sum_{m \in F'} m a_m = 0$, one has $a_m = 0$ for every $m \in F'$, which implies that F′ = 0. Consequently, $(a_m)_{m \in F}$ is null. This proves that F is free.

In other words, we have proved that the set $\mathcal{F}$ is of finite character. By corollary A.2.16 to Zorn's lemma, the set $\mathcal{F}$ admits a maximal element B. By proposition 3.7.1, the set B is a basis of M and F ⊂ B ⊂ G by construction.

c) Let B and B′ be two bases of M. Since B is free and B′ is generating, lemma 3.7.4 asserts the existence of an injection $f : B \hookrightarrow B'$. On the other hand, since B′ is free and B is generating, there is also an injection $g : B' \hookrightarrow B$. By the Cantor–Bernstein theorem (theorem A.2.1), the bases B and B′ are equipotent.   □

Lemma (3.7.4). — *Let K be a division ring and let M be a right K-vector space. Let F and G be subsets of M such that F is free and G is generating. Then there exists an injection from F into G.*

Proof. — Let Φ be the set of all pairs (F′, φ) where F′ is a subset of F and φ is an injective map from F′ to G such that the set $(F - F') \cup \varphi(F')$ is free. We endow this set Φ with the ordering for which $(F'_1, \varphi_1) \prec (F'_2, \varphi_2)$ if and only if $F'_1 \subset F'_2$ and $\varphi_2(m) = \varphi_1(m)$ for every $m \in F'_1$.

This set Φ is non-empty, because the pair $(\emptyset, \varphi)$ belongs to Φ, where φ is the unique map from the empty set to G. Let us show that Φ is inductive. So let $(F'_i, \varphi_i)_i$ be a totally ordered family of elements of Φ, let F′ be the union of all subsets F'_i and let φ be the map from F′ to G which coincides with φ_i

on F'_i. The map φ is well-defined: if an element m of F' belongs both to F'_i and F'_j, we may assume that $(F'_i, \varphi_i) \prec (F'_j, \varphi_j)$, and then $\varphi_j(m) = \varphi_i(m)$.

Let us show that (F', φ) belongs to Φ. The map φ is injective: let m, m' be distinct elements of F'; since the family (F'_i, φ_i) is totally ordered, there exists an index i such that m and m' both belong to F'_i; since φ_i is injective, $\varphi_i(m) \neq \varphi_i(m')$; then $\varphi(m) \neq \varphi(m')$.

For $m \in F - F'$, set $\varphi(m) = m$, and let $(a_m)_{m \in F}$ be an almost-null family of elements of K such that $\sum_{m \in F} \varphi(m) a_m = 0$. There exists an index i such that all elements $m \in F'$ for which $a_{m'} \neq 0$ belong to F'_i. Then, we get a linear dependence relation among the members of $(F - F'_i) \cup \varphi(F'_i)$, so that $a_m = 0$ for every $m \in F$, as was to be shown.

By Zorn's lemma (corollary A.2.13), the set Φ has a maximal element (F', φ). Let us show that $F' = F$. Otherwise, let $\mu \in F - F'$. For every $m \in F - F'$, write $\varphi(m) = m$. By the exchange property (proposition 3.7.2) applied to the free family $(\varphi(m))_{m \in F}$, to the index μ and to the generating set G, there exists an element $v \in G$ such that the family $(\varphi'(m))_{m \in F}$ is free, where $\varphi' \colon F \to M$ coincides with φ on $F - \{\mu\}$, and $\varphi'(\mu) = v$. Define $F'_1 = F' \cup \{\mu\}$ and let $\varphi_1 \cup F'_1 \to G$ be the map given by $\varphi_1(\mu) = v$ and $\varphi_1(m) = \varphi(m)$ for $m \in F'$. The map φ_1 is injective. Consequently, (F'_1, φ_1) belongs to Φ and is strictly bigger than (F', φ), contradicting the maximality hypothesis on (F', φ).

So $\varphi \colon F \to G$ is injective, establishing the lemma. □

Remark (3.7.5). — Let K be a division ring and let M be a right K-vector space. When proving the preceding theorem, we made use of the axiom of choice, by way of Zorn's lemma. This cannot be avoided in general. However, if M is finitely generated, then usual induction is enough. The first question of exercise 3.10.30 suggests a simple direct proof of lemma 3.7.4 in that case, leading to a simplification of the proof that all bases of a (finitely generated) vector space have the same cardinality.

Corollary (3.7.6). — *Let* A *be a non-null commutative ring and let* M *be an* A*-module. If* M *is free, then all bases of* M *have the same cardinality.*

This cardinality is called the *rank* of the free A-module M, and is denoted by $\mathrm{rk}_A(M)$.

Proof. — Let J be a maximal ideal of A, so that the quotient ring $K = A/J$ is a field. Let MJ be the submodule of M generated by all products ma, for $a \in J$ and $m \in M$; set $V = M/MJ$. This is an A-module. However, if $a \in J$ and $m \in M$, then $\mathrm{cl}_{MJ}(m)a = \mathrm{cl}_{MJ}(ma) = 0$; consequently, V can be viewed as an A/J-module, that is, a K-vector space. (See also §3.1.5.)

Assume that M is free and let $(e_i)_{i \in I}$ be a basis of M. Let us show that the family $(\mathrm{cl}_{MJ}(e_i))_{i \in I}$ is a basis of V. First, since the e_i generate M, this family generates V as an A-module, hence as a K-vector space. It remains to show that it is free. Let $(\alpha_i)_{i \in I}$ be an almost-null family of elements of K such that $\sum_{i \in I} \mathrm{cl}_{MJ}(e_i)\alpha_i$. For every $i \in I$ such that $\alpha_i \neq 0$, let us choose $a_i \in A$ such that $\mathrm{cl}_J(a_i) = \alpha_i$; if $\alpha_i = 0$, set $a_i = 0$. Then, the family $(a_i)_{i \in I}$ is almost-null and $\sum_{i \in I} \mathrm{cl}_{MJ}(e_i) \mathrm{cl}_J(a_i) = 0$. In other words, $m = \sum_{i \in I} e_i a_i$ belongs to MJ. Since MJ is generated by the elements of the form $e_i a$ for $i \in I$ and $a \in J$, there is a family $(b_i)_{i \in I}$ of elements of J such that $m = \sum_{i \in I} e_i b_i$. The family (e_i) being free, one has $a_i = b_i$ for every $i \in I$. Consequently, $\alpha_i = cl_J(a_i) = cl_J(b_i) = 0$ for every $i \in I$. The linear dependence relation that we have considered is thus trivial, so that the family $(e_i)_{i \in I}$ is free, as claimed.

Consequently, this family is a basis of the K-vector space V. This implies that $\dim_K(V) = \text{Card}(I)$ and $\dim_K(V)$ is the common cardinality of all bases of M. □

Theorem (3.7.7). — *Let K be a division ring and let M be a K-vector space. Every subspace N of M has a direct summand P and one has*

$$\dim(N) + \dim(P) = \dim(M).$$

Proof. — Let B_1 be a basis of N. By theorem 3.7.3, there exists a basis B of M which contains B_1. The subset $B_2 = B - B_1$ of M is then free and generates a vector subspace P of M such that $N + P = M$. Let us show that $N \cap P = 0$. So let $m \in N \cap P$; one can write m both as a linear combination of elements of B_1 and as a linear combination of elements of B_2. Subtracting these relations, we get a linear dependence relation between the members of M. If $m \neq 0$, this relation is nontrivial, contradicting the hypothesis that B is free. Hence $N \cap P = 0$ and P is a direct summand of N in M.

By definition, one has $\dim(N) = \text{Card}(B_1)$ and $\dim(M) = \text{Card}(B)$. Moreover, B_2 is free and generates P, so that $\dim(P) = \text{Card}(B_2)$. Since B_1 and B_2 are disjoint, one has

$$\dim(M) = \text{Card}(B) = \text{Card}(B_1 \cup B_2)$$
$$= \text{Card}(B_1) + \text{Card}(B_2) = \dim(N) + \dim(P),$$

as was to be shown. □

Many classical results in the theory of vector spaces over a (commutative) field remain true, with the same proof, in the more general context of vector spaces over a division ring. Let us give two examples.

Corollary (3.7.8). — *Let K be a division ring. Let V, W be K-vector spaces and let $f : V \to W$ be a K-linear map.*

(i) $\dim(\text{Im}(f)) + \dim(\text{Ker}(f)) = \dim(V)$;

(ii) *If $\dim(V) > \dim(W)$, then f is not injective;*

(iii) *Assume that $\dim(V) = \dim(W)$. Then the following properties are equivalent: (i) f is injective; (ii) f is surjective; (iii) f is bijective.*

Proof. — a) Passing to the quotient, the morphism f defines an injective K-linear map $\varphi : V/\text{Ker}(f) \to W$, which induces an isomorphism from $V/\text{Ker}(f)$ to $\text{Im}(\varphi) = \text{Im}(f)$, so that $\dim(\text{Im}(f)) = \dim(V/\text{Ker}(f))$. Let V' be a direct summand of $\text{Ker}(f)$ in V; one has $\dim(V) = \dim(\text{Ker}(f)) + \dim(V')$. Finally, the canonical surjection $cl_{\text{Ker}(f)} : V \to V/\text{Ker}(f)$ induces, by restriction, a bijective morphism from V' to $V/\mathrm{Ker}(f)$, so that $\dim(V') = \dim(V/\mathrm{Ker}(f))$. Consequently, we have

$$\dim(\mathrm{Im}(f)) + \dim(\mathrm{Ker}(f)) = \dim(V/\mathrm{Ker}(f)) + \dim(\mathrm{Ker}(f))$$
$$= \dim(V') + \dim(\mathrm{Ker}(f)) = \dim(V).$$

b) Since $\mathrm{Im}(f) \subset W$, one has $\dim(\mathrm{Im}(f)) \leq \dim(W)$. If $\dim(V) > \dim(W)$, this forces $\dim(\mathrm{Ker}(f)) > 0$, hence f is not injective.

c) If f is injective, then $\dim(\mathrm{Im}(f)) = \dim(V) = \dim(W)$, hence $\mathrm{Im}(f) = W$ and f is surjective. If f is surjective, then $\dim(\mathrm{Ker}(f)) = 0$, hence $\mathrm{Ker}(f) = 0$ and f is injective. This implies the equivalence of the three statements. □

Corollary (3.7.9). — *Let K be a division ring. Let V, W be K-vector spaces and let $f \colon V \to W$ be a K-linear map.*

(i) *The map f is injective if and only if it has a left inverse;*

(ii) *The map f is surjective if and only if it has a right inverse.*

Proof. — If a map f has a left inverse (resp. a right inverse), then it is injective (resp. surjective). It thus suffices to prove the converse assertions.

a) Assume that f is injective, so that f induces an isomorphism f_1 from V to its image $f(V)$. Let W_1 be a direct summand of $f(V)$ in W (theorem 3.7.7) and let $p \colon W \to f(V)$ be the projector with image $f(V)$ and kernel W_1. The morphism $g = f_1^{-1} \circ p \colon W \to V$ satisfies $g \circ f = \mathrm{id}_V$, so that f is left invertible.

b) Assume that f is surjective. Let V_1 be a direct summand of $\mathrm{Ker}(f)$ in V (theorem 3.7.7). Then, the morphism $f_1 = f|_{V_1} \colon V_1 \to W$ is injective and surjective (for $w \in W$, let $v \in V$ such that $f(v) = w$, and write $v = v_1 + m$, where $v_1 \in V_1$ and $m \in \mathrm{Ker}(f)$; one has $f(v_1) = f(v) = w$), hence an isomorphism. Let $j \colon V_1 \to V$ be the injection and $h = j \circ f_1^{-1}$; this is a morphism from W to V such that $f \circ h \circ f_1 = f \circ j = f_1$, hence $f \circ h = \mathrm{id}_W$ since f_1 is surjective. Consequently, f is right invertible. □

3.8. Alternating Multilinear Forms. Determinants

Definition (3.8.1). — Let A be a commutative ring, let M and N be A-modules, let p be a positive integer. A map $f \colon M^p \to N$ is said to be *multilinear*, or *p-linear*, if it is linear with respect to each variable: for every $(m_1, \ldots, m_p) \in M^p$, the map $m \mapsto f(m_1, \ldots, m_{i-1}, m, m_{i+1}, \ldots, m_p)$ is A-linear.

When $N = A$, a p-linear map from M to A is called a *p-linear form*. One also says *bilinear* for $p = 2$, *trilinear* for $p = 3$...

Definition (3.8.2). — Let A be a commutative ring, let M and N be A-modules, let p be a positive integer and let $f \colon M^p \to N$ be a multilinear map.

(i) One says that f is *symmetric* if, for every permutation $\sigma \in \mathfrak{S}_p$ and every $(m_1, \ldots, m_p) \in M^p$, one has

$$f(m_{\sigma(1)}, \ldots, m_{\sigma(p)}) = f(m_1, \ldots, m_p).$$

(ii) One says that f is *antisymmetric* if, for every permutation $\sigma \in \mathfrak{S}_p$ and every $(m_1, \ldots, m_p) \in M^p$, one has

$$f(m_{\sigma(1)}, \ldots, m_{\sigma(p)}) = \varepsilon(\sigma) f(m_1, \ldots, m_p),$$

where $\varepsilon(\sigma) \in \{\pm 1\}$ is the signature of σ.

(iii) One says that f is *alternating* if for every $(m_1, \ldots, m_p) \in M^p$ for which there exist $i \neq j$ such that $m_i = m_j$, one has $f(m_1, \ldots, m_p) = 0$.

3.8.3. — Let A be a commutative ring, let M and N be A-modules, let p be a positive integer. Let f be a p-linear alternating map from M to N; let $(m_1, \ldots, m_p) \in M^p$, let i and j be distinct elements in $\{1, \ldots, p\}$ and let us apply the alternating property to the family obtained by replacing both m_i and m_j by their sum $m_i + m_j$. Expanding, one gets (assuming $i < j$ to fix ideas)

$$\begin{aligned} 0 &= f(m_1, \ldots, m_{i-1}, m_i + m_j, \ldots, m_{j-1}, m_i + m_j, \ldots, m_p) \\ &= f(m_1, \ldots, m_{i-1}, m_i, \ldots, m_{j-1}, m_j, \ldots, m_p) \\ &\quad + f(m_1, \ldots, m_{i-1}, m_j, \ldots, m_{j-1}, m_i, \ldots, m_p), \end{aligned}$$

since the two other terms vanish. In other words, f is changed into its negative when two variables are exchanged. Now, any permutation σ of $\{1, \ldots, n\}$ can be written as a product of finitely many permutations of two elements; the number of such permutations is even or odd, according to the *signature* $\varepsilon(\sigma)$ of σ being 1 or -1. One thus gets

$$f(m_{\sigma(1)}, \ldots, m_{\sigma(n)}) = \varepsilon(\sigma) f(m_1, \ldots, m_n). \tag{3.8.3.1}$$

Conversely, let f be a p-linear antisymmetric map. If there are two indices $i \neq j$ such that $m_i = m_j$, exchanging m_i and m_j changes $f(m_1, \ldots, m_p)$ both into itself and its negative, so that $f(m_1, \ldots, m_p) = -f(m_1, \ldots, m_p)$. Consequently, $2f(m_1, \ldots, m_p) = 0$ hence $2f$ is alternating.

If 2 is regular in A or, more generally, if the multiplication by 2 in the module N is injective, then f is alternating too.

However, if $A = \mathbf{Z}/2\mathbf{Z}$, any symmetric p-linear map is antisymmetric. For example, the 2-linear form $(x_1, x_2) \mapsto x_1 x_2$ on the free A-module A is antisymmetric, but it is not alternating.

3.8.4. — The set of all alternating p-linear maps from M to N is denoted $\Lambda^p(M, N)$; it is naturally a A-module. By definition, $\Lambda^1(M, N) = \mathrm{Hom}_A(M, N)$. On the other hand, the A-module $\Lambda^0(M, N)$ identifies with N. Indeed, M^0 is a singleton so that a map $f : M^0 \to N$ is just an element of N, and every such map is linear in each variable (since there is no variable).

Let $u: M \to M'$ be a morphism of A-modules. For any $f \in \Lambda^p(M', N)$, the map $(m_1, \ldots, m_p) \mapsto f(u(m_1), \ldots, u(m_p))$ is a p-linear alternating map from M to N, denoted $\Lambda^p(u)(f)$, or simply $u^*(f)$. The map $f \mapsto \Lambda^p(u)(f)$ is a morphism of A-modules from $\Lambda^p(M', N)$ to $\Lambda^p(M, N)$. One has $\Lambda^p(\mathrm{id}_M) = \mathrm{id}_{\Lambda^p(M,N)}$, and $\Lambda^p(u \circ v) = \Lambda^p(v) \circ \Lambda^p(u)$. When N is fixed, the construction $M \mapsto \Lambda^p(M, N)$ is thus a contravariant functor in M.

Let $v: N \to N'$ be a morphism of A-modules. For any $f \in \Lambda^p(M, N)$, the map $v \circ f: (m_1, \ldots, m_p) \mapsto v(f(m_1, \ldots, m_p))$ is a p-linear alternating map from M to N', also denoted $\Lambda^p(v)(f)$, or $v_*(f)$. The map $f \mapsto \Lambda^p(v)(f)$ is a morphism of A-modules from $\Lambda^p(M, N)$ to $\Lambda^p(M, N')$. One has $\Lambda^p(\mathrm{id}_N) = \mathrm{id}_{\Lambda^p(M,N)}$, and $\Lambda^p(u \circ v) = \Lambda^p(u) \circ \Lambda^p(v)$. When M is fixed, the construction $N \mapsto \Lambda^p(M, N)$ is thus a covariant functor in N.

Proposition (3.8.5). — *Let* A *be a commutative ring. Let* M *be an A-module and let* $(m_1, \ldots, m_n)$ *be a generating family of* M. *For any A-module* P *and any integer* $p > n$, *one has* $\Lambda^p(M, P) = 0$.

Proof. — Let $f: M^p \to P$ be an alternating p-linear map. Let $(x_1, \ldots, x_p) \in M^p$. By assumption, there exist elements $a_{i,j}$ (for $1 \le i \le n$ and $1 \le j \le p$) such that $x_j = \sum a_{i,j} m_i$. Since f is p-linear, one has

$$f(x_1, \ldots, x_p) = \sum_{i_1=1}^{n} \cdots \sum_{i_p=1}^{n} a_{i_1,1} \ldots a_{i_p,p} f(m_{i_1}, \ldots, m_{i_p}).$$

Since $p > n$, in each p-tuple $(i_1, \ldots, i_p)$ of elements of $\{1, \ldots, n\}$, at least two terms are equal. The map f being alternating, one has $f(m_{i_1}, \ldots, m_{i_p}) = 0$. Consequently, $f(x_1, \ldots, x_p) = 0$ and $f = 0$. □

More generally, the above proof shows that an alternating p-linear map is determined by its values on the p-tuples of distinct elements among $(m_1, \ldots, m_n)$.

Proposition (3.8.6). — *Let* M *be a free A-module. A family* $(m_1, \ldots, m_p)$ *of elements of* M *is free if and only if there exists, for any* $a \in A - \{0\}$, *a p-linear alternating form f on* M *such that* $a f(m_1, \ldots, m_p) \ne 0$.

Proof. — Assume that the family $(m_1, \ldots, m_p)$ is bonded and let us consider a non-trivial linear dependence relation $a_1 m_1 + \cdots + a_p m_p = 0$. Up to reordering the family $(m_1, \ldots, m_p)$, we assume that $a_1 \ne 0$. Then for any alternating p-linear form f on M, one has

$$0 = f(a_1 m_1 + \cdots + a_p m_p, m_2, \ldots, m_p)$$

$$= \sum_{i=1}^{p} a_i f(m_i, m_2, \ldots, m_p)$$

$$= a_1 f(m_1, \ldots, m_p).$$

Let us now assume that the family $(m_1, \ldots, m_p)$ is free and let us show the desired property by induction on p.

The property holds for $p = 0$, for the 0-alternating form equal to the constant 1 is adequate. Let us assume that the property holds for $p-1$ and let us prove it for p. Let $a \in A - \{0\}$. Since the family $(m_1, \ldots, m_{p-1})$ is free, there exists an alternating $(p-1)$-linear form f such that $af(m_1, \ldots, m_{p-1}) \neq 0$. Then, the map

$$f' : M^p \to M, \quad f'(x_1, \ldots, x_p) = \sum_{j=1}^p (-1)^j f(x_1, \ldots, \widehat{x_j}, \ldots, x_p) x_j$$

from M^p to M is a p-linear alternating map. Assume that $af'(m_1, \ldots, m_p) = 0$. This gives a linear dependence relation among the m_i. Since the coefficient of m_p is equal to $(-1)^p af(m_1, \ldots, m_{p-1})$, this relation is non-trivial, contradicting the assumption that $(m_1, \ldots, m_p)$ is free. Consequently, $af'(m_1, \ldots, m_p) \neq 0$. In particular, there exists a linear form φ on M (for example, a suitable coordinate form in a basis of M) such that $\varphi(af'(m_1, \ldots, m_p)) \neq 0$. The composite $g = \varphi \circ f'$ is an alternating p-linear form such that $ag(m_1, \ldots, m_p) \neq 0$. $\quad\square$

Corollary (3.8.7). — *Let A be a non-zero commutative ring. Let M be a free A-module which is generated by n elements. Then the cardinality of any free family in M is at most n.*

Proof. — Let $(e_1, \ldots, e_p)$ be a free family in M. By proposition 3.8.6 applied to $a = 1$ and the family $(e_1, \ldots, e_p)$, there exists a non-zero alternating p-linear form on M. Proposition 3.8.5 then implies that $p \leq n$. $\quad\square$

Corollary (3.8.8). — *Let A be a non-zero commutative ring. Let M be a finitely generated free A-module. Then all bases of M are finite and their cardinalities are equal.*

Proof. — Let $(m_1, \ldots, m_n)$ be a generating family of M. By the preceding proposition, every free family in M is finite and has cardinality at most n. In particular, all bases of M are finite.

Let $(e_1, \ldots, e_m)$ and $(f_1, \ldots, f_p)$ be bases of M. Since $(f_1, \ldots, f_p)$ generates M and since $(e_1, \ldots, e_m)$ is free, one has $m \leq p$. Since $(e_1, \ldots, e_m)$ generates M and since $(f_1, \ldots, f_p)$ is free, one has $p \leq m$. Consequently, $m = p$. $\quad\square$

Definition (3.8.9). — Let A be a non-zero commutative ring. If M is a finitely generated free A-module, the common cardinality of all of the bases of M is called its *rank* and denoted $\mathrm{rk}_A(M)$.

If the ring A is clear from the context, one also writes $\mathrm{rk}(M)$ for $\mathrm{rk}_A(M)$.

Theorem (3.8.10). — *Let A be a non-zero commutative ring and let M be a free A-module of finite rank $n \geq 0$. For any basis $(e_1, \ldots, e_n)$ of M, there exists a unique alternating n-linear form φ on M such that $\varphi(e_1, \ldots, e_n) = 1$. This form φ is a basis of the A-module $\Lambda^n(M, A)$. In particular, $\Lambda^n(M, A)$ is a free A-module of rank 1.*

Proof. — Let $(\varphi_1, \ldots, \varphi_n)$ be the basis of $M^\vee$ dual to the basis $(e_1, \ldots, e_n)$.

It is almost obvious that there is at most one alternating linear form as in the theorem. Indeed, let φ be an alternating n-linear form on M. For any $(m_1, \ldots, m_n) \in M^n$ and any j, one has $m_j = \sum_{i=1}^n \varphi_i(m_j) e_i$. Expanding, one gets

$$f(m_1, \ldots, m_n) = \sum_{i_1=1}^n \cdots \sum_{i_n=1}^n \varphi_{i_1}(m_1) \ldots \varphi_{i_n}(m_n) f(e_{i_1}, \ldots, e_{i_n}).$$

Moreover, $f(e_{i_1}, \ldots, e_{i_n}) = 0$ if two indices i_j are equal; when they are pairwise distinct, the map $j \mapsto i_j$ is a permutation σ of $\{1, \ldots, n\}$ and

$$f(e_{i_1}, \ldots, e_{i_n}) = \varepsilon(\sigma) f(e_1, \ldots, e_n) = \varepsilon(\sigma).$$

Consequently, if such an alternating n-linear form exists at all, it is given by the (well-known) formula:

$$\varphi(m_1, \ldots, m_n) = \left(\sum_{\sigma \in \mathfrak{S}_n} \varepsilon(\sigma) \prod_{i=1}^n \varphi_i(m_{\sigma(i)}) \right) \varphi(e_1, \ldots, e_n). \qquad (3.8.10.2)$$

Let us now prove the theorem by induction on n. It suffices to show that the previous formula defines an alternating n-linear form. It is clear that it is n-linear; the alternating property is not much harder. Assume indeed that $m_j = m_k$; in formula (3.8.10.2), the terms corresponding to the permutations σ and $\sigma\tau_{j,k}$, where $\tau_{j,k}$ is the transposition exchanging j and k, only differ by their sign. If σ runs among all even permutations, σ or $\sigma\tau_{j,k}$ meet every permutation of $\{1, \ldots, n\}$ exactly once, so that the final sum is 0.

We just proved that there is an alternating n-linear form φ on M which maps $(e_1, \ldots, e_n)$ to 1, and that any alternating n-linear form f on M is of the form $a\varphi$, where $a = \varphi(e_1, \ldots, e_n)$. This concludes the proof of the theorem.□

Definition (3.8.11). — Let A be a commutative ring.

Let M be a free A-module of rank n and let $\mathscr{B} = (e_1, \ldots, e_n)$ be a basis of M. The unique alternating n-linear form on M which maps $(e_1, \ldots, e_n)$ to 1 is called the *determinant* (with respect to the basis $\mathscr{B}$); it is written $\det_{\mathscr{B}}$.

Let U be an $n \times n$ matrix with entries in A. The determinant $\det(U)$ of U is the determinant of the column vectors of U in the canonical basis of A^n. For such a matrix $U = (a_{ij})$, one thus has

$$\det(U) = \sum_{\sigma \in \mathfrak{S}_n} \varepsilon(\sigma) \prod_{i=1}^n a_{i\sigma(i)}.$$

Example (3.8.12). — Let M be a free A-module of rank n and let $\mathscr{B} = (e_1, \ldots, e_n)$ be a basis of M; write $(\varphi_1, \ldots, \varphi_n)$ for the dual basis. Let I be a subset of $\{1, \ldots, n\}$ with cardinality p; enumerate its elements in increasing order, so that $I = \{i_1, \ldots, i_p\}$ with $1 \leq i_1 < \cdots < i_p \leq n$. Let M_I be the submodule of M generated by the e_i, for $i \in I$, and let φ_I be the map from M^p to A defined by

$$\varphi_I(m_1, \ldots, m_p) = \sum_{\sigma \in \mathfrak{S}_p} \varepsilon(\sigma) \prod_{j=1}^{p} \varphi_{i_j}(m_{\sigma(j)}).$$

It is p-linear alternating. This can be shown by a direct computation. Alternatively, let $\det_{\mathscr{B}_I}$ be the determinant form on M_I with respect to its basis $\mathscr{B}_I = (e_i)_{i \in I}$ and let π_I be the projector onto M_I, whose kernel is the submodule generated by the e_i for $i \notin I$. Then

$$\varphi_I(m_1, \ldots, m_p) = \det_{\mathscr{B}_I}(\pi_I(m_1), \ldots, \pi_I(m_p)).$$

Corollary (3.8.13). — *Let* M *be a free* A-*module of rank* n *and let* p *be any integer such that* $0 \le p \le n$. *The family* (φ_I), *where* I *runs among all subsets of* $\{1, \ldots, n\}$ *with cardinality* p, *is a basis of the* A-*module* $\Lambda^p(M, A)$. *In particular, the* A-*module* $\Lambda^p(M, A)$ *is free of rank* $\binom{n}{p}$.

Proof. — Let $I = \{i_1, \ldots, i_p\}$, with $i_1 < \cdots < i_p$. It follows from the definition of φ_I that $\varphi_I(e_{i_1}, \ldots, e_{i_p}) = 1$. Let J be another subset of $\{1, \ldots, n\}$ with p elements. Write $\{j_1, \ldots, j_p\}$, with $j_1 < \cdots < j_p$. Since $J \ne I$ there exists a $k \in \{1, \ldots, p\}$ such that $j_k \notin I$ (otherwise, $J \subset I$, hence $J = I$ since they both have p elements); consequently, $\pi_I(e_{j_k}) = 0$, hence $\varphi_I(e_{j_1}, \ldots, e_{j_p}) = 0$. This implies that for any alternating p-linear form f on M,

$$f = \sum_{I = \{i_1 < \cdots < i_p\}} f(e_{i_1}, \ldots, e_{i_p})\varphi_I.$$

Indeed, the two members of the preceding equality are alternating p-linear forms that take the same value at all p-tuples of elements of the basis $\{e_1, \ldots, e_n\}$. This shows that the family (φ_I) generates $\Lambda^p(M, A)$.

Moreover, for any family (a_I) indexed by the set of p-element subsets of $\{1, \ldots, n\}$, the alternating p-linear form $\sum a_I \varphi_I$ takes the value a_I at $(e_{i_1}, \ldots, e_{i_p})$ if $I = \{i_1 < \cdots < i_p\}$. In particular, if $\sum a_I \varphi_I = 0$, then $a_I = 0$ for all I. In other words, the family (φ_I) is free. $\square$

Theorem and Definition (3.8.14). — *Let* M *be a free* A-*module of rank* n *and let* u *be an endomorphism of* M. *The endomorphism* $\Lambda^n(u)$ *of* $\Lambda^n(M, A)$ *is a homothety; its ratio is called the* determinant of u *and denoted by* $\det(u)$.

Proof. — Recall from theorem 3.8.10 that the A-module $\Lambda^n(M, A)$ is isomorphic to A: if $\mathscr{B} = (e_1, \ldots, e_n)$ is a basis of M, the determinant form $\det_{\mathscr{B}}$ is a basis of the A-module $\Lambda^n(M, A)$. Since $\Lambda^n(u)(\det_{\mathscr{B}})$ is an alternating n-linear form on M, there exists a unique element $\lambda \in A$ such that $\Lambda^n(u)(\det_{\mathscr{B}}) = \lambda \det_{\mathscr{B}}$. Let now f be any alternating n-linear form on M; there exists an element $a \in A$ such that $f = a \det_{\mathscr{B}}$. Since $\Lambda^n(u)$ is linear, one has $\Lambda^n(u)(f) = a\Lambda^n(u)(\det_{\mathscr{B}}) = a\lambda \det_{\mathscr{B}}$. Since A is commutative, $\Lambda^n(u)(f) = \lambda f$. This shows that $\Lambda^n(u)$ is a homothety of ratio λ. $\square$

3.8.15. — Let A be a commutative ring, let M be a free A-module of rank n and let u be an endomorphism of M. Let $\mathscr{B}$ be a basis of M and let U be the matrix of u in the basis $\mathscr{B}$. One has $\det(u) = \det(U)$.

Indeed, by definition, the alternating n-linear form $\Lambda^n(u)(\det_{\mathscr{B}})$ maps $(m_1, \ldots, m_n)$ to $\det_{\mathscr{B}}(u(m_1), \ldots, u(m_n))$. In particular, it maps $(e_1, \ldots, e_n)$ to $\det_{\mathscr{B}}(u(e_1), \ldots, u(e_n)) = \det(U)$. This shows that $\Lambda^n(u)(\det_{\mathscr{B}}) = (\det U)\det_{\mathscr{B}}$, so that $\det(u) = \det(U)$.

Let u and v be endomorphisms of M. One has $\Lambda^n(v \circ u) = \Lambda^n(u) \circ \Lambda^n(v)$, hence

$$\det(v \circ u) = \det(u)\det(v) = \det(v)\det(u).$$

One also has $\Lambda^n(\mathrm{id}_M) = \mathrm{id}$, so that $\det(\mathrm{id}_M) = 1$.

Assume that u is invertible, with inverse v. Then $\Lambda^n(u)$ is also invertible with inverse $\Lambda^n(v)$; this implies that $\det(u)$ is invertible with inverse $\det(v)$. In other words,

$$\det(u^{-1}) = \det(u)^{-1}.$$

Of course, the same formulas hold for matrices: if U and V are two $n \times n$ matrices, then $\det(VU) = \det(V)\det(U)$. One also has $\det(U^t) = \det(U)$, either by a direct computation, or using the fact that the transpose of a homothety is a homothety with the same ratio, so that for any endomorphism u of M, $\Lambda^n(u^t) = \Lambda^n(u)^t$ is the homothety of ratio $\det(u)$ in $\Lambda^n(M^\vee, A)$.

Let $U = (u_{i,j})$ be an $n \times n$ matrix. The *cofactor* $v_{i,j}$ of the coefficient $u_{i,j}$ of index (i, j) is the determinant of the $(n-1) \times (n-1)$ matrix obtained by removing from U the ith row and the jth column, multiplied by $(-1)^{i+j}$. The *adjugate* of the matrix U is the matrix $\tilde{U} = (v_{j,i})_{1 \leq i,j \leq n}$, the transpose of the matrix with entries the cofactors of U.

Proposition (3.8.16) (Adjugate formula). — *The following formula holds:* $U\tilde{U} = \tilde{U}U = (\det U)I_n$.

Proof. — For $i \in \{1, \ldots, n\}$, let N_i be the set $\{1, \ldots, \hat{i}, \ldots, n\} = \{1, \ldots, n\} - \{i\}$.

Let $i, j \in \{1, \ldots, n\}$. Consider the $(n-1) \times (n-1)$ matrix $U^{i,j}$ obtained by deleting the ith row and the jth column as indexed by N_i for rows and N_j for columns. If $\sigma \colon N_i \xrightarrow{\sim} N_j$ is a bijection, let $m(\sigma)$ be the number of inversions of σ, that is, the number of pairs (k, ℓ) of elements in N_i such that $k < l$ but $\sigma(k) > \sigma(\ell)$ and let $\varepsilon(\sigma) = (-1)^{m(\sigma)}$. Then, the expansion of the determinant of $U^{i,j}$ reads as

$$\det(U^{i,j}) = \sum_{\sigma \colon N_i \xrightarrow{\sim} N_j} \varepsilon(\sigma) \prod_{k \neq i} u_{k,\sigma(k)}.$$

Now, a bijection $\sigma \colon N_i \xrightarrow{\sim} N_j$ can be extended to a bijection $\bar{\sigma} \in \mathfrak{S}_n$ such that $\bar{\sigma}(i) = j$, and conversely, every such bijection $\bar{\sigma}$ is obtained exactly once. Let us compare the number of inversions of σ with that of $\bar{\sigma}$. By definition, $m(\bar{\sigma}) - m(\sigma)$ is the number of pairs $(k, \ell) \in \{1, \ldots, n\}$ such that $k < \ell$, $\bar{\sigma}(k) > \bar{\sigma}(\ell)$, and either k or ℓ is equal to i. Let a and b be the numbers of such pairs with $k = i$ and $\ell = i$ respectively. A pair (i, ℓ) counts if and only if $\ell > i$

and $\bar{\sigma}(\ell) < j$, so that a values among $\bar{\sigma}(i+1), \ldots, \bar{\sigma}(n)$ are smaller than j. Similarly, a pair (k, i) counts if and only if $k < i$ and $\bar{\sigma}(k) > j$, so that b values among $\bar{\sigma}(1), \ldots, \bar{\sigma}(i-1)$ are above j ; consequently, $i - 1 - b$ such values are smaller than j. This implies that $a + (i - 1 - b) = j - 1$, hence $a - b = j - i$. Consequently, $a + b \equiv j + i \pmod{2}$ and

$$m(\sigma) = m(\bar{\sigma}) - a - b \equiv m(\bar{\sigma}) + (i + j) \pmod{2},$$

so that

$$(-1)^{m(\sigma)} = (-1)^{m(\bar{\sigma})}(-1)^{i+j}.$$

In particular, for any $j \in \{1, \ldots, n\}$,

$$\sum_{i=1}^{n} u_{i,j}(-1)^{i+j} \det(U^{i,j}) = \sum_{i=1}^{n} \sum_{\substack{\bar{\sigma} \in \mathfrak{S}_n \\ \sigma(i)=j}} (-1)^{m(\bar{\sigma})} u_{i,j} \left(\prod_{p \neq i} u_{p,\bar{\sigma}(p)} \right)$$

$$= \sum_{\bar{\sigma} \in \mathfrak{S}_n} (-1)^{m(\bar{\sigma})} \prod_{p=1}^{n} u_{p,\bar{\sigma}(p)}$$

$$= \det(U).$$

On the other hand, for $i \neq k$, this formula shows that

$$\sum_{j=1}^{n} (-1)^{i+j} \det(U^{i,j}) u_{k,j}$$

is the determinant of the matrix U_k obtained by replacing its ith row by the kth row of U; indeed, this manipulation does not modify any of the matrices $U^{i,j}$ but replaces $u_{i,j}$ by $u_{k,j}$. Since two rows of U_k are equal, $\det(U_k) = 0$. (The definition of the determinant implies that it vanishes when two rows are equal; by transposition, it vanishes whenever two rows are equal.) Altogether, we have shown that $U\tilde{U} = \det(U)I_n$. The other formula follows by transposition: the adjugate of U^t is the transpose of $\tilde{U}$; therefore, $U^t\tilde{U}^t = \det(U^t)I_n = \det(U)I_n$, hence $\tilde{U}U = \det(U)I_n$.   □

In the course of the proof of this formula, we established the following *Laplace expansion formula* for the determinant, as a linear combination of smaller determinants:

Corollary (3.8.17). — *For every integer $i \in \{1, \ldots, n\}$, one has*

$$\det(U) = \sum_{j=1}^{n} (-1)^{i+j} u_{i,j} \det(U^{i,j}).$$

Definition (3.8.18). — The *characteristic polynomial* of a matrix $U \in \mathrm{Mat}_n(A)$ is the determinant of the matrix $XI_n - U$ with coefficients in $A[X]$.

Expansion of this determinant shows that the characteristic polynomial of U is a monic polynomial of degree n. In fact, one has

$$P_U(X) = X^n - \mathrm{Tr}(U)X^{n-1} + \cdots + (-1)^n \det(U),$$

where $\mathrm{Tr}(U) = \sum_{i=1}^n U_{ii}$ is the *trace* of U.

Remark (3.8.19). — Let $U \in \mathrm{Mat}_n(A)$ and $V \in \mathrm{GL}_n(A)$. Then the matrices U and $V^{-1}UV$ have the same characteristic polynomial. Indeed,

$$P_{V^{-1}UV}(X) = \det(XI_n - V^{-1}UV) = \det(V^{-1}(XI_n - U)V) = \det(XI_n - U) = P_U(X).$$

As a consequence, one may define the characteristic polynomial of an endomorphism u of a free finitely generated A-module M: it is the characteristic polynomial of the matrix U of u in any basis of M.

Similarly, the trace $\mathrm{Tr}(u)$ of u is defined as $\mathrm{Tr}(U)$. The map $u \mapsto \mathrm{Tr}(u)$ is a linear form on $\mathrm{End}_A(M)$. It follows from the properties of traces and determinants of matrices that $\mathrm{Tr}(uv) = \mathrm{Tr}(vu)$ and $\det(uv) = \det(vu) = \det(u)\det(v)$ for any $u, v \in \mathrm{End}_A(M)$.

Corollary (3.8.20) (Cayley–Hamilton). — *Let $U \in \mathrm{Mat}_n(A)$ be a matrix and let P_U be its characteristic polynomial. Then $P_U(U) = 0$ in $\mathrm{Mat}_n(A)$.*

Proof. — Let us write $P_U(X) = X^n + a_{n-1}X^{n-1} + \cdots + a_0$. Let $V \in \mathrm{Mat}_n(A[X])$ be the matrix defined by $V = XI_n - U$ and let $\tilde{V}$ be the adjugate of V. It is an $n \times n$ matrix with coefficients in $A[X]$, but we will rather view it as a polynomial with coefficients in $\mathrm{Mat}_n(A[X])$. Let us start from the adjugate formula (prop. 3.8.16) associated with the matrix V, namely, $P_U(X)I_n = \tilde{V} \cdot (XI_n - U)$, and let us view it as the euclidean division of the polynomial $P_U(X)I_n$ by the monic polynomial $XI_n - U$ in the noncommutative ring $\mathrm{Mat}_n(A)$ (theorem 1.3.15). The remainder is zero! By remark 1.3.16, one thus has

$$U^n + a_{n-1}U^{n-1} + \cdots + a_0I_n = 0,$$

that is, $P_U(U) = 0$. $\qquad\square$

Corollary (3.8.21). — *Let A be a commutative ring, let M be a finitely generated A-module and let u be an endomorphism of M. There exists a monic polynomial $P \in A[X]$ such that $P(u) = 0$.*

Proof. — If M is the free A-module A^n, this follows from the Cayley–Hamilton theorem (corollary 3.8.20): one may take for P the characteristic polynomial of u.

In the general case, let $(m_1, \ldots, m_n)$ be a finite family generating M. Let $p \colon A^n \to M$ be the morphism given by $p(a_1, \ldots, a_n) = \sum a_i m_i$. The morphism p is surjective. In particular, for every $i \in \{1, \ldots, n\}$, there exists an element $v_i \in A^n$ such that $p(v_i) = u(m_i)$. Let $v \colon A^n \to A^n$ be the morphism defined by $v(a_1, \ldots, a_n) = \sum a_i v_i$. One has $p \circ v = u \circ p$. Let V be the matrix of v and let P be its characteristic polynomial. By the Cayley–Hamilton theorem, one has $P(V) = 0$, hence $P(v) = 0$. Consequently, $0 = p \circ P(v) = P(u) \circ p$.

On Arthur Cayley

Arthur CAYLEY (1821–1895) was a British mathematician who worked mainly in algebra and algebraic geometry. His first paper contains the definition of what we now call the Cayley–Menger determinant and gives the volume of an $(n + 1)$-simplex (triangle, tetrahedron, etc.) of euclidean space in terms of the squares of the mutual distances between its vertices.

In fact, it is to CAYLEY (and simultaneously to GRASSMANN), that we owe, around 1844, the idea of an n-dimensional space in which points are represented by n real or complex coordinates. Although the notion of a vector space had not been invented yet, he introduces the algebra of matrices and formulates the "Cayley–Hamilton theorem", of which he proves the cases of dimension 2 and 3. However, CAYLEY introduced in 1854 the notion of an abstract group, by way of its multiplication *table*, identifying the associativity property. In 1878 he would then prove the now classic "Cayley theorem" that every such group can be viewed as a permutation group.

Invariant theory had just been founded by BOOLE when CAYLEY wrote a series of important papers. Specifically, he considers homogeneous polynomials of degree m in n variables (for example, "binary cubics" when $m = 3$ and $n = 2$) and is interested in polynomial expressions in their coefficients which are left unchanged under linear change of variables of the indetermi-

Portrait of Arthur Cayley (ca. 1860)

Photographer: Herbert Beraud
Source: Wikipedia; public domain

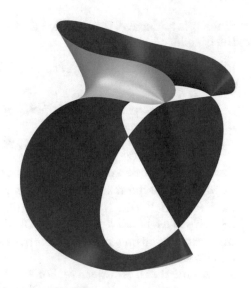

nates. The Ω-process he invented to construct such invariants would be one of the tools of HILBERT's solution of GORDAN's problem.

His name is attached to the 8-dimensional algebra of octonions which CAYLEY constructed soon after HAMILTON's discovery of quaternions. Although the algebra of octonions is not associative, its existence explains the fact from differential geometry that the 7-dimensional sphere is "parallelizable" — admits 7 independent vector fields. The real numbers, the complex numbers and the quaternions furnish a similar result for the spheres of dimensions 0, 1 and 3, but as shown by KERVAIRE, BOTT and MILNOR, it never holds otherwise.

CAYLEY's name is also attached to concepts in graph theory, such as the construction of a graph associated with a group and a generating subset of it, whose geometric structure reflect algebraic properties. He also proved the combinatorial result that there are n^{n-2} trees (connected graphs without loops) on n vertices labeled $1, 2, \ldots, n$.

In algebraic geometry, he proved the Cayley–Bacharach theorem: Consider two plane cubic curves and their nine intersection points given by BÉZOUT's theorem (see theorem 2.7.5); according to this theorem, if another cubic curve passes through eight of these points, then it has to pass through the ninth one.

He also studied cubic surfaces in 3-dimensional projective space. With SALMON, he proved that such surfaces contain lines, exactly 27 if the surface has no singularity. He also proved that the one with equation $xyz + yzw + zwx + wxy = 0$ (pictured above) is essentially the only such surface which has exactly 4 singular points.

Since p is surjective, this implies that $P(u) = 0$ and concludes the proof of the corollary.   □

Proposition (3.8.22). — *Let A be a commutative ring and let M be a free A-module of rank n.*

(i) *An endomorphism of M is surjective if and only if its determinant is invertible; it is then an isomorphism.*

(ii) *An endomorphism of M is injective if and only if its determinant is regular.*

Proof. — Let u be an endomorphism of M. Let us fix a basis $\mathscr{B} = (e_1, \ldots, e_n)$ of M, let U be the matrix of u in this basis and let $\tilde{U}$ be its adjugate. One has $\tilde{U}U = \det(U)I_n$. If $\det(U)$ is regular, the homothety of ratio $\det(U)$ is injective in M, hence u is injective. On the other hand, if $\det(U)$ is invertible, the endomorphism v of M with matrix $\det(U)^{-1}\tilde{U}$ is a left inverse of u, hence u is surjective.

Assume that u is surjective. For any $i \in \{1, \ldots, n\}$, let $f_i \in M$ be such that $u(f_i) = e_i$; let w be the unique endomorphism of M that maps e_i to f_i for every i. One has $u(w(e_i)) = e_i$ for every i, hence $u \circ w = \mathrm{id}_M$. Consequently, $\det(u)\det(w) = 1$ and $\det(u)$ is invertible.

It remains to show that $\det(u)$ is regular in A if u is injective. By hypothesis, the family $(u(e_1), \ldots, u(e_n))$ is free. Let a be a non-zero element of A. By proposition 3.8.6, there exists an alternating n-linear form f on M such that $af(u(e_1), \ldots, u(e_n)) \neq 0$. As for any alternating n-linear form, f is a multiple of the determinant form $\det_{\mathscr{B}}$ in the basis $(e_1, \ldots, e_n)$, so that $a\det(u(e_1), \ldots, u(e_n)) \neq 0$, which means exactly that $a\det(u) \neq 0$. Since a was chosen arbitrarily, this proves that $\det(u)$ is regular in A.   □

Corollary (3.8.23). — *Let M be a free A-module and let $\mathscr{B} = (e_1, \ldots, e_n)$ be a basis of M.*

A family $(x_1, \ldots, x_n)$ of elements of M is free if and only if its determinant in the given basis is regular in A.

Such a family generates M if and only if its determinant in the basis $\mathscr{B}$ is a unit of A; then, this family is even a basis of M.

Proof. — Let u be the unique endomorphism of M that maps e_i to x_i. By definition, one has $\det(u) = \det_{\mathscr{B}}(x_1, \ldots, x_n)$. For $(x_1, \ldots, x_n)$ to be free, it is necessary and sufficient that u be injective, hence that $\det(u)$ be regular. For $(x_1, \ldots, x_n)$ to generate M, it is necessary and sufficient that u be surjective, hence that $\det(u)$ be invertible; then, u is invertible and $(x_1, \ldots, x_n)$ is a basis of M.   □

3.9. Fitting Ideals

3.9.1. — Let A be a commutative ring. For every matrix $U \in \mathrm{Mat}_{n,m}(A)$ and every positive integer p, let us denote by $\Delta_p(U)$ the ideal of A generated by the determinants of all $p \times p$ matrices extracted from U. We also write this ideal as $\Delta_p(u_1, \ldots, u_m)$, where $u_1, \ldots, u_m \in A^n$ are the columns of U. It will be convenient to call any determinant of a $p \times p$ matrix extracted from U a *minor* of size p of U. By Laplace expansion of determinants, a minor of size p is a linear combination of minors of size $p-1$. This implies that $\Delta_p(U) \subset \Delta_{p-1}(U)$ for every p. If $p > \inf(m, n)$, then $\Delta_p(U) = 0$. One also has $\Delta_0(U) = A$ (the determinant of a 0×0 matrix is equal to 1); for simplicity of notation, we also set $\Delta_p(U) = A$ for $p < 0$.

Example (3.9.2). — Let $D = \mathrm{diag}(d_1, \ldots, d_{\min(n,m)}) \in \mathrm{Mat}_{n,m}(A)$ be a "diagonal matrix", by which we mean that all of its coefficients are zero, unless their row and column indices are equal. Assume moreover that d_i divides d_{i+1} for every integer i such that $1 \leq i < \min(n, m)$. For any integer $p \in \{0, \ldots, \min(n, m)\}$, *the ideal $\Delta_p(D)$ is generated by the product $d_1 \ldots d_p$.*

Let $I \subset \{1, \ldots, n\}$ and $J \subset \{1, \ldots, m\}$ be two subsets of cardinality p, let $D_{IJ} = (d_{i,j})_{\substack{i \in I \\ j \in J}}$ be the matrix obtained by extracting from D the rows whose indices belong to I and the columns whose indices belong to J. Assume that $I \neq J$ and let us show that $\det(D_{IJ}) = 0$. Indeed, $\det(D_{IJ})$ is a sum of products $d_{i_1,j_1} \ldots d_{i_p,j_p}$ (with signs), where $I = \{i_1, \ldots, i_p\}$ and $J = \{j_1, \ldots, j_p\}$. Since $I \neq J$, there must be an index k such that $i_k \neq j_k$, so that $d_{i_k,j_k} = 0$. However, if $I = J$, one such product is $\prod_{i \in I} d_i$, and all other vanish. By the divisibility assumption on the diagonal entries of D, we see that all minors of size p of M are divisible by $d_1 \ldots d_p$, and that $d_1 \ldots d_p$ is one such minor (namely, for $I = J = \{1, \ldots, p\}$). Consequently, $\Delta_p(D) = (d_1 \ldots d_p)$.

Lemma (3.9.3). — *Let $U \in \mathrm{Mat}_{n,m}(A)$. For any $P \in \mathrm{GL}_n(A)$, $Q \in \mathrm{GL}_m(A)$ and $p \in \{1, \ldots, \min(n, m)\}$, one has*

$$\Delta_p(PUQ) = \Delta_p(U).$$

Proof. — The columns of the matrix UQ are linear combinations of columns of U. By multilinearity of the determinant, each minor of size p of UQ is then a linear combination of minors of size p of U, hence $\Delta_p(UQ) \subset \Delta_p(U)$. Similarly, the rows of the matrix PU are linear combinations of rows of U and $\Delta_p(PU) \subset \Delta_p(U)$. It follows that $\Delta_p(PUQ) \subset \Delta_p(UQ) \subset \Delta_p(U)$. Since P and Q are invertible, we can write $U = P^{-1}(PUQ)Q^{-1}$, hence $\Delta_p(U) \subset \Delta_p(PUQ)$, so that, finally, $\Delta_p(U) = \Delta_p(PUQ)$.   □

3.9.4. — Let M be a finitely generated A-module and let $\varphi \colon A^n \to M$ be a surjective morphism of A-modules. For every integer $p \in \{0, \ldots, n\}$, let then $J_p(\varphi)$ be the ideal of A generated by the ideals $\Delta_p(U) = \Delta_p(u_1, \ldots, u_p)$, where $U = (u_1, \ldots, u_p)$ ranges over all subsets of $\mathrm{Ker}(\varphi)^p$. One has $J_0(\varphi) = A$. If $p < 0$, then we set $J_p(\varphi) = A$.

Lemma (3.9.5). — *Let p be an integer such that $1 \leq p \leq n$.*

a) $J_p(\varphi) \subset J_{p-1}(\varphi)$;

b) *Let $(v_i)_{i \in I}$ be a family of elements of $\mathrm{Ker}(\varphi)$ which generates $\mathrm{Ker}(\varphi)$. Then $J_p(\varphi)$ is generated by the elements $\Delta_p(v_{i_1}, \ldots, v_{i_p})$, where $i_1, \ldots, i_p \in I$;*

c) $\text{Ann}_A(M) \cdot J_{p-1}(\varphi) \subset J_p(\varphi)$.

Proof. — a) Expanding a minor of size p of U along one column, one gets an inclusion $\Delta_p(U) \subset \Delta_{p-1}(U)$. This implies the inclusion $J_p(\varphi) \subset J_{p-1}(\varphi)$.

b) One inclusion is obvious. Moreover, multilinearity of the determinants implies that if $U = (u_1, \ldots, u_p)$ is an $n \times p$-matrix with columns in $\text{Ker}(\varphi)$, then any minor involved in the definition of $\Delta_p(U)$ is a linear combination of elements of A of the form $\Delta_p(v_{i_1}, \ldots, v_{i_p})$.

c) Let $a \in \text{Ann}_A(M)$. Let $u_2, \ldots, u_p \in \text{Ker}(\varphi)$; let also $u \in A^n$ be any basis element, and let $u_1 = au$. Since $a \in \text{Ann}_A(M)$, one has $u_1 \in \text{Ker}(\varphi)$. The Laplace expansion along the first column of the determinant of any $p \times p$ submatrix of U is $\pm a$ times a minor of size $p - 1$ of U, and all such minors appear when u varies, so that $\Delta_p(u_1, \ldots, u_p) = a\Delta_{p-1}(u_2, \ldots, u_p)$. In particular, $aJ_{p-1}(\varphi) \subset J_p(\varphi)$. We thus have shown the inclusion $\text{Ann}_A(M) \cdot J_{p-1}(\varphi) \subset J_p(\varphi)$.   □

Theorem (3.9.6). — *Let A be a commutative ring, let M be a finitely generated A-module. Let $\varphi\colon A^m \to M$ and $\psi\colon A^n \to M$ be surjective morphisms. For every integer $d \geq 0$, the ideals $J_{m-d}(\varphi)$ and $J_{n-d}(\psi)$ are equal.*

Proof. — We split the proof into 5 steps. 1) *We first assume that $m = n$ and that there exists an automorphism u of A^n such that $\psi = \varphi \circ u$. Then $J_p(\varphi) = J_p(\psi)$ for every p.* Indeed, observe that for $X \in A^n$, one has $X \in \text{Ker}(\psi)$ if and only if $u(X) \in \text{Ker}(\varphi)$. Let $X_1, \ldots, X_p \in \text{Ker}(\psi)$; write $X = (X_1, \ldots, X_p)$ and $u(X) = (u(X_1), \ldots, u(X_p))$. By multilinearity of determinant, $\Delta_p(u(X)) \subset \Delta_p(X) \subset J_p(\psi)$. When X runs among all elements of $\text{Ker}(\psi)^p$, $u(X)$ runs among all elements of $\text{Ker}(\varphi)^p$ so that $J_p(\varphi) \subset J_p(\psi)$. Equality follows by symmetry.

2) Let $(e_1, \ldots, e_m)$ be the canonical basis of A^m; for every $i \in \{1, \ldots, m\}$, there exists a $u_i \in A^n$ such that $\psi(u_i) = \varphi(e_i)$, because ψ is surjective. Then, the unique morphism u from A^m to A^n such that $u(e_i) = u_i$ for every i satisfies $\varphi = \psi \circ u$. One constructs similarly a morphism $v\colon A^n \to A^m$ such that $\psi = \varphi \circ v$.

Identify A^{m+n} with $A^m \times A^n$; we define three morphisms $\theta, \theta', \theta''$ from A^{m+n} to M by $\theta(x, y) = \varphi(x) + \psi(y)$, $\theta'(x, y) = \varphi(x)$ and $\theta''(x) = \psi(y)$. They are surjective. Observe that one has

$$\theta(x, y) = \varphi(x) + \psi(y) = \varphi(x) + \varphi(v(y)) = \varphi(x + v(y)) = \theta'(x + v(y), y);$$

likewise,

$$\theta(x, y) = \theta''(x, u(x) + y).$$

Moreover, the endomorphism $(x, y) \mapsto (x + v(y), y)$ of A^{m+n} is an isomorphism (with inverse $(x, y) \mapsto (x - v(y), y)$), and so is the endomorphism $(x, y) \mapsto (x, u(x)+y)$. By the first case above, it follows that for every $p \leq m+n$, one has

$$J_p(\theta) = J_p(\theta') = J_p(\theta'').$$

3) By construction, $\psi\colon A^n \to M$ is a surjective morphism and $\theta''\colon A^m \times A^n \to M$ is the composition of the second projection with ψ. *We then prove*

that $J_p(\theta'') = J_{p-m}(\psi)$ *for any* $p \leq m + n$. This is trivial if $p \leq 0$, hence we assume $p \geq 1$. By induction on m, it suffices to treat the case $m = 1$. Then, $\mathrm{Ker}(\theta'')$ is generated by the vector $(1, 0, \ldots, 0)$ together with the vectors of the form $(0, y)$ for $y \in \mathrm{Ker}(\psi)$.

Let $z_1, \ldots, z_p$ be such vectors and let Z be the matrix with columns $(z_1, \ldots, z_p)$. If the vector $(1, 0, \ldots)$ appears twice among $z_1, \ldots, z_p$, then two columns of Z are equal; since the number of columns of Z is precisely p, this proves that any minor of size p of Z vanishes. If that vector appears exactly once, then we see by expanding a minor of size p along the corresponding column, that $\Delta_p(Z) \subset J_{p-1}(\psi)$. Finally, if all z_j are of the form $(0, y)$, then $\Delta_p(Z) \subset J_p(\psi) \subset J_{p-1}(\psi)$. This shows that $\Delta_p(Z) \subset J_{p-1}(\psi)$ hence $J_p(\theta'') \subset J_{p-1}(\psi)$. Observe moreover that all generators of $J_{p-1}(\psi)$ can be obtained as minors of size p of a suitable matrix Z: just take $z_1 = (1, 0, \ldots, 0)$ and $z_2, \ldots, z_p$ of the form $(0, y)$ with $y \in \mathrm{Ker}(\psi)$. (When $p = 1$, this gives $\Delta_p(Z) = A = J_0(\psi)$.) Consequently, $J_p(\theta'') = J_{p-1}(\psi)$.

4) *By symmetry, one has* $J_p(\theta') = J_{p-n}(\varphi)$ *for every* $p \leq m + n$.

5) The conclusion of the proof is now straightforward. Let d be an integer such that $d \geq 0$. By step 3, we have $J_{n-d}(\psi) = J_{m+n-d}(\theta'')$; by step 4, we have $J_{m-d}(\varphi) = J_{m+n-d}(\theta')$; by step 2, it follows that $J_{n-d}(\psi) = J_{m-d}(\psi) = J_{m+n-d}(\theta)$. $\qquad\square$

Definition (3.9.7). — Let A be a commutative ring and let M be a finitely generated A-module; let d be a positive integer. The dth *Fitting ideal* of M is defined as the ideal $J_{m-d}(\varphi)$, where $\varphi \colon A^m \to M$ is any surjective morphism of A-modules. It is denoted by $\mathrm{Fit}_d(M)$.

Note that, according to theorem 3.9.6, these ideals are indeed independent of the choice of φ.

Corollary (3.9.8). — *Let A be a commutative ring, let M be a finitely generated A-module and let $I = \mathrm{Ann}_A(M)$ be its annihilator.*
a) For any integer $d \geq 0$, $\mathrm{Fit}_d(M) \subset \mathrm{Fit}_{d+1}(M)$ and $I \cdot \mathrm{Fit}_{d+1}(M) \subset \mathrm{Fit}_d(M)$.
b) If M is generated by d elements, then $\mathrm{Fit}_d(M) = A$.

Proof. — This is a direct application of lemma 3.9.5. Let $\varphi \colon A^m \to M$ be a surjective morphism.

a) One has $\mathrm{Fit}_d(M) = J_{m-d}(\varphi) \subset J_{m-d-1}(\varphi) = \mathrm{Fit}_{d+1}(\varphi)$ for every $d \geq 0$. Moreover, $I \cdot \mathrm{Fit}_{d+1}(\varphi) = I \cdot J_{m-d-1}(\varphi) \subset J_{m-d}(\varphi) = \mathrm{Fit}_d(M)$.

b) Assume that m is generated by d elements and let $\psi \colon A^d \to M$ be a surjective morphism of A-modules. Then $\mathrm{Fit}_d(M) = J_0(\psi) = A$. $\qquad\square$

3.10. Exercises

Exercise (3.10.1). — Let A be a ring and let M be a right A-module.

a) Let N be a submodule of M. Show that the set $(N : M)$ of all $a \in A$ such that $ma \in N$ for every $m \in M$ is a two-sided ideal of A.

b) Let $m \in M$ and let $\text{Ann}_A(m) = \{a \in A \, ; \, ma = 0\}$ (annihilator of m). Show that it is a right ideal of A; give an example where it is not a left ideal of A. Show that the submodule generated by m is isomorphic to $A_d/\text{Ann}_A(M)$.

c) Let I be a right ideal of A, let N be a submodule of M and let $(N :_M I)$ be the set of all $m \in M$ such that $ma \in N$ for all $a \in I$. Prove that $(N :_M I)$ is a submodule of M.

Exercise (3.10.2). — Let A and B be two rings and let $f : A \to B$ be a morphism of rings.

a) Let M be a right B-module. Show that one defines a right A-module by endowing the abelian group M with the multiplication $(M, A) \to M$ given by $(m, a) \mapsto mf(a)$. This A-module shall be denoted by f^*M.

b) Let M be a right B-module. Compute the annihilator of the A-module f^*M in terms of the annihilator of M.

c) Let $u : M \to N$ be a morphism of right B-modules; show that u defines a morphism of A-modules $f^*M \to f^*N$.

d) Show that the so-defined map $\text{Hom}_B(M, N) \to \text{Hom}_A(f^*M, f^*N)$ is an injective morphism of abelian groups.

Exercise (3.10.3). — Let A and B be rings and let $f : A \to B$ be a morphism of rings. The additive group of B is endowed with the structure of a right A-module deduced from f.

a) Assume that the image of f is contained in the center of B (so that B is an A-algebra). Show that the multiplication of B is A-bilinear, namely the maps $x \mapsto xb$ and $x \mapsto bx$, for fixed $b \in B$, are A-linear.

b) Conversely, if the multiplication of B is A-bilinear, show that the image of f is contained in the center of B.

Exercise (3.10.4). — Let Λ be a commutative ring and let M and N be two A-modules.

a) Let $u \in \text{End}_A M$. Show that there is a unique structure of an A[X]-module on M for which $X \cdot m = u(m)$ and $a \cdot m = am$ for every $m \in M$ and every $a \in A$. This A[X]-module will be denoted M_u.

b) Show that the map $u \mapsto M_u$ is a bijection from $\text{End}_A(M)$ to the set of all A[X]-module structures on M such that $a \cdot m = am$ for every $m \in M$ and every $a \in A$.

c) Let $u \in \text{End}_A M$ and $v \in \text{End}_A N$. Determine the set of all morphisms from M_u to N_v.

d) If M = N, under what necessary and sufficient condition on u and v are the two modules M_u and M_v isomorphic?

e) Translate the results of the exercise when A = K is a field and $M = K^n$ is the standard K-vector space of dimension n.

Exercise (3.10.5). — *a*) Give an example of two submodules of a module whose union is not a submodule.

b) Let $(M_n)_{n \in \mathbb{N}}$ be an *increasing* family of submodules of an A-module M, meaning that $M_n \subset M_p$ if $n \leq p$. Show that its union $\bigcup_n M_n$ is a subbmodule of M.

c) Let K be a division ring, let V be a K-vector space and let $(W_i)_{1 \leq i \leq n}$ be a family of subspaces of V such that $W = \bigcup_i W_i$ is a vector subspace. If K is infinite or, more generally, if K has at least n elements show that there exists an $i \in \{1, \ldots, n\}$ such that $W = W_i$.

d) Let V_1, V_2 and V_3 be the sets of all pairs $(x, y) \in \mathbb{Z}^2$ such that x, *resp.* $x + y$, *resp.* y is even. Show that these are submodules of $\mathbb{Z}^2$, distinct from $\mathbb{Z}^2$, and that $\mathbb{Z}^2 = V_1 \cup V_2 \cup V_3$.

Exercise (3.10.6). — Let A be a ring, let M be a A-module, let $M_1, \ldots, M_r$ be submodules of M of which M is the direct sum, and let $I_1 = (0 : M_1), \ldots, I_r = (0 : M_r)$ be their annihilators. Assume that the ideals $I_1, \ldots, I_r$ are pairwise comaximal.

Set $I = \bigcap_{i=1}^r I_\alpha$ and, for $i \in \{1, \ldots, r\}$, set $J_i = \bigcap_{j \neq i} I_j$.

a) Show that for any i, I_i and J_i are comaximal two-sided ideals of A.

For every i, let N_i be the submodule of M generated by the submodules M_j, for $j \neq i$.

For any two-sided ideal J of A, let $(0 : J)$ be the submodule of M defined by $\{m \in M ; ma = 0 \text{ for every } a \in J\}$. Show the following formulas:

b) $J_i = (0 : N_i)$ and $N_i = (0 : J_i)$;

c) $N_i = MI_i$ and $MJ_i = M_i = \bigcap_{j \neq i} N_j$.

Exercise (3.10.7). — Let M be a A-module and let $m \in M$ be an element of M. Show that the following properties are equivalent:

(i) The annihilator of mA is (0) and mA has a direct summand in M;

(ii) There exists a linear form f on M such that $f(m) = 1$.

When they hold, show that $M = mA \oplus \operatorname{Ker} f$.

Exercise (3.10.8). — Let $f : M \to N$ be a morphism of A-modules.

a) Show that there exists a morphism $g : N \to M$ such that $g \circ f = \mathrm{id}_M$ if and only if f is injective and $\operatorname{Im}(f)$ has a direct summand in N.

b) Show that there exists a morphism $g : N \to M$ such that $f \circ g = \mathrm{id}_N$ if and only if f is surjective and $\operatorname{Ker}(f)$ has a direct summand in M.

Exercise (3.10.9). — Let A be a ring, let M be a right A-module and let $(M_i)_{i \in I}$ be a family of submodules of M whose sum is equal to M.

a) For the M_i to be in direct sum, it is necessary and sufficient that for every $i \in I$, the intersection of the submodules M_i and $\sum_{j \neq i} M_j$ is reduced to $\{0\}$.

b) Give an example of an A-module M and a family (M_1, M_2, M_3) of submodules of M whose sum equals M but which are not in direct sum for which $M_i \cap M_j = 0$ for every pair (i, j) with $i \neq j$.

Exercise (3.10.10). — Let A be a commutative ring and let M be an A-module.

a) Consider the following definition of a weakly torsion element: an element $m \in M$ is weakly torsion if there exists an element $a \in A - \{0\}$ such that $ma = 0$. Give an example that shows that the set of weakly torsion elements of a module may not be a submodule.

Recall that $T(M)$ is the set of all $m \in M$ for which there exists a regular element $a \in A$ such that $am = 0$.

b) Let S be the set of non-zero divisors of A. Show that $T(M)$ is the kernel of the canonical morphism from M to $S^{-1}M$.

c) Show that the quotient module $M/T(M)$ is torsion free.

d) Let $f: M \to N$ be a morphism of A-modules. Show that $f(T(M)) \subset T(N) \cap f(M)$. Give an example where the inclusion is not an equality.

Exercise (3.10.11). — Let A be a ring, let M be an A-module and let N be a submodule of M. A projector in M is an endomorphism $p \in \mathrm{End}_A(M)$ such that $p \circ p = p$.

a) If p is a projector, prove that $M = \mathrm{Ker}(p) \oplus \mathrm{Im}(p)$.

b) Show that the following conditions are equivalent:

  (i) N has a direct summand in M;

 (ii) N is the kernel of a projector in M;

 (iii) N is the image of a projector in M.

c) Let p_0 and p be two projectors in M with the same image N and let u be the map $p - p_0$. Show that u is a morphism such that $\mathrm{Im}(u) \subset N$ and $\mathrm{Ker}(u) \supset N$. Construct, by quotient, a linear map $\bar{u}: M/N \to N$.

d) Let p_0 be a given projector in M and let N be its image. Show that the map $p \mapsto \bar{u}$ defined in the preceding question induces a bijection from the set of of projectors in M with image N to the abelian group $\mathrm{Hom}_A(M/N, N)$.

Exercise (3.10.12). — Let M be a module which is finitely generated.

a) Show that any generating family of M possesses a finite subfamily which is generating.

b) If M is free, then any basis of M is finite.

Exercise (3.10.13). — Let A be an integral domain and let K be its field of fractions. Assume that $K \neq A$. Show that K is not a free A-module.

Exercise (3.10.14). — Let A be a ring and let M be a right A-module. One writes $M^{\vee} = \mathrm{Hom}_A(M, A)$ for its dual (a left A-module) and $M^{\vee\vee} = \mathrm{Hom}_A(M^{\vee}, A)$ for its *bidual*, that is, the dual of its dual (a right A-module).

a) Let $m \in M$. Show that the map

$$\lambda_m: M^{\vee} \to A, \qquad \varphi \mapsto \varphi(m)$$

is A-linear. Prove that the map $m \mapsto \lambda_m$ is a morphism of A-modules $\lambda: M \to M^{\vee\vee}$.

b) Assume that $M = A_d^n$ for some integer $n \geq 1$. Show that λ is an isomorphism.

One says that M is *reflexive* if the morphism $\lambda \colon M \to M^{\vee\vee}$ is an isomorphism.

c) Give an example where λ is not injective and an example where λ is not surjective.

Exercise (3.10.15). — Let $A = Z^{(N)}$ and $B = Z^N$ be the infinite countable direct sum and the infinite countable product of the abelian group Z respectively.

a) For $n \in N$, let e_n be the element of B whose nth term is 1, all others being 0. Prove that $(e_n)_{n \in N}$ is a basis of A, as a Z-module.

b) For every $b = (b_n) \in B$, prove that there exists a unique linear form $\varphi(b) \in A^\vee$ such that $\varphi(b)(e_n) = b_n$ for every $n \in N$. Prove that the map $b \mapsto \varphi(b)$ is an isomorphism from B to $A^\vee$.

c) For $n \in N$, let $\theta \colon B^\vee \to B$ be the map given by $\theta(f) = (f(e_n))_{n \in Z}$. Prove that θ is a morphism of Z-modules.

d) Let $f \in \operatorname{Ker}(\theta)$. Let $x \in B$; prove that there exist $y, z \in B$ such that $x = y + z$, and such that $2^n \mid y_n$ and $3^n \mid z_n$ for every $n \in N$. Prove that $2^n \mid f(y)$ and $3^n \mid f(z)$, for every $n \in Z$. Conclude that $f(x) = 0$, hence $f = 0$. This shows that θ is injective.

e) Let $f \in B^\vee$ and let $a = \theta(f)$; assume that $a \notin A$. Let (s_n) be a strictly increasing sequence of integers and let $(b_n) \in B$ be such that $2^{s_n} \mid b_n$ for all n. Prove that $f(b) \equiv \sum_{k=0}^{n-1} a_n b_n \pmod{2^{s_n}}$. Construct (s_n) and $(b_n) \in B$ so as to prevent any integer from satisfying all of these congruences. This proves that $\theta(B^\vee) \subset A$.

f) Prove that $\theta \colon B^\vee \to A$ is an isomorphism.

g) Prove that B is not a free Z-module.

Exercise (3.10.16). — Let M and N be A-modules. Let $f \colon M \to N$ be a morphism of A-modules. Its *transpose* $f^\vee$ is the map from $N^\vee$ to $M^\vee$ defined by $f^\vee(\varphi) = \varphi \circ f$, for every $\varphi \in N^\vee = \operatorname{Hom}_A(N, A)$.

a) Show that $f^\vee$ is a morphism of A-modules. Show that $(f+g)^\vee = f^\vee + g^\vee$. Show that $(f \circ g)^\vee = g^\vee \circ f^\vee$.

b) From now on, we assume that $M = N$ and that A is commutative. Let $f \in \operatorname{End}_A(M)$. Show that the set $I(f)$ of all polynomials $P \in A[X]$ such that $P(f) = 0$ is an ideal of $A[X]$.

c) Show that $I(f) \subset I(f^\vee)$.

d) If M is reflexive, show that $I(f) = I(f^\vee)$.

Exercise (3.10.17). — Let A be a commutative ring and let I be a non-zero ideal of A, viewed as a submodule of A. Show that I is a free module if and only if the ideal I is principal and is generated by a regular element.

Exercise (3.10.18). — Let A be a ring, let M and N be right A-modules and let $\varphi \colon M \to N$ be a surjective morphism.

a) If Ker(φ) and N are finitely generated, show that M is finitely generated.

In the sequel, we assume that M and N are finitely generated and that N is a free A-module.

b) Show that φ has a right inverse ψ, namely a morphism $\psi \colon N \to M$ such that $\varphi \circ \psi = \mathrm{id}_N$.

c) Show that M $\simeq$ Ker(φ) $\oplus$ Im(ψ).

d) Recover the fact that any vector subspace of a vector space has a direct summand.

e) Show that Ker(φ) is finitely generated.

Exercise (3.10.19). — Let A be a ring and let M be a right A-module which is the (internal) direct sum of an *infinite* family of non-zero submodules $(M_i)_{i \in I}$.

For $x \in M$, write $x = \sum_{i \in I} x_i$, with $x_i \in M_i$, and let $I(x)$ be the set of all $i \in I$ such that $x_i \neq 0$. This is a finite subset of I.

a) Let S be a generating subset of M. Show that I is equal to the union, for $x \in S$, of the subsets $I(x)$.

b) Show that S is infinite.

*c**)Show that $\mathrm{Card}(S) \geq \mathrm{Card}(I)$.

Exercise (3.10.20). — Let A be a commutative ring, let S be a multiplicative subset of A and let M be an A-module.

a) Let $(m_i)_{i \in I}$ be a generating family of M; show that the family $(m_i/1)$ is a generating family of the $S^{-1}A$-module $S^{-1}M$.

b) Let $(m_i)_{i \in I}$ be a linearly independent family in M. Assume that S does not contain any zero-divisors. Then show that the family $(m_i/1)$ in $S^{-1}M$ is still linearly independent.

c) Assume that A is an integral domain and that M is generated by n elements, for some integer n. Show that the cardinality of any free subset of M is at most n. (*Take for* S *the set of all non-zero elements of* A.)

Exercise (3.10.21). — Let A be a commutative ring and let S be a multiplicative subset of A. Let M be an A-module and let $i \colon M \to S^{-1}M$ be the canonical morphism.

a) Let $\mathcal{N}$ be a submodule of $S^{-1}M$ and let $N = i^{-1}(\mathcal{N})$. Prove that N satisfies the following property: for every $m \in M$, if there exists an $s \in S$ such that $sm \in N$, then $m \in N$. (One says that N is saturated with respect to S.)

b) We assume in this question that S is the set of regular elements of A. Prove that a submodule N of M is saturated with respect to S if and only if M/N is torsion-free.

c) Let N be a submodule of M which is saturated with respect to S. Prove that $N = i^{-1}(S^{-1}N)$.

d) Deduce from the preceding questions that the map $\mathcal{N} \mapsto i^{-1}(\mathcal{N})$ defines a bijection from the set of submodules of $S^{-1}M$ to the set of submodules of M which are saturated with respect to S.

Exercise (3.10.22). — Let A be a ring and let M be a right A-module. One says that a submodule N of M is *small* if for every submodule K of M such that $N + K = M$, one has $K = M$; one says that it is maximal if $N \neq M$ and if for every submodule K of M such that $N \subset K \subset M$, one has $K = N$ or $K = M$. Let J_M be the intersection of all maximal submodules of M.

a) Prove that $M = J_M$ if and only if M is not finitely generated.

b) Prove that J_M is the sum of all small submodules of M.

Exercise (3.10.23). — Let A be a ring, let J be its Jacobson radical and let M be a right A-module. Assume that there exists an A-module N such that $M \oplus N$ is a free A-module (that is, M is projective).

a) Prove that $J_M = MJ$, where J_M is the intersection of all maximal submodules of M (exercise 3.10.22).

b) Prove that $M \neq MJ$ (BASS (1960), proposition 2.7).

Exercise (3.10.24). — Let A be a ring which is not a division ring. Give examples of:

a) Non-free modules;

b) A free family with n elements of A^n which is not a basis;

c) A minimal generating set of a module which is not a basis;

d) A submodule which has no direct summand;

e) A free module possessing a non-free submodule.

Exercise (3.10.25). — *a)* Let M be a non-zero submodule of **Q** which is finitely generated.

Show that M is free and that any basis of M has cardinality 1. (*Show by induction that there exists an $a \in \mathbf{Q}$ such that $M = \mathbf{Z}a$.*)

b) Show that the **Z**-module **Q** is not finitely generated.

c) What are the maximal free subsets of **Q**?

d) Does the **Z**-module **Q** possess minimal generating subsets?

Exercise (3.10.26). — Let K be a field, let V be a right K-vector space and let A be the ring of endomorphisms of V. We assume that V is not finitely generated.

Show that the right A-module $(A_d)^2$ is isomorphic to A_d.

Exercise (3.10.27). — Let A be a ring and let $\varphi \colon (A_d)^m \to (A_d)^n$ be a morphism of right A-modules. Let Φ be the matrix of φ.

For any morphism of rings $f \colon A \to B$, one writes Φ^f for the matrix whose entries are the images by f of those of Φ.

a) Show that Φ^f is the matrix of a morphism $\varphi_0 \colon (B_d)^m \to (B_d)^n$ of right B-modules.

b) Show that if φ is an isomorphim, then φ_0 is an isomorphism too. (*Introduce the matrix Φ' of the inverse of φ, then the matrix $(\Phi')^f$.*)

c) Conclude that if A is a ring which possesses a morphism to a division ring K, then the A-module $(A_d)^m$ can only be isomorphic to the A-module $(A_d)^n$ when $m = n$. In other words, in that case, the cardinalities of all bases of a finitely generated free A-module are equal to a common integer, called the *rank* of this module.

Exercise (3.10.28). — Let K be a division ring, let V be a right K-vector space and let W be a subspace of V.

a) Show that V is the intersection of the kernels of all linear forms on V which vanish on W.

b) More generally, what is the intersection of the kernels of all linear forms on V which vanish on a given subset S of V?

c) Give an example of a ring A and of a non-zero A-module M for which every linear form is null.

d) Give an example of an integral domain A, of a free A-module M, and of a submodule N of M such that N is not equal to the intersection of the kernels of all linear forms on M which vanish on N.

Exercise (3.10.29). — Let K be a division ring, let V_1, V_2, V_3, V_4 be right K-vector spaces and let $u: V_1 \to V_2$, $v: V_3 \to V_4$ and $w: V_1 \to V_4$ be linear maps.

a) In order that there is a linear map $f: V_2 \to V_4$ such that $f \circ u = w$, it is necessary and sufficient that $\mathrm{Ker}(u) \subset \mathrm{Ker}(w)$.

b) In order that there is a linear map $g: V_1 \to V_3$ such that $v \circ g = w$, it is necessary and sufficient that $\mathrm{Im}(w) \subset \mathrm{Im}(v)$.

c) In order that there is a linear map $h: V_2 \to V_3$ such that $v \circ h \circ u = w$, it is necessary and sufficient that $\mathrm{Ker}(u) \subset \mathrm{Ker}(w)$ and $\mathrm{Im}(w) \subset \mathrm{Im}(v)$.

Exercise (3.10.30). — Let K be a division ring and let M be a right K-vector space. Assume that M is finitely generated.

a) Let $(m_1, \ldots, m_n)$ be a generating family in M and let $(v_1, \ldots, v_p)$ be a free family in M. Show by induction on n that $p \leq n$.

b) Show that all bases of M are finite and have the same cardinality.

Exercise (3.10.31). — Let A be a ring, let n be an integer and let $U \in \mathrm{Mat}_n(A)$. Let C_U be the centralizer of U in $\mathrm{Mat}_n(A)$.

a) Prove that the coefficients of the characteristic polynomial $P_U(X)$ of U belong to C_U. (*Use the adjugate formula.*)

b) Observe that U belongs to the center of C_U.

c) Applying the evaluation morphism $C_U[X] \to C_U$ at U, reprove the Cayley–Hamilton theorem 3.8.20.

Exercise (3.10.32). — Let A be a commutative ring.

a) Let f be an endomorphism of an A-module M; we assume that there exists a monic polynomial $P \in A[X]$ such that $P(f) = 0$ in $\mathrm{End}_A(M)$.

Let L be a submodule of M such that $f^{-1}(L) \subset L$. Prove that $f(L) \subset L$. (*For every integer k, let P_k be the quotient of the euclidean division of P by T_k; prove by induction on k that $P_k(f)(L) \subset L$.*)

Let M be an A-module, let N be a submodule of M and let $f : N \to M$ be a surjective morphism.

b) Assume that N is finitely generated. Prove that there exists an integer n, a surjective morphism $p : A^n \to N$ and a morphism $g : A^n \to A^n$ such that $f \circ p \circ g = p$. Applying a) with M, f and L equal to A^n, g and $\mathrm{Ker}(f \circ p)$, prove that $g(\mathrm{Ker}(f \circ p)) \subset \mathrm{Ker}(f \circ p)$. Deduce from this that f is bijective.

c) Assume that M is finitely generated. Construct a finitely generated submodule N′ of N such that $f(N') = M$. Applying b) to $f|_{N'+Am}$, for $m \in N$, prove that f is bijective.

(This is a result of ORZECH (1971); the given proof is due to GRINBERG (2016).)

Exercise (3.10.33). — a) Let M be a free finitely generated A-module and let u be an endomorphism of M. Show that the A-module $M/\mathrm{Im}(u)$ is annihilated by $\det(u)$.

b) Let M and N be free finitely generated A-modules, let $(e_1, \dots, e_m)$ be a basis of M and let $(f_1, \dots, f_n)$ be a basis of N. Let $u : M \to N$ be a linear map and let U be its matrix in the above given bases. Assume that $m \geq n$. Then show that $N/u(M)$ is annihilated by every $n \times n$ minor from U.

Exercise (3.10.34). — Let A be a commutative ring and let M be a finitely generated A-module.

a) Let I be an ideal of A. Prove that $\mathrm{Fit}_d(M/IM) = \mathrm{Fit}_d(M) + I$, and that its image in A/I coincides with the dth Fitting ideal of M/IM, viewed as an A/I-module.

b) Let B be a commutative ring and let $f : A \to B$ be a morphism of rings. Prove that $\mathrm{Fit}_d(M \otimes_A B) = \mathrm{Fit}_d(M)B$.

c) Let $f : M \to M'$ be a surjective morphism of A-modules; prove that $\mathrm{Fit}_d(M) \subset \mathrm{Fit}_d(M')$ for every integer d.

d) Let M′ be a finitely generated A-module. Let d be a positive integer; prove that

$$\mathrm{Fit}_d(M \oplus M') = \sum_{k=0}^{d} \mathrm{Fit}_k(M)\,\mathrm{Fit}_{d-k}(M').$$

Exercise (3.10.35). — Let A be a commutative ring and let M be a finitely generated A-module.

Let P be a prime ideal of A and let d be a positive integer. Prove that the following assertions are equivalent:

(i) $\mathrm{Fit}_d(M) \not\subset P$;

(ii) The A_P-module M_P can be generated by d elements;

(iii) There exists an element $a \in A - P$ such that the A_a-module M_a can be generated by d elements;

(iv) The dimension of the (A_P/P)-vector space M_P/PM_P is smaller than d.

Chapter 4.
Field Extensions

One aspect of commutative algebra is to not only consider modules over a given fixed ring, but also morphisms of (commutative) rings.

Let A and B be commutative rings. Recall that it is equivalent to give a morphism of rings $f : A \to B$ or a structure of an A-algebra on B, related by the formula $f(a) = a \cdot 1_B$.

When A and B are fields, any morphism $f : A \to B$ is injective and the study of B as an A-algebra often assumes that A is a subfield of B, f being the inclusion. The classical terminology is to say that B is a field extension of A; one reason for this terminology is that A was often assumed to be given, and B was constructed by enlarging A with new "imaginary" elements.

I will generalize this terminology here and say that the B is an extension of rings of A if it is given with the structure of an A-algebra. In fact, one can essentially reduce the study of a morphism of rings $f : A \to B$ to the study, on the one hand, of the pair consisting of B and its subring $f(A)$, and on the other hand, of the quotient ring $f(A) = A/I$, where $I = \text{Ker}(f)$. In particular, if f is injective, it induces an isomorphism from A to $f(A)$ and one is in the situation of a subring.

Often, we understate the underlying morphism $f : A \to B$, and write sentences such as "Let $A \to B$ be an extension of rings."

The main theme of this chapter is to study the ring B in terms of whether or not its elements satisfy polynomial equations with coefficients in A. If yes, one says that the extension is **algebraic**; and one says that the extension is **integral** if every element of B satisfies a monic polynomial equation with coefficients in A. As we will learn in the first sections, this property is witnessed by adequate finitely generated A-submodules of B.

The importance of integral extensions of commutative rings will be explored in chapter 9 and I rapidly focus here on algebraic extensions of fields. Because they allow us to make use of linear algebras, **finite extensions** are essential.

Although the reader has certainly heard about the notion of algebraically closed fields, I explain it here, and prove Steinitz's theorem that any field has an **algebraic closure**, and essentially only one.

© Springer Nature Switzerland AG 2021
A. Chambert-Loir, *(Mostly) Commutative Algebra*, Universitext,
https://doi.org/10.1007/978-3-030-61595-6_4

The next section is devoted to separability, *a property of algebraic extensions which makes their study considerably simpler. Fortunately, many fields only have separable extensions, and they are called* perfect.

Building on these results, I can then study finite fields. *All in all, the approach that I followed to describe them is certainly not the most economic one, and it would have been possible to prove these results earlier, at the only cost of having to redo some constructions later in greater generality.*

In the next section, I define Galois extensions *of fields, their Galois group, and show the Galois correspondence, namely a bijection between intermediate extensions of a Galois extension and subgroups of the Galois group. However, I do not discuss the classical applications of Galois's theory, such as construction with ruler and compass or the impossibility of resolutions in radicals.*

I then define norms *and* traces, *and establish their basic properties; I also show how separability is equivalent to the non-degeneracy of the trace map.*

In a final section, I discuss general extension of fields and define their transcendence degree.

4.1. Integral Elements

Definition (4.1.1). — Let $A \to B$ be an extension of commutative rings. One says that b is *integral* over A if there exists a *monic* polynomial $P \in A[X]$ such that $P(b) = 0$.

A relation of the form $b^n + a_{n-1}b^{n-1} \cdots + a_1 b + a_0 = 0$, where $a_0, \ldots, a_{n-1}$ are elements of A, is called an *integral dependence relation* for b.

Example (4.1.2). — The complex numbers $z = \exp(2i\pi/n)$, $u = (-1 + \sqrt{5})/2$ are integral over **Z**: they satisfy the relations $z^n = 1$ and $u^2 + u + 1 = 0$.

To establish the most basic properties of integral elements, we will make use of the following elementary lemma on generating families of modules.

Lemma (4.1.3). — *Let $A \to B$ be an extension of rings and let M be a B-module. Let $(b_i)_{i \in I}$ be a family of elements of B which generates B as an A-module and let $(m_j)_{j \in J}$ be a family of elements of M which generates it as a B-module.*

(i) *The family $(b_i m_j)_{(i,j) \in I \times J}$ generates M as an A-module.*

(ii) *If $(b_i)_{i \in I}$ is a basis of B as an A-module and $(m_j)_{j \in J}$ is a basis of M as a B-module, then the family $(b_i m_j)_{(i,j)}$ is a basis of M as an A-module.*

Proof. — *a)* Let $m \in M$; since the family (m_j) generates M as a B-module, there exists an almost-null family (c_j) in B such that $m = \sum_{j \in J} c_j m_j$. For every $j \in J$ such that $c_j \neq 0$, there exists an almost-null family $(a_{ij})_{i \in I}$ of elements of A such that $c_j = \sum a_{ij} b_i$, because $(b_i)_{i \in I}$ generates B as an A-module. For $j \in J$ such that $c_j = 0$, set $a_{ij} = 0$ for all $i \in I$. Then the family $(a_{ij})_{(i,j) \in I \times J}$ is almost-null and one has

$$m = \sum_{j \in J} c_j m_j = \sum_{j \in J} \sum_{i \in I} a_{ij} b_i m_j,$$

which proves that the family $(b_i m_j)$ generates M as an A-module.

b) Let (a_{ij}) be an almost-null family in A such that $\sum_{i,j} a_{ij} b_i m_j = 0$. For every $j \in J$, set $c_j = \sum_{i \in I} a_{ij} b_i$. The family $(c_j)_{j \in J}$ in B is almost-null, and one has $\sum_{j \in J} c_j m_j = 0$. Since the family (m_j) is free, one has $c_j = 0$ for all $j \in J$. Fix $j \in J$; from the relation $\sum_{i \in I} a_{ij} b_i = 0$, the freeness of the family (b_i) implies that $a_{ij} = 0$ for all i. This shows that $a_{ij} = 0$ for all i and j. This proves that the family $(b_i m_j)$ is free, as claimed. □

Corollary (4.1.4). — *Let* A → B *be an extension of rings and let* M *be a* B-*module. If* B *is finitely generated (resp. free and finitely generated) as an* A-*module, and* M *is finitely generated (resp. free and finitely generated) as a* B-*module, then* M *is finitely generated (resp. free and finitely generated) as an* A-*module.*

The next theorem gives a very useful characterization of integral elements.

Theorem (4.1.5). — *Let* A → B *be an extension of rings, let* b *be an element of* B *and let* A[b] *be the* A-*subalgebra of* B *generated by* b. *The following assertions are equivalent:*

(i) *The element* b *is integral over* A;

(ii) *The* A-*algebra* A[b] *is a finitely generated* A-*module;*

(iii) *There exists an* A[b]-*module whose annihilator is zero, and which is finitely generated as an* A-*module.*

Proof. — (i)⇒(ii). Let $P \in A[X]$ be a monic polynomial such that $P(b) = 0$. Let us write it as $P = X^n + a_1 X^{n-1} + \cdots + a_n$, for $a_1, \ldots, a_n \in A$, and let us show that the family $(1, b, \ldots, b^{n-1})$ generates A[b] as an A-module. By definition, an element of A[b] is of the form $Q(b)$, for some polynomial $Q \in A[X]$. Since P is monic, we may consider a euclidean division of Q by P, say $Q = P Q_1 + R$, where $\deg(R) < n$. Then $Q(b) = P(b) Q_1(b) + R(b) = R(b)$, hence is a linear combination of $(1, \ldots, b^{n-1})$ with coefficients in A, as required.

(ii)⇒(iii). It suffices to set $M = A[b]$. Indeed, the module M is finitely generated as an A-module, by assumption. Moreover, if $x \in A[b]$ is such that $xM = 0$, we obtain $x 1_B = x = 0$, so that the annihilator of M as an A[b]-module is trivial.

(iii)⇒(i). Let $\varphi \colon M \to M$ be the A-linear morphism defined by $\varphi(m) = bm$. Since M is finitely generated, it follows from corollary 3.8.21 to the Cayley–Hamilton theorem that there exist $n \in N$ and $a_0, \ldots, a_{n-1} \in A$ such that

$$\varphi^n + a_{n-1} \varphi^{n-1} + \cdots + a_0 = 0$$

in $\mathrm{End}_A(M)$. In particular, for every $m \in M$, one has

$$(b^n + a_{n-1} b^{n-1} + \cdots + a_0) m = 0.$$

Since the annihilator of M is trivial, this implies that $b^n + a_{n-1}b^{n-1} + \cdots + a_0 = 0$. Consequently, b is integral over A. □

Definition (4.1.6). — Let A → B be an extension of rings. The set of elements of B which are integral over A is called the *integral closure* of A in B.

Corollary (4.1.7). — *If* A → B *is an extension of rings, then the integral closure of* A *in* B *is a subalgebra of* B.

Proof. — Let $f: A → B$ be the morphism of rings which gives B the structure of an A-algebra.

Let $a \in A$. Its image $f(a)$ in B is integral over A, since it is cancelled by the polynomial $X - a$. Consequently, every element of $f(A)$ is integral over A. In particular, the zero and the unit elements of B are integral over A.

Let then b, c be elements of B which are integral over A. By theorem 4.1.5, the subalgebra $A[b]$ of B is a finitely generated A-module in B. Since c is integral over A, it is also integral over $A[b]$. By theorem 4.1.5 again, the A-subalgebra $A[b, c] = A[b][c]$ of B is a finitely generated $A[b]$-module. It then follows from corollary 4.1.4 that the subalgebra $A[b, c]$ of B is a finitely generated A-module. Corollary 4.2.2 implies that every element of $A[b, c]$ is integral over A. In particular, $b + c$ and bc are integral over A. □

Definition (4.1.8). —   (i) Let B be a ring and let A be a subring of B. One says that A is *integrally closed* in B if it coincides with its own integral closure in B.

(ii) One says that an integral domain is *integrally closed* if it is integrally closed in its field of fractions.

Theorem (4.1.9). — *A unique factorization domain is integrally closed.*

In particular, principal ideal domains and polynomial rings over a field are integrally closed.

Proof. — Let A be a unique factorization domain and let K be its field of fractions. Let $x \in K$; let us assume that x is integral over A and let us prove that $x \in A$. Let $P = X^n + a_1 X^{n-1} + \cdots + a_n$ be a monic polynomial with coefficients in A such that $P(x) = 0$. Let a, b be coprime elements of A such that $x = a/b$ and let us multiply by b^n the relation $P(x) = 0$; we obtain the relation

$$a^n + a_1 a^{n-1} b + \cdots + a_{n-1} a b^{n-1} + a_n b^n = 0.$$

Consequently, $a^n = -b(a_1 a^{n-1} + \cdots + a_n b^{n-1})$ is a multiple of b. Since b is prime to a, it is prime to a^n (proposition 2.5.17). Necessarily, b is invertible and $x \in A$. □

Proposition (4.1.10). — *Let* A *be an integral domain which is integrally closed. For any multiplicative subset* S ⊂ A *which does not contain* 0, *the ring* $S^{-1}A$ *is integrally closed.*

Proof. — Let K be the field of fractions of A, so that we have natural injections $A \subset S^{-1}A \subset K$, and K is the field of fractions of $S^{-1}A$. Let $x \in K$ be any element which is integral over $S^{-1}A$. Let

$$x^n + a_{n-1}x^{n-1} + \cdots + a_0 = 0$$

be an integral dependence relation, where $a_0, \ldots, a_{n-1}$ are elements of $S^{-1}A$. Let $s \in S$ be a common denominator of all the a_i, so that $b_i = sa_i \in A$ for every $i \in \{0, \ldots, n-1\}$. Multiplying the preceding relation by s^n, we obtain the relation

$$(sx)^n + b_{n-1}(sx)^{n-1} + \cdots + b_1 s^{n-1}(sx) + b_0 s^{n-1} = 0,$$

which shows that sx is integral over A. Since A is integrally closed in K, $sx \in A$. Consequently, $x = (sx)/s \in S^{-1}A$ and $S^{-1}A$ is integrally closed in K. □

4.2. Integral Extensions

Definition (4.2.1). — Let $A \to B$ be an extension of commutative rings. One says that B is integral over A if every element of B is integral over A.

We start from corollaries of theorem 4.1.5

Corollary (4.2.2). — *Let A be a ring and let B be a finitely generated A-algebra. Then B is integral over A if and only if B is finitely generated as an A-module.*

Proof. — First assume that B is finitely generated as an A-module and let us prove that B is integral over A. Let $b \in B$ and let M be the $A[b]$-module B. Its annihilator is 0 since for any $x \in B$, the relation $xM = 0$ implies $0 = x1_B = x$. Finally, it is finitely generated as an A-module, by assumption. Consequently, b is integral over A.

Conversely, let us assume that B is integral over A. Let $(b_1, \ldots, b_n)$ be a finite family of elements of B which generates B as an A-algebra. Let B_0 be the image of A in B; for every $i \in \{1, \ldots, n\}$, let $B_i = A[b_1, \ldots, b_i]$; observe that $B_i = B_{i-1}[b_i]$. By assumption, b_i is integral over A, hence it is integral over B_{i-1}. By theorem 4.1.5, B_i is a finitely generated B_{i-1}-module. By induction, it then follows from corollary 4.1.4 that $B = B_n$ is a finitely generated B_0-module, hence a finitely generated A-module. □

Corollary (4.2.3). — *Let $A \to B$ and $B \to C$ be extensions of rings. Assume that B is integral over A.*

If an element c of C is integral over B, then c is integral over A. In particular, if C is integral over B, then C is integral over A.

Proof. — Let $c \in C$. Let $P \in B[X]$ be a monic polynomial such that $P(c) = 0$. Let B_1 be the A-subalgebra of B generated by the coefficients of P. It is finitely generated as an A-module, because B is integral over A (corollary 4.2.2). Moreover, c is integral over B_1 by construction, so that the algebra $B_1[c]$ is finitely generated as a B_1-module. By corollary 4.1.4, $B_1[c]$ is an A-subalgebra which is finitely generated as an A-module. By theorem 4.1.5, c is integral over A.   □

Lemma (4.2.4). — *Let* A → B *be an extension of rings.*

(i) *For any multiplicative subset* S *of* A, *the ring* $S^{-1}B$ *is integral over* $S^{-1}A$.

(ii) *Let* J *be an ideal of* B *and let* I *be an ideal of* A *such that* $f(I) \subset J$. *By way of the canonical morphism* $\varphi \colon A/I \to B/J$, *the ring* B/J *is integral over* A/I.

Proof. — a) Let $x \in S^{-1}B$; let us show that x is integral over $S^{-1}A$. Let $b \in B$ and $s \in S$ be such that $x = b/s$. By assumption, there exists an integral dependence relation over A for b, say

$$b^n + a_{n-1}b^{n-1} + \cdots + a_0 = 0.$$

Then

$$x^n + \frac{a_{n-1}}{s}x^{n-1} + \cdots + \frac{a_0}{s^n} = 0$$

is an integral dependence relation for x over $S^{-1}A$.

b) Let $x \in B/J$ and let $b \in B$ be any element such that $x = cl_J(b)$. Let

$$b^n + a_{n-1}b^{n-1} + \cdots + a_0 = 0$$

be an integral dependence relation for b over A. Then

$$x^n + cl_I(a_{n-1})x^{n-1} + \cdots + cl_I(a_0) = 0$$

is an integral dependence relation for x over A/I.   □

Proposition (4.2.5). — *Let* B *be a ring and let* A *be a subring of* B *such that* B *is integral over* A.

(i) *Assume that* A *is a field. Then every regular element of* B *is invertible in* B.

(ii) *Assume that* B *is an integral domain. Then* A *is a field if and only if* B *is a field.*

Proof. — a) Let us assume that A is a field, let $b \in B$ be a regular element and let $P = X^n + a_{n-1}X^{n-1} + \cdots + a_0 \in A[X]$ be a monic polynomial of minimal degree such that $P(b) = 0$. Let $Q = X^{n-1} + a_{n-1}X^{n-2} + \cdots + a_1$, so that $P = XQ + a_0$. One has $bQ(b) = -a_0$. By the minimality assumption on n, $Q(b) \neq 0$. Since b is regular, $a_0 = -bQ(b) \neq 0$. Since $a_0 \in A$ and A is a field, this implies that b is invertible, with inverse $b' = -Q(b)/a_0$.

b) Assume that A is a field. Since B is an integral domain, $B \neq 0$; moreover, every non-zero element of B is regular, hence invertible by part a). This shows that B is a field.

Conversely, let us assume that B is a field. Since the unit elements of A and B coincide, $A \neq 0$. Let a be a non-zero element of A. Then a is invertible in B so that there exists an element $b \in$ B such that $ab = 1$. Since B is integral over A, there exists a monic polynomial $P = X^n + a_{n-1}X^{n-1} + \cdots + a_0$ with coefficients in A such that $P(b) = 0$. In other words,

$$b^n + a_{n-1}b^{n-1} + \cdots + a_0 = 0.$$

Multiplying this relation by a^{n-1}, we obtain

$$b = a^{n-1}b^n = -a_{n-1} - a_{n-2}a - \cdots - a_0 a^{n-1}.$$

In particular, $b \in$ A so that a is invertible in A. This proves that A is a field.$\square$

4.3. Algebraic Extensions

Definition (4.3.1). — Let A $\to$ B be an extension of rings. One says that an element b of B is *algebraic* over A if there exists a non-zero polynomial $P \in A[X]$ such that $P(b) = 0$.

If every element of B is algebraic over A, one says that B is algebraic over A.

A non-trivial relation of the form $a_n b^n + a_{n-1}b^{n-1} \cdots + a_1 b + a_0 = 0$, where $a_0, \ldots, a_{n-1}, a_n$ are elements of A, is called an *algebraic dependence relation* for b.

Remarks (4.3.2). —   (i) Let $b \in$ B be algebraic over A, and let $P = a_n X^n + \cdots + a_0$ be a polynomial with coefficients in A such that $P(b) = 0$. Then $a_n b$ is integral over A. Indeed, if we multiply the relation $P(b) = 0$ by a_n^{n-1}, we obtain

$$(a_n b)^n + a_{n-1}(a_n b)^{n-1} + \cdots + a_0 a_n^{n-1} = 0,$$

an integral dependence relation for $a_n b$.

(ii) Conversely, let $b \in$ B and let $a \in$ A be a non-nilpotent element such that ab is integral over A. Then b is algebraic over A. Indeed, an integral dependence relation

$$(ab)^n + c_{n-1}(ab)^{n-1} + \cdots + c_0 = 0$$

for ab gives rise to an algebraic dependence relation

$$a^n b^n + a^{n-1}c_{n-1} b^{n-1} + \cdots + ac_1 b + c_0 = 0$$

for b. Since a is not nilpotent, $a^n \neq 0$, which implies that this relation is non-trivial.

In particular, if A is a field, then an element $b \in$ B is algebraic over A if and only if it is integral over A.

(iii) The notion of being algebraic over A is not really interesting if A is not an integral domain or if some non-zero elements of A annihilate B. In fact, we shall mostly use it when A is a field.

Corollary (4.3.3). — *Let K be a field and let A be a K-algebra which is an integral domain. The set of all elements of A which are algebraic over K is a field extension of K.*

We call it the *algebraic closure* of K in A. Be cautious to distinguish this notion from that of an "absolute" algebraic closure of a field which will be defined below.

Remark (4.3.4). — Let K be a field and let A be a K-algebra. Let $a \in L$ be an element which is algebraic over K. The set of all polynomials $P \in K[T]$ such that $P(a) = 0$ is the kernel of the morphism from $K[T]$ to A given by $P \mapsto P(a)$. Consequently, the monic generator P of this ideal is the monic polynomial of minimal degree such that $P(a) = 0$; one says that it is the *minimal polynomial* of a over K.

Passing to the quotient, one gets an injective morphism $K[T]/(P) \to A$. Assume moreover that A is an integral domain. Then $K[T]/(P)$ is an integral domain as well, so that the polynomial P is irreducible.

Definition (4.3.5). — One says that a field extension $K \to L$ is *finite* if L is a finite-dimensional K-vector space. This dimension is called the *degree* of the extension, and is denoted by $[L : K]$.

Corollary (4.3.6). — *Let K be a field, let A be a K-algebra which is an integral domain and let $a_1, \ldots, a_n \in A$ be elements of A which are algebraic over K. Then the sub-algebra $K[a_1, \ldots, a_n]$ of A generated by these elements is a subfield of A, and it is a finite extension of K.*

Proof. — Let us denote this algebra by B. It follows from the previous corollary that B is a field. Then, corollary 4.2.2 implies that it is a finite extension of K. □

Proposition (4.3.7). — *Let $K \to L$ and $L \to M$ be two extensions of fields. Then the composed field extension $K \to M$ is finite if and only if both extensions $K \to L$ and $L \to M$ are finite; then one has*

$$[M : K] = [M : L][L : K].$$

Proof. — Let $(y_i)_{i \in I}$ be a basis of M as an L-vector space and let $(x_j)_{j \in J}$ be a basis of L as a K-vector space. Let us show that $(x_i y_j)_{(i,j) \in I \times J}$ is a basis of M as a K-vector space.

Indeed, let $m \in M$; one may write $m = \sum_{i \in I} \ell_i y_i$, for an almost-null family (ℓ_i) of elements of L. For every i, let $(k_{ij})_{j \in J}$ be an almost null family of elements of K such that $\ell_i = \sum_{j \in J} k_{ij} x_j$. Then the family $(k_{ij})_{(i,j) \in I \times J}$ is almost-null and $m = \sum_{(i,j) \in I \times J} k_{ij} x_j y_i$. This shows that the family $(x_i y_j)$ generates M. On the other hand, let $(k_{ij})_{(i,j)}$ be a family of elements of K such that $\sum_{(i,j)} k_{ij} x_j y_i = 0$. For every i, let us set $\ell_i = \sum_{j \in J} k_{ij} x_j \in L$. The family (ℓ_i) is almost null and $\sum_{i \in I} \ell_i y_i = 0$. Since (y_i) is free over L, one has $\ell_i = 0$ for every i. Since (x_j) is free over K, one has $k_{ij} = 0$ for every i and every j. This shows that the family $(x_j y_i)_{(i,j)\in I\times J}$ is a basis of M over K. It is finite if and only if I and J are both finite. Then,

$$[M : K] = \text{Card}(I \times J) = \text{Card}(I)\,\text{Card}(J) = [M : L][L : K].$$

Let K be a field. Recall that a polynomial $P \in K[X]$ in one indeterminate X is said to be *split* if there are elements $c, a_1, \ldots, a_n \in K$ such that $P = c(X - a_1)\ldots(X - a_n)$.

Corollary (4.3.8). — *Let K be a field. The following assertions are equivalent:*

(i) *The field K has no algebraic extension* $K \to L$ *of degree* > 1;

(ii) *Every irreducible polynomial of* $K[X]$ *has degree 1;*

(iii) *Every non-constant polynomial with coefficients in K has a root in K;*

(iv) *Every non-constant polynomial of* $K[X]$ *is split.*

Proof. — (i)$\Rightarrow$(ii). Let P be an irreducible polynomial with coefficients in K. Then the extension $K[X]/(P)$ is algebraic of degree $\deg(P)$. Consequently, $\deg(P) = 1$.

(ii)$\Rightarrow$(iii). Let P be a non-constant polynomial with coefficients in K. Let Q be an irreducible factor of P; then there exists c and a in K such that $Q = c(X - a)$; in particular, a is a root of P.

(iii)$\Rightarrow$(iv). Let $P \in K[X]$ be a non-constant polynomial; let us prove by induction on $\deg(P)$ that P is split. Let a be a root of P in K and let Q be the quotient of the euclidean division of P by $X - a$, so that $P = (X - a)Q$. Since $\deg(Q) < \deg(P)$, the polynomial Q is either constant or split, by induction. It follows that P is split.

(iv)$\Rightarrow$(i). Let $K \to L$ be an algebraic extension; let us prove that it has degree 1. We may replace K by its image in L and assume that K is a subfield of L. Let then a be an element of $L - K$. Its minimal polynomial M_a over K is irreducible in K. By assumption, it is split. This implies that $\deg(M_a) = 1$, hence $a \in K$. Consequently, $L = K$ and $[L : K] = 1$. $\square$

Definition (4.3.9). — A field K which satisfies the properties of corollary 4.3.8 is said to be *algebraically closed*.

Theorem (4.3.10) ("Fundamental theorem of algebra"). — *The field* **C** *of complex numbers is algebraically closed.*

Contrary to its name, this is not a theorem in algebra, but in analysis. Indeed, every proof of this theorem needs to use, at some point, an analytic or topological property of the field **R** of real numbers. The short proof below is due to J. R. D'ARGAND and is a variant of the maximum principle in complex analysis. A more algebraic proof is suggested in exercise 4.9.15. I also refer to the books (EBBINGHAUS ET AL, 1991; FINE & ROSENBERGER, 1997) for a historical and mathematical analysis of various proofs of this theorem.

Proof. — Let $P \in \mathbf{C}[X]$ be a non-constant polynomial; let us prove that P has a root in $\mathbf{C}$. We may assume that P is monic. Let us write $P(X) = X^n + a_{n-1}X^{n-1} + \cdots + a_0$, where $a_0, \ldots, a_{n-1} \in \mathbf{C}$, and let $M = \max(1, |a_0| + \cdots + |a_{n-1}|)$. Observe for any $z \in \mathbf{C}$ such that $|z| \geq M$, one has

$$|P(z)| = \left|z^n + a_{n-1}z^{n-1} + \cdots + a_0\right| \geq |z^n| - \left|a_{n-1}z^{n-1}\right| - \cdots - |a_0|$$
$$\geq |z|^n - M|z|^{n-1} \geq |z|^{n-1}(|z| - M).$$

Let $R = M + |P(0)|$. Since $R \geq M \geq 1$, the preceding inequality shows that $|P(z)| \geq 1 + |P(0)| > |P(0)|$ if $|z| \geq R$.

The disk D of center 0 and radius R is compact, and the function $z \mapsto |P(z)|$ is continuous on D. Consequently, there exists a point $a \in D$ where $|P|$ achieves its infimum. Necessarily, $|P(a)| \leq |P(0)|$, hence $|z| < R$.

Let us now consider the Taylor expansion of P at a, $P(a + X) = c_0 + c_1 X + \cdots + c_n X^n$. One thus has $c_0 = P(a)$. Let us argue by contradiction and assume that $c_0 \neq 0$. Since P is non-constant, there exists a least integer $m > 0$ such that $c_m \neq 0$. Let u be an mth root of $-c_0/c_m$. Then, for any small enough real number $t \geq 0$, one has $|a + ut| \leq R$ and

$$P(a + ut) = c_0 + c_m u^m t^m + \text{terms of higher order.}$$

Let c_1 be a real number such that the terms of higher order are bounded above by $c_1 t^{m+1}$. Then

$$P(a + ut) \geq c_0 - t^m(c_0 - c_1 t).$$

If $t > 0$ is small enough, then we have $0 < t^m(c_0 - c_1 t) < c_0$, so that $|P(a + ut)| < |c_0|$. We obtain a contradiction, since $\inf_{z \in D} |P(z)| = |P(a)| = |c_0|$. Consequently, $P(a) = 0$ and P has a root in $\mathbf{C}$, as was to be shown. □

Definition (4.3.11). — Let K be a field. An algebraic closure of K is an algebraic extension $K \to \Omega$ such that Ω is an algebraically closed field.

Knowing *one* example of an algebraically closed field allows us to construct other ones, as well as algebraic closures of some fields.

Proposition (4.3.12). — *Let K be a field, let E be an extension of K. Assume that E is an algebraically closed field. Let Ω be the algebraic closure of K in E. Then, Ω is algebraically closed and $K \to \Omega$ is an algebraic closure of K.*

Proof. — Since Ω is algebraic over K, by construction, we just need to prove that Ω is an algebraically closed field. Let P be a non-constant polynomial in $\Omega[X]$ and let us prove that P has a root in Ω. By assumption, P has a root, say a, in E. Then a is algebraic over Ω; since Ω is algebraic over K, corollary 4.2.3 implies that a is algebraic over K. Consequently, $a \in \Omega$. □

Example (4.3.13). — The set $\overline{\mathbf{Q}}$ of all complex numbers which are algebraic over $\mathbf{Q}$ (called algebraic numbers) is an algebraic closure of $\mathbf{Q}$.

Theorem (4.3.14) (Steinitz, 1910). — *Every field has an algebraic closure.*

Lemma (4.3.15). — *Let* K *be a field and let* $\mathcal{F}$ *be a family of polynomials in* $K[X]$. *There exists an algebraic extension* L *of* K *such that every polynomial of* $\mathcal{F}$ *is split in* L.

Proof. — We first show that for any polynomial $P \in K[X]$, there exists a finite extension $K \to L$ such that P is split in L. The proof goes by induction on $\deg(P)$. It holds if $\deg(P) = 0$. Now assume that $\deg(P) \geq 1$ and let Q be an irreducible factor of P. Let L_1 be the field $K[X]/(Q)$; this is a finite extension of K and Q has a root, say a_1, in L_1. Let P_1 be the quotient of the euclidean division of P by $X - a_1$ in $L_1[X]$. One has $\deg(P_1) = \deg(P) - 1 < \deg(P)$. By induction, there exists a finite extension L of L_1 such that P_1 is split in L. Then $P = (X - a_1)P_1$ is split in L.

If $\mathcal{F}$ is finite, this particular case suffices to establish the lemma. Indeed, let P be the product of the members of $\mathcal{F}$ and let $K \to L$ be an extension such that P is split in L. Any divisor of P, hence any element of $\mathcal{F}$, is then split in L, as was to be shown.

Let us now treat the general case. We may assume that every polynomial of $\mathcal{F}$ is monic. For $P \in \mathcal{F}$, let us set $n_P = \deg(P)$ and let $c_{P,1}, \ldots, c_{P,n_P}$ be elements of K such that

$$P(T) = T^{n_P} + c_{P,1}T^{n_P - 1} + \cdots + c_{P,n}.$$

Let us introduce the polynomial algebra $A = K[X]$ in the family $X = (X_{P,j})$ of indeterminates $X_{P,j}$, for $P \in \mathcal{F}$ and $j \in \{1, \ldots, n_P\}$. For every polynomial $P \in \mathcal{F}$, write $P(T) - \prod_{j=1}^{n_P}(T - X_{P,j}) = \sum_{j=1}^{n_P} Q_{P,j}T^{n_P - j}$, where $Q_{P,1}, \ldots, Q_{P,n_P} \in A$. Let I be the ideal of A generated by the elements $Q_{P,j}$, where P runs among $\mathcal{F}$.

Let us show that $I \neq A$. Otherwise, there would exists an almost-null family $(S_{P,j})$ in A such that

$$1 = \sum_{P \in \mathcal{F}} \sum_{j=1}^{n_P} S_{P,j} Q_{P,j}.$$

Let $\mathcal{F}_1$ be the set of all $P \in \mathcal{F}$ such that $S_{P,j} \neq 0$ for some $j \in \{1, \ldots, n_P\}$. It is finite. By the first part of the proof, there exists an algebraic extension $K \to L_1$ such that every polynomial $P \in \mathcal{F}_1$ is split in L_1. For $P \in \mathcal{F}_1$, let $a_{P,1}, \ldots, a_{P,n_P}$ be a family of elements of L_1 such that $P(T) = \prod_{j=1}^{n_P}(T - a_{P,j})$; for $P \notin \mathcal{F}_1$, set $a_{P,j} = 0$ for all j. Let $a = (a_{P,j})$. By construction, $Q_{P,j}(a) = 0$ if $P \in \mathcal{F}_1$, while $S_{P,j}(a) = 0$ if $P \notin \mathcal{F}_1$. If one evaluates the relation $1 = \sum S_{P,j}Q_{P,j}$ at a, one thus obtains $1 = \sum S_{P,j}(a)Q_{P,j}(a) = 0$, a contradiction.

By Krull's theorem (theorem 2.1.3), there exists a maximal ideal M of A such that $I \subset M$. Let $\Omega = A/M$. For every $P \in \mathcal{F}$, the polynomial P is split in Ω because the classes $a_{P,1}, \ldots, a_{P,n_P}$ of $X_{P,1}, \ldots, X_{P,n_P}$ satisfy $Q_{S,j}(a_{P,1}, \ldots, a_{P,n_P}) = 0$, hence

$$P(T) = \prod_{j=1}^{n_P} (T - a_{P,j}).$$

Moreover, as an A-algebra, Ω is generated by the elements $a_{P,j}$, and those are algebraic. Consequently, Ω is algebraic over K. □

Proof (Proof of Steinitz's theorem). — Let Ω be an algebraic extension of K such that every polynomial of K[X] is split in Ω. Such an extension exists by the preceding lemma. Let us show that Ω is an algebraic closure of K. Since it is an algebraic extension of K, it suffices to prove that Ω is algebraically closed. Let $\Omega \to L$ be an algebraic extension of Ω; let us show that it has degree 1. Replacing Ω by its image in L, we may assume that Ω is a subfield of L. Let $x \in L$; since it is algebraic over Ω and Ω is algebraic over K, the element x is algebraic over K. Let $P \in K[X]$ be its minimal polynomial over K. One has $P(x) = 0$ and P is split in Ω, by construction of this extension: this implies that there exists an $a \in \Omega$ such $x = a 1_L$. Consequently, the morphism $\Omega \to L$ is surjective and the extension $\Omega \to L$ has degree 1, as was to be shown. □

Theorem (4.3.16). — *Let K be a field, let $i: K \to L$ be an algebraic extension and let $j: K \to \Omega$ be an algebraically closed extension.*

The set Φ of morphisms of fields $f: L \to \Omega$ such that $f \circ i = j$ is non-empty. If $K \to L$ is a finite extension, one has $\mathrm{Card}(\Phi) \leq [L:K]$.

Proof. — a) We first prove the theorem under the assumption that the extension $K \to L$ is finite. Let $(a_1, \ldots, a_r)$ be a finite family of elements of L such that $L = K[a_1, \ldots, a_r]$.

For $r = 0$, one has $L = K$ and the result holds. Let us assume that $r \geq 1$ and that the result is true for any extension generated by $r - 1$ elements.

Set $L_1 = K[a_1]$; let P be the minimal polynomial of a_1 over K, so that $[L_1 : K] = \deg(P)$. Since $L_1 \simeq K[X]/(P)$, the map $j_1 \mapsto j_1(a_1)$, induces a bijection from the set of K-morphisms $f_1: L_1 \to \Omega$ to the set of roots of P in Ω. Observe that P has at least one root in Ω, and at most $\deg(P)$, so that there is at least one, and at most $\deg(P)$ morphisms $j_1: L_1 \to \Omega$ such that $f_1 \circ j = i$. By induction, each of these morphisms can be extended to L, in at least one, and in at most $[L : L_1]$ ways. This shows that $1 \leq \mathrm{Card}(\Phi) \leq [L_1 : K][L : L_1] = [L : K]$.

b) It remains to prove that Φ is non-empty in the case where the extension $K \to L$ is infinite. Let $\mathscr{F}$ be the set of pairs (L', f), where L' is a subfield of L containing $i(K)$ and $f: L \to \Omega$ is a field morphism such that $f \circ i = j$. We order the set $\mathscr{F}$ by the relation $(L'_1, f_1) \prec (L'_2, f_2)$ if $L'_1 \subset L'_2$ and $f_2|_{L'_1} = f_1$. Let us prove that the set $\mathscr{F}$ is inductive. Let then $((L'_\alpha, f_\alpha))_{\alpha \in A}$ be a totally ordered family of elements of $\mathscr{F}$. If it is empty, then it is bounded above by the pair $(i(K), f) \in \mathscr{F}$, where $f: i(K) \to \Omega$ be the unique morphism such that $f \circ i = j$. Otherwise, let L' be the union of the subfields L'_α; since the family (L'_α) is non-empty and totally ordered, L' is a subfield of L that contains $i(K)$. Moreover, there exists a unique map $f: L' \to \Omega$ such that $f|_{L'_\alpha} = f_\alpha$ for every $\alpha \in A$, and f is a morphism of fields such that $f \circ i = j$.

By Zorn's lemma (corollary A.2.13), the set $\mathscr{F}$ has a maximal element, say (L', f). We now show that $L' = L$. Otherwise, let a be any element in L such that $a \notin L'$. The extension $L' \subset L'[a]$ is finite. By part a), there exists a morphism $f' \colon L'[a] \to \Omega$ such that $f'|_{L'} = f$. Consequently, $(L'[a], f') \in \mathscr{F}$ and is strictly larger than (L', f), a contradiction.   □

Corollary (4.3.17). — *Two algebraic closures of a field are isomorphic.*

Proof. — Let $i \colon K \to \Omega$ and $j \colon K \to \Omega'$ be two algebraic closures of a field K. Let $f \colon \Omega \to \Omega'$ be any morphism of fields such that $j \circ f = i$ (theorem 4.3.16). The morphism f induces an isomorphism of fields from Ω to its image L in Ω'. In particular, L is an algebraically closed field, and it remains to prove that $L = \Omega'$. Thus let $a \in \Omega'$. Since Ω' is algebraic over K, the element a is algebraic over K, and, in particular, it is algebraic over L; since L is algebraically closed, one has $a \in L$. This concludes the proof.   □

4.4. Separability

4.4.1. — Let K be a field, let $P \in K[X]$ be a non-constant polynomial and let a be a root of P in an extension L of K. Generalizing the definition from §1.3.20, the largest integer m such that $(X - a)^m$ divides P in L[X] is called the *multiplicity* of a as a root of P.

If L' is a further extension of L, then a is also a root of P in L'; let us note that its multiplicity remains unchanged. Of course, if m is the multiplicity of a as a root of P in L, then $(X - a)^m$ still divides P in $L'[X]$, so that the multiplicity of a as a root of P in L' is at least m. Conversely, let m be the multiplicity of a as root of P in L'; one can write $P = (X - a)^m Q$, where $Q \in L'[X]$. Since the coefficients of P and $(X - a)^m$ belong to the subfield L of L', uniqueness of the euclidean division implies that the coefficients of Q belong to L; in particular, $(X - a)^m$ divides P in L[X] and the multiplicity of a as a root of P in L is at least m.

We also recall from lemma 1.3.22 that a is a root of $P^{(k)}$ of multiplicity at least $m - k$, for every integer $k \in \{0, \dots, m - 1\}$, with equality if the characteristic of K is zero, or if it is strictly larger than m (in particular, a is not a root of $P^{(m)}$).

Definition (4.4.2). — Let K be a field. One says that a polynomial $P \in K[X]$ is *separable* if it has no multiple roots in any field extension of K.

Lemma (4.4.3). — *Let K be a field and let $P \in K[X]$ be a non-zero polynomial. The following properties are equivalent:*

   (i) *The polynomial P is separable;*

   (ii) *There exists a field extension L of K in which P is split such that P has no multiple root in L;*

   (iii) *The polynomials P and P' are coprime in K[X].*

Proof. — Let $n = \deg(P)$ and write $P = a_n X^n + \cdots + a_0$ for some elements $a_0, \ldots, a_n \in K$; in particular, $a_n \neq 0$. Let $D = \gcd(P, P')$.

The implication (i)$\Rightarrow$(ii) is obvious, because there are extensions of K in which P is split, for example, an algebraic closure.

To prove (ii)$\Rightarrow$(iii), let us assume that P and P' are not coprime, so that $\deg(D) \geq 1$. Let L be a field extension of K in which P is split. Since D divides P, the polynomial D is split in L as well, hence it has a root in L, say a. One has $P(a) = P'(a) = 0$, so that a is a multiple root of P in L, contradicting (ii).

Let a be a multiple root of P in some extension L of K. Then a is a root of both P and P'. Considering a Bézout relation $D = UP + VP'$, we obtain $D(a) = 0$. In particular, P and P' are not coprime. This establishes the implication (iii)$\Rightarrow$(i).   □

Proposition (4.4.4). — *Let K be a field.*

(i) *Assume that the characteristic of K is zero. Then any irreducible polynomial in K[X] is separable.*

(ii) *Assume that the characteristic of K is a prime number p. An irreducible polynomial P in K[X] is not separable over K if and only if it is a polynomial in X^p, if and only if its derivative is 0.*

(iii) *Assume that the characteristic of K is a prime number p and that the Frobenius homomorphism of K is surjective. Then every irreducible polynomial in K[X] is separable.*

Proof. — Let $P \in K[X]$ be an irreducible polynomial which is not separable. By lemma 4.4.3, the polynomials P and P' have a common factor. Since P is irreducible, then P divides P'. Since $\deg(P') < \deg(P)$, one must have $P' = 0$.

Let $n = \deg(P)$ and write $P = a_n X^n + \cdots + a_0$, so that $a_n \neq 0$. One has $P' = na_n X^{n-1} + \cdots + 2a_2 X + a_1$. If K has characteristic 0, then $na_n \neq 0$, so that $\deg(P') = n - 1$, which contradicts the hypothesis $P' = 0$. This proves a).

Assume now that K has characteristic $p > 0$. Since $P' = 0$, one has $ma_m = 0$ for every $m \in \{0, \ldots, n\}$. If m is prime to p, then m is invertible in K, hence $a_m = 0$. Consequently, $P = \sum_m a_{pm} X^{pm} = Q(X^p)$, where $Q(X) = \sum_m a_{pm} X^m$. This proves b).

c) Let us moreover assume that the Frobenius homomorphism φ of K is surjective. Let us keep the letter φ for the Frobenius homomorphism of K[X]. For every m, let $b_m \in K$ be such that $b_m^p = a_{pm}$. Then,

$$P(X) = Q(X^p) = \sum_m b_m^p X^{pm} = \sum_m \varphi(b_m X^m) = \varphi(\sum_m b_m X^m) = R(X)^p,$$

where $R(X) = \sum_m b_m X^m$. Consequently, R divides P, contradicting the hypothesis that P is irreducible.   □

Definition (4.4.5). — Let p be a prime number and let K be a field of characteristic p. If the Frobenius morphism of K is surjective, one says that K is *perfect*.

Example (4.4.6). —   (i) A finite field is perfect. Indeed, the Frobenius homomorphism $\sigma \colon F \to F$ being injective (as is any morphism of fields), one has $\mathrm{Card}(\sigma(F)) = \mathrm{Card}(F)$. Consequently, σ is surjective.

(ii) Let F be a field of characteristic p. The field $K = F(T)$ is not perfect. Indeed, there is no rational function $f \in K$ such that $f^p = T$. Suppose by contradiction that there is such an element f. The relation $f^p = T$ shows that f is integral over the subring $F[T]$ of K. Since $F[T]$ is a unique factorization domain, it is integrally closed in its field of fractions (theorem 4.1.9), hence $f \in F[T]$. Then $1 = \deg(T) = p \deg(f)$, contradiction.

The polynomial $X^p - T \in F[X, T]$ is irreducible in $F(X)[T]$ since it has degree 1; since its content is 1, it is irreducible in $F[X][T] = F[X, T]$. Consequently, it is also irreducible in $F(T)[X] = K[X]$. However, if a is a root of $X^p - T$ in an extension L of K, one has $T = a^p$, hence $X^p - T = X^p - a^p = (X - a)^p$, so that the polynomial $X^p - T$, viewed as an element of $K[X]$, is inseparable.

Definition (4.4.7). — Let K be a field and let L be an algebraic extension of K.
One says that an element $a \in L$ is *separable* over K if its minimal polynomial is separable.
One says that L is *separable* over K if every element of L is separable over K.

Proposition (4.4.8). — *Let K be a field. If the characteristic of K is zero, or if K is a perfect field, then every algebraic extension of K is separable over K.*

Proof. — Let L be an algebraic extension of K and let $a \in L$. Let P be the minimal polynomial of a over K; it is an irreducible polynomial in $K[X]$. By proposition 4.4.4, it is separable. By definition, this says that a is separable over K, hence the algebraic extension $K \to L$ is separable.   □

Lemma (4.4.9). — *Let $K \to L$ and $L \to E$ be two algebraic extensions of fields. If an element $a \in E$ is separable over K, then it is separable over L. Consequently, if E is separable over K, then E is separable over L.*

Proof. — It suffices to prove the first statement. Let $a \in E$ be separable over K and let $P \in K[X]$ be its minimal polynomial. Since $P(a) = 0$, the minimal polynomial of a over L divides P. In particular, its roots are simple in any extension of L, so that a is separable over L.   □

Theorem (4.4.10). — *Let $i \colon K \to L$ be a finite extension of fields and let $j \colon K \to \Omega$ be an algebraically closed extension of K. Let Φ be the set of morphisms of fields $f \colon L \to \Omega$ such that $j = f \circ i$. The following conditions are equivalent:*

(i) *The field extension L is separable over K;*

(ii) *The field extension L is generated by elements which are separable over K;*

(iii) $\mathrm{Card}(\Phi) = [L : K]$.

Proof. — The implication (i)⇒(ii) is obvious.
Let $a \in L$ and let $L' = K[a]$ be the subextension it generates. We know that K-morphisms $L' \to \Omega$ correspond bijectively to the roots R in Ω of

the minimal polynomial P of a. Moreover, by theorem 4.3.16, a K-morphism $f'\colon L' \to \Omega$ can be extended to a K-morphism $f\colon L \to \Omega$ in at most $[L : L']$ ways. This implies that $\mathrm{Card}(\Phi) \leq \mathrm{Card}(R)[L : L']$. If a is not separable, then $\mathrm{Card}(R) < \deg(P) = [L' : K]$, hence $\mathrm{Card}(\Phi) < [L' : K][L : L'] = [L : K]$. This establishes the implication (iii)$\Rightarrow$(i).

Finally, let $(a_1, \ldots, a_r)$ be a finite family of separable elements of L such that $L = K[a_1, \ldots, a_r]$. Let us prove by induction on r that $\mathrm{Card}(\Phi) = [L : K]$.

This is obvious if $r = 0$, hence assume that $r \geq 1$ and set $L_1 = K[a_1]$.

If $r = 1$, then $L = L_1$, and the K-morphisms from L to Ω are in bijection with the roots in Ω of the minimal polynomial P of a_1. By assumption, these roots are simple, so that $\mathrm{Card}(\Phi) = \deg(P) = [L_1 : K]$. By the induction hypothesis, each of these K-morphisms $L_1 \to \Omega$ can be extended in exactly $[L : L_1]$ ways to a K-morphism $L \to \Omega$. This proves that $\mathrm{Card}(\Phi) = [L : L_1][L_1 : K] = [L : K]$, as was to be shown. $\square$

Corollary (4.4.11). — *Let $f\colon K \to L$ and $g\colon L \to E$ be field extensions. Assume that L is separable over K.*

If an element $a \in E$ is separable over L, then it is separable over K.

If E is separable over L, then E is separable over K.

Proof. — We first treat the case where these extensions are finite. Let $i\colon E \to \Omega$ be an algebraic closure of K. Since L is separable over K, there are exactly $[L : K]$ morphisms j from L to Ω such that $j \circ f = i$. Since E is separable over L, for each of these morphisms j, there are exactly $[E : L]$ morphisms k such that $k \circ g = j$. In particular, there are $[E : L][L : K]$ morphisms h from K to Ω such that $h \circ (g \circ f) = i$. Consequently, E is separable over K.

Let us now treat the general case. Consider an element $a \in E$ which is separable over L, and let $E_1 = L[a]$. Let P be the minimal polynomial of a over L and let $L_1 \subset L$ be the extension of K generated by the coefficients of P. By assumption, the polynomial P is separable. By construction, the minimal polynomial of a over L_1 is equal to P, which implies that a is separable over L_1, hence E_1 is separable over L_1. Moreover, L_1 is separable over K, by assumption. By the case of finite extensions, the extension $K \to E_1$ is separable. In particular, a is separable over K. $\square$

Proposition (4.4.12) (Dirichlet). — *Let G be a monoid, let Ω be a field and let Φ be a set of morphisms of monoids from G to the multiplicative monoid of Ω. Then the set Φ is a linearly independent subset of the Ω-vector space of maps from G to Ω.*

Proof. — Let us argue by contradiction and let us consider a minimal subset of Φ, say $\{\varphi_1, \ldots, \varphi_n\}$, which is linearly dependent over Ω. Let $a_1, \ldots, a_n \in \Omega$, not all zero, such that

$$a_1 \varphi_1(x) + \cdots + a_n \varphi_n(x) = 0$$

for every $x \in G$. By minimality, $a_i \neq 0$ for every $i \in \{1, \ldots, n\}$, for otherwise, the subset of Φ obtained by taking out φ_i would be linearly independent, contradicting the minimality assumption. Moreover, $n \geq 2$ (otherwise, φ_1 would be zero).

Let $y \in G$ and let us apply the relation to xy, for $x \in G$. Since $\varphi(xy) = \varphi(x)\varphi(y)$ for every $x \in G$, this gives

$$a_1\varphi_1(y)\varphi_1(x) + \cdots + a_n\varphi_n(y)\varphi_n(x) = 0,$$

for every $x \in G$. Subtracting from this relation $\varphi_n(y)$ times the first one, we obtain

$$a_1(\varphi_1(y) - \varphi_n(y))\varphi_1(x) + \cdots + a_{n-1}(\varphi_{n-1}(y) - \varphi_n(y))\varphi_{n-1}(x) = 0.$$

This holds for every $x \in G$, hence we get a linear dependence relation

$$a_1(\varphi_1(y) - \varphi_n(y))\varphi_1 + \cdots + a_{n-1}(\varphi_{n-1}(y) - \varphi_n(y))\varphi_{n-1} = 0$$

for $\varphi_1, \ldots, \varphi_{n-1}$. By the minimality assumption, $a_i(\varphi_i(y) - \varphi_n(y)) = 0$ for every i, hence $\varphi_i(y) = \varphi_n(y)$ since $a_i \neq 0$. Since y is arbitrary, we obtain $\varphi_i = \varphi_n$, a contradiction. □

Remark (4.4.13). — Let $K \to L$ be a finite extension of fields and let Ω be an algebraically closed extension of K. Every K-linear morphism of fields, $\sigma \colon L \to \Omega$, induces a morphism of groups from $L^\times$ to $\Omega^\times$, hence a morphism of monoids from $L^\times$ to Ω. Consequently, the set Φ of all K-linear field morphisms $f \colon L \to \Omega$ is a linearly independent subset of the Ω-vector space of maps from L to Ω. Observe that Φ is contained in the subspace of K-linear maps from L to Ω, which is a space of dimension $[L : K]$. Consequently, we obtain another proof of the inequality $\mathrm{Card}(\Phi) \leq [L : K]$ of theorem 4.3.16.

Theorem (4.4.14) (Primitive element theorem). — *Let E be a field and let $E \to F$ be a finite separable extension. Then there exists an element $a \in F$ such that $F = E[a]$. More precisely, if E is infinite and if $F = E[a_1, \ldots, a_d]$, there exist elements $c_2, \ldots, c_d \in E$ such that $a = a_1 + c_2 a_2 + \cdots + c_d a_d$ satisfies $F = E[a]$.*

Proof. — If E is a finite field, then F is a finite field too. By proposition 1.3.21, the multiplicative group $F^\times$ is a finite cyclic group. If a is a generator of this group, then $E[a]$ contains $F^\times$, as well as 0, so that $E[a] = F$.

We now assume that E is infinite. By induction on d, it suffices to prove the following result, which we state as an independent lemma. □

Lemma (4.4.15). — *Let $E \to F$ be a finite extension of infinite fields. Assume that there are elements $a, b \in F$ such that $F = E[a, b]$. If b is separable over E, then for all but finitely many $c \in E$, we have $F = E[a + cb]$.*

Proof. — Let P and Q be the minimal polynomials of a and b respectively; let $p = \deg(P)$ and $q = \deg(Q)$. Let Ω be an algebraic extension of F in which P and Q are split; let $a_1, \ldots, a_p$ be the roots of P in Ω, let $b_1, \ldots, b_q$ be the roots of Q in Ω, ordered so that $a_1 = a$ and $b_1 = b$. By hypothesis, the elements $b_1, \ldots, b_q$ are pairwise distinct. Consequently, for any pair (i, j) of integers, where $1 \leq i \leq p$ and $2 \leq j \leq q$, the equation $a_i + cb_j = a_1 + cb_1$ in the unknown c has at most one solution, namely $c = -(a_i - a_1)/(b_j - b_1)$. Let

$c \in E$ be any element which is not equal to any of those, set $t = a + cb$ and let us show that $F = E[t]$.

The polynomial $R(X) = P(t - cX)$ belongs to $E[t][X]$ and b is a root of R since $R(b) = P(t - cb) = P(a) = 0$. For any $j \in \{2, \dots, p\}$, observe that $t - cb_j = a_1 + cb_1 - cb_j \notin \{a_1, \dots, a_p\}$ hence $P(t - cb_j) \neq 0$ and $R(b_j) \neq 0$. Consequently, $b = b_1$ is the only common root to R and Q in Ω. Since Q is separable, all of its roots are simple, hence $\gcd(R, Q) = X - b$. Since R and Q have their coefficients in $E[t]$, it follows from remark 2.4.5 that $X - b \in E[z]$. In other words, $y \in E[t]$. Then $a = t - cb \in E[t]$ so that $E[a, b] \subset E[t] \subset E[a, b]$. It follows that $F = E[t]$.   □

We end this section with a few remarks about inseparable extensions.

Lemma (4.4.16). — *Let K be a field and let Ω be an algebraic closure of K. Let P be an irreducible monic polynomial in $K[X]$ which has only one root, say a, in Ω.*

(i) *If the characteristic of K is 0, then $a \in K$ and $P = X - a$.*

(ii) *If the characteristic of K is a prime number p, there exists a smallest integer $n \geq 0$ such that $a^{p^n} \in K$, and $P = X^{p^n} - a^{p^n}$.*

Such an element a is said to be *radicial* over K.

Proof. — Let m be the multiplicity of a as a root of P; by assumption, one has $P = (X - a)^m$.

Assume that K has characteristic zero. Being irreducible, the polynomial P is then separable so that its root a is simple. In other words, $m = 1$ and $P = X - a$.

Now assume that K has characteristic p. If P is separable, then $P = X - a$ as in the characteristic 0 case. Otherwise, we proved that P is a polynomial in X^p. Let n be the largest integer such that P is polynomial in X^{p^n}: it is the largest integer such that p^n divides m. We thus may write

$$P(X) = Q(X^{p^n}), \qquad Q(X) = X^r + b_{r-1}X^{r-1} + \cdots + b_0.$$

By assumption, Q is not a polynomial in X^p (otherwise, we could write $Q(X) = R(X^p)$, hence $P(X) = R(X^{p^{n+1}})$), contradicting the maximality of n). Consequently, Q is separable. Moreover, $b = a^{p^n}$ is the only root of Q in Ω. One thus has $Q(X) = X - b$, hence $P = X^{p^n} - a^{p^n}$.

It remains to prove that n is the smallest integer such that $a^{p^n} \in K$. Let $d \leq n$ be an integer such that $a^{p^d} \in K$. Then, $X^{p^d} - a^{p^d}$ divides P; since P is irreducible, we conclude that $p^d \geq p^n$, hence $d = n$.   □

Remark (4.4.17). — Let $K \to L$ be a finite field extension and $L_{sep} \subset L$ be the subfield of L generated by all elements of L which are separable over K. By corollary 4.4.11, the extension $K \to L_{sep}$ is separable, and it is the largest subextension of L.

If the characteristic of K is zero, then $L = L_{sep}$, hence let us assume that the characteristic of K is a prime number, p.

By corollary 4.4.11, every element of $L - L_{sep}$ is inseparable over L_{sep}. One says that L is a *radicial* extension of L_{sep}. The detailed analysis of such extensions is slightly more intricate than that of separable extensions.

Let us just show here the following two facts:

(i) *Every element of L is radicial over* L_{sep}.

Let $a \in L - L_{sep}$ and let $P = X^m + c_{m-1}X^{m-1} + \cdots + c_0$ be its minimal polynomial over L_{sep}. Since a is not separable over L and L_{sep} is separable over L, the element a is not separable over L_{sep}. In particular, P is a polynomial in X^p. Let r be the greatest integer such that there exists a polynomial $Q \in L_{sep}[X]$ with $P = Q(X^{p^r})$; by construction, Q is the minimal polynomial of a^{p^r}, and is not a polynomial in X^p. In particular, a^{p^r} is separable over L_{sep}, hence $a^{p^r} \in L_{sep}$. This implies that $Q = X - a^{p^r}$, hence $P = X^{p^r} - a^{p^r} = (X - a)^{p^r}$. This proves that a is radicial over L_{sep}.

(ii) *There exists an integer* $n \geq 0$ *such that* $[L : L_{sep}] = p^n$.

Let us consider elements $a_1, \ldots, a_n$ of $L - L_{sep}$ such that $L = L_{sep}[a_1, \ldots, a_n]$ and let us prove by induction on n that $[L : L_{sep}]$ is power of p. This holds if $n \leq 1$, by the analysis of a). Let $E = L_{sep}[a_1, \ldots, a_{n-1}]$, so that $[E : L_{sep}]$ is a power of p by induction. To simplify the notation, let us write $a = a_n$ and $r = r_n$, so that $L = E[a]$, the minimal polynomial of a over L_{sep} being $P = X^{p^r} - a^{p^r} = (X - a)^{p^r}$. Since the minimal polynomial of a over E divides P in $E[X]$, it has only one root, a. By lemma 4.4.16, this implies that a is radicial over E; in particular, its degree is a power of p. Since $[L : L_{sep}] = [L : E][E : L_{sep}]$, this implies that $[L : L_{sep}]$ is a power of p.

4.5. Finite Fields

A finite field is what it says: a field whose cardinality is finite.

Proposition (4.5.1). — *The characteristic of a finite field is a prime number p; its cardinality is a power of p.*

Proof. — Let F be a finite field. Since **Z** is infinite, the canonical morphism from **Z** to F is not injective. Its kernel is a non-zero prime ideal of **Z**, hence is generated by a prime number, say p, the characteristic of F. This morphism induces a field extension $Z/pZ \to F$, so that F is a vector space over the finite field Z/pZ. Its dimension d must be a finite integer, so that $Card(F) = p^d$. ☐

Conversely, let $q > 1$ be a power of a prime number p. We are going to show that there exists a finite field of characteristic p and cardinality q, and that any two such fields are isomorphic.

Theorem (4.5.2). — *Let p be a prime number, let f be an integer such that $f \geq 1$ and let $q = p^f$. Let Ω be a field of characteristic p in which the polynomial $X^q - X$ is split, for example, an algebraically closed field of characteristic p. There exists a unique subfield of Ω which has cardinality q: it is the set F of all roots in Ω of the polynomial $X^q - X$. Moreover, any finite field of cardinality q is isomorphic to F.*

Proof. — Let $\varphi: \Omega \to \Omega$ be the Frobenius homomorphism, defined by $x \mapsto x^p$ (see example 1.1.14). Let $\varphi_q = \sigma^f$ be the fth power of φ, so that $\varphi_q(x) = x^{p^f} = x^q$ for every $x \in \Omega$. The set F of all $x \in \Omega$ such that $x^q = \varphi_q(x) = x$ is then a subfield of Ω, as claimed. The derivative of the polynomial $P = X^q - X$ is $P' = qX^{q-1} - 1 = -1$, since Ω has characteristic p and p divides q. Consequently (see §1.3.20), P has no multiple root and $\mathrm{Card}(F) = q$.

Let K be a finite field of cardinality q. The multiplicative group $K^\times$ of K has cardinality $q-1$, hence $x^{q-1} = 1$ for every $x \in K^\times$, by Lagrange's theorem. This implies that $x^q = x$ for every $x \in K^\times$, and also for $x = 0$.

If K is a subfield of Ω, this shows that $K \subset F$, hence $K = F$ since these two fields have the same number of elements.

In the general case, we already know that K is an extension of $\mathbf{Z}/p\mathbf{Z}$ in which the polynomial $P = X^q - X$ is split, and which is generated by the roots of P. Indeed, K consists only of roots of P! It follows from theorem 4.3.16 that there exists a morphism of fields $i: K \to \Omega$. Then, $i(K)$ is a finite field of characteristic q contained in Ω, hence $i(K) = F$ and $K \simeq F$.   □

Corollary (4.5.3). — *Let F be a finite field of cardinality q and let d be an integer $\geq$ 1. There exists an irreducible monic polynomial $P \in F[X]$ such that $\deg(P) = d$. The F-algebra $F[X]/(P)$ is a finite field of cardinality q^d.*

Proof. — We may fix an extension Ω of $\mathbf{Z}/p\mathbf{Z}$ in which the polynomial $X^{q^d} - X$ is split. Since $X^{q^d} - X = X(X^{q^d-1} - 1) = X(X^{q-1} - 1)(X^{q^{d-1}} + \cdots + 1)$, $X^q - 1$ is also split in Ω. We may then assume that $F \subset \Omega$.

Then Ω has a unique subfield K of cardinality q^d. Let $x \in K^\times$ be a generator of the finite cyclic group $K^\times$ (theorem 5.5.4). The subfield $F[x]$ of K contains the multiplicative subgroup generated by x, which is $K^\times$, so that $F[x] = K$. Let P be the minimal polynomial of x over F. This is an irreducible monic polynomial with coefficients in F such that $P(x) = 0$, and $K = F[x] \simeq F[X]/(P)$. In particular, $\deg(P) = [K : F] = d$.   □

A defect of the preceding proof is that it does not give an effective way to get one's hand on a given finite field with q elements and characteristic p. However, if $q = p^d$, we know that it will be possible to find, somehow, an irreducible polynomial of degree d over $\mathbf{Z}/p\mathbf{Z}$. Such a polynomial will be a factor of $X^q - X$ over $\mathbf{Z}/p\mathbf{Z}$.

Example (4.5.4). — Let us construct a field of cardinality 8. We need to find an irreducible polynomial of degree 3 with coefficients in $\mathbf{Z}/2\mathbf{Z}$. There are $8 = 2^3$ monic polynomials of degree 3; four of them vanish at 0 and are not irreducible. The remaining four are $X^3 + aX^2 + bX + 1$ where $(a,b) \in \{0,1\}^2$; their value at $X = 1$ is $a + b$. Thus two polynomials of degree 3 with coefficients in $\mathbf{Z}/2\mathbf{Z}$ have no root, namely $X^3 + X + 1$ and $X^3 + X^2 + 1$. They are irreducible; indeed, they would otherwise have a factor of degree 1, hence a root. Consequently, $\mathbf{F}_8 \simeq \mathbf{F}_2[X]/(X^3 + X + 1)$.

4.6. Galois's Theory of Algebraic Extensions

4.6.1. — Let $K \to L$ be an extension of fields. Let $\mathrm{Aut}(L/K)$ be the set of field automorphisms of L which are K-linear. In the particularly important case where K is a subfield of L and the morphism $K \to L$ is the inclusion, an automorphism σ of L belongs to $\mathrm{Aut}(L/K)$ if and only if its restriction to K is the identity.

Observe that $\mathrm{Aut}(L/K)$ is a group with respect to composition.

Examples (4.6.2). — *a)* Let $c \colon \mathbf{C} \to \mathbf{C}$ be the complex conjugation. Then $\mathrm{Aut}(\mathbf{C}/\mathbf{R}) = \{\mathrm{id}, c\} \simeq \mathbf{Z}/2\mathbf{Z}$.

It is clear that id and c are elements of $\mathrm{Aut}(\mathbf{C}/\mathbf{R})$. Conversely, let σ be an automorphism of $\mathbf{C}$ such that $\sigma(x) = x$ for every $x \in \mathbf{R}$. Observe that $\sigma(i)^2 = \sigma(i^2) = \sigma(-1) = -1$, hence $\sigma(i) \in \{\pm i\}$. If $z = x + iy \in \mathbf{C}$, with $x, y \in \mathbf{R}$, then $\sigma(z) = \sigma(x) + \sigma(i)\sigma(y) = x + \sigma(i)y$. Consequently, $\sigma = \mathrm{id}$ if $\sigma(i) = i$ and $\sigma = c$ if $\sigma(i) = -i$.

b) Let q be a power of a prime number and let e be an integer ≥ 2. Let F be a field of cardinality q^e and let E be the subfield of F of cardinality q. Then $\mathrm{Aut}(F/E)$ is a cyclic group of order e, generated by the automorphism $\sigma_q \colon x \mapsto x^q$.

Indeed, σ_q belongs to $\mathrm{Aut}(F/E)$. For $0 \leq i \leq e - 1$, its power σ_q^i is the morphism $x \mapsto x^{q^i}$. If $0 < i < e$, the polynomial $X^{q^i} - X$ has at most q^i roots in F, so that $\sigma_q^i \neq \mathrm{id}$. This implies that the elements $\mathrm{id}, \sigma_q, \ldots, \sigma_q^{e-1}$ of $\mathrm{Aut}(F/E)$ are distinct. By the following lemma, they fill up all of $\mathrm{Aut}(F/E)$.

c) Let E be the subfield of $\mathbf{C}$ generated by $\sqrt[3]{2}$. Then $\mathrm{Aut}(E/\mathbf{Q}) = \{\mathrm{id}\}$.

Let $\alpha = \sqrt[3]{2}$, so that $E = \mathbf{Q}[\alpha]$; in particular, E is a subfield of $\mathbf{R}$. Any automorphism σ of E is determined by $\sigma(\alpha)$. Moreover, $\sigma(\alpha)^3 = \sigma(\alpha^3) = \sigma(2) = 2$, so that $\sigma(\alpha)$ is a cube root of 2 contained in E. Observe that α is the only cube root of 2 which is a real number, hence the only cube root of 2 in E. Consequently, $\sigma(\alpha) = \alpha$ and $\sigma = \mathrm{id}$.

Definition (4.6.3). — One says that a finite extension of fields $K \to L$ is Galois if $\mathrm{Card}(\mathrm{Aut}(L/K)) = [L : K]$.

Note that elements of $\mathrm{Aut}(L/K)$ coincide with K-morphisms from L to itself, since the latter are automatically surjective when the extension $K \to L$ is finite. Moreover, theorem 4.3.16 (or remark 4.4.13) asserts that $\mathrm{Card}(\mathrm{Aut}(L/K)) \leq [L : K]$. It also follows from theorem 4.4.10 that a Galois extension is separable.

When an extension $K \to L$ is Galois, its automorphism group is also denoted by $\mathrm{Gal}(L/K)$ and is called its *Galois group*.

The extension $\mathbf{R} \subset \mathbf{C}$ and the extensions of finite fields given in examples 4.6.2 are Galois extensions; the third one is not.

On Évariste Galois

Évariste GALOIS (1811–1832) is one of the iconic figures of the early nineteenth century mathematics, combining an early mathematical talent that made him read LEGENDRE, GAUSS and others at the age of 15, the suicide of his father, the victim of a political cabal, a self-consciously rebellious mind that prevented him from being admitted to the highest schools, a political involvement in Republican activist circles which brought him to jail twice, a "revolutionary" *Mémoire sur les conditions de résolubilité des équations par radicaux* which the French Science Academy refused to publish, a tragic death in a duel at the age of 20 related to a mysterious *"infâme coquette"* (despicable coquette), and an extremly moving letter to his friend Auguste CHEVALIER written the night before this duel where, in a rush, he summarized his mathematical ideas:

> *I often ventured in my life to advance propositions I was not sure of; but everything that I wrote here has been in my head for almost one year, and it is too much in my interest that I am not mistaken, so that I cannot be suspected of having stated theorems of which I wouldn't have a complete proof.*
>
> *You will ask publicly Jacobi or Gauss to give their opinion, not on the truth, but on the importance of the theorems.*
>
> *After that, there will be, so I hope, people who will find it to their advantage to decipher all this mess.*
>
> *I embrace you effusively.*

Portrait of Évariste Galois (circa 1826)

Unknown artist.
Source: Wikipedia.
Public domain

Révolutionnaire et géomètre. Postal stamp (1984) from France.
Source: `wikitimbres.fr`. Engraver: Jacques Combet
© Musée de La Poste, Paris / La Poste 2020

As pointed out by EHRHARDT (2011), the legend of GALOIS poses more questions that it solves, if even it does not bear its own contradiction of a rebellious young man considering one of the most classical problems of the mathematics of his time. GALOIS's objective was to understand resolution of equations by radicals (expressing the roots of polynomial equations with square roots, etc.), beyond ABEL's proof of the mere impossibility of solving the quintic. He did so by introducing a "group" of permutations of the roots of a given equation and showing that the equation is solvable by radicals if and only if this group is what we now call solvable. Promising later examples, he then contents himself with describing the case of an irreducible equation of prime degree.

In fact, during roughly one century, from the publication of GALOIS's *Mémoire* by LIOUVILLE in 1846 to the 1942 lectures of (ARTIN, 1998), GALOIS's theory has been read, revisited, and reshaped as long as algebra evolved from the study of equations and transformation groups to its present "abstract" form of fields and groups.

As of today, and this was already the case for the works of LIE, Galois theory has gone much further than being a mere tool for the study of field extensions, becoming a paradigm — understanding a structure through the group of its automorphisms. "Differential Galois theory", for example, replaces polynomial equations with systems of linear differential equations. GALOIS's final letter suggests that he himself envisioned such a theory.

This point of view has been pursued up to GROTHENDIECK's theories of galoisian/tannakian categories, and to the study of the "motivic Galois group".

The next lemma furnishes another, fundamental example, of a Galois extension.

Lemma (4.6.4) (Artin). — *Let* L *be a field and let* G *be a finite group of automorphisms of* L. *Let* $K = L^G$ *be the set of all* $x \in L$ *such that* $\sigma(x) = x$ *for every* $\sigma \in G$. *Then* K *is a subfield of* L, *the extension* $K \subset L$ *is finite, Galois, and* $\mathrm{Aut}(L/K) = G$.

Proof. — We leave it to the reader to prove that K is a subfield of L. Let us prove by contradiction that the extension $K \subset L$ is finite and that $[L : K] \leq \mathrm{Card}(G)$. Otherwise, let n be some integer $> \mathrm{Card}(G)$ and let $a_1, \ldots, a_n$ be elements of L which are linearly independent over K. Let us consider the system of linear equations with coefficients in L,

$$\sum_{i=1}^{n} x_i \sigma(a_i) = 0 \qquad (\sigma \in G),$$

in unknowns $x_1, \ldots, x_n$. By assumption, it has more unknowns than equations, hence it has a non-zero solution $(x_1, \ldots, x_n) \in E^n$. Let us consider one such solution for which the number of non-zero coefficients is minimal. Up to reordering $a_1, \ldots, a_n$ and $x_1, \ldots, x_n$, we assume that $x_1, \ldots, x_m$ are non-zero, but $x_{m+1} = \cdots = x_n = 0$. By linearity, we may assume $x_m = 1$, so that

$$\sum_{i=1}^{m-1} x_i \sigma(a_i) + \sigma(a_m) = 0 \qquad (\sigma \in G).$$

Let then $\tau \in G$ and apply τ to the preceding equality. One obtains

$$\sum_{i=1}^{m-1} \tau(x_i)(\tau \circ \sigma)(a_i) + (\tau \circ \sigma)(a_m) = 0.$$

Subtracting the relation corresponding to the element $\tau \circ \sigma$ of G, one gets

$$\sum_{i=1}^{m-1} (\tau(x_i) - x_i)(\tau \circ \sigma)(a_i) = 0.$$

This holds for any $\sigma \in G$, and the elements $\tau \circ \sigma$ run among all elements of G. Consequently, for any $\sigma \in G$,

$$\sum_{i=1}^{m-1} (\tau(x_i) - x_i)\sigma(a_i) = 0.$$

The n-tuple $(y_1, \ldots, y_n)$ defined by $y_i = \tau(x_i) - x_i$ is a solution of the initial system, but has $< m$ non-zero entries. By the choice of m, one has $y_i = 0$ for all i. Consequently, $\tau(x_i) = x_i$ for every i.

Since τ is arbitrary, we conclude that $x_i \in K$ for every i. In particular, $\sum_{i=1}^n x_i a_i = 0$ which contradicts the initial hypothesis that these elements were linearly independent over K.

Consequently, $[L : K] \le \mathrm{Card}(G)$. In particular, the extension $K \subset L$ is finite, and algebraic. Moreover, every element of G is an element of $\mathrm{Aut}(L/K)$, so that $\mathrm{Card}(\mathrm{Aut}(L/K)) \ge \mathrm{Card}(G)$. Since the opposite inequality always holds by theorem 4.3.16, one has $G = \mathrm{Aut}(L/K)$.   □

Let $K \subset L$ be a Galois extension, let $G = \mathrm{Gal}(L/K)$. Let $\mathscr{F}_L$ be the set of subfields of L containing K; let $\mathscr{G}_G$ be the set of subgroups of G. We order both sets by inclusion.

Theorem (4.6.5) (Fundamental theorem of Galois theory). — *Let* $K \subset L$ *be a finite Galois extension with Galois group* $G = \mathrm{Gal}(L/K)$.

(i) *For every subgroup* $H \subset G$, *the set*

$$L^H = \{x \in L ; \forall \sigma \in H, \quad \sigma(x) = x\}$$

is a subfield of L *containing* K.

(ii) *For every subfield* E *of* L *which contains* K, *the extension* L/E *is Galois.*

(iii) *The maps* $\varphi \colon H \mapsto L^H$ *from* $\mathscr{G}_G$ *to* $\mathscr{F}_K$, *and* $\gamma \colon E \mapsto \mathrm{Gal}(L/E)$ *from* $\mathscr{F}_L$ *to* $\mathscr{G}_G$, *are decreasing bijections, inverse to one another.*

Proof. — Part a) follows from lemma 4.6.4.

Let us prove b). Elements of $\mathrm{Aut}(L/E)$ are automorphisms of E which restrict to the identity on L. In particular, $\mathrm{Aut}(L/E) \subset \mathrm{Gal}(L/K)$. The inequality $\mathrm{Card}(\mathrm{Aut}(L/E)) \le [L : E]$ holds by theorem 4.3.16 and we have to prove the opposite one. Let us consider the map r from $\mathrm{Gal}(L/K)$ to $\mathrm{Hom}_K(E, L)$ given by $g \mapsto g|_E$. By construction, $\mathrm{Aut}(L/E) = r^{-1}(r(\mathrm{id}_L))$. Let Ω be an algebraic closure of L. It follows from 4.3.16 that $\mathrm{Card}(\mathrm{Hom}_K(E, L)) \le [E : K]$ and that the fibers of r have cardinality at most $[L : E]$. Since $[E : K][L : E] = \mathrm{Card}(\mathrm{Gal}(L/K))$, we see that $\mathrm{Card}(\mathrm{Hom}_K(E, L)) = [E : K]$ and all fibers of r have cardinality exactly $[L : E]$. In particular, the fiber $\mathrm{Aut}(L/E)$ of $r(\mathrm{id}_L)$ has cardinality $[L : E]$, as required.

c) Parts a) and b) show that the maps φ and γ are well-defined. If H and H′ are two subgroups of G such that $H \subset H'$, then $L^{H'} \subset L^H$, so that $\varphi(H') \subset \varphi(H)$; hence φ is decreasing. Similarly, if E and E′ are two subfields of L containing K such that $E \subset E'$, then $\mathrm{Gal}(L/E')$ is a subgroup of $\mathrm{Gal}(L/E)$, that is, $\gamma(E') \subset \gamma(E)$ and γ is decreasing.

We proved in lemma 4.6.4 that $\mathrm{Gal}(L/L^H) = H$. In other words, $\gamma \circ \varphi = \mathrm{id}$. In particular, the map φ is injective. Let E be a subfield of L which contains K; let $E' = \varphi(\gamma(E)) = L^{\mathrm{Gal}(L/E)}$. By definition of $\mathrm{Gal}(L/E)$, one has $E \subset E'$. Moreover, we know by Part b) that L/E is Galois, and by Part a) that L/E' is Galois as well, and that $\mathrm{Gal}(L/E') = \mathrm{Gal}(L/E)$. Consequently, $[L : E'] = [L : E]$. This implies that $[E' : E] = 1$, that is, $E = E'$.   □

The definition, for a finite field extension, to be a Galois extension pre-supposes that we understand its group of automorphisms. The following characterization will furnish an effective way to construct Galois extensions, via the implication (iv)$\Rightarrow$(i).

Proposition (4.6.6). — *Let* $K \subset L$ *be a finite field extension. Then the following properties are equivalent:*

(i) *The extension* $K \subset L$ *is Galois;*

(ii) *The extension* $K \subset L$ *is separable and every irreducible polynomial in* $K[X]$ *which has a root in* L *is already split in* L;

(iii) *There exists an element* $a \in L$ *such that* $L = K[a]$ *and the minimal polynomial of which is separable and split in* L;

(iv) *There exists a separable polynomial* $P \in K[X]$ *which is split in* L, *with roots* $a_1, \ldots, a_n \in L$, *such that* $L = K[a_1, \ldots, a_n]$.

Proof. — (i)$\Rightarrow$(ii). We already proved that a Galois extension is separable. Let $P \in K[X]$ be an irreducible polynomial, let $a \in L$ be a root of P in L. Let $E = K[a]$. Let $Q = \prod_{g \in G}(X - g(a))$. For every $h \in G$, the polynomial Q^h obtained by applying h to the coefficients of Q is given by

$$Q^h = \prod_{g \in G}(X - h(g(a))) = \prod_{g \in G}(X - g(a)) = Q,$$

since the map $g \mapsto hg$ is a bijection of G to itself. Consequently, the coefficients of Q belong to $L^G = K$. Moreover, $Q(a) = 0$, hence P divides Q. Since Q is split in L, it follows that P is split in L as well.

(ii)$\Rightarrow$(iii). Since the extension $K \subset L$ is separable, the Primitive element theorem (theorem 4.4.14) implies that there exists an element $a \in L$ such that $L = K[a]$. The minimal polynomial of a is separable and irreducible, hence is split in L.

(iii)$\Rightarrow$(iv). It suffices to take for P the minimal polynomial of a.

(iv)$\Rightarrow$(i). Let $E_0 = K$; for $i \in \{1, \ldots, n\}$, let $E_i = K[a_1, \ldots, a_i]$ and let M_i be the minimal polynomial of a_i over E_{i-1}. It divides P hence is separable and split in L. For every i, let Φ_i be the set of morphisms from E_i to L which restricts to the identity on K; let $r_i : \Phi_i \to \Phi_{i-1}$ be the restriction map, $\sigma \mapsto \sigma|_{E_{i-1}}$.

Given a morphism $f \in \Phi_{i-1}$, there are exactly $\deg(M_i) = [E_i : E_{i-1}]$ morphisms from E_i to L which extend f. Consequently, the fibers of the restriction map r_i all have cardinality $[E_i : E_{i-1}]$. Since Φ_0 has exactly one element, one has

$$\begin{aligned}
\mathrm{Card}(\Phi_n) &= [E_n : E_{n-1}]\,\mathrm{Card}(\Phi_{n-1}) = \ldots \\
&= [E_n : E_{n-1}] \ldots [E_1 : E_0]\,\mathrm{Card}(\Phi_0) \\
&= [L : K].
\end{aligned}$$

Moreover, any morphism $f \in \Phi_n$ is surjective, since $[f(L) : K] = \dim_K(f(L)) = \dim_K(L) = [L : K]$. Consequently, $\mathrm{Aut}(L/K) = \Phi_n$ and the extension L/K is Galois. $\square$

For future use, part of statement (ii) needs to be given a name:

Definition (4.6.7). — One says that a finite extension $K \subset L$ is *normal* if every irreducible polynomial in $K[X]$ which has a root in L is already split in L.

Lemma (4.6.8). — *Let $K \subset L$ be a finite extension. The following properties are equivalent:*

(i) *The extension $K \subset L$ is* normal;

(ii) *There exists a polynomial $P \in K[X]$ which is split in L, with roots $a_1, \ldots, a_n$, such that $L = K[a_1, \ldots, a_n]$;*

(iii) *For every extension Ω of L and any K-homomorphism $\sigma \colon L \to \Omega$, one has $\sigma(L) = L$.*

Proof. — (i)$\Rightarrow$(ii). Assume that the extension $K \subset L$ is normal and let $a_1, \ldots, a_n$ be elements of L such that $L = K[a_1, \ldots, a_n]$. For every i, let P_i be the minimal polynomial of a_i over K. It is irreducible, hence split in L. It follows that the product $P = P_1 \ldots P_n$ is split in L and L is generated by its roots.

(ii)$\Rightarrow$(iii). Let $P \in K[X]$ be a polynomial which is split in L, with roots $a_1, \ldots, a_n$, such that $L = K[a_1, \ldots, a_n]$. Let $\sigma \colon L \to \Omega$ be a K-homomorphism into an extension Ω of L. Since $\sigma(P(a_i)) = P(\sigma(a_i)) = 0$, $\sigma(a_i)$ is a root of P, hence $\sigma(a_i) \in \{a_1, \ldots, a_n\}$. In particular, $\sigma(a_i) \in L$. Since $L = K[a_1, \ldots, a_n]$, it follows that $\sigma(L) \subset L$. Since σ is injective, one has $[\sigma(L) : K] = [L : K]$, hence $\sigma(L) = L$.

(iii)$\Rightarrow$(i). Let $Q \in K[X]$ be an irreducible polynomial which has a root b in L. We need to show that Q is split in L. Let Ω be an algebraic closure of L and let $b' \in \Omega$ be a root of Q. Let $\varphi \colon K[b] \to \Omega$ be the unique K-morphism such that $\varphi(b) = b'$. The morphism φ can be extended to a K-morphism σ from L to Ω. By assumption, $\sigma(L) \subset L$. In particular, $b' = \sigma(b) \in L$. $\square$

Corollary (4.6.9). — *Let $K \subset L$ be a finite normal extension and let $G = \mathrm{Aut}(L/K)$. The extension $K \to L^G$ is radicial. In particular, if the characteristic of K is 0 or if K is perfect, then $L^G = K$ and the extension $K \subset L$ is Galois.*

Proof. — Let us prove that every element of L^G is inseparable over K. Let $a \in L^G$ and let $P \in K[T]$ be its minimal polynomial. Let $j \colon L \to \Omega$ be an algebraic closure of L. For every root b of P in Ω, there exists a unique K-morphism $f \colon K[a] \to \Omega$ such that $f(a) = b$. Then f extends to a K-morphism $\varphi \colon L \to \Omega$ and, by definition of a normal extension, the image of φ is equal to $j(L)$, so that $\varphi \in \mathrm{Aut}(L/K)$. By definition of L^G, one thus has $a = \varphi(a) = b$. This implies that a is the only root of P in Ω, and a is radicial over K.

If, moreover, the extension $K \to L$ is separable, one then has $a \in K$, so that $L^G = K$ in this case. $\square$

Remark (4.6.10). — Let $K \subset L$ be a finite extension of fields. There exists a finite extension L' of L such that the composed extension $K \to L'$ is normal. Indeed, let $a_1, \ldots, a_n$ be elements of L such that $L = K[a_1, \ldots, a_n]$. For every i, let P_i be the minimal polynomial of P_i and let P be the least common multiple of $P_1, \ldots, P_n$. If Ω is any extension of L in which P is split (for example, an algebraic closure of L), the field L' generated by the roots of P in L' is normal.

Assume moreover that the extension $K \subset L$ is separable. By the Primitive element theorem (theorem 4.4.14), there exists an element $a \in L$ such that $L = K[a]$. The minimal polynomial P of a is separable and the field L' generated by its roots in some extension Ω in which it is split is a Galois extension of K containing L. In fact, this field L' is the *smallest* Galois extension of K contained in Ω such that $L \subset L'$. Indeed, every such extension must contain the roots of P, hence L'. It is called the *Galois closure* of L in Ω.

Proposition (4.6.11). — *Let $K \subset L$ be a finite Galois extension with group $G = \mathrm{Gal}(L/K)$. Let H be a subgroup of G and let $N_G(H) = \{\sigma \in \mathrm{Gal}(L/K); \sigma H \sigma^{-1} = H\}$ be the normalizer of H in $\mathrm{Gal}(L/K)$.*

(i) *For any $\sigma \in \mathrm{Gal}(L/K)$, one has $\sigma(L^H) = L^{\sigma H \sigma^{-1}}$.*

(ii) *The morphism $N_G(H) \to \mathrm{Aut}(L^H/K)$ is surjective and its kernel is H. In particular, the extension $K \subset L^H$ is Galois if and only if H is a normal subgroup of G. In that case, $\mathrm{Gal}(L^H/K) \simeq G/H$*

Proof. — a) Let $a \in L$. That a belongs to L^H is equivalent to the relation $h(a) = a$ for every $h \in H$. Consequently, $b = \sigma(b)$ belongs to $\sigma(L^H)$ if and only if $h \circ \sigma^{-1}(b) = \sigma^{-1}(b)$ for every $h \in H$, hence if and only if $(\sigma h \sigma^{-1})(b) = b$ for every $h \in H$, hence if and only if $b \in L^{\sigma H \sigma^{-1}}$.

b) By what precedes, any element σ of $N_G(H)$ induces by restriction a K-morphism from L^H to itself. The restriction map $\sigma \mapsto \sigma|_{L^H}$ is thus a morphism of groups ρ from $N_G(H)$ to $\mathrm{Aut}(L^H/K)$. The kernel of this morphism consists of those elements $\sigma \in N_G(H)$ such that $\sigma(a) = a$ for every $a \in L^H$, hence $\mathrm{Ker}(\rho) = H$.

Let us show that ρ is surjective. Let $\sigma \in \mathrm{Aut}(L^H/K)$. Let us view σ as a K-morphism from L^H to L. Let us consider an extension $\sigma_1 : L \to \Omega$ of σ, where Ω is an algebraically closed extension of L. Since the extension $K \subset L$ is Galois, it is normal and one has $\sigma_1(L) = L$; in other words, $\sigma_1 \in \mathrm{Aut}(L/K)$ and $\sigma = \sigma_1|_{L^H}$. Since $\sigma_1(L^H) = \sigma(L^H) = L^H$, one has $\sigma_1 \in N_G(H)$ and $\sigma = \rho(\sigma_1)$.

Passing to the quotient by H, the morphism ρ induces an isomorphism $N_G(H)/H \simeq \mathrm{Aut}(L^H/K)$. In particular, the extension $K \subset L^H$ is Galois if and only if $[L^H : K] = \mathrm{Card}(N_G(H))/\mathrm{Card}(H)$. Since the extension $L^H \subset L$ is Galois of group H, one has $[L : L^H] = \mathrm{Card}(H)$; moreover, $[L : K] = \mathrm{Card}(G)$, so that $[L^H : K] = \mathrm{Card}(G)/\mathrm{Card}(H)$. Consequently, the extension $K \subset L^H$ is Galois if and only if $\mathrm{Card}(N_G(H)) = \mathrm{Card}(G)$, that is, if and only if $G = N_G(H)$, which means that H is a normal subgroup of G. $\qquad\square$

4.7. Norms and Traces

Let A be a ring and let B be an A-algebra; let us assume that B is a free finitely generated A-module. An important case of this situation arises when A is a field and B is a finite extension of A.

Since A is commutative, all bases of B have the same cardinality, equal to $n = \mathrm{rk}_A(B)$, by definition of the rank of B. For any $b \in B$, multiplication by b gives rise to an endomorphism $x \mapsto bx$ of the A-module B; one defines the characteristic polynomial of b as the characteristic polynomial P_b of this endomorphism. This polynomial is defined either as $\det(X - M_b)$, where $M_b \in \mathrm{Mat}_n(A)$ is the matrix of the multiplication by b, written in any basis of B.

Let us write

$$P_b(X) = X^n - \mathrm{Tr}_{B/A}(b)X^{n-1} + \cdots + (-1)^n N_{B/A}(b).$$

Definition (4.7.1). — The elements $\mathrm{Tr}_{B/A}(b)$ and $N_{B/A}(b)$ are called the trace and the norm of b with respect to A.

Note that $\mathrm{Tr}_{B/A}(b)$ and $N_{B/A}(b)$ are the trace and the norm of the matrix $U_b \in \mathrm{Mat}_n(A)$ of multiplication by b, written in any basis of B.

Remark (4.7.2). — Using notions that we will introduce later, an intrinsic definition is possible: consider the A[X]-module $B[X] = A[X] \otimes_A B$; it is free of rank n and multiplication by $X - b$ in B[X] induces an endomorphism of the free rank-1 module $\Lambda^n B[X]$, which is precisely multiplication by $P_b(X)$. Moreover, $\Lambda^n(B)$ is a free A-module of rank 1 and multiplication by b in B induces multiplication by $N_{B/A}(b)$ on $\Lambda^n(B)$.

Proposition (4.7.3). —   (i) $\mathrm{Tr}_{B/A}(1_B) = \mathrm{rk}_A(B)1_A$ *and* $N_{B/A}(1_B) = 1_A$;

(ii) *For any* $a \in A$ *and any* $b \in B$, $\mathrm{Tr}_{B/A}(ab) = a\,\mathrm{Tr}_{B/A}(b)$ *and* $N_{B/A}(ab) = a^{\mathrm{rk}_A(B)}N_{B/A}(b)$;

(iii) *For any* $b, b' \in B$, $\mathrm{Tr}_{B/A}(b + b') = \mathrm{Tr}_{B/A}(b) + \mathrm{Tr}_{B/A}(b')$ *and* $N_{B/A}(bb') = N_{B/A}(b)N_{B/A}(b')$;

(iv) *An element* $b \in B$ *is invertible if and only if its norm* $N_{B/A}(b)$ *is invertible in A.*

Proof. — Let us fix a basis of B as an A-module; let $n = \mathrm{rk}_A(B)$.

a) Multiplication by 1_B is the identity, hence its matrix is I_n. Consequently, $\mathrm{Tr}_{B/A}(1_B) = n1_A$ and $N_{B/A}(1_B) = 1_A$.

b) If U is the matrix of multiplication by b, the matrix of multiplication by ab is aU. Then, $\mathrm{Tr}_{B/A}(ab) = \mathrm{Tr}(aU) = a\,\mathrm{Tr}(U) = a\,\mathrm{Tr}_{B/A}(b)$ and $N_{B/A}(ab) = \det(aU) = a^n \det(U) = a^n N_{B/A}(b)$.

c) Let U' be the matrix of multiplication by b'. Since $(b + b')x = bx + b'x$ for every $x \in B$, the matrix of multiplication by $b + b'$ is $U + U'$. Then

$$\mathrm{Tr}_{B/A}(b + b') = \mathrm{Tr}(U + U') = \mathrm{Tr}(U) + \mathrm{Tr}(U') = \mathrm{Tr}_{B/A}(b) + \mathrm{Tr}_{B/A}(b').$$

Moreover, the formula $(bb')x = b(b'x)$, for $x \in B$, implies that the matrix of multiplication by bb' is UU'. Then

$$N_{B/A}(bb') = \det(UU') = \det(U)\det(U') = N_{B/A}(b)N_{B/A}(b').$$

d) Assume that b is invertible, with inverse b'. Then $bb' = 1_B$, hence multiplication by b is an isomorphism of B, whose inverse is multiplication by b'. Consequently, its determinant is invertible, that is to say, $N_{B/A}(b)$ is invertible.

Conversely, if $N_{B/A}(b)$ is invertible, then multiplication by b is invertible. In particular, there exists an element $b' \in B$ such that $bb' = 1_B$ and b is invertible.   □

Proposition (4.7.4). — *Let* $K \to L$ *be a finite separable extension, let* Ω *be an algebraic closure of* L *and let* Σ *be the set of* K-*morphisms from* L *to* Ω. *Then, for every* $x \in L$, *one has*

$$\mathrm{Tr}_{L/K}(x) = \sum_{\sigma \in \Sigma} \sigma(x) \quad and \quad N_{L/K}(x) = \prod_{\sigma \in \Sigma} \sigma(x).$$

Proof. — Let $x \in L$, let $P \in K[X]$ be its minimal polynomial, and let $x = x_1, \ldots, x_d$ be its (pairwise distinct) roots in Ω. By assumption, x is separable, hence $P(X) = (X - x_1) \ldots (X - x_d)$. Let $a_1, \ldots, a_d \in K$ be defined by the relation

$$P(X) = X^d - a_1 X^{d-1} + \cdots + (-1)^d a_d,$$

so that $a_1 = x_1 + \cdots + x_d$ and $a_d = x_1 \ldots x_d$. In the basis $\{1, x, \ldots, x^{d-1}\}$ of $K[x]$, the matrix of multiplication by x is given by

$$\begin{pmatrix} 0 & & (-1)^{d-1}a_d \\ 1 & \ddots & \vdots \\ & \ddots & 0 & -a_2 \\ & & 1 & a_1 \end{pmatrix}$$

hence is the companion matrix C_P. In particular, its trace is a_1 and its determinant is a_d. In other words,

$$\mathrm{Tr}_{K[x]/K}(x) = \sum_{i=1}^{d} x_i \quad and \quad N_{K[x]/K}(x) = \prod_{i=1}^{d} x_i.$$

Let now $\{e_1, \ldots, e_s\}$ be a basis of L over $K[x]$. The family $(e_i x^j)$, indexed by pairs (i, j) of integers such that $1 \le i \le s$ and $0 \le j < d$, is a basis of L as a K-vector space. When this basis is written in lexicographic order $(e_1, e_1 x, \ldots, e_1 x^{d-1}, e_2, \ldots, e_s x^{d-1})$, the matrix of multiplication by x is block-diagonal, with s blocks all equal to the companion matrix C_P. Consequently, $\mathrm{Tr}_{L/K}(x) = sa_1$ and $N_{L/K}(x) = a_d^s$.

Recall that for every $i \in \{1, \dots, d\}$ there is exactly one K-morphism σ_i from $K[x]$ to Ω such that $x \mapsto x_i$, and every K-morphism from $K[x]$ to Ω is one of them. Moreover, since the extension $K[x] \subset L$ is separable of degree s, each of these morphisms extends in exactly s ways to L. Therefore,

$$\sum_{\sigma \in \Sigma} \sigma(x) = s \sum_{i=1}^{d} \sigma_i(x) = s \, \mathrm{Tr}_{K[x]/K}(x) = s a_1 = \mathrm{Tr}_{L/K}(x)$$

and

$$\prod_{\sigma \in \Sigma} \sigma(x) = \left(\prod_{i=1}^{d} \sigma_i(x) \right)^s = \left(N_{K[x]/K}(x) \right)^s = a_d^s = N_{L/K}(x).$$

Corollary (4.7.5). — *If* $K \to L$ *is a Galois extension, then*

$$N_{L/K}(x) = \prod_{\sigma \in \mathrm{Gal}(L/K)} \sigma(x) \quad and \quad \mathrm{Tr}_{L/K}(x) = \sum_{\sigma \in \mathrm{Gal}(L/K)} \sigma(x)$$

for every $x \in L$.

Proof. — Let Ω be an algebraic closure of L. Since the extension $K \to L$ is Galois, it is normal and the K-morphisms from L to Ω coincide with the elements of $\mathrm{Gal}(L/K)$. Since a Galois extension is separable, the result follows from the preceding proposition. □

Corollary (4.7.6). — *Let* $K \subset L$ *be a finite extension of fields, let A be a subring of K and let B be its integral closure in L. The trace* $\mathrm{Tr}_{L/K}(x)$ *and the norm* $N_{L/K}(x)$ *of any element* $x \in B$ *are integral over A. In particular, if A is integrally closed, then* $\mathrm{Tr}_{L/K}(x)$ *and* $N_{L/K}(x)$ *belong to A, for every* $x \in B$.

Proof. — Let Ω be an algebraic closure of L, let $n = [L : K]$ and let $\sigma_1, \dots, \sigma_n \colon L \to \Omega$ be the n distinct K-morphisms from L to Ω. If $i \in \{1, \dots, n\}$ and $x \in B$, then $\sigma_i(x)$ is integral over A (it is a root of the same monic polynomial as x), so that $\mathrm{Tr}_{L/K}(x) = \sum_{i=1}^{n} \sigma_i(x)$ is integral over A. Similarly, $N_{L/K}(x) = \prod_{i=1}^{n} \sigma_i(x)$ is integral over A.

Moreover, $\mathrm{Tr}_{L/K}$ and $N_{L/K}(x)$ are elements of K. If A is integrally closed in K, this implies $\mathrm{Tr}_{L/K}(x) \in A$ and $N_{L/K}(x) \in A$. □

Theorem (4.7.7). — *Let* $K \subset L$ *be a finite separable extension. The map* $t \colon L \times L \to K$ *defined by* $t(x, y) = \mathrm{Tr}_{L/K}(xy)$ *is a non-degenerate symmetric K-bilinear form.*

Proof. — The map t is symmetric since the multiplication is commutative. The formulas

$$t(ax + a'x', y) = \mathrm{Tr}_{L/K}((ax + a'x')y) = \mathrm{Tr}_{L/K}(axy + a'x'y)$$
$$= a \, \mathrm{Tr}_{L/K}(xy) + a' \, \mathrm{Tr}_{L/K}(x'y) = at(x, y) + a't(x', y)$$

show that it is K-linear with respect to the first variable. By symmetry, it is also K-linear with the second one.

For every $y \in L$, let $\tau(y)$ be the map defined by $\tau(y)(x) = t(x, y)$. Since t is K-linear with respect to the second variable, one has $\tau(y) \in \mathrm{Hom}_K(L, K)$. Moreover, the map $\tau \colon L \to \mathrm{Hom}_K(L, K)$ is K-linear. We need to prove that τ is an isomorphism of K-vector spaces. Let $d = [L : K]$, so that $d = \dim_K(L) = \dim_K(\mathrm{Hom}_K(L, K))$. Consequently, it suffices to prove that τ is injective.

Let Ω be an algebraic closure of L and let $\sigma_1, \dots, \sigma_d$ be the d distinct K-morphisms from L to Ω. For every $x \in L$, one has

$$\mathrm{Tr}_{L/K}(x) = \sum_{j=1}^{d} \sigma_j(x).$$

Let us assume that $y \in \mathrm{Ker}(\tau)$. Then, for every $x \in L$, $\mathrm{Tr}_{L/K}(xy) = 0$, hence

$$0 = \sum_{j=1}^{d} \sigma_j(xy) = \sum_{j=1}^{d} \sigma_j(x)\sigma_j(y).$$

Since $\sigma_1, \dots, \sigma_d$ are linearly independent on L (proposition 4.4.12), this implies $\sigma_j(y) = 0$ for all j, hence $y = 0$. □

Remark (4.7.8). — Let $K \to L$ be a finite extension which is *not* separable. Let us prove that $\mathrm{Tr}_{L/K}(x) = 0$ for every $x \in L$. The characteristic of K is a prime number p. Let L_{sep} be the subfield of L generated by all elements of L which are separable over K. By remark 4.4.17, the degree $[L : L_{\mathrm{sep}}]$ is a power of p; since $L \neq L_{\mathrm{sep}}$, it is a multiple of p.

Let $x \in L$. If $x \in L_{\mathrm{sep}}$, one has

$$\mathrm{Tr}_{L/K}(x) = [L : L_{\mathrm{sep}}]] \, \mathrm{Tr}_{L_{\mathrm{sep}}/K}(x) = 0$$

since p divides $[L : K[x]]$. Otherwise, $x \in L - L_{\mathrm{sep}}$ is inseparable over K; then its minimal polynomial is a polynomial in X^p, so has no term of degree one less than its degree and $\mathrm{Tr}_{K[x]/K}(x) = 0$. Then, $\mathrm{Tr}_{L/K}(x) = [L : K[x]] \, \mathrm{Tr}_{K[x]/K}(x) = 0$.

4.8. Transcendence Degree

Proposition (4.8.1). — *Let $K \to L$ be a field extension and let $(x_i)_{i \in I}$ be a family of elements of L. The following propositions are equivalent:*

(i) *For every non-zero polynomial $P \in K[(X_i)_{i \in I}]$, one has $P((x_i)) \neq 0$;*

(ii) *The canonical morphism of K-algebras from $K[(X_i)_i]$ to L such that $X_i \mapsto x_i$ is injective;*

(iii) *There exists a K-morphism of fields $K((X_i)_i) \to L$ such that $X_i \mapsto x_i$.*

Proof. — Let $\varphi \colon K[(X_i)_i] \to L$ be the unique morphism of K-algebras such that $\varphi(X_i) = x_i$ for all $i \in I$.

Condition (i) means that $\varphi(P) \neq 0$ if $P \neq 0$; it is thus equivalent to the equality $\mathrm{Ker}(\varphi) = 0$ of (ii).

Assume (ii). Note that $K[(X_i)_{i \in I}]$ is an integral domain, and its field of fractions is $K((X_i))$. If φ is injective, then, since L is a field, it extends to a morphism of fields from $K[(X_i)]$ to L, hence (iii).

Finally, assume that (iii). Let $\psi: K((X_i)) \to L$ be a K-morphism of fields such that $\psi(X_i) = x_i$ for all i. The restriction of ψ to $K[(X_i)]$ is the morphism of K-algebras which was denoted by φ. Since ψ is injective, φ is injective too, hence (ii). $\qquad \square$

Definition (4.8.2). — If these conditions hold, one says that the family $(x_i)_{i \in I}$ is *algebraically independent* over K.

Note the similitude with proposition 3.5.2 in linear algebra, where we considered *linear* relations between elements of a module, while we now study *algebraic* relations in a ring. In fact, the theory of matroids pushes this parallel farther than mere language. Based on this analogy, a possible definition of "basis" can be made. However, in this context, "generating" families are not families that generate the extension L, but those for which L is algebraic over the field that family generates.

Definition (4.8.3). — Let $K \to L$ be a field extension. One says that a family $(x_i)_{i \in I}$ in L is a *transcendence basis* of L over K if it is algebraically independent over K and if L is algebraic over the subfield $K((x_i))$.

By abuse of language, one says that a subset B of L is algebraically independent (resp. a transcendence basis of L) over K if the family $(b)_{b \in B}$ is algebraically independent (resp. a transcendence basis of L) over K.

Theorem (4.8.4). — *Let $K \to L$ be a field extension.*

(i) *Let* B, C *be subsets of* L. *Assume that* L *is algebraic over the subfield* $K(C)$ *generated by* C *over* K *and that* B *is algebraically independent. Then* B *is a transcendence basis if and only if, for every* $c \in C - B$, *the subset* $B \cup \{c\}$ *is not algebraically independent.*

(ii) *Let* A *and* C *be subsets of* L *such that* $A \subset C$. *One assumes that* A *is algebraically independent over* K *and that* L *is algebraic over the subfield* $K(C)$ *generated by* C *over* K. *Then, there exists a transcendence basis* B *of* L *such that* $A \subset B \subset C$.

Proof. — (i) Assume that B is a transcendence basis of L. Let $c \in C - B$. By definition, L is algebraic over $K(B)$; in particular, c is algebraic over $K(B)$. Let $P_1 \in K(B)[T]$ be a non-zero polynomial in one indeterminate T such that $P_1(c) = 0$. Write $P_1 = a_n T^n + \cdots + a_0$, where $a_0, \ldots, a_n \in K(B)$. For $j \in \{0, \ldots, n\}$, there exist polynomials $f_j, g_j \in K[(X_b)_{b \in B}]$ such that $a_j = f_j((b))/g_j((b))$. Multiplying P_1 by the product of the elements $g_j((b))$, we reduce to the case where $g_j = 1$ for every j. Let $P = \sum_{j=0}^{n} f_j((X_b))T^j$; this is a non-zero polynomial in $K[(X_b)_{b \in B}, T]$. By construction, $P((b), c) = 0$. This proves that the family deduced from B by adjoining c is not algebraically independent. Since $c \notin B$, $B \cup \{c\}$ is not algebraically independent.

Conversely, assume that for every $c \in C-B$, $B \cup \{c\}$ is not algebraically independent. Let us prove that every element $c \in C$ is algebraic over $K(B)$. This is obvious if $c \in B$; otherwise, consider a non-zero polynomial $P \in K[(X_b)_{b \in B}, T]$ such that $P((b), c) = 0$. Write $P = f_n T^n + \cdots + f_0$, where $f_0, \ldots, f_n \in K[(X_b)_{b \in B}]$, and $f_n \neq 0$. Since B is algebraically independent, one has $f_n((b)) \neq 0$. Consequently, the polynomial $P_1 = P((b), T) = f_n((b))T^n + \cdots + f_0((b))$ is a non-zero element of $K(B)[T]$ and $P_1(c) = 0$. This proves that c is algebraic over $K(B)$.

Then $K(C)$ is algebraic over $K(B)$ and, since L is algebraic over $K(C)$, L is algebraic over $K(B)$. This proves that B is a transcendence basis of L.

(ii) Let $\mathscr{B}$ be the set of all subsets B of L such that $B \subset C$ and such that B is algebraically independent over K.

A polynomial in indeterminates $(X_b)_{b \in B}$ depends on only finitely many indeterminates, say $X_{b_1}, \ldots, X_{b_n}$, where $b_1, \ldots, b_n \in B$, and the condition $P((b)) = 0$ is then equivalent to the condition $P(b_1, \ldots, b_n) = 0$. Consequently, a subset B of L is algebraically independent over K if and only if each of its finite subsets is algebraically independent over K; in other words, the condition that B be algebraically independent over K is of finite character. The condition that $B \subset C$ is of finite character too (it even suffices to consider one-element subsets of B). This implies that the set $\mathscr{B}$ is of finite character. In particular, $\mathscr{B}$ is inductive, when ordered by inclusion.

Since $A \in \mathscr{B}$, by assumption, it follows from corollary A.2.16 that there exists a maximal element $B \in \mathscr{B}$ such that $A \subset B$. One has $A \subset B \subset C$, by construction, the set B is algebraically independent over K; moreover, for every $c \in C - B$, the set $B \cup \{c\}$ does not belong to $\mathscr{B}$, hence, by definition of a maximal element, it is not algebraically independent, since one has $A \subset B \cup \{c\} \subset C$. By a), the set B is a transcendence basis of L over K.   □

Applying the theorem to $A = \emptyset$ and $L = C$, we obtain:

Corollary (4.8.5). — *Any field extension has a transcendence basis.*

Theorem (4.8.6) (Exchange lemma). — *Let $K \to L$ be a field extension, let B be a transcendence basis of L over K and let C be a subset of L such that L is algebraic over $K(C)$. Then, for every $\beta \in B - C$, there exists a $\gamma \in C - B$ such that $(B - \{\beta\}) \cup \{\gamma\}$ is a transcendence basis of L over K.*

Proof. — Set $A = B - \{\beta\}$.

Since $B = A \cup \{\beta\}$ is algebraically independent, the element β is not algebraic over $K(A)$ and, in particular, L is not algebraic over $K(A)$. On the other hand, L is algebraic over $K(C)$. Consequently, there must exist a $\gamma \in C$ such that γ is not algebraic over $K(A)$. Let $B' = A \cup \{\gamma\}$ and let us show that B' is a transcendence basis of L over K.

I claim that there exists a non-zero polynomial P in indeterminates $(X_b)_{b \in B}$ and T such that $P((b), \gamma) = 0$. Indeed, since L is algebraic over $K(B)$, there exists an irreducible polynomial $R_1 \in K(B)[T]$ such that $R_1(\gamma) = 0$. Since B is algebraically independent over K, there exists a unique polynomial $R \in$

$K((X_b))[T]$, whose coefficients are rational functions in indeterminates X_b, such that $R_1 = R((b))[T]$. If we multiply R by a common denominator of the coefficients of R, we obtain a non-zero polynomial $P \in K[(X_b), T]$ such that $P((b), \gamma) = 0$, as claimed.

Since γ is not algebraic over $K(A)$, D is algebraically independent over K. We then write the polynomial P as a polynomial $\sum_{j=0}^{m} P_j X_\beta^j$ in X_β with coefficients in the indeterminates X_b, for $b \in A$, as well as T, where $m = \deg_{X_\beta}(P)$, so that $P_m \neq 0$. Since γ is not algebraic over $K(A)$, $P_m((b)_{b \in A}, \gamma) \neq 0$. Consequently, the relation

$$\sum_{j=0}^{m} P_j((b)_{b \in A}, \gamma) \beta^j = P((b), \gamma) = 0$$

shows that β is algebraic over $K(D)$. Any other element of B belongs to D, hence is algebraic over $K(D)$ as well. Therefore, $K(B)$ is algebraic over $K(D)$. Since L is algebraic over K, L is algebraic over $K(D)$. This concludes the proof that D is a transcendence basis of L over K.   □

Corollary (4.8.7). — *Let* $K \to L$ *be an extension of fields. All transcendence bases of L over K have the same cardinality.*

Proof. — Let B and B′ be two transcendence bases of L over K. According to the following lemma, there exists an injection φ from B to B′, as well as an injection φ' from B′ to B. It then follows from the Cantor–Bernstein theorem (theorem A.2.1) that B and B′ are equipotent.   □

Lemma (4.8.8). — *Let* $K \to L$ *be an extension of fields and let* A *and* B *be subsets of L such that* A *is algebraically independent over K and L is algebraic over* $K(B)$. *Then there exists an injection from A to B.*

Proof. — Let $\mathscr{F}$ be the set of all pairs (A', φ) where A′ is a subset of A and $\varphi \colon A' \to B$ is an injection such that $(A - A') \cup \varphi(A')$ is algebraically independent over K. We order the set $\mathscr{F}$ by saying that $(A'_1, \varphi_1) \prec (A'_2, \varphi_2)$ if $A'_1 \subset A'_2$ and $\varphi_2(a) = \varphi_1(a)$ for every $a \in A'_1$.

Let us prove that the set $\mathscr{F}$ is inductive.

Let $(A'_i, \varphi_i)_i$ be a totally ordered family of elements of $\mathscr{F}$. Let A′ be the union of the A'_i and let $\varphi \colon A' \to B$ be the unique function such that $\varphi(a) = \varphi_i(a)$ for any index i such that $a \in A'_i$. The map φ is injective. Indeed, let us consider two elements $a, a' \in A'$ such that $\varphi(a) = \varphi(a')$. Let i be any index such that a and a' both belong to A'_i. Then $\varphi_i(a) = \varphi(a) = \varphi(a') = \varphi_i(a')$. Since φ_i is injective, $a = a'$.

Let us show that the set $(A - A') \cup \varphi(A')$ is algebraically independent. Let P be a polynomial in indeterminates $(X_a)_{a \in A-A'}$ and $(Y_a)_{a \in A'}$ such that $P((a)_{a \notin A'}, (\varphi(a))_{a \in A'}) = 0$ and let us prove that $P = 0$. Only finitely many indeterminates Y_a, for $a \in A'$, appear in P; since the family (A'_i) of subsets of A is totally ordered, all the corresponding elements a belong to some common set A'_i. The polynomial P can then be viewed as an algebraic dependence relation for $(A - A') \cup \varphi(A'_i)$. Since this set is contained in the transcendence basis $(A - A'_i) \cup \varphi(A'_i)$, the polynomial P is zero, as claimed.

The pair (A', φ) is an element of $\mathscr{F}$ such that $(A'_i, \varphi_i) \prec (A', \varphi)$ for every i. This shows that $\mathscr{F}$ is inductive.

By Zorn's lemma (corollary A.2.13), the set $\mathscr{F}$ has a maximal element, say (A', φ). Let us prove that $A' = A$. Otherwise, let B_0 be a transcendence basis of L over K which contains $(A-A')\cup\varphi(A')$ and let β be any element of $A-A'$. By the Exchange lemma (theorem 4.8.6), there exists a $\gamma \in B$ such that the family obtained by merging $(A - A' - \{\beta\})$, $\varphi(A')$ and γ is a transcendence basis of L over K. Let $A'' = A' \cup \{\beta\}$ and let $\varphi' \colon A'' \to B$ be the map which coincides with φ on A' and such that $\varphi'(\beta) = \gamma$. By construction, the family $(\varphi'(a))_{a \in A''}$ is algebraically independent; in particular, the map φ' is injective. Moreover, $(A - A'') \cup \varphi'(A'')$ is algebraically independent. This shows that (A'', φ') belongs to $\mathscr{F}$, which contradicts the hypothesis that (A', φ) is a maximal element.   □

Definition (4.8.9). — The cardinality of any transcendence basis of L over K is called the *transcendence degree* of L over K and is denoted by $\operatorname{tr deg}_K(L)$.

Example (4.8.10). — For any integer n, the field $K(X_1, \ldots, X_n)$ of rational functions in n indeterminates has transcendence degree n over K.

Indeed, it just follows from the definition that $\{X_1, \ldots, X_n\}$ is algebraically independent over K. On the other hand, it is maximal since for every $f \in K(X_1, \ldots, X_n)$, written $f = P/Q$, with $P, Q \in K[X_1, \ldots, X_n]$, the non-zero polynomial $Q(X_1, \ldots, X_n)T - P(X_1, \ldots, X_n)$ vanishes at $(X_1, \ldots, X_n, f)$. Consequently, $\{X_1, \ldots, X_n\}$ is a transcendence basis of $K(X_1, \ldots, X_n)$ over K, and $\operatorname{tr deg}_K(L) = n$.

Example (4.8.11). — Let K be a field, let $P \in K[X_1, \ldots, X_n]$ be an irreducible polynomial and let L be the field of fractions of the integral domain $A = K[X_1, \ldots, X_n]/(P)$. Let us show that $\operatorname{tr deg}_K(L) = n - 1$.

For $j \in \{1, \ldots, n\}$, let x_j be the image of X_j in L.

Since P is irreducible, it is non-constant and its degree in at least one of the indeterminates is at least 1. For simplicity of notation, assume that $m = \deg_{X_n}(P) \geq 1$ and write $P = f_m X_n^m + \cdots + f_0$, where $f_0, \ldots, f_m \in K[X_1, \ldots, X_{n-1}]$ and $f_m \neq 0$.

Let us first prove that it is algebraically independent. Let us consider a non-zero polynomial $Q \in K[X_1, \ldots, X_{n-1}]$ such that $Q(x_1, \ldots, x_{n-1}) = 0$. This implies that Q belongs to the ideal (P) generated by P in $K[X_1, \ldots, X_n]$. Since $\deg_{X_n}(Q) = 0 < \deg_{X_n}(P)$, we get a contradiction.

In particular, $f_m(x_1, \ldots, x_{n-1}) \neq 0$. Then the expression

$$0 = P(x_1, \ldots, x_n) = f_m(x_1, \ldots, x_{n-1})x_n^m + \cdots + f_0(x_1, \ldots, x_{n-1})$$

proves that x_n is algebraic over $K(x_1, \ldots, x_{n-1})$. Since $L = K(x_1, \ldots, x_n)$, every element of L is then algebraic over $K(x_1, \ldots, x_{n-1})$.

This proves that $(x_1, \ldots, x_{n-1})$ is a transcendence basis of L; in particular, $\operatorname{tr deg}_K(L) = n - 1$.

4.9. Exercises

Exercise (4.9.1). — Let A be a ring and let $P \in A[T]$ be a monic polynomial of degree n.

a) Consider the quotient ring $B = A[T]/(P)$. Prove that B is integral over A and that there exists a $b \in B$ such that $P(b) = 0$.

b) Prove that B is a free A-module of rank n. In particular, the natural morphism $A \to B$ is injective.

c) In the ring $C = A[X_1, \ldots, X_n]$, let I be the ideal generated by the coefficients of the polynomial

$$P - \prod_{i=1}^{n} (T - X_i) \in C[T].$$

Prove that the ring C/I is a free A-module of rank $n! = n(n-1)\ldots 2 \cdot 1$. (This ring is called the *splitting algebra* of P.)

d) Assume that A is a field and let M be a maximal ideal of C. Prove that the field C/M is an extension of A in which the polynomial P is split and which is generated by the roots of P.

Exercise (4.9.2). — Let B be a ring, let A be a subring of B and let $x \in B$ be an invertible element of B.

Let $y \in A[x] \cap A[x^{-1}]$.

a) Prove that there exists an integer n such that the A-submodule $M = A + Ax + \cdots + Ax^n$ of B satisfies $yM \subset M$.

b) Prove that y is integral over A.

Exercise (4.9.3). — Let A be an integral domain and let K be its field of fractions.

a) Let $x \in K$ be integral over A. Prove that there exists an $a \in A - \{0\}$ such that $ax^n \in A$ for every $n \in \mathbf{N}$.

b) Assume that A is a noetherian ring. Let $a \in A - \{0\}$ and $x \in K$ be such that $ax^n \in A$ for every $n \in \mathbf{N}$. Prove that x is integral over A.

Exercise (4.9.4). — *a)* Let A be a commutative ring, let B be a commutative A-algebra and let $U \in \mathrm{Mat}_n(B)$ be a matrix. Let $\Delta = \det(U)$. Let $M = B^n$ be the free B-module, with canonical basis $(e_1, \ldots, e_n)$. The matrix U is viewed as an endomorphism u of M.

We assume that U is integral over A.

b) Let $V \subset M$ be the submodule generated by the elements $u^k(e_i)$, for $i \in \{1, \ldots, n\}$ and $k \in \mathbf{N}$. Prove that V is a finitely generated A-module, stable under u.

c) Construct a finitely generated submodule of $\Lambda^n(M)$ that contains $e_1 \wedge \cdots \wedge e_n$ and that is stable under $\Lambda^n(u)$. Conclude that $\det(U)$ is integral over A.

d) Introducing the ring $A[T]$ and the matrix $TI_n - U$, prove that the coefficients of the characteristic polynomial of U are integral over A. In particular, $\text{Tr}(U)$ is integral over A.

Exercise (4.9.5). — Let $f: A \to B$ be a morphism of commutative rings.

a) Assume that f is integral. Prove that the continuous map $f^*: \text{Spec}(B) \to \text{Spec}(A)$ induced by f is closed: the image of a closed subset is closed.

b) More generally, if f is integral, prove that for every A-algebra C, the continuous map $(f_C)^*: \text{Spec}(B \otimes_A C) \to \text{Spec}(C)$ is closed.

c) Let us consider the particular case $C = A[T]$. Observe that the map f_C identifies with the map $g: A[T] \to B[T]$. We assume that $g^*: \text{Spec}(B[T]) \to \text{Spec}(A[T])$ is closed.

d) Let $b \in B$ and let $Z = V(bT - 1)$ in $\text{Spec}(B[T])$. Let I be an ideal of A such that $g^*(Z) = V(I)$. Prove that $I + (T) = A[T]$. (*Consider a prime ideal* P *of* A *that contains* $I + (T)$.)

e) Prove that b is integral over A.

Exercise (4.9.6). — Let A be a commutative ring and let I be an ideal such that every element of I is nilpotent. Let $f \in A[T]$ be a polynomial, let $a \in A$ be such that $f(a) \in I$ and $f'(a)$ is a unit.

We define a sequence (a_n) of elements of A by the formula $a_{n+1} = a_n - f(a_n)f'(a)^{-1}$.

a) By induction on n, prove that, modulo I^n, a_n is the unique element of A such that $a_n \equiv a \pmod{I}$ and $f(a_n) \in I^n$.

b) Prove that $f(a_n) \in (f(a)^n)$ for all n.

c) Conclude that there exists a unique element $b \in A$ such that $f(b) = 0$ and $b \equiv a \pmod{I}$.

d) Let K be a field and let $U \in \text{Mat}_n(K)$ be a matrix. Let P be the characteristic polynomial of U and let Q be the product of its irreducible factors. Assume that Q is separable (this is automatic if K is a perfect field). Prove the Chevalley–Dunford theorem: there exists a unique pair (D, N) of commuting matrices such that D becomes diagonalizable over an algebraic closure of K, N is nilpotent and $U = D + N$. (*Solve the equation* $Q(D) = 0$ *in* $K[U]$, *with unknown* D.)

Exercise (4.9.7). — Let $d \in \mathbf{Z}$ be square-free and let $K = \mathbf{Q}[\sqrt{d}]$.

a) What is the minimal polynomial of an element $x = a + b\sqrt{d} \in K$, where $a, b \in \mathbf{Q}$?

b) Prove that the integral closure of $\mathbf{Z}$ in K is $\mathbf{Z}[(1+\sqrt{d})/2]$ if $d \equiv 1 \pmod{4}$, and is $\mathbf{Z}[\sqrt{d}]$ otherwise.

Exercise (4.9.8). — *a*) Let A be an integral domain and let $t \in A$ be such that the ring A/tA has no nilpotent element except for 0. Assume also that the ring of fractions A_t is integrally closed. Prove that A is integrally closed.

b) Prove that the ring $A = \mathbf{C}[X, Y, Z]/(XZ - Y(Y + 1))$ is integrally closed. (*Apply the first question with* $t = X$.)

Exercise (4.9.9). — Let B be a ring and let A be a subring. Assume that A is integrally closed in B.

a) Let P, Q be monic polynomials of $B[T]$ such that $PQ \in A[T]$. Prove that $P, Q \in A[T]$. (*Use exercise 4.9.1 to introduce a ring C containing B such that* $P = \prod_{i=1}^{m}(T - a_i)$ *and* $Q = \prod_{i=1}^{n}(T - b_i)$ *where* $a_1, \ldots, a_m, b_1, \ldots, b_n \in C$.)

b) Prove that $A[T]$ is integrally closed in $B[T]$. (*If* $P \in B[T]$ *is integral over* $A[T]$, *consider the monic polynomials* $Q = T^m + P$, *for large enough integers* m.)

Exercise (4.9.10). — Let $K \to L$ be a finite algebraic extension and let Ω be an algebraic closure of K.

a) Prove that the algebra $L \otimes_K \Omega$ is reduced (0 is the only nilpotent element) if and only if the extension $K \to L$ is separable.

b) In this case, prove that for every field extension M of K, the tensor product $L \otimes_K M$ is reduced.

Exercise (4.9.11). — Let E be a field and let $P, Q \in E[T]$ be two irreducible polynomials. Let Ω be an algebraic closure of E and let $x, y \in \Omega$ be such that $P(x) = Q(y) = 0$. Let $P_1, \ldots, P_r \in E[y][T]$ be pairwise distinct irreducible polynomials and $m_1, \ldots, m_r$ be strictly positive integers such that $P = \prod_{i=1}^{r} P_i^{m_i}$. Similarly, let $Q_1, \ldots, Q_s \in E[x][T]$ be pairwise distinct irreducible polynomials and $n_1, \ldots, n_s$ be strictly positive integers such that $Q = \prod_{j=1}^{s} Q_j^{n_j}$.

a) Introducing the E-algebra $R = E[X, Y]/(P(X), Q(Y))$, construct isomorphisms of E-algebras

$$R \simeq E[x][Y]/(Q_1(Y)^{n_1}) \oplus \cdots \oplus E[x][Y]/(Q_s(Y)^{n_s})$$
$$\simeq E[y][X]/(P_1(X)^{m_1}) \oplus \cdots \oplus E[y][X]/(P_r(X)^{m_r}).$$

b) What are the idempotents of R? (*If* F *is a field,* $m \geq 1$, *and* $P \in F[T]$ *is an irreducible polynomial, show that the only idempotents of* $F[T]/(P^m)$ *are 0 and 1.*)

c) Prove that $r = s$ and that there exists a permutation $\sigma \in \mathfrak{S}_r$ such that if $j = \sigma(i)$, then $E[x][Y]/(Q_j^{n_j})$ and $E[y][X]/(P_i^{m_i})$ are isomorphic as E-algebras.

d) In particular, if $j = \sigma(i)$, then $n_j \deg(Q_j) \deg(P) = m_i \deg(P_i) \deg(Q)$.

Exercise (4.9.12). — Let p be a prime number; let $E = \mathbf{F}_p$.

a) Let $n \geq 2$ be an integer prime to p and let F be the extension of E generated by a primitive nth root of unity α. Prove that $[F : E]$ is the order of p in $\mathbf{Z}/n\mathbf{Z}$.

b) Describe the action of $\mathrm{Gal}(F/E)$ on the set $\{\alpha, \alpha^2, \ldots, \alpha^{n-1}\}$. Give a simple condition of this action to factor through the alternating group on $n - 1$ elements. (*How does the Frobenius morphism* φ *act on this set? What is its signature?*)

c) Assume in the sequel that p is odd. For $a \in \mathbf{Z}$, one sets $\left(\frac{a}{p}\right) = 0$ if a is a multiple of p, $\left(\frac{a}{p}\right) = 1$ if a is a non-zero square modulo p, and $\left(\frac{a}{p}\right) = -1$ otherwise (*Legendre symbol*). Prove that $\left(\frac{a}{p}\right) \equiv a^{(p-1)/2} \pmod{p}$.

d) Let

$$\delta = \prod_{1 \le i < j \le n-1} \frac{\varphi(\alpha^j) - \varphi(\alpha^i)}{\alpha^j - \alpha^i}.$$

Prove that $\delta = ((-1)^{n(n-1)/2} n^{n-2})^{(p-1)/2} = \pm 1$.

e) Assume moreover that $n = q$ is an odd prime number. Prove that $\left(\frac{p}{q}\right)\left(\frac{q}{p}\right) = (-1)^{(p-1)(q-1)/4}$ (*Gauss's quadratic reciprocity law*).

Exercise (4.9.13). — Let K be a field and let $n \ge 2$ be an integer. Let $E = K[\alpha]$ be an extension of K generated by a primitive nth root of unity α. (In particular, the characteristic of K does not divide n.)

a) Prove that the set C of nth roots of unity in E is a cyclic group of order n, generated by α. Prove that E is a Galois extension of K.

b) Let $\sigma \in \mathrm{Gal}(E/K)$. Prove that there exists an integer d prime to n such that $\sigma(\alpha) = \alpha^d$. Prove that $\sigma(x) = x^d$ for every $x \in C$.

c) Construct a ring morphism $\varphi \colon \mathrm{Gal}(E/K) \to (\mathbf{Z}/n\mathbf{Z})^\times$ such that $\varphi(\sigma)$ is the class of d if $\sigma(\alpha) = \alpha^d$. Prove that φ is injective; in particular, $\mathrm{Gal}(E/K)$ is an abelian group.

Exercise (4.9.14). — Let K be a field and let $E = K(T)$ be the field of rational functions in one indeterminate T with coefficients in K.

a) Prove that there exists a unique K-automorphism of E, α (resp. β), such that $\alpha(T) = 1/T$ (resp. $\beta(T) = 1 - T$).

b) Let G be the subgroup of $\mathrm{Gal}(E/K)$ generated by α and β. Prove that it is isomorphic to the symmetric group $\mathfrak{S}_3$.

c) Let $F = E^G$ be the subfield of E consisting of rational functions $P \in K(T)$ such that $\alpha(P) = \beta(P) = P$. Prove that F contains

$$f = \frac{(T^2 - T + 1)^3}{T^2(T-1)^2}.$$

d) Prove that $F = K(f)$.

Exercise (4.9.15). — In this exercise, we prove that the field **C** of complex number is algebraically closed. In fact, the proof will show a bit more. So let R be a field and let $C = R(i)$ be the field obtained by adjoining an element i such that $i^2 = -1$. We make the following assumptions:

(i) The characteristic of R is zero;

(ii) Every polynomial of odd degree in R[T] has a root in R;

(iii) Every element of C is a square.

a) Prove that the field **R** of real numbers satisfies these assumptions.

b) Let E be a non-trivial finite Galois extension of R containing C. Let $G = \mathrm{Gal}(E/R)$ be its Galois group. Prove that the order of G is a power of 2. (*Let P be a 2-Sylow subgroup of G. Prove that* E^P *is an extension of* R *of odd degree. Conclude that* $E^P = R$.)

c) Let $G_1 = \mathrm{Gal}(E/C)$. Prove that $G_1 = \{1\}$ by introducing a subgroup H of index 2 in G_1, and considering the quadratic extension E^H of C. Conclude that $E = C$.

d) Prove that C is algebraically closed.

Exercise (4.9.16). — Let F be a finite field. Let $q = \mathrm{Card}(F)$.

a) For every $a \in F^\times$ and $b \in F$, prove that there exists a unique automorphism τ of $F(T)$ such that $\tau(T) = aT + b$.

b) Prove that the automorphisms of this form constitute a finite group G of cardinality $q(q - 1)$. Let A be the subgroup of G of such automorphisms with $b = 0$, and B be the subgroup of G of such automorphisms with $a = 1$.

c) Show that $F(T)^A = F(T^{q-1})$ and that $F(T)^B = F(T^q - T)$.

d) Describe the intermediate subfields between $F(T)^A$ and $F(T)$.

e) Compute the subfield $F(T)^G$. Are the extensions $F(T)^G \to F(T)^A$ and $F(T)^G \to F(T)^B$ Galois?

Exercise (4.9.17). — Let E be a field, let $P \in E[T]$ be a monic polynomial, let $n = \deg(P)$, and let $E \to F$ be a field extension such that P is split in F, and such that F is generated by the roots of P (a *splitting extension* of P). Let R be the set of roots of P in F. Let $G = \mathrm{Gal}(E/F)$.

a) Prove that the action of G on F induces an action of G on R.

b) What are the orbits of this action? In particular, prove that this action is transitive if and only if P is irreducible.

c) Let $a \in R$, and let H be the set of $\sigma \in G$ such that $\sigma(a) = a$. Prove that H is a subgroup of G. What is its index?

d) If $g \in G$, what is the fixed subfield of the subgroup gHg^{-1}? If P is irreducible, prove that $\bigcap_{g \in G} gHg^{-1} = \{\mathrm{id}_F\}$.

e) Let $E \to F$ be a finite Galois extension and let $G = \mathrm{Gal}(F/E)$. Prove that it is the splitting extension of an irreducible polynomial $P \in E[T]$ of degree n if and only if there exists a subgroup $H \subset G$ of index n such that $\bigcap_{g \in G} gHg^{-1} = \{\mathrm{id}_F\}$. What happens in the case where G is abelian?

Exercise (4.9.18). — For $n \geq 1$, let $\Phi_n \in \mathbf{C}[X]$ be the unique monic polynomials whose roots are simple, equal to the primitive nth roots of unity in **C**.

a) Prove that $\prod_{d|n} \Phi_d = X^n - 1$. Prove by induction that $\Phi_n \in \mathbf{Z}[X]$ for all n.

b) If p is a prime number, compute $\Phi_p(X)$. Prove that there are integers $a_1, \ldots, a_{p-1}$ such that $\Phi_p(1 + X) = X^{p-1} + p a_1 X^{p-2} + \cdots + p a_{p-1}$, and $a_{p-1} = 1$. Using Eisenstein's criterion (see exercice 2.8.28), prove that Φ_p is irreducible in $\mathbf{Q}[X]$.

c) Let $n \geq 2$. In the rest of the exercise, we will prove that Φ_n is irreducible. Let $\alpha \in \mathbf{C}$ be a primitive nth root of unity and let P be its minimal polynomial. Prove that $P \in \mathbf{Z}[X]$ and that P divides Φ_n in $\mathbf{Z}[X]$.

Let p be a prime number that does not divide n. Prove that there exists a $b \in \mathbf{Z}[\alpha]$ such that $P(\alpha^p) = pb$.

d) Prove that α^p is a primitive nth root of unity. If $P(\alpha^p) \neq 0$, prove by differentiating the polynomial $X^n - 1$ that $n\alpha^{p(n-1)} \in p\mathbf{Z}[\alpha]$. Derive a contradiction, hence that $P(\alpha^p) = 0$ for every prime number p which does not divide n.

e) Prove that Φ_n is irreducible in $\mathbf{Q}[X]$.

Exercise (4.9.19). — In this exercise, we give a proof of the following theorem of Wedderburn: *Any finite division ring is a field.* Let F be a finite division ring.

a) Show that any commutative subring of F is a field.

b) Let Z be the center of F. Show that Z is a subring of F. Let q be its cardinality. Show that there exists an integer $n \geq 1$ such that $\operatorname{Card} F = q^n$.

c) Let $x \in F$. Show that the set C_x of all $a \in F$ such that $ax = xa$ is a subring of F. Show that there exists an integer n_x dividing n such that $\operatorname{Card} C_x = q^{n_x}$. (Observe that left multiplication by elements of C_x endows F with the structure of a C_x-vector space.)

d) Let $x \in F^\times$; compute in terms of n_x the cardinality of the conjugacy class $C(x)$ of x in $F^\times$ (by which we mean the set of all elements of $F^\times$ of the form axa^{-1}, for $a \in F^\times$).

e) If $x \notin Z$, show that the cardinality of $C(x)$ is a multiple of $\Phi_n(q)$. (We write Φ_n for the nth cyclotomic polynomial, defined in exercise 4.9.18.)

f) Using the class equation, show that $\Phi_n(q)$ divides $q^n - q$. Conclude that $n = 1$, hence that F is a field.

Exercise (4.9.20). — For every integer $n \geq 1$, we denote by K_n the subfield of $\mathbf{C}$ generated by the nth roots of unity. We also write $\rho_n \colon \operatorname{Gal}(K_n/\mathbf{Q}) \to (\mathbf{Z}/n\mathbf{Z})^\times$ for the morphism of groups characterized by $\sigma(\omega) = \omega^{\rho(\sigma)}$ for every $\sigma \in \operatorname{Gal}(K/\mathbf{Q})$ and every nth root of unity ω.

a) Let m and n be integers ≥ 1 and let $d = \gcd(m, n)$. Prove that $K_m \cap K_n = K_d$.

b) Let p be a prime number and let $e \geq 1$ be an integer. Compute the cyclotomic polynomial Φ_{p^e}. Apply the Eisenstein's irreducibility criterion to prove that Φ_{p^e} is irreducible in $\mathbf{Q}[T]$. Conclude that $[K_{p^e} : \mathbf{Q}] = \varphi(p^e) = p^{e-1}(p - 1)$.

c) Prove that $[K_n : \mathbf{Q}] = \varphi(n)$ for every integer $n \geq 1$. Prove that the cyclotomic polynomial Φ_n is irreducible in $\mathbf{Q}[T]$.

Exercise (4.9.21). — Let E → F be a finite extension of fields.

a) Assume that this extension is separable. Prove that the set of subfields of F that contain E is finite. (*First treat the case where this extension is Galois.*)

b) Let p be a prime number, let k be a field of characteristic p, let $F = k(X, Y)$ and let $E = k(X^p, Y^p)$. For every $a \in k$, let $F_a = k(X^p, Y^p, X + aY)$. Prove that F_a is a subfield of F that contains E. Prove that $F_a \cap F_b = E$ if $a \neq b$.

Exercise (4.9.22). — Let K be an infinite field and let K → L be a finite Galois extension. We let $d = [L : K]$ and write $\sigma_1, \ldots, \sigma_d$ for the elements of Gal(L/K).

a) Let $(e_1, \ldots, e_d)$ be a family of elements in L. Prove that the determinant of the matrix $(\sigma_i(e_j))_{1 \leq i, j \leq d}$ is non-zero if and only if $(e_1, \ldots, e_d)$ is a basis of L as a K-vector space.

In the following, we fix such a basis $(e_1, \ldots, e_d)$. Let $P \in L[X_1, \ldots, X_d]$ be such that
$$P(\sigma_1(x), \ldots, \sigma_d(x)) = 0$$
for every $x \in L$.

b) Let $Q \in L[X_1, \ldots, X_d]$ be the polynomial
$$Q = P\left(\sum_{i=1}^{d} \sigma_1(e_i)X_i, \ldots, \sum_{i=1}^{d} \sigma_d(e_i)\right).$$

Prove that $Q(x_1, \ldots, x_d) = 0$ for every $(x_1, \ldots, x_d) \in K^d$. Prove that $Q = 0$.

c) Prove that $P = 0$ (*algebraic independence of $\sigma_1, \ldots, \sigma_d$*).

d) Prove that there exists a $\theta \in L$ such that $(\sigma_1(\theta), \ldots, \sigma_d(\theta))$ is a K-basis of L. (Such a basis is called a *normal basis* of L over K.)

Exercise (4.9.23). — Let F → E be a Galois extension of fields, with Galois group G = Gal(E/F).

a) Let $\alpha \in E^\times$ and let $c \colon G \to E^\times$ be the map defined by $c(\sigma) = \alpha/\sigma(\alpha)$ for every $\sigma \in G$. Prove that for every σ and τ in G, one has
$$c(\sigma\tau) = c(\sigma)\sigma(c(\tau)).$$

b) Conversely, let $c \colon G \to E^\times$ be any map satisfying this relation. Using proposition 4.4.12, prove that there exists an $x \in E$ such that
$$\alpha = \sum_{\sigma \in G} c(\sigma)\sigma(x) \neq 0.$$

Conclude that one has $c(\sigma) = \alpha/\sigma(\alpha)$ for every $\sigma \in G$.

c) Let $\chi \colon G \to F^\times$ be a group morphism. Prove that there exists an $\alpha \in E^\times$ such that $\chi(\sigma) = \alpha/\sigma(\alpha)$ for every $\sigma \in G$.

d) We assume that G is a cyclic group; let σ be a generator of G. Let $x \in F$; prove that $N_{L/K}(x) = 1$ if and only if there exists an $\alpha \in F^\times$ such that $x = \alpha/\sigma(\alpha)$. Describe the particular cases where $K \to L$ is the extension $\mathbf{R} \to \mathbf{C}$ or an extension $\mathbf{F}_q \to \mathbf{F}_{q^n}$ of finite fields.

Exercise (4.9.24). — Let K be a finite field, let p be its characteristic and let q be its cardinality.

a) For $m \in \mathbf{N}$, compute $S_m = \sum_{x \in K} x^m$.

b) For $P \in K[X_1, \ldots, X_n]$, we write $S(P) = \sum_{x \in K^n} P(x)$. If $\deg P < n(q-1)$, prove that $S(P) = 0$.

c) Let $P_1, \ldots, P_r$ be polynomials in $K[X_1, \ldots, X_n]$ such that $\sum_{i=1}^r \deg P_i < n$. Let $P = (1 - P_1^{q-1}) \ldots (1 - P_r^{q-1})$; prove that $S(P) = 0$.
Let V be the set of all $x \in K^n$ such that $P_1(x) = \cdots = P_n(x) = 0\}$. Prove that $\mathrm{Card}(V)$ is divisible by p (*the Chevalley–Warning theorem*).

d) Let $P \in K[X_1, \ldots, X_n]$ be a homogeneous polynomial of degree $d > 0$. If $d < n$, prove that there exists an $x \in K^n - \{0\}$ such that $P(x) = 0$.

Exercise (4.9.25). — This exercise is the basis of Berlekamp's algorithm for factoring polynomials over a finite field.
Let p be a prime number and let P be a nonconstant *separable* polynomial with coefficient in a finite field $\mathbf{F}_p$. Let $P = c \prod_{i=1}^r P_i$ be the factorization of P into irreducible monic polynomials, with $c \in \mathbf{F}_p^\times$. Set $n_i = \deg P_i$. Let R_P be the ring $\mathbf{F}_p[X]/(P)$.

a) Show that the ring R_{P_i} is a finite field with p^{n_i} elements.

b) For $A \in R_P$, let $\rho_i(A)$ be the remainder of a euclidean division of A by P_i. Show that the map $A \mapsto (\rho_1(A), \ldots, \rho_r(A))$ induces an isomorphism of rings $R_P \simeq \prod_{i=1}^r R_{P_i}$.

c) For $A \in R_P$, let $t(A) = A^p - A$. Show that t is a $\mathbf{F}_p$-linear endomorphism of R_P (viewed as an $\mathbf{F}_p$-vector space) which corresponds, by the preceding isomorphism, to the mapping

$$\prod_{i=1}^r \mathbf{F}_{p^{n_i}} \to \prod_{i=1}^r \mathbf{F}_{p^{n_i}}, \qquad (a_1, \ldots, a_r) \mapsto (a_1^p - a_1, \ldots, a_r^p - a_r).$$

d) Show that the kernel of t is a vector subspace of R_P, of dimension r.

e) Let a be any element in $\mathrm{Ker}(t)$. Show that there exists a monic polynomial $Q \in \mathbf{F}_p[X]$, of minimal degree, such that $Q(a) = 0$. Show that the polynomial Q is separable and split over $\mathbf{F}_p$.

f) (*continued*) If $a \notin \mathbf{F}_p$, show that Q is not irreducible. From a partial nontrivial factorization $Q = Q_1 Q_2$, show how to get a partial nontrivial factorization of P.

Exercise (4.9.26). — Let $K \subset L$ and $L \subset M$ be two finitely generated field extensions. Prove that

$$\mathrm{tr}\,\deg_K M = \mathrm{tr}\,\deg_K L + \mathrm{tr}\,\deg_L M.$$

Exercise (4.9.27). — Let k be a field and let $K \subset k(T)$ be a subfield containing k, but distinct from k.

a) Prove that the extension $K \to k(T)$ is finite. (*First prove that it is algebraic.*) Let $n = [k(T) : K]$ be its degree.

b) Prove that the minimal polynomial of T over K takes the form

$$f(X) = X^n + a_1 X^{n-1} + \cdots + a_n,$$

where $a_1, \ldots, a_n \in K$ and that there exists a $j \in \{1; \ldots; n\}$ with $a_j \notin k$.

c) Let j be such an integer, and let $u = a_j$; let $g, h \in k[T]$ be two coprime polynomials such that $u = g/h$; let $m = \sup(\deg g, \deg h)$. Prove that $m \geq n$ and that there exists a polynomial $q \in K[T]$ such that $g(X) - u h(X) = q(X) f(X)$.

d) Prove that there exist polynomials $c_0, \ldots, c_n \in k[T]$ such that $c_j/c_0 = a_j$ for all j and $\gcd(c_0, \ldots, c_n) = 1$.

e) Let $f(X, T) = c_0(T) X^n + \cdots + c_n(T) = c_0(T) f(X)$. Prove that $f(X, T)$ is irreducible in $k[X, T]$.

f) Prove that there exists a $q \in k[X, T]$ such that

$$g(T) h(X) - g(X) h(T) = q(X, T) f(X, T).$$

Conclude that $m = n$, and $K = k(u)$ (*Lüroth's theorem*).

Chapter 5.
Modules Over Principal Ideal Rings

One result that greatly simplifies undergraduate linear algebra is that vector spaces over a field have a basis. This allows us to perform computations in coordinates, as well as to representat linear maps by matrices. Over a ring which is not a field, there exist modules which are not free, and the classification of modules over general rings is much more delicate, if not impossible.

In contrast, finitely generated modules over principal ideal rings have a nice concise description using invariant factors. Explaining this description and some of its applications is the main objective of this chapter.

There are several ways to establish this result and the approach I have chosen combines algorithmic considerations related to matrix operations with the more abstract theory of Fitting ideals of chapter 3.

Matrix operations are the classical elementary modifications of matrices, permutation of rows, multiplication of a row by an invertible scalar, addition to a row of a multiple of another one, and the analogous operations on columns. Beyond their algorithmic quality, they furnish important results about matrix groups. Here again, the case of fields gives rise to the easier reduced echelon forms, and I present it first, as well as its consequences for classical linear algebra. I then pass to the Hermite and Smith normal forms for matrices with coefficients first in a euclidean ring, and then in a principal ideal domain, and explain the translation in terms of finitely modules over a principal ideal domain. I end this chapter by explaining the two classical particular cases: when the principal ideal domain is the ring of integers, we obtain a classification of finitely generated abelian groups; when it is the ring of polynomials in one indeterminate over a field, we obtain a classification (up to conjugacy) of endomorphisms of a finite-dimensional vector space, and the Jordan decomposition.

© Springer Nature Switzerland AG 2021
A. Chambert-Loir, *(Mostly) Commutative Algebra*, Universitext,
https://doi.org/10.1007/978-3-030-61595-6_5

5.1. Matrix Operations

5.1.1. Elementary Matrices — Let A be a ring. The group $GL_n(A)$ of $n \times n$ invertible matrices with entries in A contains some important elements.

We write $(e_1, \ldots, e_n)$ for the canonical basis of A^n and $(e_{i,j})$ for the canonical basis of $\mathrm{Mat}_n(A)$; that is, $e_{i,j}$ is the matrix of whose entries are all 0, except the one in row i and column j, which equals 1.

For $i, j \in \{1, \ldots, n\}$, $i \neq j$, and $a \in A$, set $E_{ij}(a) = I_n + ae_{i,j}$, where, we recall, I_n is the identity matrix. We have the relation

$$E_{ij}(a)E_{ij}(b) = E_{ij}(a + b)$$

for any $a, b \in A$. Together with the obvious equality $E_{ij}(0) = I_n$, it implies that the matrices $E_{i,j}(a)$ are invertible and that the map $a \mapsto E_{i,j}(a)$ is a morphism of groups from the additive group of A to the group $GL_n(A)$. We say that these matrices $E_{i,j}(a)$ (for $i \neq j$ and $a \in A$) are *elementary* and write $E_n(A)$ for the subgroup they generate in $GL_n(A)$.

For $\sigma \in \mathfrak{S}_n$, we let P_σ be the matrix of the linear map which sends, for every j, the vector e_j to the vector $e_{\sigma(j)}$. Explicitly, if $P_\sigma = (p_{i,j})$, one has $p_{i,j} = 1$ if $i = \sigma(j)$ and $p_{i,j} = 0$ otherwise. For all permutations $\sigma, \tau \in \mathfrak{S}_n$, one has $P_{\sigma\tau} = P_\sigma P_\tau$ and $P_{\mathrm{id}} = I_n$. Consequently, the map $\sigma \mapsto P_\sigma$ is an isomorphism of groups from $\mathfrak{S}_n$ to a subgroup of $GL_n(A)$ which we denote by W. The group W is called the *Weyl group* of GL_n; its elements are called *permutation matrices*.

Finally, for $1 \leq j \leq n$ and $a \in A$, let $D_j(a)$ be the diagonal matrix $I_n + (a - 1)e_{j,j}$. The entries of $D_j(a)$ off the diagonal are zero, those on the diagonal are 1 except for the entry in row j and column j, which equals a. For any $a, b \in A$, one has $D_j(a)D_j(b) = D_j(ab)$ and $D_j(1) = I_n$; if $a \in A^*$, then $D_j(a)$ belongs to $GL_n(A)$.

Let $GE_n(A)$ be the subgroup of $GL_n(A)$ generated by the elementary matrices $E_{i,j}(a)$, for $i \neq j$ and $a \in A$, the permutation matrices P_σ, for $\sigma \in \mathfrak{S}_n$, and the diagonal matrices $D_j(a)$, for $i \in \{1, \ldots, n\}$ and $a \in A^*$. One has $E_n(A) \subset GE_n(A)$, and the inclusion is strict in general.

5.1.2. Elementary operations — Let $M \in \mathrm{Mat}_{n,p}(A)$ be a matrix with n rows and p columns with entries in A.

Multiplying M on the right by elementary matrices from $\mathrm{Mat}_p(A)$ corresponds to the classical operations on the columns of M. Indeed, the matrix $ME_{i,j}(a)$ is obtained by adding to the jth column of M its ith column multiplied by a, an operation that we represent by writing $C_j \leftarrow C_j + C_i a$. The matrix MP_σ is obtained by permuting the columns of M: the jth column of MP_σ is the $\sigma(j)$th column of M. Finally, the matrix $MD_j(a)$ is obtained by multiplying the jth column of M by a (we write $C_j \leftarrow C_j a$).

Similarly, multiplying M on the left by elementary matrices from $\mathrm{Mat}_n(A)$ amounts to performing classical operations on the rows of M. The matrix $E_{i,j}(a)M$ is obtained by adding a times the jth row of M to its ith row (we write $R_i \leftarrow R_i + aR_j$); the ith row of M is the row of index $\sigma(i)$ of the matrix $P_\sigma M$; the rows of $D_i(a)$ are those of M, the row of index i being multiplied by a on the left (in symbols, $R_i \leftarrow aR_i$).

5.1.3. Row and column equivalences — One says that two matrices M and M' in $\mathrm{Mat}_{n,p}(A)$ are *row equivalent* if there exists a matrix $P \in \mathrm{GE}_n(A)$ such that $M' = PM$. This means precisely that there exists a sequence of elementary row operations that transforms M into M'. Row equivalence is an equivalence relation; its equivalence classes are the orbits of the group $\mathrm{GE}_n(A)$, acting on $\mathrm{Mat}_{n,p}(A)$ by left multiplication.

One says that two matrices $M, M' \in \mathrm{Mat}_{n,p}(A)$ are *column equivalent* if there exists a matrix $Q \in \mathrm{GE}_p(K)$ such that $M' = MQ$. This amounts to saying that one can pass from M to M' by a series of elementary column operations. Column equivalence is an equivalence relation, its equivalence classes are the orbits of the group $\mathrm{GE}_p(A)$ acting by right multiplication on $\mathrm{Mat}_{n,p}(A)$.

5.1.4. Reduced row echelon forms — One says that a matrix $M = (m_{i,j}) \in \mathrm{Mat}_{n,p}(A)$ is in *reduced row echelon form* if there exist an integer $r \in \{1, \ldots, \inf(n, p)\}$ and integers $j_1, \ldots, j_r$ such that the following conditions hold:

(i) $1 \le j_1 < j_2 < \cdots < j_r \le p$;

(ii) For any $i \in \{1, \ldots, r\}$ and any j such that $1 \le j < j_i$, one has $m_{i,j} = 0$;

(iii) For any $i \in \{1, \ldots, r\}$, one has $m_{i,j_i} = 1$, and $m_{k,j_i} = 0$ for any other index k;

(iv) If $i \in \{r + 1, \ldots, n\}$ and $j \in \{1, \ldots, p\}$, then $m_{i,j} = 0$.

The entries $m_{i,j_i} = 1$ are called the pivots of the matrix M, the integers $j_1, \ldots, j_r$ are called the pivot column indices; the integer r is called the *row rank* of M. In this language, the above conditions thus say that the pivot column indices are strictly increasing (condition (i)), the first non-zero entry of each of the first r rows is a pivot (condition (ii)), all entries of a pivot column other than the pivot itself are 0 (condition (iii)), and the rows of index $> r$ are zero (condition (iv)).

We mention the following important observation as a lemma.

Lemma (5.1.5). — *Let $M \in \mathrm{Mat}_{n,p}(A)$ be a matrix in reduced row echelon form. For any integer k, the matrix obtained from M by taking its first k columns is still in reduced row echelon form.*

5.1.6. Reduced column echelon form — One says that a matrix $M \in \mathrm{Mat}_{n,p}(A)$ is in *reduced column echelon form* if its transpose matrix M^t is in reduced row echelon form. Explicitly, this means that there exist an integer $s \in \{1, \ldots, \inf(n, p)\}$ and integers $i_1, \ldots, i_s$ such that the following conditions hold:

(i) $1 \le i_1 < i_2 < \cdots < i_s \le n$;

(ii) For any $j \in \{1, \ldots, s\}$ and any integer i such that $1 \le i < i_j$, one has $m_{i,j} = 0$;

(iii) For any j such that $1 \le j \le s$, one has $m_{i_j,j} = 1$ and $m_{i_j,k} = 0$ for any other $k \ne j$;

(iv) If $j \in \{s+1, \ldots, p\}$ and $i \in \{1, \ldots, n\}$, then $m_{i,j} = 0$.

In English: the first non-zero entry of each of the first s columns is equal to 1, called a pivot, the pivot row indices are increasing, all entries of a pivot row other than the pivot itself are 0, and the columns of index $> s$ are zero. The integer s is called the *column rank* of M.

5.1.7. Application to the resolution of linear systems

Let $M \in \mathrm{Mat}_{n,p}(A)$.

If M is in reduced row echelon form, then one can immediately read off the solution of the linear system $MX = 0$, with unknown $X \in A^p$, the variables corresponding to pivot column indices being expressed in terms of the other variables.

More generally, let us assume that there exists a matrix M' in reduced row echelon form which is row equivalent to M. Since each row operation can be reversed, it transforms the system $MX = 0$ into an equivalent one. Consequently, the systems $MX = 0$ and $M'X = 0$ are equivalent. Alternatively, there exists by assumption a matrix $P \in \mathrm{GE}_n(A)$ such that $M' = PM$, and since $\mathrm{GE}_n(A) \subset \mathrm{GL}_n(A)$, the condition $MX = 0$ is equivalent to the condition $M'X = 0$. The original system $MX = 0$ is now replaced by a system in reduced row echelon form.

This can also be applied to "inhomogeneous" systems of the form $MX = Y$, where $Y \in A^n$ is given. Indeed, this system is equivalent to the system $MX + yY = 0$ in the unknown $(X, y) \in A^{p+1}$, to which we add the condition $y = -1$.

Let us assume that there exists a matrix $[M'\ Y'] \in \mathrm{Mat}_{n,p+1}(A)$ in reduced row echelon form which is row equivalent to the matrix $[M\ Y]$. The system $MX = Y$ is then equivalent to the system $M'X = Y'$. Two possibilities arise. If the last column of $[M'\ Y']$ is not that of a pivot, then in the system $M'X = Y'$, the pivot variables are expressed in terms of the other variables and the entries of Y', and $y = -1$ is one of these other variables. Otherwise, when the last column of $[M'Y']$ is that of a pivot, the system $M'X = Y'$ contains the equation $0 = y$, which is inconsistent with the condition $y = -1$, so that the system $M'X = Y'$ has no solution. In that case, the system $MX = Y$ also has no solution.

We can also reason directly on M. Let $P \in \mathrm{GE}_n(A)$ be such that $M' = PM$ is in reduced row echelon form, with row rank r and pivot column indices $j_1, \ldots, j_r$. The system $MX = Y$ is equivalent to the system $M'X = PY$. The last $n - r$ rows of this system are of the form $0 = y'_i$, where $(y'_1, \ldots, y'_n)$ are the entries of $Y' = PY$. If we think of the entries $(y_1, \ldots, y_n)$ of the vector Y as indeterminates, these $n - r$ equations are linear conditions on the entries of Y which must be satisfied for the system $MX = Y$ to have a solution. When they hold, the first r rows of the system $M'X = Y'$ express the pivot variables in terms of the other variables and of the entries of $Y' = PY$.

This generalizes directly to conjunctions, that is, systems of the form $MX = Y_1, \ldots, MX = Y_q$. Assume that there exists a matrix $[M' \, Y'_1 \ldots Y'_q]$ in reduced row echelon form which is row equivalent to $[M \, Y_1 \ldots Y_q]$. If all of its pivot indices are $\leq p$, the system is solvable, otherwise, it is inconsistent.

5.2. Applications to Linear Algebra

Let K be a division ring. In this subsection, I want to show how to systematically apply row echelon forms to solve standard problems of linear algebra. In fact, we will even recover all the standard and fundamental results of linear algebra.

For any integer n, we consider K^n as a right vector space and we recall that any matrix $M \in \mathrm{Mat}_{n,p}(K)$ defines a K-linear map $X \mapsto AX$ from K^p to K^n.

We begin by proving that any matrix is row equivalent to a unique matrix in reduced row echelon form.

Proposition (5.2.1). — *Let K be a division ring. For any matrix $M \in \mathrm{Mat}_{n,p}(K)$, there exists exactly one matrix M' which is in reduced row echelon form and is row equivalent to M.*

Similarly, for any matrix $M \in \mathrm{Mat}_{n,p}(K)$, there exists one matrix M', and only one, which is in reduced column echelon form and is column equivalent to M.

Proof. — By transposition, it is enough to show the assertion for row equivalence. The first part of the proof will show the existence of a matrix equivalent to M which is in reduced row echelon form by explaining an explicit method which reduces any matrix to a matrix of this form. Technically, we show by induction on $k \in \{0, \ldots, p\}$ that we can perform elementary row operations on M so that the matrix M_k obtained by extracting the first k columns of M can be put in reduced row echelon form.

There is nothing to show for $k = 0$; so assume that $k \geq 1$ and that the assertion holds for $k - 1$. Let $A \in \mathrm{GE}_n(K)$ be such that AM_{k-1} is in reduced row echelon form, with row rank r and pivot column indices $j_1, \ldots, j_r$.

Let us then consider the matrix $(m_{i,j}) = AM_k$ in $\mathrm{Mat}_{n,k}(K)$. Its $(k - 1)$ first columns are those of AM_{k-1}.

If the entries $m_{i,k}$ for $i > r$ in the last column of AM_k are all zero, then AM_k is in reduced row echelon form, with row rank r and pivot indices $j_1, \ldots, j_r$. So the assertion holds for k in this case.

Otherwise, let i be the smallest integer $> r$ such that $m_{i,k} \neq 0$ and let us swap the rows of indices i and r. This allows us to assume that $m_{r+1,k} \neq 0$; let us then divide the row of index $r + 1$ by $m_{r+1,k}$; we are now reduced to the case where $m_{r+1,k} = 1$. Now, if, for every $i \in \{1, \ldots, n\}$ such that $i \neq r + 1$, we perform the operation $R_i \leftarrow R_i - m_{i,k}R_{r+1}$, we obtain a matrix in reduced row echelon form whose row rank is $r + 1$ and pivot column indices are $j_1, \ldots, j_r, k$. This proves the assertion for k.

By induction, the assertion holds for $k = p$, so that any $n \times p$-matrix is row equivalent to a matrix in reduced row echelon form.

To establish the uniqueness of a matrix in reduced row echelon form which is row equivalent to our original matrix, we shall make use of the interpretation through linear systems.

We want to prove that there exists at most one matrix in reduced row echelon form which is row equivalent to M. With that aim, it suffices to show that if M and M' are two matrices in reduced row echelon form which are row equivalent, then $M = M'$.

As for the existence part of the proof, let us show by induction on $k \in \{0, \ldots, p\}$ that the matrices M_k and M'_k obtained by extracting the k first columns from M and M' are equal. This holds for $k = 0$. So assume it holds for $k \in \{0, \ldots, p - 1\}$ and let us show it for $k + 1$. Let $j_1, \ldots, j_r$ be the pivot column indices of M_k. Write Y and Y' for the $(k + 1)$th columns of M and M', so that $M_{k+1} = [M_k Y]$ and $M'_{k+1} = [M'_k Y']$. We already know that $M_k = M'_k$, so it remains to show that $Y = Y'$.

Since the matrices M_{k+1} and M'_{k+1} are row equivalent, the systems $M_k X = Y$ and $M'_k X = Y'$ are equivalent. We now analyse these systems making use of the fact that both matrices M_{k+1} and M'_{k+1} are in reduced row echelon form.

The systems $M_k X = Y$ and $M'_k X = Y'$ have the same solutions. If they have one, say X, then $Y = M_k X = M'_k X = Y'$ since $M_k = M'_k$ by the induction hypothesis. Otherwise, both systems are inconsistent, which implies that $k + 1$ is a pivot column index both of M_{k+1} and of M'_{k+1}. By definition of a matrix in reduced row echelon form, Y is the column vector all of whose entries are 0 except for the $(r + 1)$st entry, which equals 1, and likewise for Y', so that $Y = Y'$.

We have thus established the induction hypothesis for $k + 1$. By induction, $M_p = M'_p$, that is $M = M'$, as was to be shown.   □

Definition (5.2.2). — Let K be a division ring and let $M \in \mathrm{Mat}_{n,p}(K)$. The *row rank* of M and the *pivot column indices* of M are the row rank and the pivot column indices of the unique matrix in reduced row echelon form which is row equivalent to M. Similarly, the *column rank* and the *pivot row indices* of M are the column rank and the pivot row indices of the unique matrix in reduced column echelon form which is column equivalent to M.

Let K be a division ring and let $M \in \mathrm{Mat}_{n,p}(K)$. Let r be its row rank and $j_1, \ldots, j_r$ be its pivot column indices.

The linear system $MX = 0$ has unique solution $X = 0$ if and only if all variables are pivot variables, that is, if and only if $r = p$ and $j_i = i$ for $1 \leq i \leq p$. This implies in particular that $n \geq p$. We therefore have proved that *a linear system which has (strictly) more unknowns than equations has a non-zero solution.* Equivalently, if $p > n$, a linear map from K^p to K^n is not injective.

Let now $Y \in K^n$ and let $[M' Y']$ be the matrix in reduced row echelon form which is row equivalent to $[M Y]$. Observe that the matrix M' is in reduced row echelon form and is row equivalent to M. Consequently, the pivot column indices of $[M Y]$ are either $j_1, \ldots, j_r$, or $j_1, \ldots, j_r, p + 1$. In the first case, the system $MX = Y$ is equivalent to the solved system $M'X = Y'$, hence has a solution; in the second case, it has no solution.

Let us show how this implies that *for $p < n$, a linear map from K^p to K^n is not surjective*. Indeed, assuming $p < n$, we have $p + 1 \leq n$ and we may find vectors $Y \in K^p$ such that the pivot column indices of the matrix $[M\,Y]$ are $j_1, \ldots, j_r, p + 1$; it suffices to choose $A \in GE_n(K)$ such that $M' = AM$ and set $Y = A^{-1}Y'$, where $Y' = (0, \ldots, 0, 1)$. In fact, one may find such vectors as soon as $r < n$. So what we have proved is: *if the row rank of M is $r < n$, then the linear map from K^p to K^n is not surjective.*

When $n = p$, the above discussion implies the following proposition.

Proposition (5.2.3). — *Let K be a division ring.*
For any matrix $M \in \mathrm{Mat}_n(K)$, the following seven conditions are equivalent:

(i) *The endomorphism of K^n defined by the matrix M is injective;*

(ii) *The row rank of M is equal to n;*

(iii) *The matrix M is row equivalent to the identity matrix I_n;*

(iv) *The matrix M belongs to $GE_n(K)$;*

(v) *The endomorphism of K^n defined by M is bijective;*

(vi) *The endomorphism of K^n defined by M is surjective;*

(vii) *The matrix M belongs to $GL_n(K)$.*

In particular, $GE_n(K) = GL_n(K)$: the linear group $GL_n(K)$ is generated by the elementary matrices.

Proof. — Let $M \in \mathrm{Mat}_n(K)$. Assuming (i), we have seen that $r = n$, hence (ii). If $r = n$, the pivot column indices are $j_i = i$ for $i \in \{1, \ldots, n\}$, so that M is row equivalent to I_n, and (iii) is proved. Observe that (iii) and (iv) are equivalent, for M is row equivalent to I_n if and only if there exists a matrix $P \in GE_n(K)$ such that $M = PI_n = P$. Let us suppose (iii); the row rank of M is equal to n and the pivot indices of any matrix $[M\,Y]$ can only be $1, \ldots, n$, implying that any system of the form $MX = Y$ has exactly one solution and (v) holds. Of course, (v) implies both (i) and (vi), and is equivalent to (vii). Finally, if the endomorphism of K^n defined by M is surjective, we have seen that $r \geq n$; since one always has $r \leq n$, we get $r = n$, so that all other properties hold. $\square$

5.2.4. Computation of a basis — Let K be a division ring, let $Y_1, \ldots, Y_p$ be vectors of K^n (viewed as a right vector space) and let V be the subspace of K^n generated by $Y_1, \ldots, Y_p$. A standard problem in linear algebra consists in determining a basis of V, or a system of linear equations defining V.

Let M be the matrix $[Y_1 \ldots Y_p]$. Observe that the solutions of the system $MX = 0$ are exactly the coefficients of linear dependence relations among the vectors Y_j. Imposing additional constraints of the form $x_j = 0$ for j in a subset J of $\{1, \ldots, p\}$, where $X = (x_1, \ldots, x_p)$, amounts to searching for the linear dependence relations among the vectors Y_j, for $j \notin J$. Let r be the row rank of M and let $j_1, \ldots, j_r$ be its pivot column indices. *I claim that the family $(Y_{j_1}, \ldots, Y_{j_r})$ is a basis of V.* Let us first show that it is linearly independent. The system $MX = 0$ is solved, expressing $x_{j_1}, \ldots, x_{j_r}$ as a linear combination of the other entries of X. If we search for relations among $Y_{j_1}, \ldots, Y_{j_r}$, this amounts to imposing $x_j = 0$ for $j \notin \{j_1, \ldots, j_r\}$, which implies $x_{j_1} = \cdots = x_{j_r} = 0$. Let us then observe that every vector Y_k is a linear combination of $Y_{j_1}, \ldots, Y_{j_r}$. To prove this, we solve the system $MX = 0$ imposing moreover that $x_j = 0$ if $j \notin \{j_1, \ldots, j_r, k\}$ and $x_k = -1$. As above, the solution of the system furnishes values for $x_{j_1}, \ldots, x_{j_r}$.

Let Y be a vector $(y_1, \ldots, y_n)$ whose entries are indeterminates, let M' be in reduced row echelon form and row equivalent to M, let $A \in \mathrm{GE}_n(K)$ be such that $M' = AM$ and set $Y' = AY$; the entries of Y' are linear combinations of the y_i. The $(n - r)$ last rows of M' are zero, so that the conditions $y_i' = 0$, for $r + 1 \le i \le n$, are exactly the linear dependence relations among $y_1, \ldots, y_n$ that are satisfied if and only $Y \in V$. We have thus obtained a system of $(n - r)$ linear equations that defines the subspace V in K^n. In practice, we obtain $[M' \, Y']$ by applying the above algorithm for a reduced row echelon form to the matrix $[M \, Y]$, but stopping this algorithm once we reach the last column.

5.2.5. Dimension of a vector subspace — Let V be a vector subspace of K^n.
Let $(Y_1, Y_2, \ldots, Y_p)$ be a family of vectors in V chosen in such a way that each of them is linearly independent from the preceding ones. For each integer m, the vectors $Y_1, \ldots, Y_m$ are linearly independent, the row rank of the matrix $[Y_1 \ldots Y_m]$ is equal to m, hence necessarily $p \le n$. This allows us to assume that p is as large as possible. Since $(Y_1, \ldots, Y_p)$ cannot be extended without violating the linear independence condition, every vector of V is a linear combination of $Y_1, \ldots, Y_p$. Then $(Y_1, \ldots, Y_p)$ is a basis of V.

The choice of such a basis $(Y_1, \ldots, Y_p)$ of V identifies it with K^p. Since a linear map from K^p to K^r can be bijective only if $p = r$, this shows that all bases of V have the same cardinality, which we call the *dimension of* V.

5.2.6. The row rank equals the column rank — It is also useful to put a
matrix into a reduced column echelon form. Indeed, let $M = [Y_1 \ldots Y_p] \in \mathrm{Mat}_{n,p}(K)$ and let $M' = [Y_1' \ldots Y_p'] \in \mathrm{Mat}_{n,p}(K)$ be the unique matrix in reduced column echelon form which is column equivalent to M. In particular, the subspace V of K^n generated by the Y_i coincides with the subspace V' generated by the Y_i'. So $\dim(V) = \dim(V')$. By definition, $\dim(V) = r$, the row rank of M. On the other hand, if s is the column rank of M, we see by direct computation that $Y_1', \ldots, Y_s'$ are linearly independent, while $Y_{s+1}' = \cdots = Y_p' = 0$, so that $(Y_1', \ldots, Y_s')$ is a basis of V'. (This is another way to obtain a basis of a subspace.) In particular, $s = \dim(V) = r$: the row rank and the column rank of a matrix coincide.

Let also $i_1, \ldots, i_s$ be the pivot column indices of M. For $i \in \{0, \ldots, n\}$ and $t \in \{1, \ldots, s\}$, the intersection $V \cap \{0\}^i \times K^{n-i}$ contains Y_t' if and only if $i < i_t$. More generally, a linear combination of $Y_1', \ldots, Y_s'$ belongs to $\{0\}^i \cap K^{n-i}$ if and only if the coefficient of Y_t' is 0 when $i_t \le i$. This shows that $\dim(V \cap (\{0\}^i \times K^{n-i})) \ge s - t + 1$ if and only if $i < i_t$.

5.3. Hermite Normal Form

5.3.1. — Let A be a principal ideal domain. We choose, once and for all, a set $\mathscr{P}(A)$ of representatives of $(A - \{0\})/A^\times$ in A, that is, a privileged generator of each non-zero ideal. We also choose, for every non-zero $a \in A$, a set $\mathscr{P}(A, a)$ of representatives of the quotient ring $A/(a)$ in A. Let us give three important examples:

(i) Assume that A is a field. Then we take $\mathscr{P}(A) = \{1\}$ and $\mathscr{P}(A, a) = \{0\}$ for every $a \in A - \{0\}$.

(ii) Assume that $A = \mathbf{Z}$, so that $\mathbf{Z}^\times = \{\pm 1\}$. We take for $\mathscr{P}(\mathbf{Z})$ the set of positive integers; for each non-zero $n \in \mathbf{Z}$, we set $\mathscr{P}(\mathbf{Z}, n) = \{0, 1, \ldots, |n|-1\}$.

(iii) Assume that K is a field and $A = K[T]$. One has $A^\times = K^\times$. We set $\mathscr{P}(K[T])$ to be the set of monic polynomials. Moreover, for every non-zero polynomial $P \in K[T]$, we define $\mathscr{P}(K[T], P)$ to be the set of polynomials Q such that $\deg(Q) < \deg(P)$.

5.3.2. Row Hermite form — Let $M = (m_{i,j}) \in \mathrm{Mat}_{n,p}(A)$. We say that the matrix M is in *row Hermite form* if there exist an integer $r \in \{1, \ldots, \inf(n, p)\}$ and integers $j_1, \ldots, j_r$, such that the following conditions hold:

(i) $1 \le j_1 < j_2 < \cdots < j_r \le p$;

(ii) For every $i \in \{1, \ldots, r\}$ and any j such that $1 \le j < j_i$, one has $m_{i,j} = 0$;

(iii) For every $i \in \{1, \ldots, r\}$, one has $m_{i,j_i} \in \mathscr{P}(A)$, $m_{k,j} \in \mathscr{P}(A, m_{i,j_i})$ if $k < i$, and $m_{k,j} = 0$ if $k > i$;

(iv) If $i \in \{r + 1, \ldots, n\}$ and $j \in \{1, \ldots, p\}$, then $m_{i,j} = 0$.

When A is a field (and the sets of representatives have been chosen as above), we recover the definition of a matrix in reduced row echelon form. The entries m_{i,j_i} are thus called the pivots of the matrix M; the integers $j_1, \ldots, j_r$ are called the pivot column indices; the integer r is called the row rank of M.

Lemma (5.3.3). — *Let $M \in \mathrm{Mat}_{n,p}(A)$ be a matrix in row Hermite form. For every integer k, the matrix deduced from M by taking its first rows (resp. its first columns) is still in row Hermite form.*

Proposition (5.3.4). — *Let A be a euclidean domain. For every matrix $M \in \mathrm{Mat}_{n,p}(A)$, there exists exactly one matrix M' which is in row Hermite form and which is row equivalent to M.*

Proof. — As for the reduced row echelon form (proposition 5.2.1), the existence part of the proof is an algorithm describing a sequence of elementary operations that transform a matrix M into a matrix in row Hermite form. We show by induction on $k \in \{0, \ldots, p\}$ that we can perform elementary row operations on M so that the matrix M_k obtained by extracting the first k columns of M is in row Hermite form. We fix a euclidean gauge δ on A.

There is nothing to prove for $k = 0$; so assume that $k \geq 1$ and that the assertion holds for $k-1$. Let $P \in \mathrm{GE}_n(\mathrm{A})$ be such that PM_{k-1} is in row Hermite form, with row rank r and pivot indices $j_1, \ldots, j_r$. Let us then consider the matrix $(m_{i,j}) = PM_k$ in $\mathrm{Mat}_{n,k}(\mathrm{A})$. Its $(k-1)$ first columns are those of PM_{k-1}.

We will now perform row operations on the rows of index $i > r$, arguing by induction on the infimum of all $\delta(m_{i,k})$, for $i > r$ such that $m_{i,k} \neq 0$.

If the entries $m_{i,k}$, for $i > r$, of the last column of PM_k are all zero, then PM_k is in row-Hermite form, with row rank r and pivot indices $j_1, \ldots, j_r$, hence the assertion holds for k.

Otherwise, let i be such that $m_{i,k}$ has the smallest gauge among the non-zero entries $m_{r+1,k}, \ldots, m_{n,k}$ and let us swap rows with indices $r+1$ and i, reducing to the case $i = r + 1$.

Then, for every integer i such that $r + 1 < i \neq n$, we consider a euclidean division $m_{i,k} = um_{r+1,k} + v$, where either $v = 0$ or $\delta(v) < \delta(m_{r+1,k})$, and perform the row operation $R_i \leftarrow R_i - uR_{r+1}$. The matrix we obtain is row equivalent to M_k, and its entries $m_{r+2,k}, \ldots, m_{n,k}$ are either zero, or their gauges are strictly smaller than that of $m_{r+1,k}$. By induction, we may thus assume that $m_{r+2,k} = \cdots = m_{n,k} = 0$.

We then divide row $r + 1$ by the unique unit $a \in \mathrm{A}^\times$ such that $m_{r+1,k}/a \in \mathscr{P}(\mathrm{A})$; we thus are reduced to the case where $m_{r+1,k} \in \mathscr{P}(\mathrm{A})$.

Now, for every integer i such that $1 \leq i \leq r$, we consider a congruence $m_{i,k} = um_{r+1,k} + v$, where $v \in \mathscr{P}(\mathrm{A}, m_{r+1,k})$, and perform the row operation $R_i \leftarrow R_i - uR_{r+1}$. The $n \times k$ matrix thus obtained is in row-Hermite form, with row rank $r + 1$ and pivot indices $j_1, \ldots, j_r, k$. This proves the assertion for k.

By induction, the assertion holds for $k = p$, so that any $n \times p$-matrix is row equivalent to a matrix in row Hermite form.

To establish the uniqueness assertion, we need to prove the following assertion: let $M, M' \in \mathrm{Mat}_{n,p}(\mathrm{A})$ be two matrices in row Hermite form which are row equivalent; then $M = M'$. We prove this assertion by induction on p, the case $p = 0$ being void, so we assume now that $p \geq 1$.

Possibly exchanging M and M', we assume the number r of pivot columns of M is greater than or equal to that for M'. Deleting the last $n - r$ rows of M and M', we then assume that all rows of M are non-zero, and write $j_1, \ldots, j_n$ for the pivot columns.

There are now two cases, depending on whether $j_n = p$ or $j_n < p$.

First assume that $j_n < p$. Then, we write $M = [M_1\, v]$ and $M' = [M_1'\, v']$, where $M_1, M_1' \in \mathrm{Mat}_{n,p-1}(\mathrm{A})$ and $v, v' \in \mathrm{A}^n$. We observe that M_1 and M_1' are in row Hermite form, and are row equivalent. By the induction hypothesis, one thus has $M_1 = M_1'$. Moreover, M_1 has n pivots as well, so that it has rank n, as a matrix with coefficients in the fraction field K of A. Consequently, there exists a $\xi \in \mathrm{K}^{p-1}$ such that $M_1\xi = v$. Let $d \in \mathrm{A} - \{0\}$ be a common denominator of the entries of ξ, and let $x = d\xi$, so that $x \in \mathrm{A}^{p-1}$; then $M_1x = dv$. The matrices $M = (M_1\, v)$ and $M' = (M_1\, v')$ being row equivalent, the systems $M_1X + yV = 0$ and $M_1X + yV' = 0$ (in unknowns $X \in \mathrm{A}^{p-1}$ and

$y \in A$) are equivalent so that $M_1 x = dv'$. Since $d \neq 0$, this implies $v = v'$ and $M = M'$.

Let us now treat the case where $j_n = p$. Now we write $M = \begin{pmatrix} M_1 & v \\ 0 & d \end{pmatrix}$ and $M' = \begin{pmatrix} M'_1 & v' \\ u' & d' \end{pmatrix}$. Again, the matrices $\begin{pmatrix} M_1 \\ 0 \end{pmatrix}$ and $\begin{pmatrix} M'_1 \\ u' \end{pmatrix}$ are in row Hermite form, and are row equivalent. By induction, they are equal, hence $M_1 = M'_1$ and $u' = 0$. Moreover, M_1 has $n-1$ pivots, so that it has rank $n-1$ as a matrix with coefficients in the fraction field K of A. As above, we may find $x \in A^{p-1}$ and $a \in A - \{0\}$ such that $M_1 x = av$. Let $\overline{x}$ be the image of x in $(A/(ad))^{p-1}$ and let $\overline{a}$ be the class of a in $A/(ad)$, so that $\xi = (\overline{x}, -\overline{a})$ is a solution of the system $M\xi = 0$ in the unknown $\xi \in (A/(ad))^p$. By row equivalence, this system is equivalent to the system $M'\xi = 0$, so that $M_1 \overline{x} = \overline{a}v'$ in $(A/(ad))^{n-1}$ and $d'\overline{a} = 0$ in $A/(ad)$. This last relation implies that d divides d'; by symmetry, d' divides d as well, so that d and d' are associates. By definition of the row Hermite form, one has $d, d' \in \mathscr{P}(A)$, hence $d = d'$.

The first relation implies that $a(v - v') \equiv 0$ in $(A/(ad))^{n-1}$, so that $v \equiv v'$ in $(A/(d))^{n-1}$. By definition of the row Hermite form again, this implies that $v = v'$ and concludes the proof by induction.   □

Theorem (5.3.5). — *Let A be a euclidean ring and let $M \in \mathrm{Mat}_{n,p}(A)$. There exist matrices $P \in E_n(A)$, $Q \in E_p(A)$ and a "diagonal matrix" D (meaning $d_{ij} = 0$ for $i \neq j$) such that $d_{i,i}$ divides $d_{i+1,i+1}$ for any integer i satisfying $1 \leq i < \inf(n,p)$ so that $M = PDQ$. Moreover, the diagonal entries $d_{1,1}, \ldots$ of D are well defined up to units.*

Proof. — Let us write δ for the euclidean gauge of A and let us show the existence of such a decomposition $M = PDQ$ by a double induction, first on $\sup(n,p)$ and then on the minimum value of the gauge δ at non-zero entries of A.

We first observe that each of the following 2×2 matrices can be deduced from the preceding by some elementary row operation:

$$\begin{pmatrix} 1 & 0 \\ 0 & 1 \end{pmatrix} \xrightarrow{R_1 \leftarrow R_1 + R_2} \begin{pmatrix} 1 & 1 \\ 0 & 1 \end{pmatrix} \xrightarrow{R_2 \leftarrow R_2 - R_1} \begin{pmatrix} 1 & 1 \\ -1 & 0 \end{pmatrix} \xrightarrow{R_1 \leftarrow R_1 + R_2} \begin{pmatrix} 0 & 1 \\ -1 & 0 \end{pmatrix}.$$

Consequently, the matrix $\begin{pmatrix} 0 & 1 \\ -1 & 0 \end{pmatrix}$ belongs to $E_2(\mathbf{Z})$. Let i and j be distinct integers in $\{1, \ldots, n\}$. Performing similar operations on the rows of indices i and j, we see that there exists, for any transposition $(i,j) \in \mathfrak{S}_n$, an element of $E_n(A)$ which exchanges up to sign the ith and the jth vector of the canonical basis of A^n and leaves the other fixed.

Now, let (i,j) be the coordinates of some non-zero coefficient of M with minimal gauge. By what precedes, we may apply elementary operations first on rows 1 and i, then on columns 1 and j, of M so as to assume that this coefficient is in position $(1,1)$. Then let $m_{1k} = m_{11} q_k + m'_{1k}$ by the euclidean division of m_{1k} by m_{11}. The column operation $C_k \leftarrow C_k - C_1 q_k$ transforms the matrix M into the matrix $M' = M E_{1k}(-q_k)$ whose entry m_{1k} is now m'_{1k}. If $m'_{1k} \neq 0$, we have $\delta(m'_{1k}) < 0$ and we conclude by induction, since the minimal value of the gauge function at non-zero entries of M has decreased. So we may assume that in the first row, only the first entry is non-zero.

By similar operations on rows, we may also assume by induction that all entries of the first column, except the first one, are zero.

Assume that some entry $m_{i,j}$ of M, with $i > 1$ and $j > 1$ is not divisible by m_{11}. Then, let us perform the row operation $R_1 \leftarrow R_1 + R_i$ (which amounts to left multiplying M by $E_{1i}(1)$); this transforms the first row of M into the row $(m_{1,1}, m_{i,2}, \ldots, m_{i,p})$. Let $m_{1,j} = m_{1,1}q + r$ be a euclidean division of $m_{1,j}$ by $m_{1,1}$. The column operation $C_j \leftarrow C_j - C_1 q$ transforms the matrix M into a new matrix whose $(1, j)$ entry is r. By hypothesis, $r \neq 0$ and $\delta(r) < \delta(m_{1,1})$. By induction, this matrix may transformed into a matrix of the form

$$\begin{pmatrix} m_{11} & 0 & \cdots & 0 \\ 0 & & & \\ \vdots & & m_{11}M' & \\ 0 & & & \end{pmatrix},$$

where $M' \in \mathrm{Mat}_{n-1,p-1}(A)$. By induction, there exist matrices $P' \in F_{n-1}(A)$, $Q' \in E_{p-1}(A)$ and a diagonal matrix $D' \in \mathrm{Mat}_{n-1,p-1}(A)$, each of whose diagonal coefficient divides the next one, such that $M' = P'D'Q'$. Let us then define the following block matrices:

$$P = \begin{pmatrix} 1 & 0 \\ 0 & P' \end{pmatrix}, \quad D = m_{11} \begin{pmatrix} 1 & 0 \\ 0 & D' \end{pmatrix}, \quad Q = \begin{pmatrix} 1 & 0 \\ 0 & Q' \end{pmatrix}.$$

We have $P_1 M Q_1 = PDQ$, hence $M = (P_1)^{-1} PDQ (Q_1)^{-1}$, which shows that M has a decomposition of the desired form.

The uniqueness assumption follows from the theory of Fitting ideals. With the notation of section 3.9, one has $\Delta_p(M) = \Delta_p(D) = (d_{1,1} \cdots d_{p,p}) = (d_{p,p})\Delta_{p-1}(D)$ for every integer p. If $\Delta_{p-1}(D) \neq 0$, this characterizes $d_{p,p}$ up to a unit; if $\Delta_{p-1}(D) = 0$, the divisibility assumption on $d_{1,1}, \ldots$ implies that $d_{p,p} = 0$. □

Corollary (5.3.6). — *Let A be a euclidean ring. One has* $E_n(A) = SL_n(A)$ *and* $GE_n(A) = GL_n(A)$.

Proof. — The two inclusions $E_n(A) \subset SL_n(A)$ and $GE_n(A) \subset GL_n(A)$ are obvious (all generators of $E_n(A)$ have determinant 1). We need to prove that any matrix $M \in GL_n(A)$ belongs to $GE_n(A)$, and that any matrix $M \in SL_n(A)$ belongs to $E_n(A)$.

So, let $M \in GL_n(A)$; let $M = PDQ$ be some decomposition with $P, Q \in E_n(A)$ and $D \in \mathrm{Mat}_n(A)$ a diagonal matrix. Observe that the diagonal entries $d_{i,i}$ of D are invertible. Consequently,

$$D = D_1(d_{1,1}) \ldots D_n(d_{n,n})$$

belongs to $GE_n(A)$ and $M \in GE_n(A)$ as well. It now follows from lemma 5.3.7 below that there exist matrices $P', Q' \in E_n(A)$, and a unit $a \in A^\times$ such that $M = P' \operatorname{diag}(a, 1, \ldots, 1)Q'$.

Assume, moreover, that $M \in SL_n(A)$. We obtain that $a = \det(M) = 1$. It follows that $M = P'Q'$, hence $M \in E_n(A)$. $\hspace{3cm}$ $\square$

Lemma (5.3.7). — *Let A be a ring and let $a_1, \ldots, a_n$ be units in A; let $a_0 = a_n a_{n-1} \ldots a_1$. There exist matrices $U, V \in E_n(A)$ such that*

$$
U \begin{pmatrix} a_1 & & & \\ & a_2 & & \\ & & \ddots & \\ & & & a_n \end{pmatrix} V = \begin{pmatrix} a_0 & & & \\ & 1 & & \\ & & \ddots & \\ & & & 1 \end{pmatrix}.
$$

Proof. — Let λ and μ be two units in A. Let us observe that each of the following matrices is deduced from the preceding one through an elementary row or column operation:

$$
\begin{pmatrix} \lambda & 0 \\ 0 & \mu \end{pmatrix} \xrightarrow{R_1 \leftarrow R_1 + \mu^{-1} R_2} \begin{pmatrix} \lambda & 1 \\ 0 & \mu \end{pmatrix} \xrightarrow{C_1 \leftarrow C_1 - C_2 \lambda} \begin{pmatrix} 0 & 1 \\ -\mu\lambda & \mu \end{pmatrix}
$$

$$
\xrightarrow{R_1 \leftarrow R_1 - R_2} \begin{pmatrix} \mu\lambda & 1 - \mu \\ -\mu\lambda & \mu \end{pmatrix} \xrightarrow{R_2 \leftarrow R_2 + R_1} \begin{pmatrix} \mu\lambda & 1 - \mu \\ 0 & 1 \end{pmatrix}
$$

$$
\xrightarrow{C_2 \leftarrow C_2 - C_1(\mu\lambda)^{-1}(1-\mu)} \begin{pmatrix} \mu\lambda & 0 \\ 0 & 1 \end{pmatrix}.
$$

Consequently, there exist matrices $U, V \in E_2(A)$ such that

$$
U \begin{pmatrix} \lambda & \\ & \mu \end{pmatrix} V = \begin{pmatrix} \mu\lambda & \\ & 1 \end{pmatrix},
$$

which proves the result for $n = 2$.

In the general case, we can perform the corresponding row and column operations for indices $n - 1$ and n, replacing a_{n-1} and a_n by $a_n a_{n-1}$ and 1. We repeat this process for indices $n - 2$ and $n - 1$, etc., until we do it for indices 2 and 1, which gives the result. $\hspace{2cm}$ $\square$

5.3.8. Principal ideal domains — We want to generalize theorem 5.3.5 to the case of any principal ideal domain A. However, elementary operations on rows and columns will not suffice anymore, the group $E_n(A)$ is not necessarily equal to $SL_n(A)$ and we need to use the full group $SL_n(A)$.

Lemma (5.3.9). — *Let A be a principal ideal domain and let a, b be two non-zero elements of A. Let d be a gcd of (a, b), let $u, v \in A$ be such that $d = au + bv$; let r and $s \in A$ be such that $a = dr$ and $b = ds$. The matrix $\left(\begin{smallmatrix} u & v \\ -s & r \end{smallmatrix}\right)$ belongs to $SL_2(A)$ and one has*

$$
\begin{pmatrix} u & v \\ -s & r \end{pmatrix} \begin{pmatrix} a & * \\ b & * \end{pmatrix} = \begin{pmatrix} d & * \\ 0 & * \end{pmatrix}.
$$

Proof. — We have $d = au + bv = d(ur + vs)$. Since $d \neq 0$, we obtain $ur + vs = 1$ so that $\left(\begin{smallmatrix} u & v \\ -s & r \end{smallmatrix} \right) \in SL_2(A)$. The rest is immediate.   □

Theorem (5.3.10). — *Let A be a principal ideal domain and let $M \in Mat_{n,p}(A)$. There exist matrices $P \in SL_n(A)$, $Q \in SL_p(A)$ and $D \in Mat_{n,p}(A)$ such that $M = PDQ$, D is "diagonal" (meaning $d_{i,j} = 0$ for $i \neq j$) and such that $d_{i,i}$ divides $d_{i+1,i+1}$ for any integer i such that $1 \leq i < \inf(n,p)$. Moreover, the diagonal entries $d_{1,1}, \ldots$ of D are well defined up to units.*

Proof. — The proof is very close to that of theorem 5.3.5, so we only indicate the modifications to be done. For any non-zero $a \in A$, let $\omega(a)$ be the *size* of a, namely the number of irreducible factors of a, counted with multiplicities. We now argue by a double induction, first on $\sup(n,p)$ and then on the minimal size of a non-zero entry of M.

As above, we may assume that $m_{1,1}$ is a non-zero element of minimal size. If all entries of the first column are divisible by $m_{1,1}$, elementary operations on rows allow us to assume that the entry $(1,1)$ is the only non-zero entry of this first column. Otherwise, it follows from lemma 5.3.9 that there exists a matrix $P_1 \in SL_n(A)$ of the form

$$\begin{pmatrix} u & 0 \ldots & v & 0 \ldots \\ 0 & 1 & & 0 \\ \vdots & & \ddots & \vdots \\ -s & & & r \\ & & & & 1 \\ & & & & & \ddots \end{pmatrix}$$

such that the $(1,1)$-entry of P_1M is a gcd of $(m_{1,1}, m_{1,j})$, the integer $j \in \{2, \ldots, p\}$ being chosen such that $m_{1,i}$ is not a multiple of $m_{1,1}$. The size of the $(1,1)$-entry of P_1M is now strictly smaller than $\omega(m_{1,1})$; by induction, there exist $P \in SL_n(A)$ and $Q \in SL_p(A)$ such that PP_1MQ is diagonal, each diagonal entry dividing the next one.

By right multiplication with analogous matrices, we may also assume that the $m_{1,1}$ is the only non-zero entry of the first row. Finally, we are reduced to the case where the matrix M takes the form

$$\begin{pmatrix} m_{11} & 0 \ldots & 0 \\ 0 & & \\ \vdots & & M' \\ 0 & & \end{pmatrix}$$

with $M' \in Mat_{n-1,p-1}(A)$. By the same argument as in the euclidean case, we may then assume that all entries of M' are divisible by $m_{1,1}$. We then conclude in the same way, applying the induction hypothesis to the matrix M' whose size is smaller than that of M.

For the proof of the uniqueness assumption, it suffices to copy the one we gave in the case of euclidean rings, see theorem 5.3.5.   □

5.4. Finitely Generated Modules Over a Principal Ideal Domain

The theory of modules over a given ring A strongly depends on its algebraic properties. The case of division rings gives rise to the particularly well-behaved theory of vector spaces. In order of complexity, the next case is that of principal ideal domains, to which we now turn.

Proposition (5.4.1). — *Let A be a principal ideal domain, let M be a free A-module of rank n and let N be a submodule of M. Then, N is a free A-module and its rank is $\leq n$.*

Proof. — It suffices to show that every submodule N of A^n is free of rank $\leq n$; let us prove this by induction on n.

If $n = 0$, then $A^n = 0$, hence $N = 0$ so that N is a free A-module of rank 0.

Assume that $n = 1$. Then, N is an ideal of A. If $N = 0$, then N is free of rank 0. Otherwise, A being a principal ideal domain, there exists a non-zero element $d \in A$ such that $N = (d)$. Since A is a domain, the map $a \mapsto da$ is an isomorphism from A to N, so that N is free of rank 1.

Let now n be an integer ≥ 2 and let us assume that for any integer $r < n$, every submodule of A^r is free of rank $\leq r$. Let N be a submodule of A^n. Let $f : A^n \to A$ be the linear form given by $(a_1, \dots, a_n) \mapsto a_n$; it is surjective and its kernel is the submodule $M_0 = A^{n-1} \times \{0\}$ of A^n. By induction, the ideal $N_1 = f(N)$ of A is free of rank ≤ 1. The submodule $N_0 = N \cap M_0$ of M_0 is isomorphic to a submodule of A^{n-1}, so is free of rank $\leq n - 1$. It then follows from proposition 3.5.5 that N is free of rank $\leq n$. □

Remark (5.4.2). — This proposition recovers the fact that when K is a field, every subspace of K^n is free of dimension $\leq n$.

It is important to observe that the hypothesis on the ring A is necessary. If A is neither a principal ideal domain nor a field, then there exists a non-zero ideal of A which is not free as a A-module. Moreover, there are noncommutative rings A possessing a submodule isomorphic to A^2. Finally, observe that it is possible for a submodule $N \subset A^n$ to have rank n while being distinct from A^n, as is witnessed by the simple example $A = \mathbf{Z}$, $n = 1$ and $N = 2\mathbf{Z} \subset \mathbf{Z}$.

The following theorem is more precise: it furnishes a basis of a free module over a principal ideal domain which is adapted to a given submodule.

Theorem (5.4.3). — *Let A be a principal ideal domain, let M be a free A-module of rank n and let N be a submodule of M. There exists a basis $(e_1, \dots, e_n)$ of M, an integer r such that $0 \leq r \leq n$ and elements $d_1, \dots, d_r$ of A such that d_i divides d_{i+1} for every integer i such that $1 \leq i < r$ and such that $(d_1 e_1, \dots, d_r e_r)$ is a basis of N.*

Moreover, the integer r and the ideals $(d_1), \dots, (d_r)$ do not depend on the choice of a particular such basis $(e_1, \dots, e_n)$.

Proof. — We may assume $M = A^n$. By the preceding proposition, the submodule N is free of some rank $p \in \{0, \ldots, n\}$. In other words, there exists a morphism $f_1: A^p \rightarrow A^n$ which induces an isomorphism from A^p to N. (In fact, the existence of a morphism $f: A^p \rightarrow A^n$ with image N is all that we will need, that is, it suffices to know that N is finitely generated; we will show below how this follows from the fact that A is a *noetherian ring*.) By theorem 5.3.10, there exist matrices $P \in GL_n(A)$, $Q \in GL_p(A)$ and $D \in Mat_{n,p}(A)$ such that $U = PDQ$, the matrix D being diagonal and each of its diagonal entries $d_1, \ldots, d_{\inf(n,p)}$ dividing the next one.

If $d_1 = 0$, set $r = 0$. Otherwise, let r be the largest integer in $\{1, \ldots, \inf(n,p)\}$ such that $d_r \neq 0$; one then has $d_{r+1} = \cdots = d_{\inf(n,p)} = 0$.

Let $u: A^p \rightarrow A^p$ be the automorphism with matrix Q, $v: A^n \rightarrow A^n$ be the automorphism with matrix P, and let $g: A^p \rightarrow A^n$ be the morphism with matrix D; one has $f = v \circ g \circ u$. Let $(\varepsilon_1, \ldots, \varepsilon_n)$ be the canonical basis of A^n. We see that $(d_1 \varepsilon_1, \ldots, d_r \varepsilon_r)$ is a basis of $Im(g) = Im(g \circ u)$, so that $(d_1 v(\varepsilon_1), \ldots, d_r v(\varepsilon_r))$ is a basis of $Im(v \circ g \circ u) = Im(f) = N$. On the other hand, $(v(\varepsilon_1), \ldots, v(\varepsilon_n))$ is a basis of A^n, because v is an automorphism. It thus suffices to set $e_i = v(\varepsilon_i)$ for $i \in \{1, \ldots, n\}$.

From this description, it follows that the Fitting ideals of the module M/N can be computed from $d_1, \ldots, d_n$, namely: $Fit_{n-k}(M/N) = \Delta_k(D) = (d_1 \ldots d_k)$ for $k \leq r$ and $Fit_{n-k}(M/N) = (0)$ for $k > r$. As in theorem 5.3.10, this allows us to recover the integer r as well as the elements $d_1, \ldots, d_r$ up to units.

We can also argue directly. Let $(e'_1, \ldots, e'_n)$ be a basis of A^n, s be an integer in $\{0, \ldots, n\}$, and $d'_1, \ldots, d'_s$ be elements of A such that $d'_i | d'_{i+1}$ for every $i < s$ and such that $(d'_1 e'_1, \ldots, d'_s e'_s)$ is a basis of N. Since any two bases of a free module over a commutative ring have the same cardinality, we have $s = r$. The matrix of the canonical injection from N to A^n in the bases $(e_1, \ldots, e_r)$ and $(e_1, \ldots, e_n)$ is equal to $D = \text{diag}(d_1, \ldots, d_r)$. Let S be the matrix expressing the basis $(e'_1, \ldots, e'_n)$ in the basis $(e_1, \ldots, e_n)$, let T be the matrix expressing the basis $(d'_1 e'_1, \ldots, d'_r e'_r)$ of N in the basis $(d_1 e_1, \ldots, d_r e_r)$ and let $D' = \text{diag}(d'_1, \ldots, d'_r)$. We have $D = S^{-1} D' T$. Since $S \in GL_n(A)$ and $T \in GL_r(A)$, it follows from the uniqueness assertion of theorem 5.3.10 that $(d_1) = (d'_1), \ldots, (d_r) = (d'_r)$.   □

Corollary (5.4.4). — *Let A be a principal ideal domain and let M be a finitely generated A-module. There exists an integer n and elements $d_1, \ldots, d_n$ of A, non-units, such that d_i divides d_{i+1} for every integer i such that $1 \leq i < n$ and such that M is isomorphic to the direct sum $\bigoplus_{i=1}^{n} A/(d_i)$; the Fitting ideals of M satisfy $Fit_p(M) = (d_1 \cdots d_{n-p})$ for every integer p such that $0 \leq p < n$, and $Fit_p(M) = A$ for $p \geq n$.*

Moreover, if m is an integer, $\delta_1, \ldots, \delta_m$ are non-units of A such that δ_j divides δ_{j+1} for every j satisfying $1 \leq j < m$ and such that M is isomorphic to $\bigoplus_{j=1}^{m} A/(\delta_j)$, then $m = n$ and $(\delta_i) = (d_i)$ for every $i \in \{1, \ldots, n\}$.

Proof. — Let $(m_1, \ldots, m_n)$ be a family of elements of M which generates M. Let $f: A^n \rightarrow M$ be the unique morphism that sends the ith vector of the canonical basis of A^n to m_i; it is surjective by construction. If N denotes its kernel, f induces an isomorphism from A^n/N to M.

Let $(e_1, \ldots, e_n)$ be a basis of A^n and let $d_1, \ldots, d_p$ be elements of A such that $(d_1 e_1, \ldots, d_p e_p)$ is a basis of N and such that d_i divides d_{i+1} for any integer i such that $1 \le i < p$ (theorem 5.4.3). Set $d_i = 0$ for $i \in \{p+1, \ldots, n\}$. The morphism φ from A^n to itself which maps $(a_1, \ldots, a_n)$ to $a_1 e_1 + \cdots + a_n e_n$ is an isomorphism, because $(e_1, \ldots, e_n)$ is a basis of A^n. Moreover, $\varphi(a_1, \ldots, a_n)$ belongs to N if and only if d_i divides a_i for every $i \in \{1, \ldots, n\}$. Passing to the quotient, we see that $M = \mathrm{Im}(f \circ \varphi)$ is isomorphic to $\bigoplus_{i=1}^{n} A/(d_i)$.

For every $i \in \{1, \ldots, n-1\}$, if d_{i+1} is invertible, then so is d_i, because d_i divides d_{i+1}. Consequently, there exists an integer $r \in \{0, 1, \ldots, n\}$ such that d_i is a unit if and only if $i \le r$. Then, $M \simeq \bigoplus_{i=r+1}^{n} A/(d_i)$; this proves the first part of the proof.

Resetting notation, we assume that $r = 0$.

By the definition of Fitting ideals, one has $\mathrm{Fit}_k(M) = J_{n-k}(f \circ \varphi)$. Moreover, the kernel of $f \circ \varphi$ is generated by the vectors $(d_1 e_1, \ldots, d_n e_n)$. Consequently, $\mathrm{Fit}_p(M) = (1)$ if $p \ge n$, $\mathrm{Fit}_p(M) = (d_1 \ldots d_{n-p})$ if $0 \le p \le n-1$, and $\mathrm{Fit}_p(M) = (0)$ if $p < 0$.

The uniqueness property follows from this description of the Fitting ideals: if $M \simeq \bigoplus_{j=1}^{m} A/(\delta_j)$, where m is an integer and $\delta_1, \ldots, \delta_m$ are non-units in A such that $\delta_1 | \delta_2 | \ldots | \delta_m$, then $(d_1 \ldots d_{n-p}) = (\delta_1 \ldots \delta_{m-p})$ for every integer p. Taking $p = n$, this implies that $m \le n$, since otherwise, the ideal $(\delta_1 \ldots \delta_{m-n})$ would be non-trivial; by symmetry, one has $n \le m$, hence $n = m$. The equalities $(d_1 \ldots d_{n-p}) = (\delta_1 \ldots \delta_{n-p})$ for every $p \in \{0, \ldots, n\}$ imply that there are units $u_1, \ldots, u_n$ such that $\delta_1 \ldots \delta_k = u_k d_1 \ldots d_k$ for all $k \in \{1, \ldots, n\}$.

Let us show that there are units $v_1, \ldots, v_n$ such that $\delta_k = v_k d_k$ for every $k \in \{1, \ldots, n\}$. This holds for $k = 1$, with $v_1 = u_1$. Assume that it holds for k. If $d_1 \ldots d_k = 0$, then $\delta_1 \ldots \delta_k = 0$; there exists $p, q \le k$ such that $d_p = \delta_q = 0$, and the divisibilities $d_k \mid d_{k+1}$ and $\delta_k \mid \delta_{k+1}$ imply that $\delta_{k+1} = d_{k+1} = 0$; in this case, we set $u_{k+1} = 1$. Otherwise, one has $d_1 \ldots d_k \neq 0$; writing

$$u_{k+1} d_1 \ldots d_{k+1} = \delta_1 \ldots \delta_{k+1} = u_k d_1 \ldots d_k \delta_{k+1},$$

we deduce that $\delta_{k+1} = u_{k+1} u_k^{-1} d_{k+1}$. This proves the result by induction.

Here is another approach.

We thus assume that an isomorphism $M \simeq \bigoplus_{i=1}^{n} A/(d_i)$ be given, where $d_1, \ldots, d_n$ are non-units in A such that d_i divides d_{i+1} for $1 \le i < n$. Let p be an irreducible element of A; by lemma 5.4.5 applied with $a = p^m$ and $b = p$, the module $p^{m-1} M / p^m M$ is an (A/pA)-vector space and its dimension is the number of indices $i \in \{1, \ldots, n\}$ such that p^m divides d_i. Using the divisibility $d_i | d_{i+1}$ for $i \in \{1, \ldots, n-1\}$, we deduce the equivalence: p^m divides d_i if and only if $\dim_{A/pA}(p^{m-1} M / p^m M) \ge m + 1 - i$. This determines the irreducible factors of the elements d_i, as well as their exponents, hence the ideals (d_i). $\square$

On Ferdinand Frobenius

Ferdinand Georg FROBENIUS (1849–1917) was a German mathematician, who made important contributions in geometry, algebra, number theory. . .

His first works were on differential equations. He solved a problem of PFAFF regarding the analytic classification of "pfaffian equations" — in modern terms, differential forms of degree 1. He also gave the necessary and sufficient condition for integrability of systems of such pfaffian equations, a theorem which is now incorporated in the definition of a foliation.

FROBENIUS then passed to questions in matrix algebra, in a slightly different language since he viewed a matrix $A = (a_{i,j})$ as the bilinear form $\sum a_{i,j} x_i y_j$, composition being given by $A \cdot B = \sum_{k=1}^{n} \frac{\partial A}{\partial x_k} \frac{\partial B}{\partial x_k}$. He gave the first general proof of the Cayley–Hamilton theorem. From there, he passed to the study of matrices with integral coefficients, classifying them up to multiplication by invertible such matrices, proving theorem 5.3.5, and applying this result to systems of linear equations in integers. He also observed, in essence, how they still held true when integers were replaced by polynomials. FROBENIUS also proved that besides the real or complex numbers, HAMILTON's quaternions are the only finite-dimensional real division algebra (theorem 1.2.7).

With STICKELBERGER, he pushed these considerations further by proving in 1878 the invariance of the factors (hence the name "invariant factors" in theorem 5.5.1) in the decomposition of a finite abelian group into a product of cyclic groups each of whose orders divides the next. That led him to study the reductions modulo prime numbers of polynomials with integral coefficients. Building on results of KRONECKER, who was interested in the number of factors of degree 1 of these reductions, FROBENIUS established in 1896 the

Portrait of Ferdinand Frobenius (circa 1886)

Photographer: Carl Günther
ETH-Bibliothek Zürich, Bildarchiv.
Public domain

Character tables of the symmetry groups of the tetrahedron, the octahedron and the icosahedron

Copied from G. Frobenius (1899), "Über die Composition der Charaktere einer Gruppe", Sitzungsberichte der Königlich Preußischen Akademie der Wissenschaften zu Berlin, p. 330–339.

In: Gesammelte Abhandlungen, Band III, edited by J-P. Serre, 1968, Springer-Verlag, p. 128

Tetraeder.
$h = 24.$

		$\chi^{(0)}$	$\chi^{(1)}$	$\chi^{(2)}$	$\chi^{(3)}$	$\chi^{(4)}$	$\chi^{(5)}$	$\chi^{(6)}$	h_m
1	X_0	1	1	1	3	2	2	2	1
2	X_1	1	1	1	3	-2	-2	-2	1
4	X_2	1	1	1	-1	0	0	0	6
R^3	X_3	1	ρ	ρ^2	0	-1	$-\rho$	$-\rho^2$	4
R^4	X_4	1	ρ^2	ρ	0	-1	$-\rho^2$	$-\rho$	4
R^2	X_5	1	ρ	ρ^2	0	1	ρ	ρ^2	4
R	X_6	1	ρ^2	ρ	0	1	ρ^2	ρ	4

Oktaeder.
$h = 48.$

		$\chi^{(0)}$	$\chi^{(1)}$	$\chi^{(2)}$	$\chi^{(3)}$	$\chi^{(4)}$	$\chi^{(5)}$	$\chi^{(6)}$	$\chi^{(7)}$	h_m
1	X_0	1	1	3	3	2	2	2	4	1
2	X_1	1	1	3	3	2	-2	-2	-4	1
3	X_2	1	1	0	0	-1	-1	-1	1	8
6	X_3	1	1	0	0	-1	1	1	-1	8
4	X_4	1	-1	1	-1	0	0	0	0	12
S^3	X_5	1	1	-1	-1	2	0	0	0	6
S^2	X_6	1	-1	-1	1	0	$\sqrt{2}$	$-\sqrt{2}$	0	6
S	X_7	1	-1	-1	1	0	$-\sqrt{2}$	$\sqrt{2}$	0	6

Ikosaeder.
$h = 120.$

		$\chi^{(0)}$	$\chi^{(1)}$	$\chi^{(2)}$	$\chi^{(3)}$	$\chi^{(4)}$	$\chi^{(5)}$	$\chi^{(6)}$	$\chi^{(7)}$	$\chi^{(8)}$	h_m
1	X_0	1	5	4	3	3	2	2	4	6	1
2	X_1	1	5	4	3	3	-2	-2	-4	-6	1
4	X_2	1	1	0	-1	-1	0	0	0	0	30
3	X_3	1	-1	1	0	0	-1	-1	1	0	20
6	X_4	1	-1	1	0	0	1	1	-1	0	20
T^2	X_5	1	0	-1	$\frac{1}{2}(1+\sqrt{5})$	$\frac{1}{2}(1-\sqrt{5})$	$\frac{1}{2}(-1+\sqrt{5})$	$\frac{1}{2}(-1-\sqrt{5})$	-1	1	12
T^3	X_6	1	0	-1	$\frac{1}{2}(1-\sqrt{5})$	$\frac{1}{2}(1+\sqrt{5})$	$\frac{1}{2}(-1-\sqrt{5})$	$\frac{1}{2}(-1+\sqrt{5})$	-1	1	12
T'	X_7	1	0	-1	$\frac{1}{2}(1-\sqrt{5})$	$\frac{1}{2}(1+\sqrt{5})$	$\frac{1}{2}(1+\sqrt{5})$	$\frac{1}{2}(1-\sqrt{5})$	1	-1	12
T	X_8	1	0	-1	$\frac{1}{2}(1+\sqrt{5})$	$\frac{1}{2}(1-\sqrt{5})$	$\frac{1}{2}(1-\sqrt{5})$	$\frac{1}{2}(1+\sqrt{5})$	1	-1	12

repartition of the "Frobenius homomorphisms" within the Galois group of the equation. His density theorem would be later refined by Čebotarev.

In 1884, Frobenius also gave a new "abstract" proof of the Sylow theorems which does not involve considerations of permutation groups.

Motivated by the study of the *group determinant*, namely the determinant of the square matrix $(X_{gh})_{g,h\in G}$ where $(X_g)_{g\in G}$ are indeterminates, Frobenius was led to his masterpiece, the theory of representations of finite groups and their characters. (The tables above, borrowed from Frobenius's 1899 paper, are the character tables of the automorphism groups of the tetrahedron, octahedron and icosahedron, respectively $\mathfrak{S}_4$, $(\mathbf{Z}/2\mathbf{Z}) \times \mathfrak{S}_4$ and $(\mathbf{Z}/2\mathbf{Z}) \times \mathfrak{A}_5$.) He applied the theory to questions of group theory, relating the character table of the symmetric group with "Young tableaux", or proving theorems about the solvability of finite groups whose cardinality has few prime factors.

Around 1910, Frobenius also studied matrices with positive coefficients, extending earlier theorems of Perron which were motivated by the theory of continued fractions. What is now called the "Perron–Frobenius theorem" is a cornerstone of the theory of Markov chains in probability theory; it has also important applications in numerical analysis.

Lemma (5.4.5). — *Let* A *be a principal ideal domain, let* d *be an element of* A *and let* $M = A/(d)$. *Let* p *be an irreducible element of* A; *for every integer* m *such that* $m \geq 1$, *let* $M_m = p^{m-1}M/p^m M$. *Then,* M_m *is an* (A/p)-*vector space; its dimension is* 1 *if* p^m *divides* d, *and is* 0 *otherwise.*

Proof. — Since the multiplication by p in the A-module M_m is zero, the canonical morphism $A \to \text{End}(M_m)$ factors through A/pA: this endows M_m with the structure of an (A/pA)-vector space. (Recall that A/pA is a field.)

Let n be the exponent of p in the decomposition of d into prime factors, that is, the supremum of all integer n such that p^n divides d. (If $d = 0$, then $n = +\infty$; otherwise, n is finite, p^n divides d but p^{n+1} doesn't.) The canonical bijection between submodules of A/dA and submodules of A containing (d) maps $p^m M$ to the ideal $(p^m, d) = (p^{\inf(m,n)})$. This furnishes an isomorphism

$$M_m = p^{m-1}M/p^m M \simeq \frac{p^{\inf(m-1,n)}A/dA}{p^{\inf(m,n)}A/dA} \simeq \frac{p^{\inf(m-1,n)}A}{p^{\inf(m,n)}A}.$$

Consequently, $M_m = 0$ if and only if $\inf(m-1, n) = \inf(m, n)$, that is, if and only if $m > n$, which amounts to saying that p^m does not divide d. Otherwise, if $m \leq n$, then $\inf(m, n) = m$, $\inf(m-1, n) = m-1$ and $M_m \simeq p^{m-1}A/p^m A$. Now, the map $A \to p^{m-1}A$ given by $a \mapsto p^{m-1}A$ induces an isomorphism from A/pA to $p^{m-1}A/p^m A$, so that $\dim_{A/pA}(M_m) = 1$ in that case. □

Definition (5.4.6). — Let A be a principal ideal domain and let M be a finitely generated A-module. Let $M \simeq \bigoplus_{i=1}^n A/(d_i)$ be a decomposition of M where $d_1, \ldots, d_n$ are non-units in A such that d_i divides d_{i+1} for any integer i such that $1 \leq i < n$.

The ideals $(d_1), \ldots, (d_n)$ are called the *invariant factors* of M. The number of those ideals which are zero is called the *rank* of M.

With these notations, we thus have $d_i = 0$ if and only if $n - r + 1 \leq i \leq n$, so that M can also be written as

$$M \simeq A^r \oplus \bigoplus_{i=1}^{n-r} A/(d_i).$$

Moreover, with the terminology of example 3.1.9, the second module $\bigoplus_{i=1}^{n-r} A/(d_i)$ in the preceding expression corresponds to the torsion submodule of M.

Corollary (5.4.7). — *A finitely generated module over a principal ideal domain is free if and only if it is torsion-free.*

Corollary (5.4.8). — *Let* A *be a principal ideal domain, let* M *be a finitely generated free* A-*module and let* N *be a submodule of* M. *For* N *to admit a direct summand in* M, *it is necessary and sufficient that the quotient* M/N *be torsion-free.*

Proof. — Assume that N has a direct summand P. Then, the canonical morphism cl from M to M/N induces an isomorphism from P onto M/N. Since

P is a submodule of the free A-module M, it is torsion-free. Consequently, M/N is torsion-free.

Conversely, let us assume that M/N is torsion-free. Since it is finitely generated and A is a principal ideal domain, the A-module M/N is free (corollary 5.4.7). Let $(f_1, \ldots, f_r)$ be a basis of M/N; for any $i \in \{1, \ldots, r\}$, let e_i be an element of M which is mapped to f_i by the canonical surjection. Let then P be the submodule of M generated by $e_1, \ldots, e_r$. Let us show that P is a direct summand of N in M.

Let $m \in N \cap P$ and write $m = \sum a_i e_i$ for some $a_1, \ldots, a_r \in A$. We have $\mathrm{cl}(m) = 0$, because $m \in N$, hence $\sum a_i f_i = 0$. Since the family (f_i) is free, it follows that $a_i = 0$ for all i. Consequently, $m = 0$ and $N \cap P = 0$.

Conversely, let $m \in M$; let $(a_1, \ldots, a_r)$ be elements of A such that $\mathrm{cl}(m) = \sum a_i f_i$. Then, $p = \sum a_i e_i$ belongs to P and $\mathrm{cl}(p) = \mathrm{cl}(m)$. It follows that $\mathrm{cl}(m - p) = 0$, hence $m - p \in N$, which proves that $m \in P + N$. This shows that $M = P + N$.

We have thus shown that P is a direct summand of N, as claimed. □

Lemma and Definition (5.4.9). — *Let A be a principal ideal domain and let M be an A-module. For every irreducible element $p \in A$, the set $M(p)$ of all $m \in M$ for which there exists an integer $n \geq 0$ with $p^n m = 0$ is a submodule of M, called the p-primary component of M.*

Proof. — One has $0 \in M(p)$. Let $m, m' \in M(p)$; we want to show that $m + m' \in M(p)$. Let n, n' be positive integers such that $p^n m = p^{n'} m' = 0$; set $k = \sup(n, n')$. Then, $p^k(m + m') = p^{k-n}(p^n m) + p^{k-n'}(p^{n'} m') = 0$, so that $m + m' \in M(p)$. Finally, let $m \in M(p)$ and let $a \in A$; if $n \geq 0$ is such that $p^n m = 0$, we see that $p^n(am) = a(p^n m) = 0$, hence $am \in M(p)$. This shows that $M(p)$ is a submodule of M. □

Let p and q be two irreducible elements of A. If there exists a unit $u \in A$ such that $q = up$, it is clear from the definition that $M(p) \subset M(q)$, hence $M(p) = M(q)$ by symmetry. Otherwise, it will follow from proposition 5.4.10 below that $M(p) \cap M(q) = 0$.

We fix a set $\mathscr{P}$ of irreducible elements in A such that any irreducible element of A is equal to the product of a unit by a unique element of $\mathscr{P}$. In particular, when p varies in $\mathscr{P}$, the ideals (p) are maximal and pairwise distinct.

Proposition (5.4.10) (Primary decomposition). — *Let A be a principal ideal domain, let M be an A-module. Then the torsion submodule of M decomposes as the direct sum $T(M) = \bigoplus_{p \in \mathscr{P}} M(p)$.*

Proof. — Let us first prove these modules $M(p)$ are in direct sum. Let thus $(m_p)_{p \in \mathscr{P}}$ be an almost-null family of elements of M, where $m_p \in M(p)$ for every $p \in \mathscr{P}$, and assume that $\sum m_p = 0$. We want to prove that $m_p = 0$ for every $p \in \mathscr{P}$. Let I be the subset of $\mathscr{P}$ consisting of those p for which $m_p \neq 0$; by hypothesis, it is a finite subset. For every $p \in I$, let n_p be a positive integer such that $p^{n_p} m_p = 0$. For any $p \in I$, let

$$a_p = \prod_{\substack{q \in I \\ q \neq p}} q^{n_q}.$$

By construction, $a_p m_q = 0$ for every $q \in I$ such that $q \neq p$, and also for every $q \in \mathscr{P} - I$. Consequently, $a_p m_p = a_p(\sum_q m_q) = 0$. However, let us observe that a_p and p^{n_p} are coprime. Since A is a principal ideal domain, there exist $u, v \in A$ such that $u a_p + v p^{n_p} = 1$; we get

$$m_p = (u a_p + v p^{n_p}) m_p = u(a_p m_p) + v(p^{n_p} m_p) = 0.$$

We thus have shown that $m_p = 0$ for every $p \in \mathscr{P}$; in other words, the modules $M(p)$ are in direct sum, as claimed.

By definition, one has $M(p) \subset T(M)$ for every p, hence $\bigoplus_p M(p) \subset T(M)$. Conversely, let $m \in T(M)$ and let a be any non-zero element of A such that $am = 0$; let $a = u \prod_{p \in I} p^{n_p}$ be the decomposition of a into irreducible factors, where I is a finite subset of $\mathscr{P}$ and u is a unit. Let us prove by induction on $\mathrm{Card}(I)$ that m belongs to $\bigoplus_p M(p)$. This is obvious if $\mathrm{Card}(I) \leq 1$. Otherwise, fix $q \in I$ and let $b = a/q^{n_q}$; the elements b and q^{n_q} are coprime, hence there exist $c, d \in A$ such that $cb + dq^{n_q} = 1$, hence $m = cbm + dq^{n_q}m$. Since $q^{n_q}(bm) = am = 0$, one has $bm \in M_q$. On the other hand, $bq^{n_q}m = am = 0$, hence $q^{n_q}m \in \bigoplus_p M_p$ because b has fewer irreducible factors than a. By induction, it follows that m belongs to the submodule $\bigoplus_p M(p)$ of M, hence $T(M) = \bigoplus_p M(p)$, as claimed. $\square$

Remark (5.4.11). — Let A be a principal ideal domain and let M be a finitely generated torsion A-module. Let $p \in \mathscr{P}$ be an irreducible element of A and let $M(p)$ be the corresponding p-primary component of M. By the primary decomposition, we see that $M(p)$ is isomorphic to a quotient of M, namely, the quotient by the direct sum of the other primary components. Consequently, $M(p)$ is finitely generated. Let $m_1, \dots, m_r$ be elements of $M(p)$ which generate $M(p)$; for every $i \in \{1, \dots, r\}$, let $k_i \in \mathbf{N}$ be such that $p^{k_i} m_i = 0$ and set $k = \sup(k_1, \dots, k_r)$; then $p^k m_i = 0$ for every i, hence $p^k m = 0$ for every $m \in M(p)$.

The invariant factors of $M(p)$ are of the form $(p^{n_1}), \dots, (p^{n_s})$, where $n_1 \leq \cdots \leq n_s$ are positive integers. Their knowledge, for every $p \in \mathscr{P}$, determines the invariant factors of M, and conversely. We shall see explicit examples in the next section.

5.5. Application: Finitely Generated Abelian Groups

The main examples of principal ideal rings are $\mathbf{Z}$ and $K[X]$ (where K is a field). In this section we consider the case of the ring $\mathbf{Z}$; the case of a polynomial ring will be the subject of the next section.

Recall that a $\mathbf{Z}$-module is nothing but an abelian group, so that finitely generated $\mathbf{Z}$-modules are just finitely generated abelian groups. Moreover, any ideal of $\mathbf{Z}$ has a unique positive generator. In the case $A = \mathbf{Z}$, corollary 5.4.4 gives the following theorem:

Theorem (5.5.1). — *Let G be a finitely generated abelian group. There exists an integer $r \geq 0$ and a family $(d_1, \ldots, d_s)$ of integers at least equal to 2, both uniquely determined, such that d_i divides d_{i+1} for $1 \leq i < s$ and such that*

$$G \simeq \mathbf{Z}^r \oplus (\mathbf{Z}/d_1\mathbf{Z}) \oplus \cdots \oplus (\mathbf{Z}/d_s\mathbf{Z}).$$

The integer r is the rank of G, the integers $d_1, \ldots, d_s$ are called its *invariant factors.*

Theorem 5.5.1 furnishes a "normal form" for any finitely generated abelian group, which allows us, in particular, to decide whether two such groups are isomorphic or not.

However, one must pay attention that the divisibility condition is satisfied. To determine the invariant factors of a group, written as a direct sum of cyclic groups, the simplest procedure consists in using the Chinese remainder theorem twice: first, decompose each cyclic group as the direct sum of its primary components; then, collect factors corresponding to distinct prime numbers, beginning with those of highest exponents.

Example (5.5.2). — Let us compute the invariant factors of the abelian groups $G_1 = (\mathbf{Z}/3\mathbf{Z}) \oplus (\mathbf{Z}/5\mathbf{Z})$ and $G_2 = (\mathbf{Z}/6\mathbf{Z}) \oplus (\mathbf{Z}/4\mathbf{Z})$.

Since 3 and 5 are coprime, $(\mathbf{Z}/3\mathbf{Z}) \oplus (\mathbf{Z}/5\mathbf{Z})$ is isomorphic to $\mathbf{Z}/15\mathbf{Z}$, by the Chinese remainder theorem. Consequently, the group G_1 has exactly one invariant factor, namely 15.

The integers 6 and 4 are not coprime, their gcd being 2. Since $6 = 2 \cdot 3$, and 2 and 3 are coprime, it follows from the Chinese remainder that

$$G_2 \simeq (\mathbf{Z}/6\mathbf{Z}) \oplus (\mathbf{Z}/4\mathbf{Z}) \simeq (\mathbf{Z}/2\mathbf{Z}) \oplus (\mathbf{Z}/3\mathbf{Z}) \oplus (\mathbf{Z}/4\mathbf{Z})$$
$$\simeq ((\mathbf{Z}/2\mathbf{Z}) \oplus (\mathbf{Z}/4\mathbf{Z})) \oplus ((\mathbf{Z}/3\mathbf{Z})).$$

Observe that the preceding expression furnishes the primary decomposition of the group G_2: the factors $(\mathbf{Z}/2\mathbf{Z}) \oplus (\mathbf{Z}/4\mathbf{Z})$ and $(\mathbf{Z}/3\mathbf{Z})$ being respectively the 2-primary and 3-primary components of G_2. We will now collect factors corresponding to distinct prime numbers, and we begin with those whose exponents are maximal, namely $\mathbf{Z}/3\mathbf{Z}$ and $\mathbf{Z}/4\mathbf{Z}$. We then obtain

$$G_2 \simeq ((\mathbf{Z}/2\mathbf{Z})) \oplus ((\mathbf{Z}/4\mathbf{Z}) \oplus (\mathbf{Z}/3\mathbf{Z})) \simeq (\mathbf{Z}/2\mathbf{Z}) \oplus (\mathbf{Z}/12\mathbf{Z}).$$

This shows that the invariant factors of G_2 are 2 and 12.

Using this result we can also make the list of all finite abelian groups whose cardinality is a given integer g. Indeed, it suffices to make the list of all families of integers $(d_1, \ldots, d_s)$ such that $d_1 \geq 2$, d_i divides d_{i+1} if $1 \leq i < s$, and $g = d_1 \ldots d_s$. Again, the computation is easier if one first considers the primary decomposition of the abelian group.

Example (5.5.3). — Let us determine all abelian groups G of cardinality 48. Since $48 = 2^4 \times 3$, such a group will have a primary decomposition of the form $G = G_2 \oplus G_3$, where G_2 has cardinality 2^4 and G_3 has cardinality 3. In particular, we have $G_3 = \mathbf{Z}/3\mathbf{Z}$. Let us now find all possible 2-primary components. This amounts to finding all families $(d_1, \dots, d_s)$ of integers such that $d_1 \geq 2$, d_i divides d_{i+1} if $1 \leq i < s$ and $d_1 \dots d_s = 2^4$. All of the d_i must be powers of 2, say of the form 2^{n_i}, for integers $n_1, \dots, n_s$ such that $1 \leq n_1 \leq n_2 \leq \cdots \leq n_s$, and one has $\sum_{i=1}^{s} n_i = 4$. The list is the following:

- $d_1 = 2, d_2 = 2, d_3 = 2, d_4 = 2$;
- $d_1 = 2, d_2 = 2, d_3 = 4$;
- $d_1 = 2, d_2 = 4$, but then $d_3 \leq 2$ so that this case does not happen;
- $d_1 = 2, d_2 = 8$;
- $d_1 = 4, d_2 = 4$;
- $d_1 = 8$, but then $d_2 \leq 2$, so that this case does not happen neither;
- $d_1 = 16$.

We thus find that up to isomorphy, there are only five abelian groups of cardinality 16:

$$(\mathbf{Z}/2\mathbf{Z})^4, \quad (\mathbf{Z}/2\mathbf{Z})^2 \oplus (\mathbf{Z}/4\mathbf{Z}), \quad (\mathbf{Z}/2\mathbf{Z}) \oplus (\mathbf{Z}/8\mathbf{Z}), \quad (\mathbf{Z}/4\mathbf{Z})^2, \quad (\mathbf{Z}/16\mathbf{Z}).$$

The abelian groups of cardinality 48 are obtained by taking the product of one of these five groups with $\mathbf{Z}/3\mathbf{Z}$. We collect the factor $\mathbf{Z}/3\mathbf{Z}$ with the 2-primary factor with highest exponent and obtain the following list:

$$(\mathbf{Z}/2\mathbf{Z})^3 \oplus (\mathbf{Z}/6\mathbf{Z}),$$
$$(\mathbf{Z}/2\mathbf{Z})^2 \oplus (\mathbf{Z}/12\mathbf{Z}),$$
$$(\mathbf{Z}/2\mathbf{Z}) \oplus (\mathbf{Z}/24\mathbf{Z}),$$
$$(\mathbf{Z}/4\mathbf{Z}) \oplus (\mathbf{Z}/12\mathbf{Z}),$$
$$(\mathbf{Z}/48\mathbf{Z}).$$

Let us conclude with an application to field theory that has already been proved, by another method, in proposition 1.3.21

Theorem (5.5.4). — *Let K be a (commutative) field and let G be a finite subgroup of the multiplicative group of K. Then K is a cyclic group. In particular, the multiplicative group of a finite field is cyclic.*

Proof. — Let n be the cardinality of G. We may assume that $G \neq \{1\}$. By theorem 5.5.1, there exist an integer $r \geq 1$ and integers $d_1, \dots, d_r \geq 2$ such that d_i divides d_{i+1} for every $i \in \{1, \dots, r-1\}$ and such that $G \simeq (\mathbf{Z}/d_1\mathbf{Z}) \times \cdots \times (\mathbf{Z}/d_r\mathbf{Z})$. In particular, $x^{d_r} = 1$ for every $x \in G$.

Since a non-zero polynomial with coefficients in a field has no more roots than its degree, we conclude that $n \leq d_r$. Since $n = d_1 \dots d_r$, this implies that $r = 1$ and $d_1 = n$. Consequently, $G \simeq \mathbf{Z}/n\mathbf{Z}$ is a cyclic group of order n. $\quad\square$

5.6. Application: Endomorphisms of a Finite-Dimensional Vector Space

Let K be a field and let $A = K[X]$ be the ring of polynomials in one indeterminate.

5.6.1. — Recall from remark 3.2.6 (see also exercise 3.10.4) that an endomorphism u of a K-vector space V endows it with the structure of a K[X]-module, given by $P \cdot x = P(u)(x)$ for any polynomial $P \in K[X]$ and any vector $x \in V$. Explicitly, if $P = \sum_{n=0}^{d} a_n X^n$, one has

$$P \cdot x = \sum_{n=0}^{d} a_n u^n(x).$$

Let V_u be the K[X]-module so defined. Conversely, let M be a K[X]-module. Forgetting the action of X, we get a K-vector space V. The action of X on V is a K-linear endomorphism u, and $M = V_u$.

In order for the theory of K[X]-modules to reflect faithfully that of K-vector spaces endowed with endomorphisms, we need to understand this correspondence from a categorical point of view, that is, at the level of morphisms.

Let W be a K-vector space and let v be an endomorphism of W. A morphism of K[X]-modules from V_u to W_v is an additive map $f : V \to W$ such that $f(P \cdot x) = P \cdot f(x)$ for every $x \in V$ and every $P \in K[X]$. Taking $P = \lambda \in K$, this gives $f(\lambda x) = \lambda f(x)$: the map f is a morphism of K-vector spaces. Taking $P = X$, this gives $f(u(x)) = v(f(x))$, hence $f \circ u = v \circ f$. Conversely, if $f : V \to W$ is a morphism of K-vector spaces such that $f \circ u = v \circ f$, then $f \circ P(u) = P(v) \circ f$ for every $P \in K[X]$. (Indeed, the set of polynomials that satisfy this relation is a subring of K[X] that contains K and X, hence is the whole of K[X].) Consequently, $P \cdot f(x) = P(v)(f(x)) = f(P(u)(x)) = f(P \cdot x)$ for every $P \in K[X]$ and every $x \in V$, so that f is a morphism of K[X]-modules.

In particular, isomorphisms $f : V_u \to W_v$ of K[X]-modules are precisely isomorphisms $f : V \to W$ of K-vector spaces such that $f \circ u = v \circ f$. In the particular case $V = W$, we conclude that the K[X]-modules V_u and V_v are isomorphic if and only if there exists an $f \in GL(V)$ such that $u = f^{-1} \circ v \circ f$, that is, if and only if the endomorphisms u and v are conjugate under $GL(V)$.

Definition (5.6.2). — One says that a non-zero K[X]-module M is *cyclic* if there exists a non-zero polynomial $P \in K[X]$ such that $M \simeq K[X]/(P)$.

We remark that the cyclic K[X]-module $M = K[X]/(P)$ is finite-dimensional as a K-vector space and its dimension is equal to $\deg(P)$. Indeed, euclidean division by P shows that any polynomial of K[X] is equal modulo P to a *unique* polynomial of degree $< \deg(P)$. Consequently, the elements $cl(1), cl(X), \ldots, cl(X^{\deg(P)-1})$ form a basis of M.

The terminology "cyclic" is chosen in analogy with the notion of a cyclic group: both can be viewed as modules (of finite length) generated by one element.

Lemma (5.6.3). — *Let V be a K-vector space, let u be an endomorphism of V. The K[X]-module V_u is cyclic if and only if there exists a vector $x \in V$ and an integer $n \geq 1$ such that the family $(x, u(x), \ldots, u^{n-1}(x))$ is a basis of V.*

Such a vector $x \in V$ is called a *cyclic vector* for the endomorphism u.

Proof. — Assume that the K[X]-module V_u is cyclic and fix an isomorphism φ of K[X]/(P) with V_u, for some non-zero polynomial $P \in K[X]$. Let $n = \deg(P)$. The image of $cl(1)$ by φ is an element x of V. Since $(1, cl(X), \ldots, cl(X^{n-1}))$ is a basis of K[X]/(P) as a K-vector space, the family $(x, u(x), \ldots, u^{n-1}(x))$ is a basis of V.

Conversely, let $x \in V$ be any vector such that $(x, u(x), \ldots, u^{n-1}(x))$ is a basis of V. Write $u^n(x) = \sum_{p=0}^{n-1} a_p u^p(x)$ in this basis, for some elements $a_0, \ldots, a_{n-1} \in K$. Let $\Pi = X^n - \sum_{p=0}^{n-1} a_p X^p \in K[X]$. Then, the map $f : K[X] \to V$, $P \mapsto P(u)(x)$, is a surjective morphism of K[X]-modules. Let us show that $\mathrm{Ker}(f) = (\Pi)$. Indeed, one has

$$f(\Pi) = u^n(x) - \sum_{p=0}^{n-1} a_p u^p(x) = 0,$$

so that $\Pi \in \mathrm{Ker}(f)$; it follows that $(\Pi) \subset \mathrm{Ker}(f)$. Let then $P \in \mathrm{Ker}(f)$ and let R be the euclidean division of P by Π; one has $\deg(R) < n$ and $f(R) = f(P) = 0$. We get a linear dependence relation between $x, u(x), \ldots, u^{n-1}(x)$; since this family is a basis of V, this dependence relation must be trivial, which gives $R = 0$. This shows that $\mathrm{Ker}(f) = (\Pi)$, so that f induces, passing to the quotient, an injective morphism $\varphi : K[X]/(P) \to V$. Since f is surjective, φ is surjective as well, hence is an isomorphism. This concludes the proof of the lemma. □

Remark (5.6.4). — If V_u is a cyclic module and $x \in V$ is a cyclic vector, the matrix of u in the basis $(x, \ldots, u^{n-1}(x))$ is of the form

$$\begin{pmatrix} 0 & & & & a_0 \\ 1 & 0 & & & a_1 \\ & \ddots & \ddots & & \vdots \\ & & \ddots & 0 & \vdots \\ & & & 1 & a_{n-1} \end{pmatrix}.$$

It is the *companion matrix* C_Π of the polynomial

$$\Pi = X^n - a_{n-1}X^{n-1} - \cdots - a_1 X - a_0.$$

Let us prove that Π is both the minimal and characteristic polynomial of the matrix C_Π (and of the endomorphism u).

Since $x, \ldots, u^{n-1}(x)$ are linearly independent, one has $Q(u)(x) \neq 0$ for every non-zero polynomial Q of degree $< n$. Moreover, one has $u^n(x) = a_0 x + a_1 u(x) + \cdots + u^{n-1}(x)$, so that $\Pi(u)(x) = 0$. Since u^k and $\Pi(u)$ commute, it follows that $\Pi(u)(u^k(x)) = u^k(\Pi(u)(x)) = 0$ for every $k \in \{0, \ldots, n-1\}$. Consequently, $\Pi(u) = 0$. Since $\deg(\Pi) = n$, this implies that Π *is the minimal polynomial of u, and of C_Π.*

The characteristic polynomial of u, or of C_Π, is a monic polynomial of degree n. By the Cayley–Hamilton theorem, it is a multiple of Π. It is thus equal to Π since both are monic and have degree n. One can also compute it explicitly by induction on n: expanding the determinant along the first row, one has

$$
P_{C_\Pi}(X) = \det \begin{pmatrix} X & & & & -a_0 \\ -1 & X & & & -a_1 \\ & -1 & \ddots & & \vdots \\ & & \ddots & X & -a_{n-2} \\ & & & -1 & X - a_{n-1} \end{pmatrix}
$$

$$
= X \det \begin{pmatrix} X & & & & -a_1 \\ -1 & X & & & -a_2 \\ & -1 & \ddots & & \vdots \\ & & \ddots & X & -a_{n-2} \\ & & & -1 & X - a_{n-1} \end{pmatrix} + (-1)^n a_0 \det \begin{pmatrix} -1 & X & & \\ & -1 & \ddots & \\ & & \ddots & X \\ & & & -1 \end{pmatrix}
$$

$$
= X(X^{n-1} - a_{n-1}X^{n-2} - \cdots - a_1) + (-1)^n a_0 (-1)^{n-1}
$$

$$
= \Pi(X).
$$

Applied to the $K[X]$-module defined by an endomorphism of a finite-dimensional K-vector space, corollary 5.4.4 thus gives the following theorem.

Theorem (5.6.5). — *Let K be a field. Let V be a finite-dimensional K-vector space and let u be an endomorphism of V. There exists a unique family $(P_1, \ldots, P_r)$ of monic, non-constant, polynomials of $K[X]$ such that P_i divides P_{i+1} for any integer i such that $1 \leq i < r$ and such that, in some basis of V, the matrix of u takes the following block-diagonal form*

$$
\begin{pmatrix} C_{P_1} & & & \\ & C_{P_2} & & \\ & & \ddots & \\ & & & C_{P_r} \end{pmatrix}.
$$

The polynomials $(P_1, \ldots, P_r)$ are called the *invariant factors* of the endomorphism u. It is apparent from the matrix form above that P_r is the *minimal polynomial* of u, while $P_1 \ldots P_r$ is its *characteristic polynomial*.

Corollary (5.6.6). — *Two endomorphisms of a finite-dimensional vector space are conjugate if and only if they have the same invariant factors.*

Applying the theory of Fitting ideals, we also get the following algorithm for computing the invariant factors of a matrix.

Proposition (5.6.7). — *Let K be a field and let U be a matrix in $\mathrm{Mat}_n(K)$. For any integer p such that $1 \le p \le n$, let $\Delta_p \in K[X]$ be the (monic) gcd of the $p \times p$ minors of the matrix $XI_n - U$. One has $\Delta_1 \mid \Delta_2 \mid \cdots \mid \Delta_n$. Let $m \in \{0, \ldots, n\}$ be such that $\Delta_p = 1$ for $p \le m$ and $\Delta_p \ne 1$ for $p \in \{m+1, \ldots, n\}$. Then, there exist monic polynomials $Q_1, \ldots, Q_{n-m} \in K[X]$ such that*

$$Q_1 = \Delta_{m+1}, \quad Q_1 Q_2 = \Delta_{m+2}, \quad \ldots, \quad Q_1 \ldots Q_{n-m} = \Delta_n.$$

The invariant factors of U are the polynomials $Q_1, \ldots, Q_{n-m}$.

Proof. — Let $A = K[X]$. The endomorphism U of K^n endows K^n with the structure of an A-module $(K^n)_U$ defined by $P \cdot v = P(u)(v)$ for $P \in A$ and $v \in K^n$. The proof of the proposition consists in computing in two different ways the Fitting ideals of $(K^n)_U$, once by proving that $(K^n)_U$ is isomorphic to $A^n / \mathrm{Im}(XI_n - U)$, and once by the formula of corollary 5.4.4, $\mathrm{Fit}_k((K^n)_U) = (P_1 \ldots P_{r-k})$ for all integers $k \ge 0$, where $P_1, \ldots, P_r$ are the invariant factors of U.

Let $(e_1, \ldots, e_n)$ be the canonical basis of the free A-module A^n; let also $(f_1, \ldots, f_n)$ be the canonical basis of K^n. The matrix $U \in \mathrm{Mat}_n(K)$ induces an endomorphism of K^n, as well as an endomorphism of A^n; we denote both of them by u. By the universal property of free modules, there is a unique morphism ψ of A-modules from A^n to $(K^n)_u$ such that $\psi(e_i)$ is the ith vector f_i of the canonical basis of K^n, and a unique morphism θ of K-vector spaces from K^n to A^n such that $\theta(f_i) = e_i$.

By construction, the matrix of the morphism

$$\varphi \colon A^n \to A^n, \qquad e_i \mapsto Xe_i - u(e_i)$$

is equal to $XI_n - U$. For any $i \in \{1, \ldots, n\}$, one has

$$\psi(\varphi(e_i)) = \psi(Xe_i - u(e_i)) = u(\psi(e_i)) - \psi(u(e_i)) = 0$$

since ψ is a morphim of $K[X]$-modules. Consequently, $\psi \circ \varphi = 0$, that is, $\mathrm{Im}(\varphi) \subset \mathrm{Ker}(\psi)$. Passing to the quotient, ψ induces a morphism $\tilde{\psi}$ from $A^n / \mathrm{Im}(\varphi)$ to $(K^n)_u$. Let $\tilde{\theta}$ be the composition of θ with the projection to $A^n / \mathrm{Im}(\varphi)$. For any $i \in \{1, \ldots, n\}$, one has

$$\tilde{\theta}(f_i) = \mathrm{cl}(\theta(f_i)) = \mathrm{cl}(e_i),$$

and
$$\tilde{\theta}(u(f_i)) = \mathrm{cl}(u(e_i)) = \mathrm{cl}(X \cdot e_i) = X \cdot \mathrm{cl}(e_i),$$

so that $\tilde{\theta}$ is a morphism of A-modules. For any i, one has

$$\tilde{\theta} \circ \tilde{\psi}(\mathrm{cl}(e_i)) = \tilde{\theta} \circ \psi(e_i) = \tilde{\theta}(f_i) = \mathrm{cl}(e_i);$$

since the vectors $\mathrm{cl}(e_i)$ generate the A-module $A^n/\mathrm{Im}(\varphi)$, the composition $\tilde{\theta} \circ \tilde{\psi}$ is the identity of $A^n/\mathrm{Im}(\varphi)$. On the other hand, one has

$$\tilde{\psi} \circ \tilde{\theta}(f_i) = \tilde{\psi}(\mathrm{cl}(e_i)) = \psi(e_i) = f_i$$

for each i, so that $\tilde{\psi} \circ \tilde{\theta}$ is the identity of $(K^n)_u$. Finally, $\tilde{\psi}$ is an isomorphism and $(K^n)_u$ is isomorphic to $A^n/\mathrm{Im}(\varphi)$.

Since the matrix of φ, as an endomorphism of A^n, is equal to $XI_n - U$, we see that for all integers p, one has

$$(\Delta_p) = J_p(XI_n - U) = \mathrm{Fit}_{n-p}((K^n)_u).$$

On the other hand, the $K[X]$-module associated to a companion matrix C_P is isomorphic to $K[X]/(P)$, so that $(K^n)_u$ has an alternative presentation as the quotient of A^r by the image of the diagonal matrix V with diagonal entries $(P_1, \ldots, P_r)$, as in theorem 5.6.5. By example 3.9.2, one then has

$$\mathrm{Fit}_{n-p}((K^n)_u) = J_{r-n+p}(V) = \begin{cases} (1) & \text{if } r - n + p \leq 0; \\ (P_1 \ldots P_{r-n+p}) & \text{if } 1 \leq r - n + p \leq r; \\ (0) & \text{if } r - n + p > r. \end{cases}$$

We thus obtain

$$(\Delta_p) = \begin{cases} (1) & \text{if } p \leq n - r; \\ (P_1 \ldots P_{p+r-n}) & \text{if } n - r + 1 \leq p \leq n; \\ (0) & \text{if } p > n. \end{cases}$$

This implies that $\Delta_1 = \cdots = \Delta_{n-r} = 1$ and $\Delta_{n-r+p} = (P_1 \ldots P_p)$ for $1 \leq p \leq r$ and concludes the proof of the proposition, with $m = n - r$. $\qquad\square$

All the preceding theory has the following application, which is important in subsequent developments of algebra. Note that it would not be so easy to prove without the theory of invariant factors, especially if the field K is finite.

Corollary (5.6.8). — *Let K be a field and let U and V be two matrices in $\mathrm{Mat}_n(K)$. Let $K \to L$ be an extension of fields; assume that U and V are conjugate as matrices of $\mathrm{Mat}_n(L)$, that is to say, there exists a matrix $P \in \mathrm{GL}_n(L)$ such that $V = P^{-1}UP$. Then, U and V are conjugate over K: there exists an invertible matrix $Q \in \mathrm{GL}_n(K)$ such that $V = Q^{-1}UQ$.*

Proof. — Let $(P_1, \ldots, P_r)$ be the family of invariant factors of U, viewed as a matrix in $\mathrm{Mat}_n(K)$. Therefore, there exists a basis of $W = K^n$ in which the matrix of U has a block-diagonal form, the blocks being the companion matrices with characteristic polynomials $P_1, \ldots, P_r$. The matrix that expresses this new basis of W also gives a basis of L^n in which the matrix of U is exactly the same block-diagonal matrix of companion matrices. In other words, when one "extends the scalars" from K to L, the invariant factors of U are left unchanged. Since, as matrices with coefficients in K, the invariant factors of U coincide with those of V, the matrices U and V have the same invariant factors as matrices in $\mathrm{Mat}_n(K)$, hence are conjugate.   □

Theorem (5.6.9) (Jordan decomposition). — *Let K be an algebraically closed field. Let V be a finite-dimensional K-vector space and let u be an endomorphism of V. There exists a basis of V in which the matrix of u is block-diagonal, each block being of the form ("Jordan matrix")*

$$J_n(\lambda) = \begin{pmatrix} \lambda & 1 & & 0 \\ & \lambda & \ddots & \\ & & \ddots & 1 \\ 0 & & & \lambda \end{pmatrix} \in \mathrm{Mat}_n(K),$$

where λ is an eigenvalue of u. Moreover, two endomorphisms u and u' are conjugate if and only if, for every $\lambda \in K$, the lists of sizes of Jordan matrices (written in increasing order) corresponding to λ for u and u' coincide.

Proof. — Let us first write down the primary decomposition of the $K[X]$-module V_u. Since K is algebraically closed, the irreducible monic polynomials in $K[X]$ are the polynomials $X - \lambda$, for $\lambda \in K$. For $\lambda \in K$, let V_λ be the $(X - \lambda)$-primary component of V_u; it is the set of all $v \in V$ such that there exists an integer $n \in \mathbf{N}$ with $(X - \lambda)^n \cdot v = 0$, that is, $(u - \lambda\,\mathrm{id})^n(v) = 0$. In other words, V_λ is the subspace of V known as the characteristic subspace of u associated to the eigenvalue λ. (It is indeed non-zero if and only if λ is an eigenvalue.) By construction, V_λ is a $K[X]$-submodule of V_u, which means that this subspace of V is stable under the action of u.

The $(X - \lambda)$-primary component V_λ of V is isomorphic, as a $K[X]$-module, to a direct sum

$$\bigoplus_{i=1}^{s} K[X]/((X - \lambda)^{n_i}),$$

for some unique increasing family $(n_1, \ldots, n_s)$ of integers. Let us observe the following lemma:

Lemma (5.6.10). — *The $K[X]$-module corresponding to the Jordan matrix $J_n(\lambda)$ is isomorphic to $K[X]/(X - \lambda)^n$.*   □

Proof. — Let $W = K^n$ and let w be the endomorphism of W given by the matrix $J_n(\lambda)$. Let $(e_1, \ldots, e_n)$ be the canonical basis of K^n; one has $w(e_m) = \lambda e_m + e_{m-1}$ if $m \in \{2, \ldots, n\}$, and $w(e_1) = \lambda e_1$. In other words, if we view W

as a K[X]-module W_w via w, we have $(X - \lambda) \cdot e_m = e_{m-1}$ for $2 \leq m \leq n$ and $(X - \lambda) \cdot e_1 = 0$; in other words, $(X - \lambda)^m \cdot e_n = e_{n-m}$ for $m \in \{0, \ldots, n - 1\}$ and $(X - \lambda)^n \cdot e_n = 0$.

There is a unique morphism φ of K[X]-modules from K[X] to W_w such that $\varphi(1) = e_n$; it is given by $\varphi(P) = P(X) \cdot \varphi(1) = P(w)(e_n)$. The above formulas show that the vectors e_m belong to the image of φ, hence φ is surjective. Moreover, these formulas also show that, $(X_\lambda)^n \in \mathrm{Ker}(\varphi)$, so that φ defines a morphism $\tilde{\varphi}$ from $K[X]/(X - \lambda)^n$ to W_v. The morphism $\tilde{\varphi}$ has the same image as φ, hence is surjective. Since, $K[X]/(X - \lambda)^n$ has dimension n as a K-vector space, as well as W, the morphism $\tilde{\varphi}$ must be an isomorphism. $\quad\square$

Let us now return to the proof of the Jordan decomposition. Thanks to the lemma, we see that the subspace V_λ has a basis in which the matrix of u (restricted to V_λ) is block-diagonal, the blocks being Jordan matrices $J_{n_1}(\lambda), \ldots, J_{n_s}(\lambda)$. Moreover, the polynomials $(X - \lambda)^{n_i}$ (ordered by increasing degrees) are the invariant factors of V_λ. This shows the existence of the Jordan decomposition.

Conversely, the Jordan blocks determine the invariant factors of each primary component of V_u, so that two endomorphisms u and u' are conjugate if and only if for each $\lambda \in K$, the lists of sizes of Jordan blocks corresponding λ for u and u' coincide. $\quad\square$

5.7. Exercises

Exercise (5.7.1). — *a*) Show that the permutation matrix P associated to the transposition $(1, 2) \in \mathfrak{S}_2$ belongs to the subgroup of $GL_2(\mathbf{Z})$ generated by $E_2(\mathbf{Z})$ together with the matrices $D_i(-1)$ for $i \in \{1, 2\}$.

b) Does it belong to the subgroup $E_2(\mathbf{Z})$?

c) Prove that the Weyl group of $GL_n(\mathbf{Z})$ is contained in the subgroup generated by $E_n(\mathbf{Z})$ together with the matrices $D_i(-1)$.

Exercise (5.7.2). — Let A be a commutative ring.

a) Show that $E_n(A)$ is a normal subgroup of the group $GE_n(A)$; show also that $GE_n(A) = E_n(A) \cdot \{D_1(a)\}$ so that $GE_n(A)$ is a semidirect product of $E_n(A)$ by $A^\times$.

b) Consider the same question, replacing $GE_n(A)$ by $GL_n(A)$ and $E_n(A)$ by $SL_n(A)$.

Exercise (5.7.3). — Let K be a division ring and let $D = D(K^\times)$ be the derived group of the multiplicative group $K^\times$, that is, the quotient of $K^\times$ by its subgroup generated by all commutators $[a, b] = aba^{-1}b^{-1}$, for $a, b \in K^\times$. Write π for the canonical projection from $K^\times$ to D.

Let n be an integer and let V be a right K-vector space of dimension n. Let $B(V)$ be the set of bases of V and let $\Omega(V)$ be the set of all maps $\omega \colon B(V) \to D$ satisfying the following properties:

$$\omega(v_1 a_1, \dots, v_n a_n) = \omega(v_1, \dots, v_n) \pi(a_1 \dots a_n)$$

and

$$\omega(v_1, \dots, v_{i-1}, v_i + v_j, v_{i+1}, \dots, v_n) = \omega(v_1, \dots, v_n)$$

for $(v_1, \dots, v_n) \in B(V)$, $a_1, \dots, a_n \in K^\times$ and $i, j \in \{1, \dots, n\}$ such that $i \neq j$.

a) For $\omega \in \Omega(V)$, $(v_1, \dots, v_n) \in B(V)$ and $\sigma \in \mathfrak{S}_n$, prove that

$$\omega(v_{\sigma(1)}, \dots, v_{\sigma(n)}) = \pi(\varepsilon(\sigma))\omega(v_1, \dots, v_n).$$

b) For $\beta = (v_1, \dots, v_n) \in B(V)$, let $\beta_i = (v_1, \dots, v_{i-1}, v_{i+1}, \dots, v_n)$. Let W be a hyperplane of V and let $v \in V - W$. Let $\varphi \in \Omega(W)$. Prove that there exists a unique $\omega \in \Omega(V)$ such that

$$\omega(v_1, \dots, v_{n-1}, v) = \varphi(v_1, \dots, v_{n-1})$$

for every $(v_1, \dots, v_{n-1}) \in B(W)$.

c) Let $(v_1, \dots, v_n) \in B(V)$ and let $a \in D$. Prove that there exists a unique $\omega \in \Omega(V)$ such that $\omega(v_1, \dots, v_n) = a$.

d) Let $U \in GL(V)$. Prove that there exists a unique element $\delta(U) \in D$ such that $\omega(U(v_1), \dots, U(v_n)) = \delta(U)\omega(v_1, \dots, v_n)$ for every $(v_1, \dots, v_n) \in B(V)$. Prove that the map $\delta: GL(V) \to D$ is a morphism of groups.

When K is commutative, then $D = K^\times$, and the morphism $\delta: GL_n(K) \to K^\times$ is the determinant. For this reason, the map δ is called the *noncommutative determinant*. This construction is due to J. Dieudonné.

e) Let $U \in GL_2(K)$ have matrix $\left(\begin{smallmatrix} a & b \\ c & d \end{smallmatrix}\right)$. If $a \neq 0$, then $\delta(U) = \pi(ad - aca^{-1}b)$; if $a = 0$, then $\delta(U) = \pi(-cb)$.

Exercise (5.7.4). — Let N be a positive integer.

a) Show that the group $SL_n(\mathbf{Z}/N\mathbf{Z})$ is generated by its elementary matrices.

b) Deduce that the canonical morphism $SL_n(\mathbf{Z}) \to SL_n(\mathbf{Z}/N\mathbf{Z})$ (reduction mod. N) is surjective.

c) Is the analogous morphism $GL_n(\mathbf{Z}) \to GL_n(\mathbf{Z}/N\mathbf{Z})$ surjective?

Exercise (5.7.5). — a) Let $u \in \mathbf{Z}^n$ be a vector whose entries are coprime. Show by induction on the smallest non-zero entry of u that there exists a matrix $M \in SL_n(\mathbf{Z})$ whose first column is u.

b) Let M be a matrix with n rows and p columns with entries in a principal ideal domain A. Assume that $p \leq n$.

Show that it is possible to complete M into a matrix $P \in GL_n(A)$ (that is, there exists such a matrix P whose first p columns give M) if and only if the ideal $\Delta_p(M)$ generated by all minors of size p of M is equal to (1).

Exercise (5.7.6). — Let A be a principal ideal domain and let K be its fraction field.

a) Let $u = (a_1, \ldots, a_n) \in A^n$. Let $a \in A$ be a generator of the ideal generated by $a_1, \ldots, a_n$. Show that there exists a matrix $M \in E_n(A)$ such that $Mu = (a, 0, \ldots, 0)$.

b) Show that the group $E_n(A)$ acts transitively on the set of lines of K^n.

c) Let $u_1 = (1, 2)$ and $u_2 = (2, 1)$ in $\mathbf{Q}^2$ and let $D_1 = \mathbf{Q}u_1$ and $D_2 = \mathbf{Q}u_2$ be the lines in $\mathbf{Q}^2$ generated by u_1 and u_2. Show that there does not exist a matrix $M \in GL_2(\mathbf{Z})$ such that $M(D_1)$ and $M(D_2)$ are the two coordinate axes.

d) Show that there does not exist a matrix $P \in GL_2(k[X, Y])$ which maps the line generated by the vector $(X, Y) \in k(X, Y)^2$ to the line generated by the vector $(1, 0)$.

Exercise (5.7.7). — Let A be a principal ideal domain and let K be its fraction field.

a) Let X be a non-zero vector in K^n. Show that there exists a matrix $M \in GL_n(A)$ whose first column is proportional to X.

b) Show that each square matrix $M \in \mathrm{Mat}_n(K)$ can be written as a product PB where $P \in GL_n(A)$ and B is an upper-triangular matrix with entries in K. (Argue by induction.)

c) *Numerical example:* In the case $A = \mathbf{Z}$, give such an explicit decomposition for the matrix
$$M = \begin{pmatrix} 1/2 & 1 & -1/4 \\ 2/5 & 2 & 2/3 \\ 3/4 & 1/7 & -1 \end{pmatrix}.$$

Exercise (5.7.8). — Let A be a principal ideal domain and let M be a free finitely generated A-module.

a) For any $m \in M$, show that the following properties are equivalent:

   (i) The vector m belongs to some basis of M;
   (ii) There exists a linear form $f \in M^\vee$ such that $f(m) = 1$;
   (iii) In any basis of M, the coordinates of M are coprime;
   (iv) There exists a basis of M in which the coordinates of m are coprime;
   (v) For any $a \in A$ and $m' \in M$ such that $m = am'$, one has $a \in A^\times$;
   (vi) For any $a, a' \in A$ and $m' \in M$ such that $am = a'm'$ and $a \neq 0$, then a' divides a.

Such a vector is said to be primitive.

b) Show that any non-zero vector is a multiple of a primitive vector.

c) *Example :* $A = \mathbf{Z}$, $M = \mathbf{Z}^4$, $m = (126, 210, 168, 504)$.

Exercise (5.7.9). — Let A be a principal ideal domain and let L, M be two finitely generated A-modules. Show that $\mathrm{Hom}_A(L, M)$ is a finitely generated A-module.

Exercise (5.7.10). — Le A be a principal ideal domain and let M be a finitely generated A-module. Let $(d_1), \ldots, (d_n)$ be its invariant factors, where $d_1, \ldots, d_n$ are non-units and d_i divides d_{i+1} for $1 \le i < n$.

a) Let m be a vector of M whose annihilator equals (d_n). Show that the submodule Am of M generated by m admits a direct summand in M.

b) Assume that A $=$ **Z** and M $=$ $(\mathbf{Z}/p\mathbf{Z}) \oplus (\mathbf{Z}/p^2\mathbf{Z})$. Give a necessary and sufficient condition on a vector $m \in$ M for the subbmodule Am to possess a direct summand.

Exercise (5.7.11). — Let $q(x, y) = ax^2 + 2bxy + cy^2$ be a positive definite quadratic form with real coefficients. Let $m = \inf_{(x,y) \in \mathbf{Z}^2 - \{0\}} q(x, y)$.

a) Show that there exists a non-zero vector $e_1 \in \mathbf{Z}^2$ such that $m = q(e_1)$.

b) Show that there exists a vector $e_2 \in \mathbf{Z}^2$ such that (e_1, e_2) is a basis of $\mathbf{Z}^2$.

c) Deduce from the inequality $q(e_2 + ne_1) \ge q(e_1)$ that $m \le 2\sqrt{(ac - b^2)/3}$.

Exercise (5.7.12). — Let A be a commutative ring and let $I_1, \ldots, I_n$ be ideals of A such that $I_1 \subset I_2 \subset \cdots \subset I_n \subsetneq$ A. Let M $= \bigoplus_{i=1}^{n} A/I_i$.

a) Let J be a maximal ideal of A containing I_n. Endow the A-module M/JM with the structure of an (A/J)-vector space. Show that its dimension is equal to n.

b) Show that any generating family of M has at least n elements.

c) Assume that A is a principal ideal domain. Let M be a finitely generated A-module and let $(d_1), \ldots, (d_r)$ be its invariant factors. Show that any generating set in M has at least r elements.

Exercise (5.7.13). — The goal of this exercise is to provide another proof of theorem 5.4.3 that does not use matrix operations.

Let A be a principal ideal domain, let M be a free finitely generated A-module and let N be a submodule of M.

a) Show that there exists a linear form f on M such that the ideal I $= f(N)$ of A is maximal among the ideals of this form (that is, there does not exist a linear form $g \in M^\vee$ such that $f(N) \subsetneq g(N)$).
Let M$'$ be the kernel of f and let N$' =$ N $\cap$ M$'$.

b) Show that for any linear form f' on M$'$ and any vector $m \in$ N$'$, one has $f'(m) \in$ I.

c) Using the same induction method as the one used for proving proposition 5.4.1, show that there exists a basis $(e_1, \ldots, e_n)$, an integer $r \in \{0, \ldots, n\}$ and elements $d_1, \ldots, d_r$ of A such that d_i divides d_{i+1} for any integer $i < r$ and such that $(d_1e_1, \ldots, d_re_r)$ is a basis of N.

Exercise (5.7.14). — Let A be a principal ideal domain and let M be a finitely generated A-module.

a) Justify the existence of an integer s, elements $m_1, \ldots, m_s$ of M, elements $d_1, \ldots, d_s$ of A such that d_i divides d_{i+1} for $1 \le i < s - 1$ such that (d_i) is the annihilator of m_i and M $= \bigoplus_{i=1}^{s} A m_i$.

b) Let $i \in \{1, \ldots, s\}$. Show that there exists a $u_i \in \mathrm{End}_A(M)$ such that

$$u_i(m_1) = \cdots = u_i(m_{s-1}) = 0, \quad u_i(m_s) = m_i.$$

c) Let u be any element in the center of $\mathrm{End}_A(M)$. Show that there exists an $a \in A$ such that $u(m) = am$ for every $m \in M$.

d) Let $u: M \to M$ be an additive map such that $u \circ v = v \circ u$ for every $v \in \mathrm{End}_A(M)$. Show that there exists $a \in A$ such that $u(m) = am$ for every $m \in M$.

Exercise (5.7.15). — a) Give the list of all abelian groups of order 16 (up to isomorphism).

b) Give the list of all abelian groups of order 45 (up to isomorphism).

Exercise (5.7.16). — Let M be the set of triples $(x, y, z) \in \mathbf{Z}^3$ such that $x + y + z$ is even.

a) Show that M is a submodule of $\mathbf{Z}^3$; prove that it is free of rank 3 by exhibiting 3 linearly independent vectors in M.

b) Construct a basis of M.

c) Exhibit a linear map $f: M \to \mathbf{Z}/2\mathbf{Z}$ such that $M = \mathrm{Ker}(f)$. Conclude that $\mathbf{Z}^3/M$ is isomorphic to $\mathbf{Z}/2\mathbf{Z}$.

Exercise (5.7.17). — Let L be the set of vectors $(x, y, z) \in \mathbf{Z}^3$ such that

$$x - 3y + 2z \equiv 0 \pmod 4 \quad \text{and} \quad x + y + z \equiv 0 \pmod 6.$$

a) Show that L is a free $\mathbf{Z}$-submodule of $\mathbf{Z}^3$. What is its rank?

b) Construct a morphism f from $\mathbf{Z}^3/L$ to $(\mathbf{Z}/4\mathbf{Z}) \times (\mathbf{Z}/6\mathbf{Z})$ such that $L = \mathrm{Ker}(f)$. Prove that f is surjective; conclude that $\mathbf{Z}^3/L$ is isomorphic to $(\mathbf{Z}/4\mathbf{Z}) \times (\mathbf{Z}/6\mathbf{Z})$.

c) Compute integers d_1, d_2, d_3 and a basis (e_1, e_2, e_3) of $\mathbf{Z}^3$ such that $d_1|d_2|d_3$ and such that $(d_1 e_1, d_2 e_2, d_3 e_3)$ is a basis of L.

Exercise (5.7.18). — a) Let G be a finite abelian group; let n be the smallest positive integer such that $nG = 0$. Show that there exists an element $g \in G$ of exact order n.

b) Let K be a field. Recover the fact that a finite subgroup of K^* is cyclic.

c) Let G be the multiplicative subgroup $\{\pm 1, \pm i, \pm j, \pm k\}$ of Hamilton's quaternions $\mathbf{H}$. Prove that G is not cyclic (it suffices to remark that G is not commutative!). This shows that in the previous question, the commutativity assumption on K is crucial.

Exercise (5.7.19). — a) Let G, H be finitely generated abelian groups such that $G \times G \simeq H \times H$. Show that $G \simeq H$.

b) Let G, G′, H be finitely generated abelian groups such that $G \times H \simeq G′ \times H$. Show that $G \simeq G′$.

c) Show that the finite generation hypothesis is necessary by exhibiting abelian groups for which the two previous results fail.

Exercise (5.7.20). — If G is a finite abelian group, one writes G^* for the set of all group morphisms from G to $\mathbf{C}^*$. An element of G^* is called a *character* of G.

a) Show that G^* is a group with respect to pointwise multiplication. Show also that it is finite.

b) In this question, we assume that $G = \mathbf{Z}/n\mathbf{Z}$, for some positive integer n. If $\chi \in G^*$, show that $\chi(1)$ is an nth root of unity. Prove that the map $\chi \mapsto \chi(1)$ from G^* to $\mathbf{C}^*$ is an isomorphism from G^* to the group of nth roots of unity in $\mathbf{C}^*$.

Construct an isomorphism from G to G^*.

c) If $G = H \times K$, show that G^* is isomorphic to $H^* \times K^*$.

d) Show that for any finite abelian group, G^* is isomorphic to G.

Exercise (5.7.21). — Let G be a finite abelian group. For any function $f : G \rightarrow \mathbf{C}$ and any character $\chi \in G^*$,

$$\hat{f}(\chi) = \sum_{g \in G} f(g)\chi(g).$$

This defines a function $\hat{f}$ on G^*, called the Fourier transform of f.

a) Compute $\hat{f}$ when $f = \mathbf{1}$ is the constant function $g \mapsto 1$ on G.

b) For any $g \in G$, show that

$$\sum_{\chi \in G^*} \chi(g) = \begin{cases} \text{Card}(G) & \text{if } g = 0; \\ 0 & \text{otherwise.} \end{cases}$$

c) Let $f : G \rightarrow \mathbf{C}$ be a function. Show that for any $g \in G$, one has the following *Fourier inversion formula*:

$$f(g) = \frac{1}{\text{Card}(G)} \sum_{\chi \in G^*} \hat{f}(\chi)\chi(g)^{-1}.$$

d) Show also the *Plancherel formula*:

$$\sum_{\chi \in G^*} \left|\hat{f}(\chi)\right|^2 = \text{Card}(G) \sum_{g \in G} |f(g)|^2 .$$

Exercise (5.7.22). — Let A be a matrix in $\text{Mat}_n(\mathbf{Z})$, viewed as an endomorphism of $\mathbf{Z}^n$. Let $M = \mathbf{Z}^n / A(\mathbf{Z}^n)$. Show that M is a finite abelian group if and only if $\det(A) \neq 0$. In that case, show also that $\text{Card}(M) = |\det(A)|$.

Exercise (5.7.23). — Let G be an abelian group and let H be a subgroup of G. Assume that H is divisible: for any $h \in H$ and any strictly positive integer n, there exists an $h' \in H$ such that $nh' = h$.

a) Let K be a subgroup of G such that $K \cap H = \{0\}$ and $K + H \neq G$. Let $g \in G - (K + H)$ and let $K' = K + \mathbf{Z}g$. Show that $K' \cap H = \{0\}$ and $K' \supsetneq K$.

b) Using Zorn's lemma, show that there is a maximal subgroup $K \subset G$ such that $K \cap H = \{0\}$. Using the previous question, show that $G = K + H$.

c) Show that $G \simeq H \times (G/H)$.

Exercise (5.7.24). — Let K be a field, let V be a finite-dimensional K-vector space and let $u \in \mathrm{End}_K(V)$. Show that an endomorphism of V which commutes with every endomorphism which commutes with u is a polynomial in u. (*Introduce the* K[X]*-module structure on* V *which is defined by u.*)

Exercise (5.7.25). — Let k be a field. Determine all conjugacy classes of matrices $M \in \mathrm{Mat}_n(k)$ satisfying the indicated properties.

a) The characteristic polynomial of M is $X^3(X - 1)$.

b) The minimal polynomial of M is $X(X - 1)$.

c) $(M - I_n)^2 = 0$.

Chapter 6.
Noetherian and Artinian Rings. Primary Decomposition

This chapter explores more advanced results in the theory of modules.

Despite its apparent simplicity, Nakayama's lemma *allows one to efficiently transfer results which are initially proved by reasoning modulo a maximal ideal, or modulo the Jacobson radical. It is a very important tool in commutative algebra.*

The next three sections are devoted to three different finiteness conditions for modules, all of them phrased in terms of chains of submodules, *that is, totally ordered sets of submodules: we say that a module has finite length if the lengths of all of these chains are bounded above, that it is noetherian if any non-empty chain has a maximal element, and artinian if any non-empty chain has a minimal element.*

Modules of finite length *are probably the closest to finite-dimensional vector spaces. The length function is an integer-valued function that behaves similarly to the dimension, and it is as useful.*

The noetherian property, *named in honor of Emmy Noether, is of particular importance in commutative algebra, as it often guarantees that given modules are finitely generated. As applications of this notion, I present two fundamental finiteness theorems of Hilbert. First, the* finite basis theorem *asserts that polynomial rings over a noetherian ring are polynomial rings. Using a method due to Emil Artin and John Tate, I also prove that when a finite group acts on a finitely generated algebra, the subalgebra of invariant elements is itself finitely generated; this theorem is one of the starting results in the theory of invariants.*

The role of the artinian property, *named in honor of Emil Artin, is more subtle and will be less obvious in this book. (Its strength would require the introduction of other techniques, such as completion, which are not studied here.) Nevertheless, I prove Akizuki's theorem that asserts that a ring is artinian if and only if it has finite length; in the case of commutative rings, artinian rings can be also be characterized by the fact that they are noetherian and that all of their prime ideals are maximal.*

The last two sections consider commutative rings only and have an implicit geometric content in algebraic geometry. I define the support *of a module, and its* associated ideals. *My exposition of associated ideals is slightly different from the classical one, which is limited to modules over noetherian rings; contrary to a plausible expectation, I feel that it makes the first results easier, and their proofs more natural.*

© Springer Nature Switzerland AG 2021
A. Chambert-Loir, *(Mostly) Commutative Algebra*, Universitext,
https://doi.org/10.1007/978-3-030-61595-6_6

I then discuss the existence of a primary decomposition, *a theorem due to Emmy Noether and Emmanuel Lasker that gives an approximation of the decomposition into powers of primes. I conclude the chapter by proving Krull's intersection theorem.*

6.1. Nakayama's Lemma

Recall (definition 2.1.5) that the Jacobson radical J of a ring A is the intersection of all maximal left ideals of A; it is a two-sided ideal of A and $1 + a$ is invertible for every element $a \in J$ (lemma 2.1.6).

Theorem (6.1.1) (Nakayama's lemma). — *Let A be a ring and let J be its Jacobson radical. Let M be a finitely generated A-module such that M = MJ. Then M = 0.*

Proof. — We prove the theorem by induction on the number of elements of a generating set of M.

If M is generated by 0 elements, then $M = 0$.

Let then $n \geq 1$ and assume that the theorem holds for any A-module which is generated by $(n - 1)$ elements. Let M be an A-module which is generated by n elements, say $e_1, \ldots, e_n$, and such that $M = MJ$. Let $N = M/e_1 A$. Then, N is generated by the classes of $e_2, \ldots, e_n$ and one has $N = NJ$. Consequently, $N = 0$. In other words, $M = e_1 A$.

Since $M = MJ$, there is an $a \in J$ such that $e_1 = e_1 a$, hence $e_1(1 - a) = 0$. Since $a \in J$, $1 - a$ is right invertible in A and $e_1 = 0$. This implies that $M = 0$.□

Corollary (6.1.2). — *Let A be a ring and let J be its Jacobson radical. Let M be a finitely generated A-module and let N be a submodule of M such that M = N + MJ. Then, M = N.*

Proof. — Setting $P = M/N$, the hypothesis $M = N + MJ$ implies that $P = PJ$. Consequently, $P = 0$ and $M = N$.   □

Corollary (6.1.3). — *Let A be a ring and let J be its Jacobson radical. Let $U \in \mathrm{Mat}_n(A)$ be a matrix and let $\overline{U}$ be its image in $\mathrm{Mat}_n(A/J)$. If $\overline{U}$ is right invertible (resp. left invertible, resp. invertible), then U is right invertible (resp. left invertible, resp. invertible).*

Proof. — Let $M = A_d^n$ and let u be the endomorphism of M represented by the matrix U in the canonical basis $(e_1, \ldots, e_n)$ of A_d^n. If $\overline{U}$ is right invertible, the endomorphism $\overline{u}$ of $(A/J)_d^n$ it defines is surjective, so that, for every $m \in M$, there exist m' and $m'' \in M$ such that $m = u(m') + m''$ and all coordinates of m'' belong to J. In other words, $M = u(M) + MJ$. By the preceding corollary, $M = u(M)$, so that u is surjective. For every $i \in \{1, \ldots, n\}$, choose $m_i \in M$ such that $u(m_i) = e_i$ and let v be the endomorphism of M such that $v(e_i) = m_i$. Then, $u \circ v(e_i) = u(m_i) = e_i$ for every i so that $u \circ v = \mathrm{id}_M$. In particular, u is right invertible. The matrix V of v is then a right inverse to U.

The case of left invertible matrices is proven analogously, or by considering the transposed matrices, or by working in the opposite ring; the case of invertible matrices follows.   □

Corollary (6.1.4). — *Let* A *be a commutative ring and let* I *be an ideal of* A. *Let* M *be a finitely generated* A-*module and let* N *be a submodule of* M *such that* M = N + IM. *Then, there exists an element* $a \in I$ *such that* $(1 + a)M \subset N$.

Proof. — Let $S = 1 + I = \{1 + a \, ; \, a \in I\}$; it is a multiplicative subset of A. The equation $M = N + MI$ implies the equality $S^{-1}M = S^{-1}N + S^{-1}M \cdot S^{-1}I$ of $S^{-1}A$-modules. For every $a \in S^{-1}I$, $1 + a$ is invertible in $S^{-1}A$. Indeed, write $a = u/(1+v)$, for some $u, v \in I$; then one has $1+a = (1+(1+v)u)/(1+v)$, which is invertible since $1 + (1 + v)u \in 1 + I$. Consequently, $S^{-1}I$ is contained in the Jacobson radical of $S^{-1}A$. Since $S^{-1}M$ is finitely generated (corollary 3.6.9), the preceding corollary implies that $S^{-1}M = S^{-1}N$.

It follows that for every $m \in M$, there exists an $a \in S$ such that $ma \in N$. More precisely, let $(m_1, \dots, m_n)$ be a generating family of M; for every $i \in \{1, \dots, n\}$, let $a_i \in S$ be such that $m_i a_i \in N$, and set $a = a_1 \dots a_n$. Then $m_i a \in N$ for every $i \in \{1, \dots, n\}$, hence $Ma \subset N$.   □

Remark (6.1.5). — One can also prove the corollary by a similar induction to the one performed in the proof of Nakayama's lemma (theorem 6.1.1).

Corollary (6.1.6). — *Let* A *be a commutative ring, let* M *be a finitely generated* A-*module and let* $u \in \mathrm{End}_A(M)$ *be an endomorphism of* M. *If* u *is surjective, then* u *is an isomorphism.*

Proof. — Let us use u to view M as an A[X]-module, setting $P \cdot m = P(u)(m)$ for every $P \in A[X]$ and every $m \in M$ (remark 3.2.6). Then M is finitely generated as an A[X]-module. Since u is surjective, one has $M = u(M) = X \cdot M = (X)M$. By the preceding corollary, there exists a polynomial $P \in A[X]$ such that $(1 - XP(X)) \cdot M = 0$. In other words, $0 = m - u(P(u)(m))$ for every $m \in M$, so that $\mathrm{id}_M = u \circ P(u)$. Since $P(u) \circ u = u \circ P(u) = (XP)(u)$, this shows that $P(u)$ is the inverse of u, hence u is an isomorphism.   □

6.2. Length

Definition (6.2.1). — Let A be a ring. One says that an A-module is *simple* if it is non-zero and if its only submodules are 0 and itself.

Examples (6.2.2). —   a) The module 0 is not simple.

b) Let A be a ring and let I be a right ideal of A. The bijection $V \mapsto \mathrm{cl}_I^{-1}(V)$ between A-submodules of A_d/I and submodules of A_d containing I preserves the containment relation. Consequently, the A-module A_d/I is simple if and only if $I \neq A$ and the only right ideals of A which contain I are I and A. In other words, A_d/I is a simple right A-module if and only if I is a maximal right ideal of A.

c) Let M be a simple A-module and let m be a non-zero element of M. The set mA of all multiples of M is a non-zero submodule of M. Since M is simple, one has $mA = M$.

The set I of all $a \in A$ such that $ma = 0$ is a right ideal of A and the map $a \mapsto ma$ induces an isomorphism from A_d/I to M. Consequently, I is a maximal right ideal of A.

d) Assume that A is a division ring. Then, a simple A-module is nothing but a 1-dimensional A-vector space.

e) Let A be a ring and let I be a two-sided ideal of A. In the identification between A-modules annihilated by I and (A/I)-modules, A-submodules correspond to (A/I)-submodules. Consequently, an A-module which is annihilated by I is a simple A-module if and only if it is simple as an (A/I)-module.

f) Let M be an A-module and let N be a submodule of M. The submodules of M/N correspond to the submodules of M containing N. Consequently, M/N is a simple A-module if and only if N is a maximal submodule of M, in the sense that $M \neq N$ and the only submodules of M containing N are N and M.

Proposition (6.2.3) (Schur lemma). — *Let A be a ring and let $u \colon M \to N$ be a non-zero morphism of A-modules.*

   a) *If M is simple, then u is injective.*

   b) *If N is simple, then u is surjective.*

   c) *If M and N are both simple, then u is an isomorphism.*

Proof. — The image of u is a non-zero submodule of N; if N is simple, then $\mathrm{Im}(u) = N$ and u is surjective. The kernel of u is a submodule of M, distinct from M; if M is simple, then $\mathrm{Ker}(u) = 0$ and u is injective. If both M and N are simple, then u is bijective, hence an isomorphism.   □

Corollary (6.2.4). — *The ring of endomorphisms of a simple A-module is a division ring.*

Proof. — Let M be a simple A-module. By the proposition, any non-zero element of $\mathrm{End}_A(M)$ is an isomorphism, hence is invertible in $\mathrm{End}_A(M)$. Moreover, $M \neq 0$, so that $\mathrm{id}_M \neq 0$ and $\mathrm{End}_A(M)$ is not the zero ring. This proves that $\mathrm{End}_A(M)$ is a division ring.   □

6.2.5. — Let A be a ring and let M be an A-module. A *chain* of submodules of M is a set of submodules of M which is totally ordered by inclusion.

As in general ordered sets (see §A.1.4.4), the elements of a non-empty finite chain of submodules of M can be uniquely enumerated as a finite sequence $(M_0, \ldots, M_n)$ of submodules of M such that $M_0 \subsetneq M_1 \cdots \subsetneq M_n$. The length of such a chain is then defined as n.

In the sequel, we will make no distinction between the set $\{M_0, \ldots, M_n\}$ and the sequence $(M_0, \ldots, M_n)$.

Definition (6.2.6). — Let A be a ring and let M be an A-module. The *length* of M is the supremum of the lengths of all finite chains of submodules of M. It is denoted by $\ell_A(M)$ or $\ell(M)$.

Examples (6.2.7). —   *a*) The null A-module has length 0, and conversely.

b) Simple A-modules are exactly the A-modules of length 1.

Indeed, let M be an A-module. If M is simple, then the only chains of submodules of M are (0), (M) (both of length 0) and $(0, M)$ (of length 1), so that M has length 1. Conversely, if $\ell_A(M) = 1$, then $M \neq 0$ (otherwise, the only chain of submodules of M would be (0), of length 0) and if there existed a submodule N such that $N \neq 0$ and $N \neq M$, the chain $(0, N, M)$ would imply that $\ell_A(M) \geq 2$.

c) Assume that A is a division ring. The expression "chain of submodules" is equivalent to "strictly increasing sequence of vector subspaces". In such a sequence, the dimension increases at least by 1 at each inclusion, and exactly by 1 if the sequence cannot be enlarged. This implies that $\ell_A(M) = \dim_A(M)$: the length of M is its dimension as a vector space.

d) The length of the **Z**-module **Z** is infinite, as witnessed by the arbitrarily long chains of ideals: $(2^n \mathbf{Z}, 2^{n-1}\mathbf{Z}, \ldots, \mathbf{Z})$.

e) Let I be a two-sided ideal and let M be an A-module annihilated by I. Then every submodule of M is also annihilated by I, hence can be viewed as an A/I-module, and conversely. Consequently, the length of M as an A-module is equal to its length as an A/I-module.

f) A chain $(M_0, M_1, \ldots, M_n)$ of submodules is *maximal* if it cannot be made longer by inserting a module at the beginning, between two elements, or at the end of the sequence without destroying its strictly increasing character. Therefore, such a chain is maximal if and only if $M_0 = 0$, M_i/M_{i-1} is a simple A-module for $1 \leq i \leq n$, and $M_n = M$.

If an A-module has finite length, then any chain can be extended to a maximal one. A chain of length $\ell_A(M)$ must be maximal; and we shall prove below (theorem 6.2.11) that all maximal chains have length $\ell_A(M)$, but this is not *a priori* obvious.

Lemma (6.2.8). — *Let A be a ring, let M be a right A-module and let* M', M'', N *be submodules of M such that* $M' \subset M''$, $M' \cap N = M'' \cap N$ *and* $M' + N = M'' + N$. *Then* $M' = M''$.

Proof. —   Let $m \in M''$. Since $M'' \subset M'' + N$ and $M'' + N = M' + N$, there exist $m' \in M'$ and $n \in N$ such that $m = m' + n$. Then, $n = m - m' \in M'' \cap N$, hence $n \in M' \cap N$. This implies that $m = m' + n$ belongs to M'. Consequently, $M'' \subset M'$, hence $M'' = M'$.   □

Proposition (6.2.9). — *Let A be a ring, let M be an A-module and let N be a submodule of N. Assume that among the modules M, N and M/N, two have finite length. Then the length of the other one is finite too and one has*

$$\ell_A(M) = \ell_A(N) + \ell_A(M/N).$$

Proof. —   Let $(N_0, N_1, \ldots, N_a)$ be a chain of submodules of N of length a. Let $(P_0, \ldots, P_b)$ be a chain of submodules of M/N of length b; for each $i \in \{0, \ldots, b\}$, there exists a unique submodule $M_i \subset M$ containing N such that $P_i = M_i/N$ and $(M_0, \dots, M_b)$ is a chain in M. The union of these two chains furnishes a chain of submodules of M, $(N_0, N_1, \dots, N_a, M_1, \dots, M_b)$, of length $a + b$. This implies the inequality $\ell(M) \geq \ell(N) + \ell(M/N)$, with the usual convention $\infty = n + \infty$. In particular, if M has finite length, then so have N and M/N.

Conversely, let us assume that N and M/N have finite length; we want to prove that M has finite length and that $\ell(M) = \ell(N) + \ell(M/N)$. Let thus $(M_0, \dots, M_a)$ be a chain of submodules of M. Applying lemma 6.2.8 to $M' = M_i$ and $M'' = M_{i+i}$, we see that at least one of the two inclusions

$$M_i \cap N \subset M_{i+1} \cap N \quad \text{and} \quad M_i + N \subset M_{i+1} + N$$

is strict. Consequently, removing the inclusions which are equalities in the sequences

$$M_0 \cap N \subset M_1 \cap N \subset \cdots \subset M_a \cap N$$

and

$$(M_0 + N)/N \subset (M_1 + N)/N \subset \cdots \subset (M_a + N)/N,$$

we obtain two chains of submodules in N and in M/N whose lengths add up to a, at least. In particular, $\ell(N) + \ell(M/N) \geq \ell(M)$. This concludes the proof of the proposition. □

Corollary (6.2.10). — *A direct sum $\bigoplus_{i=1}^{n} M_i$ of a family $M_1, \dots, M_n$ of right A-modules of finite lengths has finite length.*

Theorem (6.2.11) (Jordan–Hölder). — *Let A be a ring and let M be a right A-module. Let $M_0 \subsetneq M_1 \subsetneq \dots M_m$ and $N_0 \subsetneq N_1 \subsetneq \dots N_n$ be maximal chains of submodules of A. Then $m = n = \ell_A(M)$ and there exists a permutation σ of $\{1, \dots, n\}$ such that for each $i \in \{1, \dots, n\}$, the module M_i/M_{i-1} is isomorphic to $N_{\sigma(i)}/N_{\sigma(i)-1}$.*

In particular, if M possesses a maximal chain $(M_0, M_1, \dots, M_n)$ of submodules of M, then $\ell_A(M) = n$. Moreover, up to reordering, the simple A-modules M_i/M_{i-1}, for $1 \leq i \leq n$, do not depend on the chosen maximal chain.

On the other hand, if M does not possess a maximal chain, then it admits arbitrarily long chains, hence $\ell_A(M)$ is infinite.

Proof. — For $1 \leq i \leq m$ and $0 \leq j \leq n$, let us set $M_{i,j} = M_{i-1} + M_i \cap N_j$. It is a submodule of M such that $M_{i-1} \subset M_{i,j} \subset M_i$; moreover, $M_{i,0} = M_{i-1}$ and $M_{i,n} = M_i$. Consequently, there exists a smallest integer $\sigma(i) \in \{1, \dots, n\}$ such that $M_{i,\sigma(i)} = M_i$; then, $M_{i-1} + M_i \cap N_{\sigma(i)} = M_i$ and $M_i \cap N_{\sigma(i)-1} \subset M_{i-1}$. In fact, $\sigma(i)$ is the only integer j such that $M_{i,j}/M_{i,j-1} \neq 0$.

Similarly, let us set $N_{j,i} = N_{j-1} + N_j \cap M_i$, for $1 \leq j \leq n$ and $0 \leq i \leq m$. For any j, there exists a smallest integer $\tau(j) \in \{1, \dots, m\}$ such that $N_{j,i} = N_j$; it is the unique integer i such that $N_{j,i}/N_{j,i-1} \neq 0$.

By Zassenhaus's lemma below, for any i, j such that $1 \leq i \leq m$ and $1 \leq j \leq n$, we have isomorphisms

$$\frac{M_{i,j}}{M_{i,j-1}} = \frac{M_{i-1} + M_i \cap N_j}{M_{i-1} + M_i \cap N_{j-1}}$$

$$\simeq \frac{M_i \cap N_j}{(M_{i-1} \cap N_j) + (M_i \cap N_{j-1})}$$

$$\simeq \frac{N_{j-1} + M_i \cap N_j}{N_{j-1} + M_{i-1} \cap N_j}$$

$$= \frac{N_{j,i}}{N_{j,i-1}}$$

of A-modules. In particular, $M_{i,j}/M_{i,j-1}$ is non-zero precisely when $j = \sigma(i)$, hence if and only if $N_{j,i}/N_{j,i-1}$ is non-zero. Considering $j = \sigma(i)$, we get $\tau(\sigma(i)) = i$; setting $i = \tau(j)$, we get $\sigma(\tau(j)) = j$. This implies that σ is a bijection from $\{1, \ldots, m\}$ to $\{1, \ldots, n\}$, with reciprocal bijection τ; in particular, $m = n$. Moreover, if $j = \sigma(i)$, $M_{i,j} = M_i$, $M_{i,j-1} = M_{i-1}$, while $N_{j,i} = N_j$ and $N_{j,i-1} = N_{i-1}$; the above isomorphisms imply that the A-modules M_i/M_{i-1} and $N_{\sigma(i)}/N_{\sigma(i)-1}$ are isomorphic, hence the theorem. $\qquad\square$

Lemma (6.2.12) (Zassenhaus). — *Let M be an A-module, let $N' \subset N$ and $P' \subset P$ be submodules of M. There exist isomorphisms of A-modules*

$$\frac{N' + (N \cap P)}{N' + (N \cap P')} \xleftarrow{\sim} \frac{N \cap P}{(N' \cap P) + (N \cap P')} \xrightarrow{\sim} \frac{P' + (N \cap P)}{P' + (N' \cap P)},$$

respectively induced by the inclusions $N \cap P \to N' + (N \cap P)$ and $N \cap P \to P' + (N \cap P)$.

Proof. — Let $f \colon N \cap P \to (N' + (N \cap P))/(N' + (N \cap P'))$ be the linear map defined as the composition of the injection j from $N \cap P$ into $N' + (N \cap P)$ and of the canonical surjection π from $N' + (N \cap P)$ to its quotient by $N' + (N \cap P')$. Let us show that f is surjective. Indeed, let $x \in N' + (N \cap P)$; one may write $x = n_1 + n_2$, with $n_1 \in N'$ and $n_2 \in N \cap P$. One has $\pi(n_1) = 0$, hence $\pi(x) = \pi(n_2) = f(n_2)$ and $\pi(x)$ belongs to the image of f.

Let now $x \in N \cap P$ be any element such that $f(x) = 0$. By assumption, one may write $x = n_1 + n_2$, where $n_1 \in N'$ and $n_2 \in N \cap P'$. In particular, $n_2 \in P$ so that $n_1 = x - n_2 \in P$. It follows that $n_1 \in N' \cap P$ and x belongs to the sum of the submodules $N' \cap P$ and $N \cap P'$ of $N \cap P$. Conversely, these submodules are both contained in $\mathrm{Ker}(f)$, hence $\mathrm{Ker}(f) = (N' \cap P) + (N \cap P')$. Passing to the quotient, f induces an isomorphism

$$\frac{N \cap P}{(N' \cap P) + (N \cap P')} \simeq \frac{N' + (N \cap P)}{N' + (N \cap P')},$$

so that the first two A-modules in the lemma are indeed isomorphic. Exchanging the roles of N', N and of P', P, the second and the third A-modules are isomorphic too, QED. $\qquad\square$

Remark (6.2.13). — Let K be a division ring and let M be a K-vector space.

a) Let us explain how the concept of length allows us to recover the main results about the dimension of vector spaces.

If M is a simple K-module, then, for any non-zero element x of M, xK is a non-zero submodule of M, hence xK = M. Consequently, (x) is a basis of M. Conversely, assume that M is a K-module generated by one element x; then, for any non-zero $y \in$ M, there exists an element $a \in$ K such that $y = xa$. Necessarily, $a \neq 0$, so that $x = ya^{-1}$ belongs the submodule generated by y, hence M = yK. The simple K-modules are the K-modules generated by a single non-zero element.

Assume that M is finitely generated, say by a family $(x_1, \dots, x_n)$. Let us also assume that this family is minimal, so that it is a basis of M. Then, for $i \in \{0, \dots, n\}$, define $M_i = \mathrm{Span}(x_1, \dots, x_i)$. We have thus defined a family of subspaces of M such that $M_0 \subset M_1 \subset \cdots \subset M_n$. If $M_i = M_{i-1}$, then $x_i \in \mathrm{Span}(x_1, \dots, x_{i-1})$, contradicting the hypothesis that the family $(x_1, \dots, x_n)$ is minimal among those generating M. Therefore, M_i/M_{i-1} is a non-zero vector space generated by one element (the class of x_i), hence is a simple K-module. Consequently, the Jordan–Hölder theorem implies the two following results:

- $\ell_K(M) = n$;
- Any minimal generating family of M has exactly n elements.

This reproves the fact that any two bases of M have the same cardinality.

b) Assume that M is finite-dimensional. Let E = $(e_1, \dots, e_n)$ and F = $(f_1, \dots, f_n)$ be two bases of M.

For $0 \leq i \leq n$, define $M_i = \mathrm{Span}(e_1, \dots, e_i)$ and $N_i = \mathrm{Span}(f_1, \dots, f_i)$. The proof of the Jordan–Hölder theorem furnishes a (unique) permutation σ of $\{1, \dots, n\}$ such that $M_{i-1} + M_i \cap N_{\sigma(i)-1} = M_{i-1}$ and $M_{i-1} + M_i \cap N_{\sigma(i)} = M_i$, for every $i \in \{1, \dots, n\}$. For any i, let x_i be a vector belonging to $M_i \cap N_{\sigma(i)}$ but not to M_{i-1}. For every i, one has $\mathrm{Span}(x_1, \dots, x_i) = M_i$; it follows that X = $(x_1, \dots, x_n)$ is a basis of M; moreover, there exists a matrix B_1, in upper triangular form, such that X = EB_1.

Set $\tau = \sigma^{-1}$. Set $y_j = x_{\tau(j)}$ for $j \in \{1, \dots, n\}$; one has $y_j \in N_j \cap M_{\tau(i)}$ and $y_j \notin N_{j-1}$. Similarly, one thus has $\mathrm{Span}(x_{\tau(1)}, \dots, x_{\tau(j)}) = N_j$ for every j. Consequently, there exists a matrix B_2, still in upper triangular form, such that $(x_{\tau(1)}, \dots, x_{\tau(n)}) = FB_2$. Let P_τ be the permutation matrix associated to τ, we have $(x_{\tau(1)}, \dots, x_{\tau(n)}) = (x_1, \dots, x_n)P_\tau$. This implies that $FB_2 = EB_1 P_\tau$, hence F = $EB_1 P_\tau B_2^{-1}$. Therefore, the matrix A = $B_1 P_\tau P_2^{-1}$ that expresses the coordinates of the vectors of F in the basis E is the product of an upper-triangular matrix, a permutation matrix and another upper-triangular matrix.

In the group $\mathrm{GL}(n, K)$, let B be the subgroup consisting of upper-triangular matrices, and let W be the subgroup consisting of permutation matrices. We have proved that $\mathrm{GL}(n, K) = BWB$: this is called the *Bruhat decomposition*.

Proposition (6.2.14). — *Let* A *be a commutative ring, let* S *be a multiplicative subset of* A *and let* M *be an* A-*module of finite length. Then* $S^{-1}M$ *is an* $S^{-1}A$-*module of finite length, and* $\ell_{S^{-1}A}(S^{-1}M) \leq \ell_A(M)$.

Proof. — Let $N = S^{-1}M$ and let $(N_0, N_1, \ldots, N_n)$ be a chain of submodules of N. For every i, let M_i be the inverse image of N_i in M by the canonical morphism $M \to S^{-1}M$. By construction, we have inclusions of submodules $M_0 \subset M_1 \subset \cdots \subset M_n$. More precisely, since $S^{-1}M_i = N_i$ for every i (see proposition 3.6.7), this is even a chain of submodules, hence $\ell_A(M) \geq n$. By definition, the length $\ell_{S^{-1}A}(S^{-1}M)$ is the supremum of those integers n, hence $\ell_A(M) \geq \ell_{S^{-1}A}(S^{-1}M)$. $\qquad\square$

6.3. The Noetherian Property

Definition (6.3.1). — Let A be a ring. One says that an A-module M is *noetherian* if every non-empty family of submodules of M possesses a maximal element.

Proposition (6.3.2). — *Let* A *be a ring and let* M *be an* A-*module. The following properties are equivalent:*

(i) *The module* M *is noetherian;*

(ii) *Every submodule of* M *is finitely generated;*

(iii) *Every increasing family of submodules is stationary* (ascending chain condition).

Proof. — (i)⇒(ii) Let us assume that M is noetherian, that is, any non-empty family of submodules of M admits a maximal element and let us show that every submodule of M is finitely generated.

Let N be a submodule of M and let us consider the family $\mathscr{S}_N$ of all finitely generated submodules of N. This family is non-empty because the null module 0 belongs to $\mathscr{S}_N$. By hypothesis, $\mathscr{S}_N$ possesses a maximal element, say, N′. By definition, the A-module N′ is a finitely generated submodule of N and no submodule P of N such that $N' \subsetneq P$ is finitely generated.

For every $m \in N$, the A-module $P = N' + Am$ satisfies $N' \subset P \subset N$ and is finitely generated; by maximality of N′, one has $P = N'$, hence $m \in N'$. This proves that N′ = N, hence N is finitely generated.

(ii)⇒(iii). Let us assume that every submodule of M is finitely generated and let us consider an increasing sequence $(M_n)_{n \in \mathbf{N}}$ of submodules of M. Let $N = \bigcup M_n$ be the union of these modules M_n. Since the family is increasing, N is a submodule of M. By hypothesis, N is finitely generated. Consequently, there exists a finite subset $S \subset N$ such that $N = \langle S \rangle$. For every $s \in S$, there exists an integer $n_s \in \mathbf{N}$ such that $s \in M_{n_s}$; then, $s \in M_n$ for any integer n such that $n \geq n_s$. Let us set $\nu = \sup(n_s)$, so that $S \subset M_\nu$. It follows that $N = \langle S \rangle$ is contained in M_ν. Finally, for $n \geq \nu$, the inclusions $M_\nu \subset M_n \subset N \subset M_\nu$

for $n \geq \nu$ show that $M_n = M_\nu$. We have shown that the sequence (M_n) is stationary.

(iii)$\Rightarrow$(i) Let us assume that any increasing sequence of submodules of M is stationary and let $(M_i)_{i \in I}$ be a family of submodules of M indexed by a non-empty set I. Assuming by contradiction that this family has no maximal element, we are going to construct from the family $(M_i)_{i \in I}$ a *strictly* increasing sequence of submodules of M. Let $i \in I$; since M_i is not a maximal element of the family, there exists a $j \in J$ such that $M_i \subsetneq M_j$. Let us thus choose a function $f : I \to I$ such that $M_i \subsetneq M_{f(i)}$ for every $i \in I$. Let then $i_0 \in I$ and let us consider the sequence (i_n) of elements of I defined by $i_n = f(i_n)$ for $n > 0$. By construction, the sequence $(M_{i_n})_{n \in \mathbf{N}}$ of submodules of M is strictly increasing, contradicting (iii). (*In fact, the equivalence (i)$\Leftrightarrow$(iii) has nothing to do with modules and is valid in any ordered set.*) $\square$

Proposition (6.3.3). — *Let A be a ring, let M be an A-module and let N be a submodule of M. Then M is noetherian if and only if both N and M/N are noetherian.*

Proof. — Let us assume that M is noetherian. Since every submodule of N is also a submodule of M, every submodule of N is finitely generated, and N is noetherian. Let $\mathscr{P}$ be a submodule of M/N and let $P = cl^{-1}(\mathscr{P})$ be its inverse image by the canonical morphism $cl : M \to M/N$. By hypothesis, P is finitely generated, so that $\mathscr{P} = cl(P)$ is the image of a finitely generated module, hence is finitely generated too. This shows that M/N is noetherian.

Let us assume that N and M/N are noetherian. Let P be a submodule of M. By assumption, the submodule $P \cap N$ of N is finitely generated, as well as the submodule $cl(P)$ of M/N. Moreover, $cl(P) \simeq P/(P \cap N)$. By proposition 3.5.5 applied to the module P and its submodule $P \cap N$, P is finitely generated. $\square$

Corollary (6.3.4). — *The direct sum $M_1 \oplus \cdots \oplus M_n$ of a finite family of noetherian A-modules $M_1, \ldots, M_n$ is noetherian.*

Definition (6.3.5). — One says that a ring A is a *left noetherian ring* if the left A-module A_s is noetherian. One says that A is a *right noetherian ring* if the right A-module A_d is noetherian.

When a commutative ring A is left noetherian, it is also right noetherian, and vice versa; one then simply says that A is *noetherian*.

Remark (6.3.6). — Let A be a ring. The submodules of the left A-module A_s are its left ideals. Consequently, the ring A is left noetherian if and only if one of the following (equivalent) properties holds:

(i) Every non-empty family of left ideals of A possesses a maximal element;

(ii) Every left ideal of A is finitely generated;

(iii) Every increasing family of left ideals of A is stationary.

In particular, a principal ideal domain (all of whose ideals are generated by *one* element) is a noetherian ring. We had already observed this fact in lemma 2.5.6 in the course of the proof that a principal ideal domain is a unique factorization domain.

Corollary (6.3.7). — *Assume that the ring A is left noetherian (resp. right noetherian) and let n be an integer. Every submodule of the left A-module A_s^n (resp. of the right A-module A_d^n) is finitely generated.*

Corollary (6.3.8). — *Assume that the ring A is left noetherian (resp. right noetherian). Then, for any two-sided ideal I of A, the quotient ring A/I is left noetherian (resp. right noetherian).*

Proof. — Indeed, every left ideal of A/I is of the form J/I for some left ideal J of A containing I. By hypothesis, J is finitely generated, so that J/I is finitely generated too. The case of right ideals is analogous. $\square$

Remark (6.3.9). — Let A be a ring and let M be a right A-module. Let $I = \mathrm{Ann}_A(M)$ be the annihilator of M; it is a two-sided ideal of A. *If M is noetherian, then the quotient ring A/I is right noetherian.*

Indeed, let $(m_1, \dots, m_n)$ be a finite generating family of M. The map $f: A_d \to M^n$ given by $f(a) = (m_1 a, \dots, m_n a)$ is a morphism of right A-modules, and its kernel is equal to I. Consequently, f defines an injective morphism $\varphi: A_d/I \to M^n$. Since M is noetherian, M^n is noetherian too (corollary 6.3.4). Consequently, A_d/I is a noetherian right A-module. Since its submodules are the right ideals of A/I, this proves that A/I is right noetherian.

Proposition (6.3.10). — *Let A be a commutative ring and let S be a multiplicative subset of A. If M is a noetherian A-module, then $S^{-1}M$ is a noetherian $S^{-1}A$-module.*

Proof. — Let N be a submodule of $S^{-1}M$. By proposition 3.6.7, there exists a submodule N of M such that $N = S^{-1}N$. Since M is a noetherian A-module, N is finitely generated. Consequently, N is a finitely generated $S^{-1}A$-module. This shows that $S^{-1}M$ is a noetherian $S^{-1}A$-module. $\square$

Corollary (6.3.11). — *Let A be a commutative noetherian ring and let S be a multiplicative subset of A. The fraction ring $S^{-1}A$ is noetherian.*

Proof. — By proposition 6.3.10, $S^{-1}A$ is a noetherian $S^{-1}A$-module. Consequently, $S^{-1}A$ is a noetherian ring. $\square$

Theorem (6.3.12) (Hilbert's finite basis theorem). — *For any commutative noetherian ring A, the ring $A[X]$ is noetherian.*

Proof. — Let I be an ideal of $A[X]$; let us prove that I is finitely generated. For $n \geq 0$, let I_n be the set of coefficients of X^n, in all polynomials of degree $\leq n$ belonging to I; explicitly, I_n is the set of all $a \in A$ for which there exists a polynomial $P \in I$ such that $P = aX^n +$ terms of lower degree (which we will write $P = aX^n + \dots$). Observe that I_n is an ideal of A. Indeed, if $a, b \in I_n$, there are polynomials $P, Q \in I$ of degree $\leq n$ such that $P = aX^n + \cdots$ and $Q = bX^n + \cdots$. Then, for any $u, v \in A$, $uP + vQ = (ua + vb)X^n + \cdots$; since I is an ideal of A, $uP + vQ \in I$, hence $ua + vb \in I_n$. Moreover, one may write the polynomial 0 as $0 = 0X^n + \cdots$, so that $0 \in I_n$.

For any integer n, one has $I_n \subset I_{n+1}$; indeed, if $a \in I_n$ and $P \in I$ is such that $P = aX^n + \cdots$, then $XP = aX^{n+1} + \cdots$, so that $a \in I_{n+1}$. Since A is a noetherian ring, the sequence (I_n) is stationary. Let $v \in \mathbf{N}$ be such that $I_n = I_v$ for any integer $n \geq v$.

The ideals $I_0, I_1, \ldots, I_v$ are finitely generated; let us choose, for every integer $n \leq v$, a generating family $(a_{n,i})_{1 \leq i \leq r(n)}$ of I_n, as well as polynomials $P_{n,i} \in I_n$ such that $P_{n,i} = a_{n,i} X^n + \cdots$. Let J be the ideal of A[X] generated by the polynomials $P_{n,i}$ for $0 \leq n \leq v$ and $1 \leq i \leq r(n)$. The ideal J is finitely generated and, by construction, one has $J \subset I$. We shall now show the converse inclusion by induction on the degree of a polynomial $P \in I$.

If $\deg(P) = 0$, then P is constant and belongs to I_0, so that $P \in J$. Assume now that every polynomial in I whose degree is $< n$ belongs to J. Let $P \in I$ be a polynomial of degree n and let a be its leading coefficient. Let $m = \inf(n, v)$; if $n \leq v$, then $m = n$, hence $a \in I_m$; otherwise, $m = v$ and $a \in I_v$ since $I_n = I_v$ by definition of v. Consequently, there are elements $c_i \in A$, for $1 \leq i \leq r(m)$ such that $a = \sum_i c_i a_{m,i}$; let Q be the polynomial $Q = P - X^{n-m} \sum_i c_i P_{m,i}$. By construction, $\deg(Q) \leq n$; moreover, the coefficient of X^n in Q is equal to $a - \sum_i c_i a_{m,i} = 0$, so that $\deg(Q) < n$. By induction, $Q \in I$, so that $Q \in J$. It follows that $P \in J$.

We thus have proved that $I = J$. In particular, I is finitely generated.   □

Corollary (6.3.13). — *Let* A *be a commutative noetherian ring and let* n *be a positive integer. The ring* $A[X_1, \ldots, X_n]$ *is noetherian. In particular, for any field* K, *the ring* $K[X_1, \ldots, X_n]$ *is noetherian.*

Corollary (6.3.14). — *Let* A *and* B *be commutative rings. Assume that* A *is a noetherian ring and that* B *is a finitely generated* A*-algebra. Then,* B *is a noetherian ring.*

Proof. — By hypothesis, there are elements $b_1, \ldots, b_n \in B$ which generate B as an A-algebra, so that the canonical morphism of A-algebras from $A[X_1, \ldots, X_n]$ to B which maps X_i to b_i is surjective. This shows that B is a quotient of the noetherian ring $A[X_1, \ldots, X_n]$. Consequently, B is noetherian.   □

Theorem (6.3.15) (Hilbert). — *Let* K *be a field, let* A *be a finitely generated commutative* K*-algebra and let* G *be a finite subgroup of* $\mathrm{Aut}_K(A)$. *Then, the set* A^G *of all* $a \in A$ *such that* $g(a) = a$ *for any* $g \in G$ *is a finitely generated* K*-algebra.*

It is precisely to prove a theorem of this form, where the algebra A is $C[X_1, \ldots, X_n]$ and the group G is $SL_n(\mathbf{C})$, acting on A by linear transformations of the indeterminates, see HILBERT (1890), that Hilbert proved the finite basis theorem for polynomial rings over a field or over $\mathbf{Z}$. The terminology of "noetherian rings" only emerged in the 1940s, in homage to the paper (NOETHER, 1921) on primary decomposition in which she made this concept prominent.

The proof of theorem 6.3.15 requires three steps.

Lemma (6.3.16). — *The set* A^G *is a subalgebra of* A.

Proof. — We need to prove that

- The elements 0 and 1 belong to A^G;
- If a and b belong to A^G, so do $a + b$ and ab;
- If a belongs to A^G and $\lambda \in K$, then λa belongs to A^G.

Recall that every $g \in G$ is an automorphism of A as a K-algebra, so that $g(0) = 0$, $g(1) = 1$, $g(a + b) = g(a) + g(b)$, $g(ab) = g(a)g(b)$ for $a, b \in A$, $g(\lambda a) = \lambda g(a)$ for $a \in A$, $g \in G$ and $\lambda \in K$. In particular, we see that $0 \in A^G$ and $1 \in A^G$. Moreover, if $a, b \in A^G$, then these formulas imply that $g(a + b) = g(a) + g(b) = a + b$ and $g(ab) = g(a)g(b) = ab$ for every $g \in G$, so that $a + b$ and ab belong to A^G. Finally, if $a \in A^G$ and $\lambda \in K$, then $g(\lambda a) = \lambda g(a) = \lambda a$ for every $g \in G$, hence $\lambda a \in A^G$. □

Lemma (6.3.17). — *Under the hypotheses of theorem 6.3.15, the ring A is a finitely generated A^G-module.*

Proof. — Since A is finitely generated as a K-algebra, we may choose elements $a_1, \ldots, a_r \in A$ such that $A = K[a_1, \ldots, a_r]$.

For $i \in \{1, \ldots, r\}$, let us consider the polynomial

$$P_i(X) = \prod_{g \in G}(X - g(a_i))$$

in $A[X]$. Let G act on $A[X]$ by $g(P) = \sum g(c_n)X^n$ if $P = \sum c_n X^n$ and $g \in G$. Let us write

$$P_i(X) = X^n + b_1 X^{n-1} + \cdots + b_n$$

where $b_1, \ldots, b_n \in A$ and $n = \mathrm{Card}(G)$. For any $h \in G$, one then has

$$h(P_i(X)) = h\left(\prod_{g \in G}(X - g(a_i))\right) = \prod_{g \in G}(X - h(g(a_i))) = \prod_{g \in G}(X - g(a_i)) = P_i(X)$$

so that $b_1, \ldots, b_n$ are invariant under h. It follows that $P_i \in A^G[X]$, hence $b_1, \ldots, b_n \in A^G$. By construction, one has $P_i(a_i) = 0$, which is a monic polynomial relation for a_i, with coefficients in A^G. In the terminology of definition 4.1.1, we have just proved that the elements a_i are integral over A^G; we now proceed to show that A is a finitely generated A^G-module. Indeed, let us prove that A is generated, as an A^G-module, by the n^r products of the form $\prod_{i=1}^r a_i^{n_i}$, where, for every i, $0 \le n_i \le n-1$. Let A' be the A^G-submodule of A generated by these elements.

Since A is generated by all products $\prod_{i=1}^r a_i^{n_i}$ with $n_1, \ldots, n_r \in \mathbf{N}$, as a K-module, henceforth as an A^G-module, it suffices to prove that each such product belongs to A'. Let $X^{n_i} = Q_i(X)P_i(X) + R_i(X)$ be the euclidean division of X^{n_i} by the monic polynomial P_i in $A^G[X]$ (theorem 1.3.15). Since $P_i(a_i) = 0$, we have $a_i^{n_i} = R_i(a_i)$ for all i, hence

$$\prod_{i=1}^r a_i^{n_i} = \prod_{i=1}^r R_i(a_i).$$

If we expand this last product, we see that it belongs to A'. This shows that $A' = A$, hence that A is finitely generated as an A^G-module. □

Lemma (6.3.18) (Artin–Tate). — *Let K, B, A be three commutative rings such that $K \subset B \subset A$. Assume that K is a noetherian ring and that A is a finitely generated K-algebra as well as a finitely generated B-module. Then B is a finitely generated K-algebra.*

Proof. — Let $(x_1, \ldots, x_r)$ be a finite family of elements of A which generates A as a K-algebra; let $(a_1, \ldots, a_n)$ be a finite family of elements of A which generates A as a B-module. We assume, as we may, that $x_1 = 1$ and $a_1 = 1$. Then, every element of A can be written both as a polynomial in $x_1, \ldots, x_r$ with coefficients in K and as a linear combination of $a_1, \ldots, a_n$ with coefficients in B. If we apply this remark to $x_1, \ldots, x_r$ and to the products $x_i x_j$ (for $1 \le i, j \le r$), we obtain elements $b_{i,\ell}$ and $c_{i,j,\ell} \in B$ such that

$$x_i = \sum_{\ell=1}^{n} b_{i\ell} a_\ell$$

for $i \in \{1, \ldots, r\}$ and

$$a_i a_j = \sum_{\ell=1}^{n} c_{i,j,\ell} a_\ell$$

for $i, j \in \{1, \ldots, n\}$.

Let B_0 be the K-subalgebra of B generated by all the elements $b_{i,\ell}$ and $c_{i,j,\ell}$; it is a finitely generated K-algebra, in particular a noetherian ring.

Let A_0 be the B_0-submodule of A generated by the elements a_i. It is a K-submodule of A; moreover, all products $a_i a_j$ belong to A_0 by construction, so that A_0 is stable under the multiplication law of A; and $1 \in A_0$. Consequently, A_0 is a subalgebra of A. Still by construction, the elements $x_1, \ldots, x_r$ belong to A_0, hence $A_0 = A$. This proves that A is a finitely generated B_0-module.

Since B_0 is a noetherian ring, the ring A is noetherian as a B_0-module, so that every B_0-submodule of A is finitely generated. Since B is such a B_0-submodule, we obtain that B is finitely generated as a B_0-module; it is in particular finitely generated as a B_0-algebra.

Recall that B_0 is finitely generated as a K-algebra. By corollary 1.3.7, this implies that B is also finitely generated as a K-algebra, as was to be shown. □

Proof (Proof of theorem 6.3.15). — Set $B = A^G$, so that $K \subset B \subset A$. By the second lemma, A is finitely generated as a B-module; by hypothesis, it is finitely generated as a K-algebra. The Artin–Tate lemma (lemma 6.3.18) now implies that B is finitely generated as a K-algebra. □

Example (6.3.19). — Let us assume that $A = K[X_1, \ldots, X_n]$ and G is the symmetric group $\mathfrak{S}_n$ acting on A by permuting the indeterminates X_i. Then, $A^{\mathfrak{S}_n}$ is the algebra of *symmetric polynomials*. In this case, it is generated by the elementary symmetric polynomials $S_0, S_1, \ldots, S_n$ defined, for $1 \le j \le n$, by

$$S_j = \sum_{i_1 < \cdots < i_j} X_{i_1} \ldots X_{i_j}.$$

For example, $S_0 = 1, S_1 = X_1 + \cdots + X_n$ and $S_n = X_1 \ldots X_n$.

In fact, we prove a more precise result: *Let K be any commutative ring, and let $P \in K[X_1, \ldots, X_n]$ be a symmetric polynomial; there exists a unique polynomial $Q \in K[T_1, \ldots, T_n]$ such that $P = Q(S_1, \ldots, S_n)$; more precisely, if $cT_1^{m_1} \ldots T_n^{m_n}$ is a monomial that appears in Q, then $m_1 + 2m_2 + \cdots + nm_n \leq \deg(P)$.* (We say that the "weighted degree" of Q is at most that of P.)

Let us now prove this result by induction, first on n, then on $\deg(P)$. When $n = 1$, one has $S_1 = X_1$ and $A^{\mathfrak{S}_1} = A = K[X_1]$. Assume that the result holds for $(n-1)$. Writing $S'_0, \ldots, S'_{n-1}$ for the symmetric polynomials in the variables $X_1, \ldots, X_{n-1}$, we observe the relations $S_0 = S'_0$, $S_1 = S'_1 + X_n S'_0, S_2 = S'_2 + X_n S'_1, \ldots, S_{n-1} = S'_{n-1} + X_n S'_{n-2}$, and $S_n = X_n S'_{n-1}$. Consider a symmetric polynomial $P \in A[X_1, \ldots, X_n]$. Then $P(X_1, \ldots, X_{n-1}, 0)$ is a symmetric polynomial in $A[X_1, \ldots, X_{n-1}]$, hence it can be expressed as a polynomial in $S'_1, \ldots, S'_{n-1}$. Thus let $Q_1 \in A[T_1, \ldots, T_n]$ be such that $P(X_1, \ldots, X_{n-1}, 0) = Q_1(S'_1, \ldots, S'_{n-1})$, where the weighted degree of Q_1 is at most that of $P(X_1, \ldots, X_{n-1}, 0)$, hence at most that of P. Let us then consider the polynomial $P_1 = P - Q_1(S_1, \ldots, S_n)$; this is a symmetric polynomial of degree $\leq \deg(P)$ such that $P_1(X_1, \ldots, X_{n-1}, 0) = 0$, so that all monomials appearing in P_1 are multiples of X_n and X_n divides P_1. By symmetry, $X_1, \ldots, X_{n_1}$ divide P_1, which means that every monomial appearing in P_1 is divisible by the product $X_1 \ldots X_n = S_n$. There exists a polynomial $P_2 \in K[X_1, \ldots, X_n]$ such that $P_1 = X_1 \ldots X_n P_2$, and P_2 is necessarily symmetric (because $X_1 \ldots X_n$ is symmetric and regular), and its degree is $\leq \deg(P) - n$. By induction, there exists a polynomial $Q_2 \in A[T_1, \ldots, T_n]$ of weighted degree $\leq \deg(P) - n$ such that $P_2 = Q_2(S_1, \ldots, S_n)$. Let us set $Q = Q_1 + T_n Q_2$; its weighted degree is $\leq \deg(P)$ and $Q(S_1, \ldots, S_n) = Q_1(S_1, \ldots, S_{n-1}) + S_n Q_2(S_1, \ldots, S_n) = (P - P_1) + X_1 \ldots X_n P_2 = P$.

This proves the existence; let us now establish the uniqueness of such a polynomial. Let $Q \in K[T_1, \ldots, T_n]$ be such that $P = Q(S_1, \ldots, S_n) = 0$. Then $0 = P(X_1, \ldots, X_{n-1}, 0) = Q(S'_1, \ldots, S'_{n-1}, 0)$. By induction, the polynomial $Q(T_1, \ldots, T_{n-1}, 0)$ vanishes, which means that there exists a polynomial $Q_1 \in K[T_1, \ldots, T_n]$ such that $Q = T_n Q_1$. Then $0 = Q(S_1, \ldots, S_n) = S_n Q_1(S_1, \ldots, S_n)$. Since $S_n = X_1 \ldots X_n$ is a regular element of $K[X_1, \ldots, X_n]$, one has $Q_1(S_1, \ldots, S_n) = 0$. By induction on the degree of Q, one has $Q_1 = 0$, hence $Q = T_n Q_1 = 0$.

6.4. The Artinian Property

Definition (6.4.1). — Let A be a ring and let M be a right A-module. One says that M is *artinian* if every non-empty family of submodules of M possesses a minimal element.

On Emil Artin

Emil ARTIN (1898–1962) was an Austrian mathematician who worked in algebra, number theory, and topology. He became a professor in Hamburg in 1925. Revoked from this position by the Nazis in 1937, ARTIN and his family took refuge in the United States; he would come back to Germany in 1958.

DEDEKIND had defined for number fields an analogue of the RIEMANN zeta function, and ARTIN's dissertation concerned an analogue of this function for quadratic extensions of the field of rational functions over a finite field k, conjecturing a version of the Riemann hypothesis which would be only proved by WEIL in 1948.

In the following years, he and SCHREIER introduced "real closed fields", building an algebraic theory of the real numbers. They characterized them as the only fields whose algebraic closure is a finite extension. He also applied this theory to solve HILBERT's 17th problem, that a rational function in $\mathbf{R}(X_1, \ldots, X_n)$ which is positive everywhere can be written as a sum of squares of rational functions.

In the middle of the 1920s, ARTIN's lectures had a tremendous influence on the development of algebra, as can be witnessed by the subtitle of VAN DER WAERDEN (1930), acknowledging both him and NOETHER, and his notes on Galois theory, which shaped the theory as we now teach it.

Emil Artin, 1935, at his Hamburg-Langenhorn home, reading

Photographed by Natascha Artin-Brunswick
Courtesy of Tom Artin

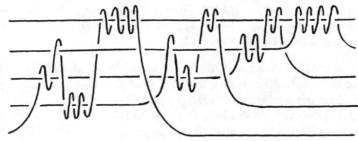

Braids drawing Copied from E. Artin (1947), "Theory of braids",
Annals of Mathematics **48** (1), p. 101–126
By courtesy of the Department of Mathematics at Princeton University

The descending chain condition that Artin introduced allowed him to give a more general treatment of Wedderburn's simple rings. (The terminology of *artinian rings* emerged in the 1950s.)

From these years, we also owe him fundamental papers in algebraic number theory. He gave a new construction of L-functions for number fields, associated with representations of Galois groups; a theorem of Brauer asserts their meromorphy, but it is conjectured that they should be holomorphic and satisfy a functional equation, in analogy with the Riemann or Dedekind zeta functions. Artin also gave the proof of the general reciprocity law of class field theory, a vast generalization of Gauss's quadratic reciprocity law. The book *Class field theory* that he would publish with Tate in 1967 introduces the "algebraic" approach to that subject, based on group cohomology.

Artin also worked in topology. With Fox, he constructed a "wild arc" in 3-dimensional space, wild in the sense that its complementary subset is not simply connected.

In 1925, he had initiated the theory of braids, a variant of the theory of knots, which are diagrams of non-crossing strings such as the one on the left, considered up to "isotopy". He identified the group they form with the fundamental group of a space and used this relation to give the now classic presentation of this group. Braids became an important chapter of topology, with applications to representation theory and statistical physics. In 2001, Bigelow and Krammer proved that the n-strings braid group is isomorphic to a subgroup of $GL_n(\mathbf{R})$.

Definition (6.4.2). — Let A be a ring. One says that the ring A is a *right artinian* ring if it is artinian as a right A-module, and that it is *left artinian* if it is artinian as a left A-module.

If the ring A is commutative, then it is equivalent for A to be left artinian or to be right artinian; we then say that A is an *artinian ring*.

In some sense, the artinian property is symmetric to the noetherian property. However, we shall see in theorem 6.4.11 that for a *ring*, the property of being artinian is substantially stronger than that of being noetherian. But for the moment, let us observe a few consequences which are quite analogous to those obtained from the noetherian property.

Remark (6.4.3). — *An A-module M is artinian if and only if every decreasing sequence of submodules of M is stationary (descending chain condition).* This is a general result about ordered sets, which is proved essentially in the same way as the equivalence of (i) and (iii) in proposition 6.3.2.

Let us indeed assume that M is artinian and let (M_n) be a decreasing sequence of submodules. By assumption, it has a minimal element, say M_m, and one has $M_n = M_m$ for $n \geq m$.

In the other direction, let $(M_i)_{i \in I}$ be non-empty family of submodules of M that has no minimal element. Let $i_0 \in I$; then M_{i_0} is not a minimal element of $(M_i)_{i \in I}$, hence there exists an index $i_1 \in I$ such that $M_{i_1} \subsetneq M_{i_0}$. One then constructs by induction a sequence $(i_n)_{n \in \mathbf{N}}$ of elements of I such that the sequence $(M_{i_n})_{n \in \mathbf{N}}$ is strictly decreasing.

Proposition (6.4.4). — *Let A be a ring. Let M be an A-module and let N be a submodule of M. Then M is artinian if and only if N and M/N are artinian modules.*

Proof. — a) Let $\mathrm{cl}_N \colon M \to M/N$ be the canonical surjection. Let us first assume that N and M/N are artinian. Let (M_n) be a decreasing sequence of submodules of M. Then $(M_n \cap N)$ is a decreasing sequence of submodules of N, hence is stationary; similarly, $(\mathrm{cl}_N(M_n))$ is a decreasing sequence of submodules of M/N, so is stationary too. Consequently, there exists an integer $p \in \mathbf{N}$ such that $M_n \cap N = M_{n+1} \cap N$ and $\mathrm{cl}_N(M_n) = \mathrm{cl}_N(M_{n+1})$ for $n \geq p$. Applying lemma 6.2.8 to the submodules M_n, M_{n+1}, N of M, we conclude that $M_n = M_{n+1}$ for $n \geq p$: the sequence (M_n) is stationary.

Conversely, let us assume that M is artinian. A decreasing sequence of submodules of N is also a decreasing sequence of submodules of M, hence is stationary. Consequently, N is artinian. Let then (P_n) be a decreasing sequence of submodules of M/N. The sequence $(\mathrm{cl}_N^{-1}(P_n))$ of submodules of M is decreasing, hence is stationary since M is artinian. Since cl_N is surjective, one has $P_n = \mathrm{cl}_N(\mathrm{cl}_N^{-1}(P_n))$, hence the sequence (P_n) is also stationary. □

Corollary (6.4.5). — *The direct sum $M_1 \oplus \cdots \oplus M_n$ of a finite family of artinian A-modules $M_1, \ldots, M_n$ is artinian.*

Proposition (6.4.6). — *Assume that A is commutative and let S be a multiplicative subset of A. If M is an artinian A-module, then $S^{-1}A$ is an artinian $S^{-1}A$-module.*

Proof. — Let us assume that M is an artinian A-module and let $i\colon M \to S^{-1}M$ be the canonical morphism. Let (P_n) be a decreasing sequence of submodules of $S^{-1}M$. The sequence $(i^{-1}(P_n))$ of submodules of M is decreasing, hence is stationary. By proposition 3.6.7, one has $P_n = S^{-1}(i^{-1}(P_n))$ for any integer n. This implies that the sequence (P_n) is stationary.   □

Corollary (6.4.7). — *Let A be a commutative ring. If A is artinian, then $S^{-1}A$ is artinian for any multiplicative subset S of A.*

Theorem (6.4.8). — *Let A be a ring. An A-module has finite length if and only if it is both artinian and noetherian.*

Proof. — Let us assume that M has finite length and let us consider a sequence (M_n) of submodules of M which is either increasing or decreasing. The sequence $(\ell_A(M_n))$ of their lengths is then increasing or decreasing, according to the case; it is bounded from below by 0, and bounded from above by $\ell_A(M)$. Consequently, it is stationary. This implies that the sequence (M_n) is stationary. (Recall that if $P \subset Q$ are two modules of the same finite length, then $\ell_A(Q/P) = \ell_A(Q) - \ell_A(P) = 0$; consequently, $Q/P = 0$ and $Q = P$.) Considering decreasing sequences, we obtain that M is artinian; considering increasing sequences, we conclude that M is noetherian.

Let us now assume that M is artinian and that its length is infinite, and let us prove that M is not noetherian.

Let us set $M_0 = 0$. By assumption, $M \neq 0$, so that the set of non-zero submodules of M is non-empty. Since M is artinian, it admits a minimal element M_1; this module M_1 is a non-zero submodule of M such that any strict submodule of M_1 is null; in other words, M_1 is simple. Since $\ell_A(M_1) = 1$ and $\ell_A(M) = \infty$, one has $M_1 \neq M$ and we may construct similarly a submodule M_2 of M containing M_1 such that M_2/M_1 is simple, hence $\ell_A(M_2) = \ell_A(M_1) + 1 = 2$. We thus construct by induction an increasing sequence (M_n) of submodules of M such that M_{n+1}/M_n is simple and $\ell_A(M_n) = n$ for every integer n. In particular, the sequence (M_n) is not stationary, hence M is not noetherian.   □

Proposition (6.4.9). — *Let A be a right artinian ring and let J be the Jacobson radical of A. There exists an integer n such that $J^n = 0$.*

Proof. — Since A is right artinian, the decreasing sequence (J^n) of right ideals of A is stationary. Consequently, there exists an integer $n \geq 0$ such that $J^n = J^{n+1}$. Let us assume by contradiction that $J^n \neq 0$.

The set of right ideals I of A such that $J^n I \neq 0$ is non-empty (for example, $I = A$ is such a right ideal). Since A is right artinian, there exists a minimal such ideal, say I. Let x be any element of I such that $J^n x \neq 0$; then xA is a right ideal of A such that $J^n(xA) \neq 0$; moreover, $xA \subset I$. By minimality of I, one has $I = xA$. In particular, I is finitely generated. Moreover, $J^n JI = J^{n+1}I = J^n I \neq 0$. Since $JI \subset I$, the minimality of I implies that $I = JI$. It then follows from Nakayama's lemma (theorem 6.1.1) that $I = 0$, in contradiction with the hypothesis that $J^n I \neq 0$.   □

Lemma (6.4.10). — *Let* A *be a ring and let* n *be a positive integer. Let* M *be an artinian* A*-module. Assume that* M *is the sum of its submodules of length* $\leq n$. *Then* M *has finite length.*

Proof. — Let (M_i) be a family of submodules of M such that $\ell_A(M_i) \leq n$ for every i, and $M = \sum_i M_i$. Let us show that $\ell_A(M) < \infty$ by induction on n.

Let us first treat the case $n = 1$. If the length of M is infinite, we may construct by induction on m a sequence $i(m)$ such that $M_{i(m)}$ is not included in the sum of the modules $M_{i(k)}$ for $k < m$ (otherwise, one would have $M = \sum_{k<m} M_{i(k)}$ and $\ell_A(M)$ would be finite). Since all modules M_i are simple or zero, one has $M_{i(m)} \cap \sum_{k<m} M_{i(k)} = 0$ and the modules $M_{i(k)}$ are in direct sum in their union $N = \sum_{k\in\mathbf{N}} M_{i(k)}$. For every m, set $N_m = \bigoplus_{k\geq m} M_{i(k)}$. The sequence (N_m) is strictly decreasing, and this contradicts the hypothesis that M is artinian.

Let us now assume that $n \geq 2$ and that the result holds for $n-1$. For every i, let M_i' be a submodule of M_i such that $\ell_A(M_i') \leq n-1$ and $\ell_A(M_i/M_i') \leq 1$. Set $M' = \sum_i M_i'$; this is a submodule of M; by the induction hypothesis, $\ell_A(M') < \infty$. Moreover, M/M' is the sum of the modules $M_i/(M' \cap M_i)$ whose lengths are ≤ 1. By the case $n = 1$, $\ell_A(M/M') < \infty$. By proposition 6.4.4, one has $\ell_A(M) < \infty$, as was to be shown.   □

Theorem (6.4.11) (Akizuki–Hopkins–Levitzki). — *Let* A *be a right artinian ring. Then the ring* A *is right noetherian and has finite length as a right* A*-module.*

Proof (Proof[1]). — It suffices to prove that $\ell_A(A_d)$ is finite.

Let $\mathscr{S}$ be the set of right ideals I of A such that A/I has finite length. Since $\mathscr{S}$ is non-empty (it contains A) and A is right artinian, we may consider a minimal element I of $\mathscr{S}$. Let $n = \ell_A(A_d/I)$.

Let x be an element of A such that $\ell_A(xA) < \infty$, and let $J = \{a \in A \,;\, xa = 0\}$ be its right annihilator, so that J is a right ideal of A and $xA \simeq A_d/J$. I claim that $A_d/(I \cap J)$ has finite length. First of all, $(I+J)/I$ has finite length, as a submodule of A_d/I. Then, the composition of the inclusion morphism $J \to I+J$ and of the projection $I + J \to (I + J)/I$ is surjective, and its kernel is $I \cap J$, so that $J/(I \cap J)$ has finite length. This module is a submodule of $A_d/(I \cap J)$, with quotient A_d/J, which has finite length, by the assumption on x. By proposition 6.2.9, we conclude that $\ell_A(A/(I \cap J)) < \infty$. Since $I \cap J \subset I$, the minimality of I implies that $I \cap J = I$, hence $I \subset J$. In particular, $\ell_A(xA) = \ell_A(A/J) \leq \ell_A(A/I) \leq n$.

Let K be the sum of all right ideals xA of A such that $\ell_A(xA) < \infty$. By lemma 6.4.10, $\ell_A(K) < \infty$. Moreover, every right ideal K' such that $\ell_A(K') < \infty$ is contained in K. Indeed, if $x \in K'$, then $\ell_A(xA) \leq \ell_A(K') < \infty$, hence $x \in K$. In other words, K is the largest right ideal of A which is of finite length.

Let us assume by contradiction that $K \neq A$. Let K' be a right ideal of A which is minimal among the ideals containing K and distinct from K. Such an ideal exists because A is right artinian. Necessarily, the right A-module K'/K is simple, hence has length 1. Consequently, $\ell_A(K') < \infty$, which contradicts the definition of K. This shows that $K = A$. In particular, $\ell_A(A_d) < \infty$.   □

---

[1] This proof is borrowed from Bourbaki (2012), §1, n° 2, théorème 1.

Corollary (6.4.12). — *Let A be a right artinian ring and let M be right A-module. The following conditions are equivalent:*

(i) *The module M is finitely generated;*

(ii) *The module M has finite length;*

(iii) *The module M is artinian;*

(iv) *The module M is noetherian.*

Proof. — Without any hypothesis on the ring A, (ii) implies (iii) and (iv), and (iv) implies (i).

Assume that M is finitely generated, so that there exists an integer n and a surjective morphism $A_d^n \to M$ of A-modules. Then $\ell_A(M) \le \ell_A(A_d^n) = n \, \ell_A(A_d)$ is finite. This proves the implication (i)$\Rightarrow$(ii).

It remains to show that if M is artinian, then M has finite length. Let $x \in M$ and let $J = \{a \in A; xa = 0\}$, so that $xA \simeq A/J$. Since A is right artinian, its length is finite (theorem 6.4.11) and $\ell_A(xA) = \ell_A(A/J) \le \ell_A(A_d)$. This shows that M is generated by its submodules of lengths $\le \ell_A(A_d)$. By lemma 6.4.10, $\ell_A(M)$ is finite.   □

Lemma (6.4.13). — *Let A be a commutative artinian ring.*

a) If A is an integral domain, then A is a field.

b) Every prime ideal of A is maximal.

c) The set of maximal ideals of A is finite.

Proof. — a) Let us assume that A is a domain. Let $a \in A - \{0\}$ and let us consider the decreasing sequence $(a^n A)_{n \ge 0}$ of principal ideals. Since A is artinian, there exists an integer n such that $a^n A = a^{n+1} A$, hence there exists an element $b \in A$ such that $a^n = a^{n+1}b$. Consequently, $a^n(1 - ab) = 0$. Since A is a domain and $a \ne 0$, one has $1 - ab = 0$, hence a is invertible.

b) Let P be a prime ideal of A. Then, A/P is an integral domain. Moreover, it is artinian as an A-module, because it is a quotient of the A-module A. Since the A-submodules of A/P coincide with its (A/P)-submodules, we see that the ring A/P is artinian. By *a)*, it is a field, hence P is a maximal ideal.

c) Let us argue by contradiction and let (M_n) be a sequence of pairwise distinct maximal ideals of A. Let us consider the decreasing sequence of ideals

$$M_1 \supset M_1 M_2 \supset \cdots \supset M_1 \ldots M_n \supset \cdots$$

For every $i \le n$, M_i is not contained in M_{n+1}, so there is an element $a_i \in M_i$ such that $a_i \notin M_{n+1}$. The product $a = a_1 \ldots a_n$ belongs to $M_1 \ldots M_n$; since M_{n+1} is a prime ideal, a does not belong to M_{n+1}. This shows that the preceding sequence of ideals is not stationary, contradicting the hypothesis that A is artinian.   □

Theorem (6.4.14) (Akizuki). — *Let A be a commutative ring. The following conditions are equivalent:*

(i) *The ring A is artinian;*

   (ii) *The ring A is noetherian and all of its prime ideals are maximal ideals;*

   (iii) *The A-module A has finite length.*

With the terminology introduced in chapter 9, condition (ii) can be rephrased as "A is noetherian and $\dim(A) = 0$".

Proof. — By theorem 6.4.8, condition (iii) is equivalent to the fact that A is both artinian and noetherian. This shows the implication (iii)⟹(i).

   We also have shown in theorem 6.4.11 that an artinian ring is noetherian; moreover, by lemma 6.4.13, all of its prime ideals are maximal, so that (i)⟹(ii).

   It remains to show that under assumption (ii), the ring A has finite length.

   Let $\mathscr{S}$ be the set of ideals of A such that the length of A/I is infinite. If A does not have finite length, then $I = (0)$ belongs to $\mathscr{S}$, hence $\mathscr{S}$ is non-empty. Since A is noetherian, the set $\mathscr{S}$, ordered by inclusion, has a maximal element I. The ideal I is such that the length of A/I is infinite, while the length of A/J is finite for any ideal J such that $J \supsetneq I$. The ring A/I is noetherian, and its prime ideals, being in correspondance with the prime ideals of A containing I, are maximal ideals. We may thus replace A by A/I and assume that $I = (0)$.

   Let us now show that A is an integral domain. Since $\ell_A(A) = +\infty$, we have $A \neq 0$. Let then a, b be two non-zero elements of A such that $ab = 0$. By assumption, A/aA has finite length. Moreover, the map $x \mapsto bx$ vanishes on aA, hence induces a surjective morphism from A/aA to bA; in particular, bA has finite length. Since A/bA has finite length, this implies that A has finite length, a contradiction. We have shown that A is an integral domain.

   In particular, (0) is a prime ideal of A. By assumption, (0) is maximal, so that A is a field. In particular, A is a simple module. In particular, its length is finite. This contradiction shows that the hypothesis that $\ell_A(A)$ is infinite is false.   □

Remark (6.4.15). — We shall establish in the next section (theorem 6.5.15) that for any noetherian commutative ring A, there is an integer n, a sequence $(I_0, \ldots, I_n)$ of ideals of A and a sequence $(P_1, \ldots, P_n)$ of prime ideals of A such that

$$0 = I_0 \subset I_1 \subset \cdots \subset I_{n-1} \subset I_n = A$$

and $I_k/I_{k-1} \simeq A/P_k$ for every integer $k \in \{1, \ldots, n\}$. Under assumption (ii) of theorem 6.4.14, all quotients I_k/I_{k-1} are simple A-modules. This gives another proof that A has finite length.

6.5. Support of a Module, Associated Ideals

In this section, all rings are tacitly assumed to be commutative.

Definition (6.5.1). — Let A be a commutative ring and let M be an A-module. The *support* of M is the subset of $\mathrm{Spec}(A)$ consisting of all prime ideals P of A such that $M_P \neq 0$. It is denoted by $\mathrm{Supp}_A(M)$, or $\mathrm{Supp}(M)$ if no confusion can arise regarding the ring A.

Let us expand the definition. Let M be an A-module and let P be a prime ideal. The condition $M_P \neq 0$ means that there exists an $m \in M$ such that $m/1 \neq 0$ in M_P, hence $sm \neq 0$ for any $s \in A-P$. In other words, *a prime ideal P of A belongs to* $\mathrm{Supp}_A(M)$ *if and only if there exists an* $m \in M$ *whose annihilator* $\mathrm{Ann}_A(m)$ *is contained in* P. This can also be rewritten as

$$\mathrm{Supp}_A(M) = \bigcup_{m \in A} V(\mathrm{Ann}_A(m)).$$

Example (6.5.2). — Let A be a commutative ring and let I be an ideal of A. For $m \in A/I$, one has $I \subset \mathrm{Ann}_A(m)$, hence $V(\mathrm{Ann}_A(m)) \subset V(I)$, so that $\mathrm{Supp}_A(M) \subset V(I)$. Conversely, $\mathrm{Ann}_A(\mathrm{cl}_I(1)) = I$, hence $V(I) \subset \mathrm{Supp}_A(M)$. This proves that $\mathrm{Supp}_A(A/I) = V(I)$, the set of all prime ideals of A which contain I.

Proposition (6.5.3). — *Let* A *be a commutative ring, let* M *be an A-module and let* N *be a submodule of* M. *Then* $\mathrm{Supp}_A(M) = \mathrm{Supp}_A(N) \cup \mathrm{Supp}_A(M/N)$.

Proof. — Recall (exactness of localization, proposition 3.6.6) that for any prime ideal P of A, N_P identifies with a submodule of M_P and one has an isomorphism $(M/N)_P \simeq M_P/N_P$. Consequently, $M_P = 0$ if and only if $N_P = (M/N)_P = 0$. In other words,

$$P \notin \mathrm{Supp}_A(M) \qquad \Leftrightarrow \qquad P \notin \mathrm{Supp}_A(N) \cup \mathrm{Supp}_A(M/N),$$

that is to say, $\mathrm{Supp}_A(M) = \mathrm{Supp}_A(N) \cup \mathrm{Supp}_A(M/N)$.   □

Theorem (6.5.4). — *Let* A *be a commutative ring and let* M *be an A-module.*

a) *If* $M \neq 0$, *then* $\mathrm{Supp}_A(M) \neq \emptyset$;

b) *Let* P *be a prime ideal in* $\mathrm{Supp}_A(M)$; *then every prime ideal containing* P *belongs to* $\mathrm{Supp}_A(M)$;

c) *One has* $\mathrm{Supp}_A(M) \subset V(\mathrm{Ann}_A(M))$: *any prime ideal of* A *belonging to* $\mathrm{Supp}_A(M)$ *contains* $\mathrm{Ann}_A(M)$;

d) *If* M *is finitely generated, then* $\mathrm{Supp}_A(M) = V(\mathrm{Ann}_A(M))$ *is the set of all prime ideals of* A *which contain* $\mathrm{Ann}_A(M)$.

Proof. — a) Let us assume that $M \neq 0$ and let $m \in M - \{0\}$. In other words, $\mathrm{Ann}_A(m) \neq A$, hence $V(\mathrm{Ann}_A(m)) \neq \emptyset$ (this follows from Krull's theorem, theorem 2.1.3, see remark 2.2.6, b)). Since $\mathrm{Supp}_A(M)$ contains $V(\mathrm{Ann}_A(m))$, this implies that $\mathrm{Supp}_A(M) \neq \emptyset$.

b) Indeed, every subset of $\mathrm{Spec}(A)$ which is a union of closed subsets satisfies this property. Let P and Q be prime ideals of Q such that $P \subset Q$ and $P \in \mathrm{Supp}_A(M)$. Let $m \in M$ be such that $P \in V(\mathrm{Ann}_A(m))$. Then $Q \supset P \supset \mathrm{Ann}_A(m))$, hence $Q \in V(\mathrm{Ann}_A(m))$, hence $Q \in \mathrm{Supp}_A(M)$.

c) For every $m \in M$, one has $\mathrm{Ann}_A(M) \subset \mathrm{Ann}_A(m)$, hence $V(\mathrm{Ann}_A(m)) \subset V(\mathrm{Ann}_A(M))$. Consequently, $\mathrm{Supp}_A(M) \subset V(\mathrm{Ann}_A(M))$.

d) Let $(m_1, \ldots, m_n)$ be a finite family of elements of M which generates M. Let us observe that

$$\mathrm{Ann}_A(M) = \mathrm{Ann}_A(m_1) \cap \cdots \cap \mathrm{Ann}_A(m_n).$$

Indeed, the inclusion $\subset$ is obvious; conversely, if for every i, one has $a \in \mathrm{Ann}_A(m_i)$, then $am = 0$ for every linear combination m of the m_is, hence $am = 0$ for every $m \in M$. In particular,

$$V(\mathrm{Ann}_A(M)) = V\big(\mathrm{Ann}_A(m_1) \cap \cdots \cap \mathrm{Ann}_A(m_n) \big)$$
$$= V(\mathrm{Ann}_A(m_1)) \cup \cdots \cup V(\mathrm{Ann}_A(m_n)),$$

by remark 2.2.6, *a)*.

Using *c)*, we thus have

$$\mathrm{Supp}_A(M) \subset V(\mathrm{Ann}_A(m_1)) \cup \cdots \cup V(\mathrm{Ann}_A(m_n)),$$

and the other inclusion follows from the definition of $\mathrm{Supp}_A(M)$, hence the desired equality.   □

Remark (6.5.5). — Let M be an A-module. If M is finitely generated, then assertion *d)* of the theorem says that its support is the closed subset $V(\mathrm{Ann}_A(M))$ of $\mathrm{Spec}(A)$. In general, $\mathrm{Supp}_A(M)$ is not necessarily closed. For example, let us consider $A = \mathbf{Z}$ and $M = \bigoplus_{n \geq 1}(\mathbf{Z}/n\mathbf{Z})$. Since every element of M is torsion, one has $M_{(0)} = 0$, hence $(0) \notin \mathrm{Supp}_\mathbf{Z}(M)$. On the other hand, for every prime number p, the module M contains $\mathbf{Z}/p\mathbf{Z}$ as a submodule, whose support is $\{(p)\}$, so that $(p) \in \mathrm{Supp}_\mathbf{Z}(M)$. Consequently, $\mathrm{Supp}_\mathbf{Z}(M) = \{(2), (3), \ldots\} = \mathrm{Spec}(\mathbf{Z}) - \{(0)\}$ and this set is not closed in $\mathrm{Spec}(\mathbf{Z})$.

By assertion *b)*, the support of an A-module is a union of closed subsets, hence it contains the closure of any of its points. One says that it is *closed under specialization.*

Definition (6.5.6). — Let A be a commutative ring and let M be an A-module. One says that a prime ideal P of A is *associated* to M if there exists an element $m \in M$ such that P is minimal among all prime ideals of A containing $\mathrm{Ann}_A(m)$. The set of all associated prime ideals of M is denoted $\mathrm{Ass}_A(M)$.

Example (6.5.7). — Let A be a commutative ring, let P be a prime ideal of A. Then $\mathrm{Ass}_A(A/P) = \{P\}$.

Let $\mathrm{cl}_P \colon A \to A/P$ be the canonical surjection. Let $x \in A/P$. If $x = 0$, then $\mathrm{Ann}_A(x) = A$ and no prime ideal of A contains $\mathrm{Ann}_A(x)$. Let us assume that $x \neq 0$ and let us prove that $\mathrm{Ann}_A(x) = P$. The inclusion $P \subset \mathrm{Ann}_A(x)$ is obvious. Conversely, let $a \in A$ be such that $x = \mathrm{cl}_P(a)$; one has $a \notin P$; consequently, if $b \in \mathrm{Ann}_A(x)$, the equality $bx = 0$ means $ba \in P$, hence it implies $b \in P$ by the definition of a prime ideal. In that case, P is the only minimal prime ideal containing $\mathrm{Ann}_A(x)$. This implies the claim.

Theorem (6.5.8). — *Let A be a commutative ring and let M be an A-module.*

a) *Multiplication by an element $a \in A$ is injective in M if and only if a does not belong to any element of $\mathrm{Ass}_A(M)$.*

b) *An element $a \in A$ belongs to every element of $\mathrm{Ass}_A(M)$ if and only if $M_a = 0$.*

c) *In particular, $\mathrm{Ass}_A(M) = \varnothing$ if and only if $M = 0$.*

Proof. — a) Let $P \in \mathrm{Ass}_A(M)$ and let $a \in P$. Let $m \in M$ be such that P is minimal among the prime ideals containing $\mathrm{Ann}_A(m)$. By lemma 2.2.15, a is a zero-divisor in the ring $A/\mathrm{Ann}_A(m)$. In other words, there exists an element $b \in A - \mathrm{Ann}_A(m)$ such that $ab \in \mathrm{Ann}_A(m)$. Consequently, $bm \neq 0$ and $abm = 0$. In particular, multiplication by a is not injective in M.

Conversely, let $a \in A$ and $m \in M$ be such that $am = 0$ but $m \neq 0$. One thus has $\mathrm{Ann}_A(m) \neq A$, so that there exists a minimal prime ideal P that contains $\mathrm{Ann}_A(m)$ (lemma 2.2.13). Since $a \in \mathrm{Ann}_A(m)$, one has $a \in P$.

b) Let us assume that $M_a = 0$. Let $m \in M$; since $M_a = 0$, there exists an integer $n \geq 1$ such that $a^n m = 0$. Consequently, $a^n \in \mathrm{Ann}_A(m)$ and any prime ideal containing $\mathrm{Ann}_A(m)$ contains a. In particular, a belongs to every element of $\mathrm{Ass}_A(M)$.

Conversely, let us assume that $M_a \neq 0$ and let $m \in M$ be such that $m/1 \neq 0$ in M_a. The ideal $\mathrm{Ann}_A(m)$ does not contain any power a^n, hence generates an ideal I of A_a such that $I \neq A_a$. By lemma 2.2.13 (applied to the ideal I of the ring A_a), there exists a prime ideal Q of A_a which contains I and is minimal among those prime ideals. Let P be the inverse image of Q in A, so that $Q = PA_a$. Then, the ideal P is prime and does not contain a; it is also a minimal prime ideal among those containing $\mathrm{Ann}_A(m)$; in particular, $P \in \mathrm{Ass}_A(M)$ and $a \notin P$.

c) It suffices to apply b) to $a = 1$.   □

Proposition (6.5.9). — *Let A be a commutative ring, let M be an A-module and let N be a submodule of M. Then,*

$$\mathrm{Ass}_A(N) \subset \mathrm{Ass}_A(M) \subset \mathrm{Ass}_A(N) \cup \mathrm{Ass}_A(M/N).$$

Proof. — It follows from the definition that any prime ideal P which is associated to N is also associated to M. Let now P be a prime ideal which is associated to M, and let $m \in M$ be such that P is minimal among the prime ideals containing $\mathrm{Ann}_A(m)$. Let $\mathrm{cl}_N \colon M \to M/N$ be the canonical surjection. One has $\mathrm{Ann}_A(\mathrm{cl}_N(m)) \supset \mathrm{Ann}_A(m)$. If P contains $\mathrm{Ann}_A(\mathrm{cl}_N(m))$, then P is also minimal among the prime ideals containing $\mathrm{Ann}_A(\mathrm{cl}_N(m))$, hence $P \in \mathrm{Ass}_A(M/N)$. Let us thus assume that P does not contain $\mathrm{Ann}_A(\mathrm{cl}_N(m))$, and let $b \in \mathrm{Ann}_A(\mathrm{cl}_N(m))$ such that $b \notin P$. Let $a \in \mathrm{Ann}_A(bm)$; then $abm = 0$, hence $ab \in \mathrm{Ann}_A(m)$, hence $ab \in P$; since $b \notin P$, this implies $a \in P$, hence the inclusions $\mathrm{Ann}_A(m) \subset \mathrm{Ann}_A(bm) \subset P$. It follows that the prime ideal P is minimal among those containing $\mathrm{Ann}_A(bm)$. This implies $P \in \mathrm{Ass}_A(N)$ since, by choice of b, we have $bm \in N$.   □

Proposition (6.5.10). — *Let A be a commutative ring, let S be a multiplicative subset of A and let M be an A-module. A prime ideal P disjoint from S is associated to the A-module M if and only if $S^{-1}P$ is associated to the $S^{-1}A$-module $S^{-1}M$.*

Viewing $\mathrm{Spec}(S^{-1}A)$ as a subset of $\mathrm{Spec}(A)$ via proposition 2.2.7, this can be rephrased as the equality $\mathrm{Ass}_{S^{-1}A}(S^{-1}M) = \mathrm{Ass}_A(M) \cap \mathrm{Spec}(S^{-1}A)$.

Proof. — Let a prime ideal P, disjoint from S, be associated to M. Let $m \in M$ be such that P is minimal among the prime ideals containing $\mathrm{Ann}_A(m)$. The inverse image in A of $\mathrm{Ann}_{S^{-1}A}(m/1)$ contains $\mathrm{Ann}_A(m)$. Moreover, let $a \in A$ be such that $(a/1)(m/1) = 0$ in $S^{-1}M$; then, there exists an $s \in S$ such that $sam = 0$, hence $sa \in \mathrm{Ann}_A(m)$ and *a fortiori* $sa \in P$. Since P is disjoint from S, we get $a \in P$ and $a/1 \in S^{-1}P$. Consequently, $S^{-1}P$ is minimal among the prime ideals of $S^{-1}A$ containing $\mathrm{Ann}_{S^{-1}A}(m/1)$, hence, $S^{-1}P \in \mathrm{Ass}_{S^{-1}A}(S^{-1}M)$.

In the other direction, assume that $S^{-1}P$ is associated to $S^{-1}M$. Let $m \in M$, $s \in S$ be such that $S^{-1}P$ is minimal among the prime ideals of $S^{-1}A$ containing $\mathrm{Ann}_{S^{-1}A}(m/s)$. Let $a \in \mathrm{Ann}_A(m)$; then $am/s = 0$, hence $a/1 \in \mathrm{Ann}_{S^{-1}A}(m/s)$ and in particular, $a/1 \in S^{-1}P$. Therefore, there exists a $t \in S$ such that $ta \in P$; since S and P are disjoint, $a \in P$. In other words, $\mathrm{Ann}_A(m) \subset P$.

Let us show that P is minimal among the prime ideals of A containing $\mathrm{Ann}_A(m)$. Let Q be a prime ideal of A such that $\mathrm{Ann}_A(m) \subset Q \subset P$. Then, $\mathrm{Ann}_{S^{-1}A}(m/1) \subset S^{-1}Q \subset S^{-1}P$. This implies that $S^{-1}Q = S^{-1}P$. Since P and Q are disjoint from S and prime, we get $Q = P$, as was to be shown.   □

Corollary (6.5.11). — *Let A be a commutative ring and let M be an A-module.*

a) *A prime ideal of A belongs to the support $\mathrm{Supp}_A(M)$ if and only if it contains an ideal of $\mathrm{Ass}_A(M)$.*

b) *$\mathrm{Ass}_A(M) \subset \mathrm{Supp}_A(M)$, and both sets have the same minimal elements.*

Proof. — a) Let $m \in M$. Let P be a prime ideal of A. By definition of the support of a module, $P \in \mathrm{Supp}_A(M)$ if and only if $M_P \neq 0$. By theorem 6.5.8, this is itself equivalent to $\mathrm{Ass}_{A_P}(M_P) \neq \emptyset$ Finally, proposition 6.5.10 implies that this holds if and only if there exists an ideal $Q \in \mathrm{Ass}_A(M)$ such that $Q \subset P$.

b) This is essentially a reformulation. Let $P \in \mathrm{Ass}_A(M)$; by a), since $P \supset P$, one has $P \in \mathrm{Supp}_A(M)$.

Let us assume that P is minimal in $\mathrm{Ass}_A(M)$, and let us prove that P is also minimal in $\mathrm{Supp}_A(M)$. Let $Q \in \mathrm{Supp}_A(M)$ be such that $Q \subset P$. By a), there exists an associated ideal $P' \in \mathrm{Ass}_A(M)$ such that $Q \supset P'$; then $P' \subset P$, hence $P = P'$ by minimality, hence $P = Q$.

In the other direction, let $P \in \mathrm{Supp}_A(M)$ be a minimal element of the support of M and let us prove that P belongs to $\mathrm{Ass}_A(M)$. Let $m \in M$ be such that $P \supset \mathrm{Ann}_A(m)$. If Q is a prime ideal of A such that $\mathrm{Ann}_A(m) \subset Q \subset P$, then $Q \in \mathrm{Supp}_A(M)$, hence $P = Q$ by minimality of P. In other words, P is minimal among the prime ideals of A that contain $\mathrm{Ann}_A(m)$. This proves that $P \in \mathrm{Ass}_A(M)$.   □

Proposition (6.5.12). — *Let A be a commutative ring, let M be an A-module and let P be a prime ideal of A which is finitely generated. Then P is associated to M if and only if there exists an element $m \in M$ such that $P = \mathrm{Ann}_A(m)$.*

Proof. — If $P = \mathrm{Ann}_A(m)$, then P is obviously a minimal prime containing $\mathrm{Ann}_A(m)$, hence $P \in \mathrm{Ass}_A(M)$.

To prove the converse assertion, we first treat the case where A is a local ring and P is its maximal ideal. Let $m \in M$ be such that P is the only minimal prime ideal containing $\mathrm{Ann}_A(m)$. By proposition 2.2.11 applied to $A/\mathrm{Ann}_A(m)$, every element of P is nilpotent modulo $\mathrm{Ann}_A(m)$: for every $a \in P$, there exists an integer $n \geq 1$ such that $a^n \in \mathrm{Ann}_A(m)$. Since P is finitely generated, there exists an integer $n \geq 1$ such that $P^n \subset \mathrm{Ann}_A(m)$. Let us choose a minimal such integer n. Then $P^{n-1} \cdot m \neq 0$, hence we may choose $a \in P^{n-1}$ such that $am \neq 0$; then $P \subset \mathrm{Ann}_A(am) \subset P$, hence $P = \mathrm{Ann}_A(am)$.

Let us now treat the general case. By proposition 6.5.10, one has $PA_P \in \mathrm{Ass}_{A_P}(M_P)$. Moreover, the ideal PA_P is finitely generated. By the case of a module over a local ring, there exists an element of M_P, say m/a, with $m \in M$ and $a \in A - P$, with annihilator PA_P. Then $\mathrm{Ann}_{A_P}(m/1) = PA_P$.

Let us now assume that $P = (a_1, \ldots, a_n)$ and let us construct a sequence $(m_0, \ldots, m_n)$ of multiples of m such that for all $k \in \{0, \ldots, n\}$, $PA_P = \mathrm{Ann}_{A_P}(m_k/1)$ and $a_1 m_k = \cdots = a_k m_k = 0$. Let us set $m_0 = m$. Assume $m_0, \ldots, m_{k-1}$ are defined. Since $a_k(m_{k-1}/1) = 0$, there exists a $b \in A - P$ such that $ba_k m_{k-1} = 0$; set $m_k = bm_{k-1}$. Since $b \notin P$, one has $\mathrm{Ann}_{A_P}(m_k/1) = \mathrm{Ann}_{A_P}(m_{k-1}/1) = PA_P$. This constructs the desired sequence by induction. Finally, m_n is an element of M such that $PA_P = \mathrm{Ann}_{A_P}(m_n/1)$ and $P \subset \mathrm{Ann}_A(m_n)$. It follows that $P = \mathrm{Ann}_A(m_n)$: if $am_n = 0$, then $a(m_n/1) = 0$, hence $a \in PA_P$, hence $a \in P$.   □

Corollary (6.5.13). — *Let A be a noetherian ring and let M be an A-module. A prime ideal P of A is associated to M if and only if there exists an element $m \in M$ such that $P = \mathrm{Ann}_A(m)$.*

Proof. — Let M be a noetherian A-module and let P be a prime ideal which is associated with M. Let $m \in M$ be such that P is a minimal prime ideal among those containing $I = \mathrm{Ann}_A(m)$. The morphism $a \mapsto am$ from A to M induces an isomorphism from A/I to the submodule Am of M. Since M is a noetherian A-module, so is its submodule Am. Since the A-submodules of A/I coincide with the ideals of A/I, the ring A/I is noetherian (see also remark 6.3.9). Moreover, P/I is a minimal prime ideal among those containing $(0) = \mathrm{Ann}_{A/I}(1)$, so that $P/I \in \mathrm{Ass}_{A/I}(A/I)$ and there exists an $x \in A/I$ such that $P/I = \mathrm{Ann}_{A/I}(x)$ (corollary 6.5.13). Let $a \in A$ be such that $x = \mathrm{cl}_I(a)$. For every $b \in P$, one has $ba \in I$, hence $bam = 0$; conversely, if $b \in A$ satisfies $bam = 0$, then $bx = 0$, hence $b \in P$. This proves that $P = \mathrm{Ann}_A(am)$.   □

Remark (6.5.14). — There is a conflicting terminology in the literature between two notions of associated primes. Books where the emphasis lies on noetherian rings, such as MATSUMURA (1986), usually prefer the one given by corollary 6.5.13, and BOURBAKI (1989b) call *weakly associated primes* those of definition 6.5.6. In more advanced works on commutative algebra, the latter definition is preferred; I also believe that the notion and the exposition of the main results are more natural. However, the following results require the hypothesis that M is noetherian.

Theorem (6.5.15). — *Let A be a commutative ring and let M be a noetherian A-module. There exist a finite family $(M_0, \ldots, M_n)$ of submodules of M and a finite family $(P_1, \ldots, P_n)$ of prime ideals of A such that*

$$0 = M_0 \subset M_1 \subset \cdots \subset M_n = M$$

and $M_i/M_{i-1} \simeq A/P_i$ for every $i \in \{1, \ldots, n\}$.

Proof. — Let us begin with a remark. Let N be a submodule of M such that $N \neq M$. Applying corollary 6.5.13 to the noetherian A-module M/N, there exists an $m \in M$ and a prime ideal P of A such that $(Am+N)/N$ is isomorphic to A/P.

We set $M_0 = 0$ and apply this remark inductively, constructing an increasing sequence $(M_1, M_2, \ldots)$ of submodules of M, and a sequence $(P_1, P_2, \ldots)$ of prime ideals of A such that M_i/M_{i-1} is isomorphic to A/P_i for every i. Since M is noetherian, this strictly increasing sequence must be finite and the process must stop at some level n. Then, $M_n = M$, hence the theorem. □

Corollary (6.5.16). — *Let A be a commutative ring and let M be a noetherian A-module. Then $\mathrm{Ass}_A(M)$ is a finite set.*

Proof. — Let us consider a composition series

$$0 = M_0 \subset M_1 \subset \cdots \subset M_n = M,$$

and a sequence $(P_1, \ldots, P_n)$ of prime ideals of A, as given by theorem 6.5.15, so that $A/P_i \simeq \mathrm{Ann}_A(M_i/M_{i-1})$ for $1 \leq i \leq n$.

By proposition 6.5.9, $\mathrm{Ass}(M) \subset \mathrm{Ass}(M_{n-1}) \cup \mathrm{Ass}(A/P_n)$. We had also explained in example 6.5.7 that $\mathrm{Ass}(A/P_n) = \{P_n\}$. Consequently, $\mathrm{Ass}(M) \subset \mathrm{Ass}(M_{n-1}) \cup \{P_n\}$. By induction on n, we obtain

$$\mathrm{Ass}_A(M) \subset \{P_1; \ldots; P_n\}.$$

In particular, $\mathrm{Ass}_A(M)$ is a finite set.   □

Corollary (6.5.17). — *Let A be a commutative ring and let M be a noetherian A-module. Then, M has finite length if and only if all of its associated ideals are maximal ideals.*

Proof. — Let us assume that M has finite length. Let then

$$0 = M_0 \subset M_1 \subset \cdots \subset M_n = M$$

be a Jordan–Hölder composition series, so that for every i, there is a maximal ideal P_i of A such that $M_i/M_{i-1} \simeq A/P_i$. By the same proof as for the preceding corollary, $\text{Ass}_A(M)$ is contained in $\{P_1; \ldots; P_n\}$. In particular, all associated prime ideals of M are maximal ideals.

Conversely, let us assume that all associated prime ideals of M are maximal and let us consider a composition series

$$0 = M_0 \subset M_1 \subset \cdots \subset M_n = M,$$

and prime ideals $P_1, \ldots, P_n$, as given by theorem 6.5.15, so that for every $i \in \{1, \ldots, n\}$, one has $M_i/M_{i-1} \simeq A/P_i$. By proposition 6.5.3,

$$\text{Supp}_A(M) = \{P_1; \ldots; P_n\}.$$

By corollary 6.5.11, the minimal elements of $\text{Supp}_A(M)$ are associated prime ideals, hence are maximal ideals of A, by assumption. This implies that $P_1, \ldots, P_n$ are maximal ideals of A. In particular, M_i/M_{i-1} is a simple A-module, for every i, and M has finite length.   □

6.6. Primary Decomposition

Definition (6.6.1). — Let A be a commutative ring.

One says that an A-module M is *coprimary* if $\text{Ass}_A(M)$ has exactly one element. If $\text{Ass}_A(M) = \{P\}$, one also says that M is P-coprimary.

One says that a submodule N of an A-module is *primary* if M/N is coprimary. If M/N is P-coprimary, one says that N is P-primary.

Lemma (6.6.2). — *Let A be a commutative ring and let M be a non-zero A-module.*

a) *The module M is coprimary if and only if for every $a \in A$, either the multiplication by a is injective in M, or the fraction module M_a is 0.*

b) *Assume that M is coprimary. If M is finitely generated, then $\sqrt{\text{Ann}_A(M)}$ is the unique associated prime ideal of M.*

Proof. — a) By theorem 6.5.8, the second condition is equivalent to saying that the union of the associated prime ideals of M (the set of elements a such that multiplication by a is not injective in M) coincides with the intersection of the associated prime ideals of M (the set of elements a such that $M_a = 0$). Since $M \neq 0$, it possesses associated ideals. Consequently, the union of the associated prime ideals contains the intersection of the prime ideals. If there are at least two distinct associated prime ideals, say P and Q, then $P \cup Q$ strictly contains $P \cap Q$, so that that containment is strict unless there is exactly one associated prime ideal, that is to say, unless M is coprimary.

b) Let P be the unique associated prime ideal of M. Let $a \in A$. By a), either $a \notin P$ and the multiplication by a is injective on M, or $a \in P$ and one has $M_a = 0$. Since M is finitely generated, the condition $M_a = 0$ is equivalent to the existence of an integer n such that $a^n M = 0$, that is, $a \in \sqrt{\text{Ann}_A(M)}$. In the first case, the multiplication by every power a^n of a is still injective on M, so that $a \notin \sqrt{\mathrm{Ann}_A(M)}$, since $M \neq 0$. This proves that $\sqrt{\mathrm{Ann}_A(M)} = P$. $\quad\square$

Examples (6.6.3). — Let A be a commutative ring.

a) Let I be an ideal of A. The ideal I is a primary submodule of A if and only if for any $a, b \in A$ such that $ab \in I$ and $b \notin I$, there exists an integer $n \geq 1$ such that $a^n \in I$. One then simply says that I is a *primary ideal*. If this holds, then $\sqrt{I}$ is a prime ideal of A.

b) If I is a prime ideal of A, then I is primary. This follows either from the characterization in a), or from the computation of $\mathrm{Ass}_A(A/I)$ done in example 6.5.7.

c) Assume that A is a principal ideal domain. Let us show that primary ideals of A other than (0) are the ideals of the form (p^n), for some irreducible element p and some integer $n \geq 1$. Let I be a primary ideal distinct from (0), let $p \in A$ be such that $\sqrt{I} = (p)$. Necessarily, p is the only prime divisor of I, so that there exists an integer $n \geq 1$ such that $I = (p^n)$. Conversely, let us prove that such ideals (p^n) are (p)-primary. Let $a, b \in A$ be such that $ab \in I$ but $b \notin I$. Then, p^n divides ab but p^n does not divide b; necessarily, p divides a, so that $a^n \in I$.

In particular, any ideal of A, say $(p_1^{n_1} \ldots p_r^{n_r})$ can be written as an intersection of the primary ideals $(p_i^{n_i})$. *The primary decomposition, due to Lasker and Noether, generalizes this fact to all noetherian rings.*

d) *Let I be an ideal of A such that $P = \sqrt{I}$ is a maximal ideal of A. Then I is P-primary.* (This generalizes the previous example.) Let $a, b \in A$ be such that $ab \in I$ but $a \notin P$. Since P is maximal, the image of a in A/P is invertible and there exists a $c \in A$ such that $x = 1 - ac \in P$. Since $P = \sqrt{I}$, there exists an integer $n \geq 1$ such that $x^n = (1 - ac)^n \in I$. Expanding the power, there is a $y \in A$ such that $x^n = 1 - yac$. Then, $b = (x^n + yac)b = x^n b + yabc \in I$.

It is however simpler to apply the general theory. (Is it really?) Observe that a prime ideal of A contains I if and only if it contains $\sqrt{I}$. Consequently, the support of A/I is equal to $\{P\}$. Since $\mathrm{Ass}_A(A/I)$ and $\mathrm{Supp}_A(A/I)$ have the same minimal elements (corollary 6.5.11), one has $\mathrm{Ass}_A(A/I) = \{P\}$ and I is a P-primary ideal.

Lemma (6.6.4). — *Let A be a ring, let M be an A-module and let N, N′ be two submodules of M.*

a) *One has $\mathrm{Ass}_A(M/(N \cap N')) \subset \mathrm{Ass}_A(M/N) \cup \mathrm{Ass}_A(M/N')$.*

b) *Let P be a prime ideal of A such that N, N′ are P-primary; then $N \cap N'$ is P-primary.*

c) *Let P be a prime ideal of A such that N is P-primary. If $N \cap N' \neq N'$, then $P \in \mathrm{Ass}_A(M/(N \cap N'))$.*

Proof. — a) Let us set $V = M/(N \cap N')$ and let V′ be the submodule $N'/(N \cap N')$ of V, so that V/V' is isomorphic to M/N'. In particular,

$Ass_A(V/V') = Ass_A(M/N')$. Then, $N' \cap N$ is the kernel of the composition of the inclusion map $N' \to M$ with the canonical surjection $M \to M/N$, so that $V' = N'/(N \cap N')$ is isomorphic to a submodule of M/N. By proposition 6.5.9, $Ass_A(V') \subset Ass_A(M/N)$ and $Ass_A(V) \subset Ass_A(V') \cup Ass_A(V/V')$. In particular, $Ass_A(V) \subset Ass_A(M/N) \cup Ass_A(M/N')$, as was to be shown.

b) Let P be a prime ideal of A and assume, moreover, that both N and N' are P-primary submodules of M. By *a*), we have $Ass_A(V) \subset \{P\}$. Since $V \neq 0$, $Ass_A(V) \neq \emptyset$; consequently, $Ass_A(V) = \{P\}$, which means that $N \cap N'$ is a P-primary submodule of M.

c) By assumption, $Ass_A(M/N) = \{P\}$. Moreover, $N' \cap N \neq N'$, so that $V' \neq 0$. Being a non-empty subset of $Ass_A(M/N) = \{P\}$, one has $Ass_A(V') = \{P\}$. Since $Ass_A(V)$ contains $Ass_A(V')$, the lemma is proved. □

Theorem (6.6.5) (Lasker–Noether). — *Let A be a commutative ring and let M be a noetherian A-module. Any submodule of M is the intersection of a finite family of primary submodules of M.*

Proof. — Assume that the proposition does not hold. Since the module M is noetherian, there is a largest submodule N of M which cannot be written as the intersection of finitely many primary submodules of M. Observe that $N \neq M$ (for M is the intersection of the empty family) and N is not a primary submodule of M (for one could write $N = N$). By lemma 6.6.2, there exists an $a \in A$ such that $N_a \neq M_a$ and $m \in M - N$ such that $am \in N$.

For every integer k, let P_k be the submodule of M consisting of those $p \in M$ such that $a^k p \in N$. By assumption, $P_k \neq M$: otherwise, we would have $a^k M \subset N$, hence $M_a = N_a$.

The family (P_k) is an increasing family of submodules of M, hence it is stationary, because M is noetherian. Let $k \in \mathbf{N}$ such that $P_k = P_{k+1}$. It is clear that $N \subset (N + Am) \cap (N + a^k M)$. Conversely, let $n \in (N + Am) \cap (N + a^k M)$; let us write $n = n' + bm = n'' + a^k p$, where $n', n'' \in N$, $b \in A$, and $p \in M$. Then, $an = an' + abm \in N$, hence $a^{k+1}p = a(n - n'') \in N$. Consequently, $p \in P_{k+1}$; by the choice of k, it follows that $p \in P_k$, so that $a^k p \in N$ and $n \in N$. This shows that $N = (N + Am) \cap (N + a^k M)$.

On the other hand, $m \notin N$, so that $N \subsetneq N + Am$; moreover, $N_a \neq M_a$, so that $a^k M \neq N$, and $N \subsetneq N + a^k M$. By the choice of N, the submodules $N + Am$ and $N + a^k M$ can be written as the intersections of finitely many primary submodules of M. Then so can N, a contradiction. □

Definition (6.6.6). — Let A be a ring and let M be an A-module and let N be a submodule of M. A *primary decomposition* of N in M is a finite sequence $(N_1, \ldots, N_n)$ of primary submodules of M such that $N = N_1 \cap \cdots \cap N_n$.

A primary decomposition as above is said to be *minimal* if the following two properties also hold:

(i) For every $i \neq j$, $Ass_A(M/N_i) \neq Ass_A(M/N_j)$;
(ii) For every j, $N \neq \bigcap_{i \neq j} N_i$.

On Emmy Noether

Emmy NOETHER (1882–1935) was a German mathematician. The only female member of the historical notices of this book, she worked at a time when women were excluded from academic positions. She had to work without pay, or under the name of HILBERT. In 1933, the Nazis revoked her position at Göttingen and she left to Bryn Mawr College in the USA where she died soon after.

Her first works belonged to invariant theory, a generalization of GORDAN's methods to forms in n variables, but she adopted HILBERT's point of view, giving in 1915 a short proof of the finiteness of invariants for a finite group (theorem 6.3.15).

At that time, she also proved the fundamental theorem in mathematical physics that conservation laws are associated with "continuous" symmetries of a physical system; for example, the time-invariance of the Lagrangian corresponds with the conservation of energy.

We also owe her an important result in inverse Galois theory. The problem is to decide whether a given finite group G is the Galois group of a finite extension of the rationals, and her paper from 1917 furnishes one of the rare methods at our disposal, relating that question to another question ("Noether's problem") of deciding whether a finitely generated field extension of the rationals is isomorphic to a field of rational functions.

While STEINITZ had elaborated the general theory of fields in 1910, the general theory of rings emerged more slowly and most theorems were only proved in the context of polynomials; this was for example the case of

Portrait of Emmy Noether (before 1910)

Unknown photographer
Source: Wikipedia
Public domain.

quadratischen Ungleichungen von Newton. — E. Noether: Ableitung der Elementarteilertheorie aus der Gruppentheorie. Die Elementarteilertheorie gibt bekanntlich für Moduln aus ganzzahligen Linearformen eine Normalbasis von der Form $(c_1 y_1, c_2 y_2, \ldots, c_r y_r)$, wo jedes e durch das folgende teilbar ist; die e sind dadurch bis aufs Vorzeichen eindeutig festgelegt. Da jede Abelsche Gruppe mit endlich vielen Erzeugenden dem Restklassensystem nach einem solchen Modul isomorph ist, ist dadurch der Zerlegungssatz dieser Gruppen als direkte Summe größter zyklischer mitbewiesen. Es wird nun umgekehrt der Zerlegungssatz rein gruppentheoretisch direkt gewonnen, in Verallgemeinerung des für endliche Gruppen üblichen Beweises, und daraus durch Übergang vom Restklassensystem zum Modul selbst die Elementarteilertheorie abgeleitet. Der Gruppensatz erweist sich so als der einfachere Satz; in den Anwendungen des Gruppensatzes — z. B. Bettische und Torsionszahlen in der Topologie — ist somit ein Zurückgehen auf die Elementarteilertheorie nicht erforderlich. — *3. Februar 1925.* E. Landau: Konforme Abbildung. — *10. Fe-*

Facsimile of a short note by Emmy Noether: Derivation of the theory of elementary divisors from the theory of groups (27 January 1925). In the last sentence, she suggests the now classic group theoretic interpretation of the Betti numbers of torsion numbers from topology.
Jahresbericht Deutschen Math. Verein. (2. Abteilung) 34 (1926), p. 104.
Source: Göttinger Digitalisierungszentrum

LASKER's results on primary decomposition. In 1921, NOETHER showed that these theorems went on to hold provided that one assumes that HILBERT's finite basis theorem holds, that is, that the ring is "noetherian" (theorem 6.6.5). In 1927, she also proved the classic characterization of Dedekind rings (theorem 9.6.9). With those papers, the general theory of "abstract" commutative rings was finally on foot; modern algebra was born.

She then studied generalizations of HAMILTON's quaternions, which were known as "hypercomplex numbers", unifying and generalizing existing results, notably due to WEDDERBURN, as well as identifying their relations with the representation theory of finite groups, proving the fundamental theorems of the theory in the form in which they are still expressed today.

In his obituary address, VAN DER WAERDEN formulated the following maxim which, he says, guided NOETHER in all of her works:

> *All relations between number, functions and operations can only become clear, generalizable, and truly fruitful, when they are are separated from their particular objects and reduced to the terms of a general conceptual context.*

This maxim should allow the reader to appreciate NOETHER's fundamental, but extremely brief, contribution to algebraic topology, which is reproduced on the right. Indeed, the "Betti numbers" studied by topologists were defined as the invariant factors of some matrix, but she suggested that what should be studied are the appropriate homology modules, and that was it!

Remark (6.6.7). — Let us restate the definition in the important case of ideals. *Let* A *be a ring, let* I *be an ideal of* A. A primary decomposition of I *in* A *is an expression of the form* $I = Q_1 \cap \cdots \cap Q_n$, *where* $Q_1, \ldots, Q_n$ *are primary ideals of* A. *Such a decomposition is said to be* minimal *if, moreover,*

   (i) *For every* $i \neq j$, $\sqrt{Q_i} \neq \sqrt{Q_j}$;
   (ii) *For every* j, $I \neq \bigcap_{i \neq j} Q_i$.

Corollary (6.6.8). — *Let* A *be a ring and let* M *be a noetherian* A-*module. Every submodule of* M *possesses a minimal primary decomposition.*

Proof. — Let N be a submodule of M. By theorem 6.6.5, we know that N admits a primary decomposition: there is an integer n and primary submodules $N_1, \ldots, N_n$ of M such that $N = N_1 \cap \cdots \cap N_n$. Let us choose such a decomposition for which n is minimal and let us prove that we have indeed a minimal primary decomposition.

Let P be a prime ideal of A which is associated to at least two modules M/N_j. By lemma 6.6.4, the intersection of all modules N_i such that $\mathrm{Ass}_A(M/N_i) = \{P\}$ is a P-primary submodule of M. This gives rise to a new primary decomposition of N, with strictly less factors. By the minimality assumption on n, this implies that the associated prime ideals of the N_i are pairwise distinct.

Now, if we can remove one of the modules from the list without changing the intersection, we get a new primary decomposition with one less factor. By the minimality assumption on n, this cannot hold.   □

Theorem (6.6.9). — *Let* A *be a ring, let* M *be an* A-*module and let* N *be a submodule of* M. *Let* $(N_1, \ldots, N_n)$ *be a minimal primary decomposition of* N; *for every* i, *let* P_i *be the unique prime ideal associated to* M/N_i.

   a) $\mathrm{Ass}_A(M/N) = \{P_1, \ldots, P_n\}$.
   b) *If* P_i *is minimal among the associated primes of* M/N, *then*

$$N_i = M \cap N_{P_i} = \{m \in M \, ; \, \exists a \notin P_i, \, am \in N\}.$$

Proof. — *a)* By induction on n, it follows from lemma 6.6.4 that $\mathrm{Ass}_A(M/N) \subset \{P_1, \ldots, P_n\}$. This inclusion does not depend on the fact that the given primary decomposition is minimal. Let us now show that for every i, P_i is an associated prime ideal to M/N. Let $N' = \bigcap_{j \neq i} N_j$, so that $N = N_i \cap N'$. Since the given decomposition is minimal, $N \neq N'$. Since M/N_i is P_i-primary, it then follows from lemma 6.6.4 that $\mathrm{Ass}_A(M/N)$ contains P_i, as required.

b) Let $j \in \{1, \ldots, n\}$ be such that $j \neq i$. By proposition 6.5.10, the associated prime ideals of $(M/N_j)_{P_i}$ are the prime ideals of the form PA_{P_i} where $P \in \mathrm{Ass}(M/N_j) = \{P_j\}$ is a prime ideal of A contained in P_i. By minimality of P_i, P_j is not contained in P_i, so that $\mathrm{Ass}_{A_{P_i}}((M/N_j)_{P_i}) = \emptyset$. By theorem 6.5.8, $(M/N_j)_{P_i} = 0$ and $(N_j)_{P_i} = M_{P_i}$. Consequently,

$$N_{P_i} = (N_1)_{P_i} \cap \cdots \cap (N_n)_{P_i} = (N_i)_{P_i}.$$

Let now $N'_i = M \cap (N_i)_{P_i}$. It is obvious that $N_i \subset N'_i$. Conversely, let $m \in N'_i$, so that there exists an $a \notin P_i$ such that $am \in N_i$. Since $\mathrm{Ass}_A(M/N_i) = \{P_i\}$, multiplication by a is injective on M/N_i, hence $m \in N_i$. □

Theorem (6.6.10) (Krull's intersection theorem). — *Let* A *be a commutative ring, let* M *be a noetherian* A*-module and let* I *be a finitely generated ideal of* A*. Let* $N = \bigcap_{n \geq 0} I^n M$. *Then* $N = IN$ *and there exists an* $a \in I$ *such that* $(1 + a)N = 0$.

Proof. — We first prove that $N = IN$. With that aim, let us consider a primary decomposition of the submodule IN of M, namely, $IN = N_1 \cap \cdots \cap N_m$; for $j \in \{1, \ldots, m\}$, let P_j be the unique prime ideal of A which is associated to M/N_j. Let us assume, by contradiction, that $N \neq IN$ and let $x \in N - IN$. Let $j \in \{1, \ldots, m\}$ be such that $x \notin N_j$.

Let $a \in I$; one has $ax \in IN \subset N_j$, so that the multiplication by a on M/N_j is not injective. By the definition of a primary submodule, one has $(M/N_j)_a = 0$. Since M is finitely generated, there exists an integer $n \geq 1$ such that $a^n M \subset N_j$.

Since the ideal I is finitely generated, there exists an integer $k \geq 1$ such that $I^k M \subset N_j$. Now, using that $x \in N = \bigcap_{n \geq 0} I^n M$, we see that $x \in I^k M \subset N_j$, a contradiction which establishes that $N = IN$.

Since M is noetherian, its submodule N is a finitely generated A-module. By Nakayama's lemma (corollary 6.1.4), there exists an element $a \in I$ such that $(1 + a)N = 0$. □

Corollary (6.6.11). — *Let* A *be a noetherian commutative ring and let* I *be an ideal of* A*. Under each of the following hypotheses, one has* $\bigcap_{n \geq 0} I^n = 0$:

(i) *Every element of* $1 + I$ *is regular;*

(ii) *The ideal* I *is contained in the Jacobson radical of* A;

(iii) *One has* $I \neq A$ *and the ring* A *is an integral domain.*

Proof. — Let $J = \bigcap_{n \geq 0} I^n$. By Krull's intersection theorem 6.6.10, there exists an $a \in I$ such that $(1 + a)J = 0$. Under each of the hypotheses (i), (ii), (iii), the element $1 + a$ of A is regular: this is obvious in case (i); in case (ii), $1 + a$ is invertible (lemma 2.1.6); in case (iii), $1 + a \neq 0$. Consequently, $J = 0$. □

6.7. Exercises

Exercise (6.7.1). — Show that in Nakayama's lemma, one cannot omit the hypothesis that the module is finitely generated. (*Consider the* **Z**-*module* **Q**.)

Exercise (6.7.2). — Let A be a ring and let I be a finitely generated two-sided ideal of A such that $I = I^2$. Show that there exists a central element $e \in A$ such that $e = e^2$ and $I = eA = Ae$. (*Apply Nakayama's lemma to find an element* $a \in I$ *such that* $I(1 + a) = 0$.)

Exercise (6.7.3). — Let A be a commutative ring and let V be a finitely generated A-module. For any maximal ideal M of A, define

$$d_V(P) = \dim_{A_P/PA_P}(V_P/PV_P).$$

a) Let M be a maximal ideal of A. Prove that $d_V(M) = \dim_{A/M}(V/MV)$.

b) Let P be a prime ideal of A. Show that the A_P-module V_P is generated by $d_V(P)$ elements.

c) Show that there exists an $a \in A - P$ such that the A_a-module V_a is generated by $d_V(P)$ elements.

d) Let Q be a prime ideal of A such that $a \notin Q$. Show that $d_V(Q) \leq d_V(P)$. (In other words, the function $P \mapsto d_V(P)$ from Spec(A) to **N** is upper-semicontinuous, when Spec(A) is endowed with the Zariski topology.)

Exercise (6.7.4). — Let A be a ring, let M, N be A-modules and let $f : M \to N$ be a morphism. Assume that M has finite length.

a) Show that Ker(f) and Im(f) have finite lengths and that $\ell_A(\mathrm{Im}(f)) + \ell_A(\mathrm{Ker}(f)) = \ell_A(M)$.

b) Assume that $\ell_A(N) > \ell_A(M)$; prove that f is not injective.

c) Assume that $\ell_A(N) > \ell_A(M)$; prove that f is not surjective.

d) Assume that N = M (so that f is an endomorphism). Show that the following conditions are equivalent: (i) f is bijective; (ii) f is injective; (iii) f is surjective.

Exercise (6.7.5). — Let A be a ring, let M be an A-module and let $(S_i)_{i \in I}$ be a family of simple submodules of M such that $M = \sum_{i \in I} S_i$. (One says that M is a *semisimple* A-module.)

a) Let N be a submodule of M. Using Zorn's lemma show that there exists a maximal subset J of I such that N and the submodules S_j, for $j \in J$, are in direct sum in their sum. (You may restrict to the case where I is finite, if you wish to avoid Zorn's lemma.) Show that $M = N \oplus \left(\bigoplus_{j \in J} S_j \right)$.

b) In particular, any submodule of M has a direct summmand in M.

c) Show that there exists a subset $J \subset I$ such that M is isomorphic to the module $\bigoplus_{j \in J} S_j$.

Exercise (6.7.6). — Let K be a field.

a) For any non-zero polynomial $P \in K[X]$, show that $K[X]/(P)$ is a $K[X]$-module of finite length equal to the number of irreducible factors of P (repeated according to their multiplicities). What happens if K is algebraically closed?

b) Let $A = K[X_1, \ldots, X_n]$. Show that an A-module M has finite length if and only if $\dim_K(M) < \infty$. (*Use Hilbert's Nullstellensatz.*)

c) If, moreover, K is algebraically closed, prove that $\ell_A(M) = \dim_K(M)$.

Exercise (6.7.7). — Let A be a ring and let M be a non-zero right A-module which is finitely generated.

a) Show that the set of all submodules of M which are distinct from M is inductive.

b) Show that for any submodule N of M such that N ≠ M, there exists a submodule P of M such that N ⊂ P ⊂ M and M/P is a simple A-module.

c) Assume that A possesses a unique maximal right ideal I. Show that $\operatorname{Hom}_A(M, A/I) \neq 0$.

d) Set $A = \mathbf{Z}$ and $M = \mathbf{Q}$. Show that there does not exist any submodule P in M such that M/P is simple.

Exercise (6.7.8). — Let A be a ring and let J be its Jacobson radical. Let M be a finitely generated right A-module such that M ≠ 0. Let N be a maximal submodule of M (exercise 6.7.7).

Show that MJ ⊂ N; conclude that M ≠ MJ. (This gives another proof of Nakayama's lemma.)

Exercise (6.7.9). — Let A be a principal ideal domain.

a) Let a be any non-zero element of A. Prove that the A-module $A/(a)$ has finite length and compute its length in terms of a decomposition of a as a product of irreducible elements.

b) Using the Jordan–Hölder theorem, give a second proof of the uniqueness property of decomposition of a as a product of irreducible elements.

Exercise (6.7.10). — Let A be a ring. One says that an A-module M is *indecomposable* if M ≠ 0 and if there does not exist non-zero submodules N, N′ of M such that M = N ⊕ N′.

a) Let M be an A-module such that $\operatorname{End}_A(M)$ is a local ring. Prove that M is indecomposable.

b) Let M be an indecomposable A-module of finite length. Prove that $\operatorname{End}_A(M)$ is a local ring.

c) Let M be an A-module of finite length. Prove that there exists a finite family $(N_1, \dots, N_r)$ of indecomposable submodules of M such that $M = N_1 \oplus \cdots \oplus N_r$.

d) Let M be an A-module and let N, N′ be submodules of M such that $M = N \oplus N'$. Assume that N is indecomposable. Prove that for every $u \in \operatorname{End}_A(M)$, either u or $\operatorname{id}_M - u$ induces an isomorphism from N to a direct summand of N′ in M.

e) Let M be an A-module of finite length. Let $(N_1, \dots, N_r)$ and $(N'_1, \dots, N'_s)$ be finite families of indecomposable submodules of M such that $M = N_1 \oplus \cdots \oplus N_r = N'_1 \oplus \cdots \oplus N'_s$. Prove that $r = s$ and that there exists a bijection $\sigma \in \mathfrak{S}_r$ such that $N'_j \simeq N_{\sigma(j)}$ for every $j \in \{1, \dots, r\}$. (*Krull–Remak–Schmidt theorem.*)

Exercise (6.7.11). — Let A be a local noetherian commutative ring and let M be its maximal ideal. Let I be an ideal of A. Show that A/I has finite length if and only if there exists an integer $n \geq 0$ such that $M^n \subset I$.

Exercise (6.7.12). — Let A be a ring, let (I_n) be an increasing family of left ideals of A. Let $I = \bigcup I_n$. Show that I is a left ideal of A. If I is finitely generated, show that the sequence (I_n) is stationary.

Exercise (6.7.13). — Let A be a ring, let I, J be two-sided ideals of A such that $I \cap J = (0)$. Show that A is left noetherian if and only if both A/I and A/J are left noetherian.

Exercise (6.7.14) (Examples of non-noetherian rings). — Show that the following commutative rings are not noetherian.

a) The ring $k[X_1, X_2, \ldots, X_n, \ldots]$ of polynomials in infinitely many indeterminates with coefficients in a non-zero commutative ring k;

b) The ring $C^0(\mathbf{R}, \mathbf{R})$ of real-valued continuous functions on $\mathbf{R}$;

c) The ring $C^\infty(\mathbf{R}, \mathbf{R})$ of real-valued indefinitely differentiable functions on $\mathbf{R}$. Show however that the ideal of functions vanishing at 0 is a principal ideal;

d) The subring of $\mathbf{C}[X, Y]$ generated as a submodule by $\mathbf{C}$ and the ideal (X).

Exercise (6.7.15). — Let $\mathscr{F}$ be the set of all polynomials $P \in \mathbf{Q}[X]$ such that $P(n) \in \mathbf{Z}$ for every $n \in \mathbf{Z}$.

a) Show that $\mathscr{F}$ is a $\mathbf{Z}$-subalgebra of $\mathbf{Q}[X]$.

b) Let $f: \mathbf{Z} \to \mathbf{Z}$ be a function. Assume that $f(0) \in \mathbf{Z}$ and that there exists a $P \in \mathscr{F}$ such that $f(n) = P(n + 1) - P(n)$ for every $n \in \mathbf{Z}$. Show that there exists a unique element $Q \in \mathscr{F}$ such that $f(n) = Q(n)$ for every $n \in \mathbf{Z}$.

c) Show that the polynomials $\binom{X}{0} = 1$, $\binom{X}{1} = X$, $\binom{X}{2} = X(X - 1)/2$, $\ldots$, $\binom{X}{p} = X(X - 1) \ldots (X - p + 1)/p!$, $\ldots$ form a basis of $\mathscr{F}$ as a $\mathbf{Z}$-module.

d) Show that the ring $\mathscr{F}$ is not noetherian.

Exercise (6.7.16). — Let A be a ring, let M be a noetherian right A-module and let $\varphi: M \to M$ be an endomorphism of M. Show that there exists an integer $n \geq 1$ such that $\mathrm{Ker}(\varphi^n) \cap \mathrm{Im}(\varphi^n) = (0)$.

Exercise (6.7.17). — Let A be a commutative ring which is not noetherian. Let $\mathscr{F}$ be the set of all ideals of A which are not finitely generated, so that $\mathscr{F} \neq \emptyset$.

a) Prove that $\mathscr{F}$ is an inductive set with respect to the inclusion relation. Let P be a maximal element of $\mathscr{F}$. In the rest of the exercise, we shall prove that P is a (non-finitely generated) prime ideal of A. We argue by contradiction. Let $a, b \in A$ be such that $ab \in P$ while $a \notin P$ and $b \notin P$.

b) Prove that there exist $u_1, \ldots, u_m \in P$ and $v_1, \ldots, v_n \in A$ such that $P + (a) = (u_1, \ldots, u_m, a)$ and $(P : (a)) = (v_1, \ldots, v_n)$.

c) Prove that $P = (u_1, \ldots, u_m, av_1, \ldots, av_n)$. Derive from this contradiction that P is a prime ideal of A.

This proves that a commutative ring is noetherian if and only if every prime ideal is finitely generated, a theorem of I. S. Cohen.

Exercise (6.7.18). — Let A be a commutative ring. Assume that for every maximal ideal M of A, the fraction ring A_M is noetherian and that, for every non-zero $a \in A$, the set of maximal ideals of A containing a is finite.

Let I be a non-zero ideal of A. Let $M_1, \ldots, M_n$ be the maximal ideals of A that contain I.

a) Prove that there exist elements $a_1, \ldots, a_m \in I$ such that every maximal ideal of A containing $(a_1, \ldots, a_m)$ is one of the M_j.

b) For every $i \in \{1, \ldots, n\}$, prove that there exist an integer m_i and elements $b_{i,j}$ (for $j \in \{1, \ldots, m_i\}$) such that $b_{i,j}/1, \ldots, b_{i,m_i}/1$ generate the ideal IA_{M_i} of A_{M_i}.

c) Prove that I is finitely generated.

d) Prove that A is noetherian.

Exercise (6.7.19). — Let R be a commutative noetherian ring and let I be an ideal of R; let $(a_1, \ldots, a_n)$ be a generating family of I. Let $J = \bigcap_{k \geq 1} I^k$ and let $b \in J$.

a) Show that for every integer $k \geq 1$, there exists a polynomial $P_k \in R[T_1, \ldots, T_n]$, homogeneous of degree k, such that $b = P_k(a_1, \ldots, a_n)$.

b) Show that there exist an integer $m \geq 1$ and, for every $k \in \{1, \ldots, m\}$, a polynomial $Q_k \in R[T_1, \ldots, T_n]$ which is homogeneous of degree k, such that $P_{m+1} = Q_m P_1 + \cdots + Q_1 P_m$.

c) Prove that there exists an $a \in I$ such that $b = ab$. In particular, one has $J = IJ$. (This proof of Krull's intersection theorem is due to PERDRY (2004).)

Exercise (6.7.20). — Let A be a ring which satisfies the two conditions:

(i) For every increasing sequence $(J_n)_n$ of finitely generated ideals of A, there exists an $m \in \mathbf{N}$ such that $J_m = J_{m+1}$;

(ii) All finitely generated ideals of A are finitely presented.

a) Prove that for every integer n, every finite generated submodule of A^n is finitely presented. (*This only uses (ii). Adapt the proof of corollary 6.3.7.*)

b) Prove that for every increasing sequence $M_1, M_2, \ldots$ of finitely generated A-modules, there exists an m such that $M_m = M_{m+1}$.

c) Let $I = (f_1, \ldots, f_r)$ be a finitely generated ideal of A[X]; let n be an integer such that $n \geq \sup(\deg(f_1), \ldots, \deg(f_r))$. Construct a finitely generated submodule M of A[X], generated by polynomials of degree $\leq n$, such that $I = \langle M \rangle$ and such that for every $f \in M$ such that $\deg(f) < M$, then $Xf \in M$.

d) For every integer $m \in \mathbf{N}$, let J_m be the set of all coefficients of X^m of elements $f \in I$ such that $\deg(f) \leq m$. Prove that (J_m) is an increasing sequence of finitely generated ideals of A and that $J_m = J_n$ for $m \geq n$.

e) Prove that the ideal I of A[X] is finitely presented (condition (ii)).

f) Prove that the ring A[X] satisfies (i) and (ii).

These results of SEIDENBERG *(1974) are the basis of a constructivist point of view on noetherian rings.*

Exercise (6.7.21). — One says that a commutative ring R is graded if there is a family $(R_n)_{n \geq 0}$ of subgroups of $(R, +)$ such that $R = \bigoplus_{n=0}^{\infty} R_n$ and $R_n \cdot R_m \subset R_{n+m}$ for all $n, m \geq 0$.

a) Show that R_0 is a subring of R and that $I = \bigoplus_{n \geq 1} R_n$ is an ideal of R.

b) Assume that R_0 is a noetherian ring and that R is finitely generated as an R_0 algebra. Show that R is noetherian.

c) Conversely, let us assume that R is a noetherian ring. Show that R_0 is noetherian. Show that there is an integer $r \geq 0$, elements $x_1, \ldots, x_r \in R$ and integers $n_1, \ldots, n_r$ such that $x_i \in R_{n_i}$ for every i and such that $I = (x_1, \ldots, x_r)$. Show by induction on n that $R_n \subset R_0[x_1, \ldots, x_r]$. Conclude that R is a finitely generated R_0-algebra.

Exercise (6.7.22). — Let $R = \oplus R_n$ be a commutative noetherian graded ring. One says that an R-module M is graded if there is a family $(M_n)_{n \geq 0}$ of subgroups of M such that $M = \bigoplus_{n \geq 0} M_n$ and such that $R_n \cdot M_n \subset M_{n+m}$ for all $m, n \geq 0$.

Let M be a graded R-module.

a) Observe that M_n is an R_0-module. If M is finitely generated, show that for each n, M_n is a finitely generated R_0-module.

b) Assume that R_0 is an artinian ring. Let $P_M(t)$ be the power series with integer coefficients given by

$$P_M(t) = \sum_{n=0}^{\infty} \ell_{R_0}(M_n) t^n.$$

Assume that there is an integer r, elements $x_1, \ldots, x_r \in R$ and integers $n_1, \ldots, n_r$ such that $x_i \in R_{n_i}$ for every i and such that $R = R_0[x_1, \ldots, x_r]$. Prove by induction on r that there exists a polynomial $f_M \in \mathbf{Z}[t]$ such that

$$f_M(t) = P_M(t) \prod_{i=1}^{r} (1 - t^{n_i}).$$

c) Assume moreover that $n_i = 1$ for every i. Show that there is a polynomial $\varphi_M \in \mathbf{Q}[t]$ such that

$$\ell_{R_0}(M_n) = \varphi_M(n)$$

for every large enough integer n.

Exercise (6.7.23). — Let A be a noetherian commutative ring and let I be an ideal of A.

a) Let B be the set of all polynomials $P \in A[T]$ of the form $P = \sum a_n T^n$, where $a_n \in I^n$ for every n. Show that B is a noetherian ring. (*Use exercise 6.7.21.*)

b) Let M be a finitely generated A-module and let N be a submodule of M. Let $M[T] = M^{(\mathbf{N})}$ be considered as an A[T]-module via $T \cdot (m_0, m_1, \dots) = (0, m_0, m_1, \dots)$. Let P (resp. Q) be the subset of M[T] consisting of sequences (m_n) such that $m_n \in I^n M$ (resp. such that $m_n \in N \cap I^n M$) for every $n \in \mathbf{N}$. Prove that P is a B-submodule of M[T], and that Q is B-submodule of P.

c) Prove that there exists an integer $m \in \mathbf{N}$ such that $N \cap I^n M = I^{n-m}(N \cap I^m M)$ for every integer $n \geq m$. (A result known as the *Artin–Rees lemma.*)

d) Let $f \colon M \to N$ be a morphism of finitely generated A-modules. Prove that there exists an integer m such that

$$f^{-1}(I^n N) = \mathrm{Ker}(f) + I^{n-m} \cdot f^{-1}(I^m M) \quad \text{and} \quad f(M) \cap I^n M \subset f(I^{n-m} M)$$

for every integer $n \geq m$.

Exercise (6.7.24). — Let A be a noetherian commutative ring and let I be an ideal of A. Let M be a finitely generated A-module and let $N = \bigcap_{n \in \mathbf{N}} I^n M$.

a) Prove that there exists an $a \in I$ such that $(1+a)N = 0$. (*Use exercise 6.7.23 and Nakayama's lemma.*)

b) Assuming that I is contained in the Jacobson radical of A, prove that $N = 0$.

c) Let P be a prime ideal of A containing I. Prove that there exists an $a \in A - P$ such that $aN = 0$.

d) Assume that A is an integral domain, $I \neq A$, and M is torsion free. Prove that $N = 0$.

Exercise (6.7.25). — In this exercise, we describe all **Z**-submodules of **Q**.

For any non-zero $a \in \mathbf{Q}$, let $v_p(a)$ be the exponent of p in the decomposition of a as a product of prime factors; we also set $v_p(0) = +\infty$. Let

$$V = \prod_{p \text{ prime}} (\mathbf{Z} \cup \{-\infty, +\infty\})$$

and $v \colon \mathbf{Q} \to V$ be the map given by $v(a) = (v_p(a))_p$. The set V is endowed with the product ordering: for two families $a = (a_p)_p$ and $b = (b_p)_p$ in V, say that $a \leq b$ if $a_p \leq b_p$ for every prime number p.

a) Let x and y be two elements of **Q** such that $x\mathbf{Z} \subset y\mathbf{Z}$; show that $v(y) \leq v(x)$.

b) Show that any subset of V admits a least upper bound and a greatest lower bound in V.

c) Let M be a submodule of $\mathbf{Q}$. Assume that M is finitely generated. Show that there exists an $m \in \mathbf{Q}$ such that $M = m\mathbf{Z}$. Show that the element $v(m)$ of V does not depend on the choice of a generator $m \in M$; it will be denoted $v(M)$. Then, prove that
$$M = \{x \in \mathbf{Q}\,;\, v(x) \geq v(M)\}.$$

d) For an arbitrary submodule of $\mathbf{Q}$, set $v(M) = \inf_{x \in \mathbf{Q}} v(x)$. Conversely, for any $u \in V$, set $M_u = \{x \in \mathbf{Q}\,;\, v(x) \geq u\}$. Show that $M_u = 0$ if and only if $\sum_p \sup(0, u_p) = +\infty$; explicitly, $M_u = 0$ if and only if one of the two following conditions holds:

(i) There exists a prime number p such that $u_p = +\infty$;

(ii) The set of all prime numbers p such that $u_p > 0$ is infinite.

e) Show that the map $M \mapsto v(M)$ induces a bijection from the set of all non-zero submodules of $\mathbf{Q}$ to the subset V^0 of V consisting of families $u \in V$ such that $\sum_p \sup(0, u_p) < \infty$.

f) For $u \in V^0$, show that M_u is finitely generated if and only if $\sum_p |u_p| < \infty$, that is, if and only if:

(i) For every p, $u_p \neq -\infty$;

(ii) The set of all primes p such that $u_p < 0$ is finite.

If this holds, show that M_u is generated by one element.

g) For $u \in V^0$, show that M_u contains $\mathbf{Z}$ if and only if $u \leq 0$, namely $u_p \leq 0$ for every p. If this holds, show that $M_u/\mathbf{Z}$ is artinian if and only if the set of all primes p such that $u_p < 0$ is finite.

h) Let $u \in V^0$ be such that $u \leq 0$. Show that $M_u/\mathbf{Z}$ has finite length if and only if

(i) For every primer number p, $u_p \neq -\infty$;

(ii) The set of all primes p such that $u_p < 0$ is finite.

More precisely, show that $\ell_{\mathbf{Z}}(M_u/\mathbf{Z}) = \sum_p |u_p|$.

Exercise (6.7.26). — Let A be a ring, let M be an artinian A-module and let u be an endomorphism of M.

a) Assume that u is injective. Show that u is bijective.

b) Give an example where u is surjective but not bijective.

Exercise (6.7.27). — Let A be a ring, let M be an artinian A-module and let φ be an endomorphism of M.

a) Show that there exists an integer $n \geq 0$ such that $\operatorname{Im}(\varphi^p) = \operatorname{Im}(\varphi^n)$ for every integer p such that $p \geq n$.

b) Prove that $\operatorname{Ker}(\varphi^n) + \operatorname{Im}(\varphi^n) = M$.

c) Assume moreover that M has finite length. Using exercise 6.7.16, show that the preceding sum is a direct sum.

d) Conclude that if M has finite length and is indecomposable, then φ is either nilpotent or an automorphism (*Fitting's lemma*).

Exercise (6.7.28). — Let A be a commutative ring.

a) Let V be an A-module of finite length. Show that the canonical morphism $V \to \prod_{M \in \mathrm{Max}(A)} V_M$ (sending $x \in M$ to the family of fractions $x/1$, Max(A) being the set of all maximal ideals of A) is an isomorphism of A-modules.

b) Assume that A is an artinian ring. Show that the canonical morphism $A \to \prod_{M \in \mathrm{Max}(A)} A_M$ is an isomorphism of rings: *Any commutative artinian ring is a product of local rings.*

Exercise (6.7.29). — Let A be a ring and let M be a right A-module. Let $A' = \mathrm{End}_A(M)$ be the ring of endomorphisms of M. Then M is seen as a left A'-module and we set $A'' = \mathrm{End}_{A'}(M)$.

a) For $a \in A$, let $\rho_M(a)$ be the homothety of ratio a in M. Prove that $\rho_M(a) \in A''$. Prove that the map $\rho_M \colon A \to A''$ is a morphism of rings.

One says that the A-module M is *balanced* if the morphism ρ_M is an isomorphism.

b) Let $n \geq 1$ be an integer such that M^n is balanced. Prove that M is balanced. (*Represent endomorphisms of* M^n *by matrices with entries in* $\mathrm{End}_A(M)$.)

c) Let M be a right A-module. For every $e \in A \oplus M$, prove that there exists a $u \in \mathrm{End}_A(A_d \oplus M)$ such that $u(1, 0) = e$. Prove that $A_d \oplus M$ is balanced.

d) Let M be a right A-module such that A is a quotient of M^n, for some $n \geq 1$ (one says that M is *generator*). Prove that M is balanced.

e) Assume that the ring A is simple. Prove that every non-zero right ideal of A is generator, hence balanced (RIEFFEL (1965)).

Exercise (6.7.30). — Let A be a simple artinian ring. Following HENDERSON (1965) and NICHOLSON (1993), we prove in this exercise a theorem of Wedderburn according to which A is isomorphic to a matrix ring over a division ring.

a) Prove that A contains a right ideal I which is a simple right A-module.

b) Prove that $I = A \cdot I$ and deduce from this that $I^2 \neq 0$.

c) Prove that there exists an idempotent element $e \in A$ such that $I = eA$ and such that $D = eAe$ is a division ring.

d) For $a \in A$, the homothety $\rho_I(a)$ of ratio a in I is an element of $\mathrm{End}_D(I)$; prove that the map $\rho_I \colon A \to \mathrm{End}_D(I)$ is an isomorphism. (*Use that* $1 \in AeA$.)

e) Prove that I is finite-dimensional as a left D-vector space. (*Observe that the endomorphisms of* I *of finite rank form a non-trivial two-sided ideal.*) Conclude that there exists an integer $n \geq 1$ such that $A \simeq \mathrm{Mat}_n(D)$.

Exercise (6.7.31). — Let A be a commutative ring and let M be a finitely generated A-module. Let I be an ideal of A. Prove that $\mathrm{Supp}_A(M/IM) = V(I) \cap V(\mathrm{Ann}_A(M))$.

Exercise (6.7.32). — Let A be a commutative ring and let $U \in \mathrm{Mat}_{m,n}(A)$. Let $u: A^n \to A^m$ be the associated morphism of A-modules and let $M = \mathrm{Coker}(u)$.

a) Let P be a prime ideal of A. Prove that $P \in \mathrm{Supp}_A(M)$ if and only if $\det(V) \in P$ for every $m \times m$ submatrix V of A.

b) Prove that $\mathrm{Fit}_0(M)$ is finitely generated and $V(\mathrm{Fit}_0(M)) = \mathrm{Supp}_A(M)$.

c) Prove that $\mathrm{Spec}(A) - \mathrm{Supp}_A(M)$ is a quasi-compact subset of $\mathrm{Spec}(A)$.

Exercise (6.7.33). — a) Let A be a ring. Let M and N be A-modules such that $M \subset N$. Show that $\mathrm{Ass}_A(M) \subset \mathrm{Ass}_A(N)$.

b) Give examples that show that, in general, there is no relation between the associated prime ideals of a module and those of a quotient.

c) Let M be an A-module, let M_1 and M_2 be submodules of M such that $M = M_1 \oplus M_2$. Prove that $\mathrm{Ass}_A(M) = \mathrm{Ass}_A(M_1) \cup \mathrm{Ass}_A(M_2)$.

d) Assume only that $M = M_1 + M_2$. How can you relate $\mathrm{Ass}_A(M)$ to $\mathrm{Ass}_A(M_1)$ and $\mathrm{Ass}_A(M_2)$?

Exercise (6.7.34). — Let A be a commutative ring and let $x \in A$. Assume that x is regular but not a unit. Show that for any integer $n \geq 1$, A/xA and $A/x^n A$ have the same associated prime ideals.

Exercise (6.7.35). — Let A be a commutative ring and let I be an ideal of A such that $I \neq A$. Show that I is primary if and only if every element of A/I is either nilpotent or regular.

Exercise (6.7.36). — Let A be a commutative ring, let $f: M' \to M$ be a morphism of A-modules and let N be a submodule of M'. If N is primary, show that $f^{-1}(N)$ is a primary submodule of M. Give two proofs, one relying on the manipulation of associated prime ideals, the other on lemma 6.6.2.

Exercise (6.7.37). — Let A be a commutative ring.

a) Let M be a maximal ideal of A. For any integer $n \geq 1$, show that M^n is an M-primary ideal of A.

b) Let P be a prime ideal of A, let $n \geq 1$ be an integer and let $P^{(n)}$ be the inverse image in A of the ideal $P^n A_P$ of A_P. Show that $P^{(n)}$ is a P-primary ideal of A.

c) Prove that P^n is P-primary if and only if $P^n = P^{(n)}$.

d) Give an example of a prime ideal P such that P^n is not a P-primary ideal.

Exercise (6.7.38). — Let k be a field and let $A = k[X_1, X_2, \dots]$ be the ring of polynomial in infinitely many indeterminates $X_1, \dots$. Let $M = (X_1, \dots)$ be the ideal generated by the indeterminates, $I = (X_1^2, \dots)$ be the ideal generated by their squares and $J = (X_1, X_2^2, \dots)$ be the ideal generated by the elements X_n^n.

a) Show that the radicals of I and J are equal to M and that M is a maximal ideal. In particular, I and J are primary ideals.

b) Show that there is no integer n such that $M^n \subset J$.

c) Show that there is no element $a \in A$ such that $M = \{x \in A \, ; \, ax \in I\}$.

d) Prove that $\mathrm{Ass}_A(A/I) = \{M\}$ but that there does not exist any $f \in A/I$ such that $M = \mathrm{Ann}_A(f)$.

Exercise (6.7.39). — Let A be a commutative ring and let M be an A-module. Show that the canonical morphism from M to $\prod_{P \in \mathrm{Ass}_A(M)} M_P$ is injective.

Exercise (6.7.40). — Let A be a commutative ring, let M be an A-module and let S be a multiplicative subset of A. Let N be the kernel of the canonical morphism from M to $S^{-1}M$.

a) Show that a prime ideal P of A belongs to $\mathrm{Ass}_A(N)$ if and only if it belongs to $\mathrm{Ass}_A(M)$ and $P \cap S \neq \emptyset$.

b) Show that a prime ideal P of A belongs to $\mathrm{Ass}_A(M/N)$ if and only if it belongs to $\mathrm{Ass}_A(M)$ and $P \cap S = \emptyset$.

Exercise (6.7.41). — Let A be a commutative noetherian ring and let M, N be finitely generated A-modules.

a) Assume that A is local, with maximal ideal P. Prove that $M \neq 0$ if and only if there exists a non-zero morphism $f : M \to A/P$.

b) We still assume that A is local, with maximal ideal P. Prove that $P \in \mathrm{Ass}_A(\mathrm{Hom}(M, N))$ if and only if $M \neq 0$ and $P \in \mathrm{Ass}_A(N)$.

c) Prove that $\mathrm{Ass}_A(\mathrm{Hom}(M, N)) = \mathrm{Supp}(M) \cap \mathrm{Ass}_A(N)$.

Exercise (6.7.42). — Let $f : A \to B$ be a morphism of commutative rings. Let N be a B-module and let M be an A-submodule of N.

a) Let $P \in \mathrm{Ass}_A(M)$; prove that there exists a prime ideal $Q \in \mathrm{Ass}_B(N)$ such that $P = f^{-1}(Q)$.

b) Let $Q \in \mathrm{Ass}_B(N)$, let N' be a Q-primary submodule of N and let $M' = N' \cap M$. Assume that $M' \neq M$. Prove that M' is a P-primary submodule of M, where $P = f^{-1}(Q)$.

Exercise (6.7.43). — Let A be a commutative von Neumann ring.

a) Prove that every prime ideal of A is associated to A.

b) Let P be a prime ideal of A which is of the form $\mathrm{Ann}_A(a)$, for some $a \in A$. Prove that P is finitely generated. (*If E is a field, the von Neumann ring E^N shows that proposition 6.5.12 does not hold without the hypothesis that A is noetherian.*)

Exercise (6.7.44). — Let A be a commutative ring, let S be a multiplicative subset of A and let M be an A-module. Let N be the kernel of the canonical morphism $i : M \to S^{-1}M$.

a) Prove that $\mathrm{Ass}_A(M/N)$ is the set of primes $P \in \mathrm{Ass}_A(M)$ which are disjoint from S.

b) Prove that $\mathrm{Ass}_A(N)$ is the set of primes $P \in \mathrm{Ass}_A(M)$ which meet S.

Exercise (6.7.45). — Let A be a commutative ring.

a) Let M be an A-module, let $P \in \mathrm{Ass}_A(M)$ be an associated ideal of M and let $m \in M$ be such that P is minimal among all prime ideals containing $\mathrm{Ann}_A(m)$. Prove that for every $a \in P$, there exist $b \in A - P$ and $n \in \mathbf{N}$ such that $a^n bm = 0$. (*First treat the case where A is local with maximal ideal P.*)

b) If A is reduced, then $\mathrm{Ass}_A(A)$ is the set of minimal prime ideals of A. (*First treat the case where A is local and and $P \in \mathrm{Ass}_A(A)$ is its maximal ideal.*)

c) Assume that A_P is reduced for every prime ideal P of A. Prove that A is reduced.

d) Assume that for every $P \in \mathrm{Ass}_A(A)$, the local ring A_P is a field. Prove that for every prime ideal P of A, the local ring A_P is reduced. (*Considering a minimal counterexample, prove that there exists $a \in P$ which is not a zero divisor and introduce the ring $A_P[1/a]$.*) Conclude that A is reduced.

Exercise (6.7.46). — Let A be the ring of continuous functions on $[-1; 1]$.

a) Is the ring A an integral domain? Is it reduced?

b) Let I be the ideal of A consisting of functions f such that $f(0) = 0$. Show that I is not finitely generated. Show that $I = I^2$.

c) Show that the ideal (x) is not primary. (In fact, its radical is not prime.)

d) Prove that $\bigcap_n (x^n) \neq 0$.

Exercise (6.7.47). — Let $A = \mathscr{C}([0; 1], \mathbf{R})$ be the ring of real-valued, continuous functions on $[0; 1]$. The support $S(f)$ of a function $f \in A$ is the closure of $\{x ; f(x) \neq 0\}$.

a) Let $f \in A$; what is $\mathrm{Ann}_A(f)$? Prove in particular it is a radical ideal and that $A/\mathrm{Ann}_A(f) \simeq \mathscr{C}(S(f); \mathbf{R})$.

b) Prove that $\mathrm{Ann}_A(f)$ is not a prime ideal.

c) Prove that the zero ideal (0) has no primary decomposition in A.

Exercise (6.7.48). — Let k be a field and let A be the ring $k[X, Y, Z]/(XY - Z^2)$. One writes $x = \mathrm{cl}(X)$, etc. Let P be the ideal $(x, z) \subset A$.

a) Show that P is a prime ideal.

b) Show that P^2 is not a primary ideal. (*Consider the multiplication by y in A/P^2.*)

c) Show that $(x) \cap (x^2, y, z)$ is a minimal primary decomposition of P^2.

Exercise (6.7.49). — Let E be a field and let I be the ideal $(X) \cap (X, Y)^2$ of the polynomial ring $E[X, Y]$.

a) Prove that $I = (X^2, XY)$.

Let $A = E[X, Y]/I$ be the quotient ring and let x, y be the classes of X, Y modulo I. Let n be an integer such that $n \geq 1$.

b) Prove that the ideal (y^n) of A is (x, y)-primary and that $0 = (x) \cap (y^n)$ is a minimal primary decomposition of the ideal 0 in A.

c) Let $a \in E^\times$. Prove that the ideal $(x + ay^n)$ is (x, y)-primary and that $0 = (x + ay^n)$ is an (x, y)-primary ideal and that $0 = (x) \cap (x + ay^n)$ is also a minimal primary decomposition of 0.

Exercise (6.7.50). — Let A be a commutative noetherian ring.

a) Let I be an ideal of A and let $a \in I$ be a regular element. Let P be a prime ideal of A such that $P \in \text{Ass}_A(A/I)$. Prove that there exist a P-primary ideal Q and an ideal J of A such that $I = Q \cap J$ and $P \notin \text{Ass}_A(A/J)$.

b) Prove that J contains a regular element x.

c) Prove that $I : (x) = Q : (x)$ and that this ideal is P-primary.

d) Prove that there exists an integer $n \geq 0$ such that $P^n \not\subset I : (x)$ and $P^{n+1} \subset I : (x)$. Prove that P^n contains a regular element y such that $y \notin I : (x)$.

e) Prove that $P = I : (xy)$.

f) Let $a \in A$ be a regular element and let P be a prime ideal of A such that $P \in \text{Ass}_A(A/(a))$. Let $b \in A$ be a regular element such that $b \in P$; prove that $P \in \text{Ass}_A(A/(b))$. (*Find regular elements* $x, y \in A$ *such that* $bx = ay$ *and* $P = (a) : (x)$; *then prove that* $P = (b) : (y)$.)

Exercise (6.7.51). — Let K be a field and let $n \in \mathbf{N}$. Let $A = K[T_1, \ldots, T_n]$.

a) Let $a_1, \ldots, a_n$ be integers. Prove that the ideal $(T_1^{a_1}, \ldots, T_n^{a_n})$ of A is primary and compute its radical.

b) Let J be a monomial ideal of A and let f, g be two monomials. If f and g are coprime, then $J + (fg) = (J + (f)) \cap (J + (g))$.

c) Let J be the monomial ideal (X^2YZ, Y^2Z, YZ^3) of $K[X, Y, Z]$. Compute a minimal primary decomposition of J.

d) Let J be a monomial ideal of A. Show that it admits a minimal primary decomposition which consists of monomial ideals.

Chapter 7.
First Steps in Homological Algebra

In this chapter, we explain a few elementary notions in homological algebra. *At the heart of algebraic topology, where it has striking applications (such as Brouwer's theorem, for example), the algebraic formalism of homological algebra is now spread over many fields of mathematics — algebraic topology, commutative algebra, algebraic geometry, representation theory for example — and its applications range as far as robotics!*

Algebraic topology and differential calculus naturally furnish sequences of modules linked by morphisms $f_0 \colon M_0 \to M_1$, $f_1 \colon M_1 \to M_2$, etc., such that the composites $f_1 \circ f_0$, $f_2 \circ f_1$, etc. of two successive morphisms vanish. One important example of such a situation, the de Rham complex *of an open subset U of $\mathbf{R}^2$, is given in the last section of this chapter, where elements of M_0, M_1, M_2 respectively are functions, vector fields, and functions on U, and f_0 and f_1 correspond to gradient and curl. The relation $f_p \circ f_{p-1} = 0$ means that $\mathrm{Im}(f_{p-1}) \subset \mathrm{Ker}(f_p)$, and a good reason for the inclusion is the equality, in which case one says that the sequence is* exact. *The game of homological algebra starts by systematically associating with such a sequence its* homology modules, *defined as $\mathrm{Ker}(f_p)/\mathrm{Im}(f_{p-1})$, which quantify in what respect that inclusion is not an equality.*

Two classes of modules have particularly good properties, that of projective *and* injective *modules. I define them in sections 7.3 and 7.4. While their role in more advanced homological algebra is very important, it is barely touched here, and I only prove the theorem of Kaplansky that projective modules over a local ring are free, and the theorem of Baer that every module is a submodule of an injective module.*

Although one may start with exact sequences, algebraic constructions tend to lose this property and I initiate the study of the exactness of some functors, *such as the functor of homomorphisms. This is the occasion for introducing* adjoint functors.

© Springer Nature Switzerland AG 2021
A. Chambert-Loir, *(Mostly) Commutative Algebra*, Universitext,
https://doi.org/10.1007/978-3-030-61595-6_7

7.1. Diagrams, Complexes and Exact Sequences

The notion of an exact sequence allows one to combine many algebraic properties in a diagram which is quite simple to write down as well as to read.

7.1.1. — We have already seen a few diagrams in this book, such as the one on p. 29 that immediately follows and illustrates theorem 1.5.3.

Such a *diagram* starts from a *quiver*, that is a collection of vertices and arrows linking one vertex to another; moreover, the vertices are labeled by an object from a category, and an arrow linking one vertex to another is labeled by a morphism from the object that labels its source to the object that labels its target.

Some arrows might be drawn using a dashed line; this usually reflects that the existence or the uniqueness of the corresponding morphisms is the conclusion of a mathematical statement that the diagram illustrates.

A path in a diagram is a finite sequence of consecutive arrows, meaning that the target of an arrow is the source of the next one. Given such a path, one may compose the morphisms by which the arrows are labeled. A diagram is said to be *commutative* if for any two paths linking one vertex to another, the two morphisms that they define coincide.

Definition (7.1.2). — Let A be a ring. A *complex* of A-modules is a diagram

$$M_1 \xrightarrow{f_1} M_2 \xrightarrow{f_2} \cdots \to M_{n-1} \xrightarrow{f_{n-1}} M_n$$

where $M_1, \ldots, M_n$ are A-modules and $f_1, \ldots, f_{n-1}$ are homomorphisms such that $f_i \circ f_{i-1} = 0$ for $i \in \{2, \ldots, n-1\}$.

One says that such a diagram is an *exact sequence* if $\mathrm{Ker}(f_i) = \mathrm{Im}(f_{i-1})$ for every $i \in \{2, \ldots, n-1\}$.

Observe that the condition $f_i \circ f_{i-1} = 0$ means that $\mathrm{Im}(f_{i-1}) \subset \mathrm{Ker}(f_i)$. Consequently, an exact sequence is a complex.

Sometimes, complexes are implicitly extended indefinitely by adding null modules and null morphisms.

Definition (7.1.3). — Let

$$M_1 \xrightarrow{f_1} M_2 \to \cdots \to M_{n-1} \xrightarrow{f_{n-1}} M_n$$

be a complex of A-modules. Its homology modules are the modules $H_2, \ldots, H_{n-1}$ defined by $H_i = \mathrm{Ker}(f_i)/\mathrm{Im}(f_{i-1})$ for $2 \le i \le n-1$.

As a consequence of this definition, a complex is an exact sequence if and only if its homology modules are zero.

If the complex is extended by null modules, we get also $H_0 = \mathrm{Ker}(f_1)$ and $H_n = M_n/\mathrm{Im}(f_{n-1})$.

Example (7.1.4). — Let A be a ring and let $f : M \to N$ be a morphism of A-modules.

The diagram $0 \to M \xrightarrow{f} N \to 0$ is a complex, and its homology modules are $\mathrm{Ker}(f)$ and $N/\mathrm{Im}(f) = \mathrm{Coker}(f)$ (see example 3.4.3). In particular, this diagram is an exact sequence if and only if f is an isomorphism.

In general, note that $\mathrm{Ker}(f)$ and $\mathrm{Coker}(f)$ sit in an exact sequence

$$0 \to \mathrm{Ker}(f) \to M \xrightarrow{f} N \to \mathrm{Coker}(f) \to 0.$$

Proposition (7.1.5). — *A diagram of* A-*modules*

$$0 \to N \xrightarrow{i} M \xrightarrow{p} P \to 0$$

is an exact sequence if and only if

(i) *The morphism i is injective;*

(ii) $\mathrm{Ker}(p) = \mathrm{Im}(i)$;

(iii) *The morphism p is surjective.*

Then, i induces an isomorphism from N *to the submodule* $i(N)$ *of* M, *and p induces an isomorphism of* $M/i(N)$ *with* P.

Such diagrams are very important in practice, and called *short exact sequences*.

Proof. — It suffices to write down all the conditions of an exact sequence. The image of the map $0 \to N$ is 0; it has to be the kernel of i, which means that i is injective. The next condition is $\mathrm{Im}(i) = \mathrm{Ker}(p)$. Finally, the image of p is equal to the kernel of the morphism $P \to 0$, which means that p is surjective. The rest of the proof follows from the factorization theorem: if p is surjective, it induces an isomorphism from $M/\mathrm{Ker}(p)$ to P; if i is injective, it induces an isomorphism from N to $i(N) = \mathrm{Ker}(p)$. □

Lemma (7.1.6). — *Let* A *be a ring and let*

$$0 \to N \xrightarrow{i} M \xrightarrow{p} P \to 0$$

be a short exact sequence of A-*modules.*

a) *The map* $q \mapsto \mathrm{Im}(q)$ *is a bijection from the set of right inverses of p (morphisms* $q : P \to M$ *such that* $p \circ q = \mathrm{id}_P$) *to the set of direct summands of* $i(N)$ *in* M.

b) *The map* $j : \mathrm{Ker}(j)$ *is a bijection from the set of left inverses of i (morphisms* $j : M \to N$ *such that* $j \circ i = \mathrm{id}_N$) *to the set of direct summands of* $i(N)$ *in* M.

c) *The morphism p has a right inverse if and only if the morphism i has a left inverse, if and only if the submodule* $i(N) = \mathrm{Ker}(p)$ *of* M *has a direct summand.*

Definition (7.1.7). — An exact sequence which satisfies the three equivalent properties of lemma 7.1.6, c), is said to be *split*.

Proof. — *a*) Let $q \colon P \to M$ be a right inverse of p, let $Q = q(P)$ be the image of q and let us show that Q is a direct summand of $i(N)$ in M. It suffices to prove that for any $m \in M$, there is a unique pair (x, y) such that $x \in N$, $y \in P$ and $m = i(x) + q(y)$. If $m = i(x) + q(y)$, then $p(m) = p(i(x)) + p(q(y)) = y$ since $p \circ i = 0$ and $p \circ q = \mathrm{id}_P$. Then, $i(x) = m - q(p(m))$. Observe that $p(m - q(p(m))) = p(m) - (p \circ q)(p(m)) = 0$; by the definition of an exact sequence, $\mathrm{Im}(i) = \mathrm{Ker}(p)$, so that $m - q(p(m)) \in \mathrm{Im}(i)$. Since i is injective, there is a unique $x \in N$ such that $i(x) = m - q(p(m))$. Then, $y = p(m)$ satisfies $m - i(x) = q(p(m)) = q(y)$. Moreover, if $m = i(x) + q(y) = i(x') + q(y')$, then $i(x - x') = q(y' - y)$; then $0 = p \circ i(x - x') = p \circ q(y' - y) = y' - y$, hence $y' = y$; then $i(x) = i(x')$, hence $x = x'$ since i is injective.

Conversely, let $Q \subset M$ be a direct summand of $i(N)$ and let us prove that the morphism $p|_Q \colon Q \to P$ is an isomorphism. Its kernel is $Q \cap \mathrm{Ker}(p) = Q \cap i(N) = 0$, so that it is injective. On the other hand, let $x \in P$; since p is surjective, there exists an element $m \in M$ such that $p(m) = x$; let us write $m = y + n$, with $n \in i(N)$ and $y \in Q$; one has $x = p(m) = p(y)$, so that $p|_Q$ is surjective. Let $q \colon P \to Q$ be the inverse of this isomorphism; one has $q \circ p|_Q = \mathrm{id}_Q$ and $p \circ q = \mathrm{id}_P$; in particular, q is a right inverse of p.

These constructions $q \mapsto Q$ and $Q \mapsto q$ are inverse to one another; this proves *a*).

b) Let Q be the kernel of j. Let us show that Q is a direct summand of $i(N)$. If $x \in Q \cap i(N)$, one may write $x = i(y)$ for some $y \in N$, so that $0 = j(x) = j(i(y)) = y$. Consequently, $y = 0$ and $x = 0$. Moreover, let $x \in M$ and set $y = x - i(j(x))$. Then, $j(y) = j(x) - j(i(j(x))) = 0$ so that $y \in Q$. Therefore, $x = i(j(x)) + y$ belongs to $i(N) + Q$. Hence $M = Q \oplus i(N)$.

Conversely, let Q be a direct summand of $i(N)$ in M. Since i is injective, any element of M can be written uniquely as $x + i(y)$, for some $x \in Q$ and some $y \in N$. Let $j \colon M \to N$ be the map such that $j(m) = y$ if $m = x + i(y)$ with $x \in Q$ and $y \in N$. One checks directly that j is a morphism. Finally, for any $y \in N$, the decomposition $m = i(y) = 0 + i(y)$ shows that $j(m) = y$, so that $j \circ i = \mathrm{id}_N$.

Again, these two constructions $j \mapsto Q$ and $Q \mapsto j$ are inverse to one another, which proves *b*).

Finally, assertion *c*) follows from *a*) and *b*).   □

Examples (7.1.8). —   *a*) Any subspace of a vector space has a direct summand. Consequently, every exact sequence of modules over a division ring is split.

b) Let $A = \mathbf{Z}$, let $M = \mathbf{Z}$, let $N = 2\mathbf{Z}$ and let $P = M/N = \mathbf{Z}/2\mathbf{Z}$. The natural injection i of N to M and the canonical surjection from M to P give rise to an exact sequence

$$0 \to 2\mathbf{Z} \xrightarrow{i} \mathbf{Z} \xrightarrow{p} \mathbf{Z}/2\mathbf{Z} \to 0.$$

This exact sequence is not split. Indeed, if q were a right inverse to p, its image would be a submodule of $\mathbf{Z}$ isomorphic to $\mathbf{Z}/2\mathbf{Z}$. However, there is no non-zero x element of $\mathbf{Z}$ such that $2x = 0$ while there is such an element in $\mathbf{Z}/2\mathbf{Z}$.

7.2. The "Snake Lemma". Finitely Presented Modules

Theorem (7.2.1) (Snake lemma). — *Let* A *be a ring. Let us consider a diagram of morphisms of* A-*modules*

$$
\begin{array}{ccccccc}
N & \xrightarrow{\;i\;} & M & \xrightarrow{\;p\;} & P & \longrightarrow & 0 \\
\downarrow{f} & & \downarrow{g} & & \downarrow{h} & & \\
0 & \longrightarrow & N' & \xrightarrow{\;i'\;} & M' & \xrightarrow{\;p'\;} & P'.
\end{array}
$$

Let us assume the two rows are exact sequences, and that the two squares are commutative, meaning that $i' \circ f = g \circ i$ *and* $p' \circ g = h \circ p$.

(i) *There is a unique morphism of* A-*modules* $\partial\colon \mathrm{Ker}(h) \to \mathrm{Coker}(f)$ *such that* $\partial(p(y)) = \mathrm{cl}(x')$ *for* $y \in p^{-1}(\mathrm{Ker}(h))$ *and* $x' \in N'$ *such that* $g(y) = i'(x')$.

(ii) *These morphisms induce an exact sequence*

$$
\mathrm{Ker}(f) \xrightarrow{\;i_*\;} \mathrm{Ker}(g) \xrightarrow{\;p_*\;} \mathrm{Ker}(h) \xrightarrow{\;\partial\;} \mathrm{Coker}(f) \xrightarrow{\;(i')_*\;} \mathrm{Coker}(g) \xrightarrow{\;(p')_*\;} \mathrm{Coker}(h);
$$

(iii) *If* i *is injective, then* i_* *is injective;*

(iv) *If* p' *is surjective, then* $(p')_*$ *is surjective.*

Proof. — The proof needs a number of steps. In particular, assertion *a*) is proved in step 7; assertions *c*) and *d*) are proved in steps 3 and 5.

1) We first show that $i(\mathrm{Ker}(f)) \subset \mathrm{Ker}(g)$ and $p(\mathrm{Ker}(g)) \subset \mathrm{Ker}(h)$. Indeed, let $x \in N$ be such that $f(x) = 0$; then $g(i(x)) = (g \circ i)(x) = (i' \circ f)(x) = i'(f(x)) = 0$, so that $i(x) \in \mathrm{Ker}(g)$. Similarly, for any $y \in \mathrm{Ker}(g)$, one has $h(p(y)) = (h \circ p)(y) = (p' \circ g)(y) = p'(g(y)) = 0$, hence $p(y) \in \mathrm{Ker}(h)$.

Consequently, the morphisms i and p induce by restriction morphisms $i_*\colon \mathrm{Ker}(f) \to \mathrm{Ker}(g)$ and $p_*\colon \mathrm{Ker}(g) \to \mathrm{Ker}(h)$.

2) One has $i'(\mathrm{Im}\, f) \subset \mathrm{Im}\, g$ and $p'(\mathrm{Im}\, g) \subset \mathrm{Im}\, h$. Indeed, let $x' \in \mathrm{Im}(f)$ and let $x \in N$ be such that $x' = f(x)$. Then $i'(x') = (i' \circ f)(x) = (g \circ i)(x) = g(i(x))$, which shows that $i'(x')$ belongs to $\mathrm{Im}(g)$. Similarly, let $y' \in \mathrm{Im}(g)$ and let $y \in M$ be such $y' = g(y)$. One has $p'(y') = (p' \circ g)(y) = (h \circ p)(y) = h(p(y))$, hence $p'(y')$ belongs to $\mathrm{Im}(h)$. Consequently, the kernel of the composition

$$
N' \xrightarrow{\;i'\;} M' \to M'/\mathrm{Im}\, g = \mathrm{Coker}(g)
$$

contains $\mathrm{Im}(f)$. Passing to the quotient by $\mathrm{Im}(f)$, we obtain a morphism $(i')_*\colon \mathrm{Coker}(f) = N'/\mathrm{Im}(f) \to \mathrm{Coker}(g)$.

In the same way, for every $x \in M$, one has $p'(g(x)) = h(p(x)) \in \mathrm{Im}(h)$, so that the kernel of the composition

$$
M' \xrightarrow{\;p'\;} P' \to P'/\mathrm{Im}(h) = \mathrm{Coker}(h)
$$

contains $\mathrm{Im}(g)$. We thus deduce from p' a morphism $(p')_*\colon \mathrm{Coker}(g) \to \mathrm{Coker}(h)$.

3) Assume that i is injective. Then the morphism i_* is injective: if $i_*(x) = 0$, then $i(x) = 0$ hence $x = 0$.

4) Since p_* is the restriction of p to $\mathrm{Ker}(g)$, and since $p \circ i = 0$, we have $p_* \circ i_* = 0$, hence $\mathrm{Im}\, i_* \subset \mathrm{Ker}\, p_*$. Conversely, let $y \in \mathrm{Ker}\, p_*$, that is, $y \in \mathrm{Ker}(g)$ and $p(y) = 0$. Since the first row of the diagram is an exact sequence, $y \in \mathrm{Im}(i)$ and there exists an element $x \in N$ such that $y = i(x)$. Then $0 = g(y) = g(i(x)) = (g \circ i)(x) = (i' \circ f)(x) = i'(f(x))$. Since i' is injective, $f(x) = 0$ and $x \in \mathrm{Ker}(f)$. This implies that $y = i(x) \in i(\mathrm{Ker}(f)) = i_*(\mathrm{Ker}(f))$.

5) Assume that p' is surjective. Then the morphism p'_* is surjective: let $\zeta' \in \mathrm{Coker}(h)$, and let $z' \in P'$ be such that $\zeta' = \mathrm{cl}(z')$. Since p' is surjective, there exists an element $y' \in M'$ such that $z' = p'(y')$. By definition of p'_*, $\zeta' = p'_*(\mathrm{cl}(y'))$, hence $\zeta' \in \mathrm{Im}(p'_*)$.

6) One has $p'_* \circ i'_* = 0$. Indeed, the definition of i'_* implies that for any $x' \in N'$, $i'_*(\mathrm{cl}(x')) = \mathrm{cl}(i'(x'))$, hence

$$p'_*(i'_*(\mathrm{cl}(x'))) = p'_*(\mathrm{cl}(i'(x'))) = \mathrm{cl}(p'(i'(x'))) = 0.$$

Conversely, if $p'_*(\mathrm{cl}(y')) = 0$, then $\mathrm{cl}(p'(y')) = 0$, so that $p'(y') \in \mathrm{Im}(h)$. Let us thus write $p'(y') = h(z)$ for some $z \in P$. Since p is surjective, there exists an element $y \in M$ such $z = p(y)$ and $p'(y') = h(p(z)) = p'(g(z))$. Consequently, $y' - g(z)$ belongs to $\mathrm{Ker}(p')$, hence is of the form $i'(x')$ for some $x' \in N'$. Finally,

$$\mathrm{cl}(y') = \mathrm{cl}(g(z) + i'(x')) = \mathrm{cl}(i'(x')) = i'_*(x'),$$

which proves that $\mathrm{Ker}(p'_*) = \mathrm{Im}(i'_*)$.

7) Let us now define the morphism $\partial \colon \mathrm{Ker}(h) \to \mathrm{Coker}(f)$. Let $y \in p^{-1}(\mathrm{Ker}(h))$; then $p(y) \in \mathrm{Ker}(h)$, hence $p'(g(y)) = h(p(y)) = 0$; one thus has $g(y) \in \mathrm{Ker}(p') = \mathrm{Im}(i')$, since the bottom row of the diagram is exact; since i' is injective, there exists a unique element $x' \in N'$ such that $g(y) = i'(x')$. Let then $\partial_1(y) \in \mathrm{Coker}(f)$ be the class of x' modulo $\mathrm{Im}(f)$. This defines a map $\partial_1 \colon p^{-1}(\mathrm{Ker}(h)) \to \mathrm{Coker}(f)$. This map ∂_1 is a morphism of A-modules. This may be proved by a direct elementary computation, but we simply observe that ∂_1 is the composition of the morphism $g|_{p^{-1}(\mathrm{Ker}(h))}$ (which lands into $\mathrm{Ker}(p')$) with the inverse of the isomorphism $N' \to \mathrm{Ker}(p') = \mathrm{Im}(i')$ induced by i', with the projection to $\mathrm{Coker}(f)$.

The surjective morphism $p \colon M \to P$ induces, by restriction, a surjective morphism from $p^{-1}(\mathrm{Ker}(h)) \to \mathrm{Ker}(h)$ whose kernel is $\mathrm{Ker}(p)$. Let us show that there exists a unique morphism $\partial \colon \mathrm{Ker}(h) \to \mathrm{Coker}(f)$ such that $\partial \circ p = \partial_1$. By the factorization theorem, it suffices to prove that $\mathrm{Ker}(p) \subset \mathrm{Ker}(p_1)$. Let $y \in \mathrm{Ker}(p) \subset p^{-1}(\mathrm{Ker}(h))$ and $x' \in N'$ be such that $i'(x') = g(y)$. Since the top row of the diagram is exact, there exists an $x \in N$ such that $y = i(x)$; then $g(y) = g(i(x)) = i'(f(x))$, so that $x' = f(x)$ and $\partial_1(y) = \mathrm{cl}(x') = 0$. Consequently, $\mathrm{Ker}(\partial_1)$ contains $\mathrm{Ker}(p)$, as was to be shown.

Conversely, if $\partial' \colon \mathrm{Ker}(h) \to \mathrm{Coker}(f)$ satisfies the conditions of assertion a), one has $\partial' \circ p = \partial_1 = \partial \circ p$, hence $\partial = \partial'$ since p induces a surjective morphism from $p^{-1}(\mathrm{Ker}(h))$ to $\mathrm{Ker}(h)$.

8) Let us show that $\mathrm{Im}(p_*) = \mathrm{Ker}(\partial)$. Let $z \in \mathrm{Im}(p_*)$. There exists an element $y \in \mathrm{Ker}(g)$ such that $p(y) = z$. In other words, $g(y) = 0$; with the notation of the preceding paragraph, we may take $x' = 0$, hence $\partial(z) = 0$ and $z \in \mathrm{Ker}(\partial)$.

Conversely, let $z \in \mathrm{Ker}(\partial)$. Keeping the same notation, we have $x' \in \mathrm{Im}(f)$, so that there is an $x \in N$ such that $x' = f(x)$, hence $g(y) = i'(x') = g(i(x))$ and $y - i(x) \in \mathrm{Ker}(g)$. It follows that $z = p(y) = p(y - i(x)) \in p(\mathrm{Ker}(g)) = \mathrm{Im}(p_*)$, as required.

9) Finally, let us show that $\mathrm{Im}(\partial) = \mathrm{Ker}(i'_*)$. Let $z \in N$; with the same notation, $i'(\partial(z)) = i'(\mathrm{cl}(x')) = \mathrm{cl}(i'(x')) = \mathrm{cl}(g(y)) = 0$, hence $\partial(z) \in \mathrm{Ker}(i'_*)$ and $\mathrm{Im}(\partial) \subset \mathrm{Ker}(i'_*)$.

In the other direction, let $\xi' \in \mathrm{Ker}(i'_*)$. There is an element $x' \in N'$ such that $\xi' = \mathrm{cl}(x')$, so that $i'_*(\xi') = \mathrm{cl}(i'(x'))$. This shows that $i'(x') \in \mathrm{Im}(g)$. Let $y \in M$ be such that $i'(x') = g(y)$; the definition of ∂ shows that $\partial(p(y)) = \mathrm{cl}(x') = \xi'$ so that $\mathrm{Ker}(i'_*) \subset \mathrm{Im}(\partial)$.

This concludes the proof of the theorem. $\qquad\qquad\square$

Corollary (7.2.2). — *Let us retain the hypotheses of theorem 7.2.1.*

(i) *If f and h are injective, then g is injective. If f and h are surjective, then g is surjective.*

(ii) *If f is surjective and g is injective, then h is injective.*

(iii) *If g is surjective and h is injective, then f is surjective.*

Proof. — a) Assume that f and h are injective. The exact sequence given by the snake lemma begins with $0 \xrightarrow{i_*} \mathrm{Ker}(g) \xrightarrow{p_*} 0$. Necessarily, $\mathrm{Ker}(g) = 0$. If f and h are surjective, the exact sequence ends with $0 \xrightarrow{i'_*} \mathrm{Coker}(g) \xrightarrow{p'_*} 0$, so that $\mathrm{Coker}(g) = 0$ and g is surjective.

b) If f is surjective and g is injective, one has $\mathrm{Ker}(g) = 0$ and $\mathrm{Coker}(f) = 0$. The middle of the exact sequence can thus be rewritten as $0 \xrightarrow{p_*} \mathrm{Ker}(h) \xrightarrow{\partial} 0$, so that h is injective.

c) Finally, if g is surjective and h is injective, we have $\mathrm{Ker}(h) = 0$, $\mathrm{Coker}(g) = 0$, hence an exact sequence $0 \xrightarrow{\partial} \mathrm{Coker}(f) \xrightarrow{i'_*} 0$, which implies that $\mathrm{Coker}(f) = 0$ and f is surjective. $\qquad\square$

Definition (7.2.3). — Let A be a ring. Let M be a right A-module. A presentation of M is an exact sequence

$$F \xrightarrow{\psi} E \xrightarrow{\varphi} M \to 0,$$

where E and F are *free* A-modules.

One says that this presentation is finite if moreover F and G are finitely generated.

If an A-module M has a finite presentation, one also says that M is finitely presented.

Remark (7.2.4). — Let M be an A-module and let $F \xrightarrow{\psi} E \xrightarrow{\varphi} M \to 0$ be a presentation of M. By definition of an exact sequence, the morphism φ is surjective. Let $(e_i)_{i \in I}$ be a basis of E. Then $(\varphi(e_i))_{i \in I}$ is a generating family in M. Let $(f_j)_{j \in J}$ be a basis of F. The family $(\psi(f_i))_{j \in J}$ generates $\mathrm{Im}(\psi)$. By the definition of an exact sequence, we see that this family generates $\mathrm{Ker}(\varphi)$.

Reversing the argument, we can construct a presentation of an A-module M as follows. First choose a generating family $(m_i)_{i \in I}$ in M. Let then $\varphi \colon A^{(I)} \to M$ be the A-linear map defined by $\varphi((a_i)) = \sum_{i \in I} m_i a_i$. Let $(n_j)_{j \in J}$ be a generating family of $\mathrm{Ker}(\varphi)$ and let $\psi \colon A^{(J)} \to A^{(I)}$ be the A-linear map defined by $\psi((a_j)) = \sum n_j a_j$. Then $A^{(J)} \xrightarrow{\psi} A^{(I)} \xrightarrow{\varphi} M \to 0$ is a presentation of M.

Lemma (7.2.5). — *Let A be a ring and let M be a right A-module. Let $0 \to P \xrightarrow{q} N \xrightarrow{p} M \to 0$ be an exact sequence. If M is finitely presented and N is finitely generated, then P is finitely generated.*

Proof. — Let $F \xrightarrow{\psi} E \xrightarrow{\varphi} M \to 0$ be a finite presentation of M. Let $(e_i)_{i \in I}$ be a finite generating family of E and let $(f_j)_{j \in J}$ be a finite generating family of F. For every $i \in I$, choose an element $u_i \in N$ such that $p(u_i) = \varphi(e_i)$; such an element exists because p is surjective. Let $u \colon E \to N$ be the unique morphism of A-modules such that $u(e_i) = u_i$ for every $i \in I$. One has $p \circ u = \varphi$.

Let $j \in J$. One has $p(u(\psi(f_j))) = \varphi(\psi(f_j)) = 0$. Consequently, $u(\psi(f_j)) \in \mathrm{Ker}(p)$. By definition of an exact sequence, the morphism q induces an isomorphism from P to $\mathrm{Im}(q)$. Consequently, there exists a unique element $v_j \in P$ such that $u(\psi(f_j)) = q(v_j)$. Let then $v \colon A^{(J)} \to P$ be the unique morphism such that $v(f_j) = v_j$ for every $j \in J$. One has $u \circ \psi = q \circ v$.

We have constructed a commutative diagram

$$
\begin{array}{ccccccc}
F & \xrightarrow{\psi} & E & \xrightarrow{\varphi} & M & \longrightarrow & 0 \\
\downarrow{v} & & \downarrow{u} & & \downarrow{\mathrm{id}_M} & & \\
0 \longrightarrow & P & \xrightarrow{q} & N & \xrightarrow{p} & M & \longrightarrow 0
\end{array}
$$

the two rows of which are exact sequences. By the snake lemma (theorem 7.2.1), we deduce from this diagram an exact sequence

$$\mathrm{Ker}(\mathrm{id}_M) \to \mathrm{Coker}(v) \to \mathrm{Coker}(u) \to \mathrm{Coker}(\mathrm{id}_M).$$

Since id_M is an isomorphism, one has $\mathrm{Ker}(\mathrm{id}_M) = \mathrm{Coker}(\mathrm{id}_M) = 0$ so that $\mathrm{Coker}(v)$ and $\mathrm{Coker}(u)$ are isomorphic. Since $\mathrm{Coker}(u) = N/\mathrm{Im}(u)$ is a quotient of the finitely generated module N, it is finitely generated. Consequently, $\mathrm{Coker}(v) = P/\mathrm{Im}(v)$ is finitely generated.

On the other hand, $\mathrm{Im}(v)$ is a quotient of the finitely generated A-module F, hence $\mathrm{Im}(v)$ is finitely generated.

By proposition 3.5.5, the module P is finitely generated, as was to be shown. □

7.2.6. — We now let A be a commutative ring and explore the behavior of homomorphisms with respect to localization. Let S be a multiplicative subset of A. Let M and N be A-modules. Every morphism $f: M \to N$ gives rise to a morphism $S^{-1}f: S^{-1}M \to S^{-1}N$ of $S^{-1}A$-modules, characterized by $(S^{-1}f)(m/s) = f(m)/s$ for $m \in M$ and $s \in S$. The map $f \mapsto S^{-1}f$ from $\mathrm{Hom}_A(M, N)$ to $\mathrm{Hom}_{S^{-1}A}(S^{-1}M, S^{-1}N)$ is A-linear, and its target is an $S^{-1}A$-module; consequently, it extends to a morphism $\delta_{M,N}: S^{-1}\mathrm{Hom}_A(M, N) \to \mathrm{Hom}_{S^{-1}A}(S^{-1}M, S^{-1}N)$ of $S^{-1}A$-modules.

Proposition (7.2.7). — *Let A be a commutative ring, let S be a multiplicative subset of A and let M and N be A-modules. If M is a finitely presented A-module, then the morphism $\delta_{M,N}$ is an isomorphism.*

Proof. — We first treat the case where M is a free finitely generated A-module. Let $(e_1, \ldots, e_m)$ be a basis of M. Then every morphism $f: M \to N$ is determined by the image of the e_i, and the map $\mathrm{Hom}_A(M, N) \to N^m$ given by $f \mapsto (f(e_1), \ldots, f(e_m))$ is an isomorphism of A-modules. Moreover, $S^{-1}M$ is a free $S^{-1}A$-module, with basis $(e_1/1, \ldots, e_m/1)$, and the map $\mathrm{Hom}_{S^{-1}A}(S^{-1}M, S^{-1}N) \to (S^{-1}N)^m$ given by $f \mapsto (f(e_1/1), \ldots, f(e_m/1))$ is an isomorphism of $S^{-1}A$-modules. Given these isomorphism, the morphism $\delta_{M,N}$ identifies as the morphism from $S^{-1}(N^m)$ to $S^{-1}N)^m$ given by $(n_1, \ldots, n_m)/s \mapsto (n_1/s, \ldots, n_m/s)$. Since this morphism is an isomorphism, this implies that $\delta_{M,N}$ is an isomorphism if M is a free finitely generated A-module.

Let now $F \xrightarrow{u} E \xrightarrow{p} M \to 0$ be a finite presentation of M. By left exactness of the functor $\mathrm{Hom}_A(\bullet, N)$, it induces an exact sequence

$$0 \to \mathrm{Hom}_A(M, N) \xrightarrow{p^*} \mathrm{Hom}_A(E, N) \xrightarrow{u^*} \mathrm{Hom}_A(F, N)$$

of A-modules. By exactness of localization, it then induces an exact sequence

$$0 \to S^{-1}\mathrm{Hom}_A(M, N) \xrightarrow{S^{-1}p^*} S^{-1}\mathrm{Hom}_A(E, N) \xrightarrow{S^{-1}u^*} S^{-1}\mathrm{Hom}_A(F, N)$$

of $S^{-1}A$-modules.

The given presentation of M also induces a finite presentation $S^{-1}F \xrightarrow{S^{-1}u} S^{-1}E \xrightarrow{S^{-1}p} S^{-1}M \to 0$ and, similarly, an exact sequence

$$0 \to \mathrm{Hom}_{S^{-1}A}(S^{-1}M, S^{-1}N) \xrightarrow{p^*} \mathrm{Hom}_{S^{-1}A}(S^{-1}E, S^{-1}N) \xrightarrow{u^*} \mathrm{Hom}_{S^{-1}A}(S^{-1}F, S^{-1}N)$$

of $S^{-1}A$-modules.

Then the morphisms $\delta_{M,N}$, $\delta_{E,N}$ and $\delta_{F,N}$ sit within a commutative diagram

$$
\begin{array}{ccccccc}
0 & \longrightarrow & S^{-1}\mathrm{Hom}_A(M, N) & \xrightarrow{S^{-1}p^*} & S^{-1}\mathrm{Hom}_A(E, N) & \xrightarrow{S^{-1}u^*} & S^{-1}\mathrm{Hom}_A(F, N) \\
& & \downarrow{\delta_{M,N}} & & \downarrow{\delta_{E,N}} & & \downarrow{\delta_{F,N}} \\
0 & \longrightarrow & \mathrm{Hom}_{S^{-1}A}(S^{-1}M, S^{-1}N) & \xrightarrow{(S^{-1}p)^*} & \mathrm{Hom}_{S^{-1}A}(S^{-1}E, S^{-1}N) & \xrightarrow{(S^{-1}u)^*} & \mathrm{Hom}_{S^{-1}A}(S^{-1}F, S^{-1}N)
\end{array}
$$

the two rows of which are exact sequences. We also know that $\delta_{E,N}$ and $\delta_{F,N}$ are isomorphisms, because E and F are free and finitely generated.

To deduce from this that $\delta_{M,N}$ is an isomorphism, we will apply the snake lemma. However, the morphism $S^{-1}u^*$ is not surjective *a priori*, hence we replace the module $S^{-1}\operatorname{Hom}_A(F, N)$ by the image of $S^{-1}u^*$ and the morphism $\delta_{F,N}$ by its restriction $\delta'_{F,N}$. The snake lemma (theorem 7.2.1) then furnishes a morphism ∂ sitting in an exact sequence

$$0 \to \operatorname{Ker}(\delta_{M,N}) \to \operatorname{Ker}(\delta_{E,N}) \to \operatorname{Ker}(\delta'_{F,N})$$
$$\xrightarrow{\partial} \operatorname{Coker}(\delta_{M,N}) \to \operatorname{Coker}(\delta_{E,N}) \to \operatorname{Coker}(\delta'_{F,N}).$$

The morphism $\delta_{E,N}$ is an isomorphism, hence $\operatorname{Ker}(\delta_{E,N}) = \operatorname{Coker}(\delta_{E,N}) = 0$; the morphism $\delta'_{F,N}$ is injective, hence $\operatorname{Ker}(\delta'_{F,N}) = 0$. The preceding exact sequence thus becomes

$$0 \to \operatorname{Ker}(\delta_{M,N}) \to 0 \to 0 \xrightarrow{\partial} \operatorname{Coker}(\delta_{M,N}) \to 0$$

so that $\operatorname{Ker}(\delta_{M,N}) = \operatorname{Coker}(\delta_{M,N}) = 0$. This proves that $\delta_{M,N}$ is an isomorphism, as was to be shown.   □

7.3. Projective Modules

Definition (7.3.1). — One says that an A-module P is *projective* if every short exact sequence
$$0 \to N \to M \to P \to 0$$
is split.

Proposition (7.3.2). — *Let A be a ring and let P be an A-module. The following properties are equivalent:*

(i) *The module P is projective;*

(ii) *The module P is isomorphic to a direct summand of a free A-module;*

(iii) *For all A-modules M and N, any surjective morphism $p \colon M \to N$ and any morphism $f \colon P \to N$, there exists a morphism $g \colon P \to M$ such that $f = p \circ g$;*

(iv) *For every A-module M, every surjective morphism $f \colon M \to P$ has a right inverse.*

The third property of the theorem is often taken as the definition of a projective module. It can be summed up as a diagram

$$
\begin{array}{ccc}
 & & M \\
 & \overset{g}{\nearrow} & \downarrow{\scriptstyle p} \\
P & \underset{f}{\longrightarrow} & N
\end{array}
$$

where the solid arrows represent the given maps, and the dashed arrow is the one whose existence is asserted by the property.

Proof. — (i)⇒(ii). Let us assume that P is projective. We need to construct a free A-module L and submodules Q and Q′ of L such that L = Q ⊕ Q′ and P ≃ Q. Let S be a generating family of P and let L = $A^{(S)}$ be the free A-module on S; let $(e_s)_{s \in S}$ be the canonical basis of L (see example 3.5.3). Let $p : L \to P$ be the unique morphism such that $p(e_s) = s$ for every $s \in S$; it is surjective (proposition 3.5.2). Let N = Ker(p); we thus have a short exact sequence

$0 \to N \to L \xrightarrow{p} P \to 0$. Since P is projective, this short exact sequence is split and N has a direct summand Q in L, namely the image of any right inverse of p (lemma 7.1.6). In particular, P ≃ Q and L = Q ⊕ N. This shows that P is (isomorphic) to a direct summand of free module.

(ii)⇒(iii). Let us assume that there exists an A-module Q such that L = P ⊕ Q is a free A-module. Let S be a basis of L.

Let M and N be A-modules and let $p : M \to N$ and $f : P \to N$ be morphisms of A-modules, p being surjective. We need to show that there exists a morphism $g : P \to M$ such that $p \circ g = f$.

Let us define a morphism $\varphi : L \to N$ by $\varphi(x + y) = f(x)$ for $x \in P$ and $y \in Q$. By construction, $\varphi|_P = f$. Since p is surjective, there exists for every $s \in S$ an element $m_s \in M$ such that $p(m_s) = \varphi(s)$. Then the universal property of free A-modules implies that there exists a unique morphism $\gamma : L \to M$ such that $\gamma(s) = m_s$ for every $s \in S$, in particular, $p \circ \gamma(s) = p(m_s) = \varphi(s)$. Since S is a basis of L, $p \circ \gamma = \varphi$. The restriction $g = \varphi|_P$ of φ to P satisfies $p \circ g = \varphi|_P = f$.

(iii)⇒(iv). Let $p : M \to P$ be a surjective morphism of A-modules. Let us apply the hypothesis of (iii) to N = P and $f = \mathrm{id}_P$. There exists a $g : P \to M$ such that $p \circ g = \mathrm{id}_P$; in other words, p has a right inverse.

(iv)⇒(i). Let $0 \to N \xrightarrow{i} M \xrightarrow{p} P \to 0$ be a short exact sequence of A-modules. By assumption, p has a right inverse. It then follows from lemma 7.1.6 that this exact sequence is split.   □

Corollary (7.3.3). — *Every free A-module is projective.*

Remark (7.3.4). — Let P be a projective A-module. Let $(e_i)_{i \in I}$ be a generating family of elements of P and let $f : A^{(I)} \to P$ be the surjective morphism given by $f((a_i)) = \sum a_i e_i$. Let $g : P \to A^{(I)}$ be a right inverse of f. For every i, let $\varphi_i : P \to A$ be the morphism such that $\varphi_i(x)$ is the ith coordinate of $g(x)$, for every $x \in P$. Then $(\varphi_i)_{i \in I}$ is a family of elements of $P^\vee$, for every $x \in I$, the family $(\varphi_i(x))$ has finite support and one has $x = f(g(x)) = f((\varphi_i(x))_i) = \sum_{i \in I} \varphi_i(x) e_i$.

In this case, the family (φ_i) plays more or less the role of a "dual basis" of the generating family (e_i).

Conversely, let P be an A-module and let us assume that there exist a family $(e_i)_{i \in I}$ in P and a family $(\varphi_i)_{i \in I}$ in $P^\vee$ such that, for every $x \in I$, the family $(\varphi_i(x))$ has finite support and $x = \sum_{i \in I} \varphi_i(x) e_i$. Let us prove that P is projective. As above, we consider the morphism $f : A^{(I)} \to P$ such that

$f((a_i)) = \sum a_i e_i$ and the map $g\colon P \to A^{(I)}$ given by $g(x) = (\varphi_i(x))_i$. By assumption, one has $f \circ g(x) = x$ for every $x \in P$, so that f is surjective and has a right inverse. This implies that $0 \to \mathrm{Ker}(f) \to A^{(I)} \xrightarrow{f} P \to 0$ is a split exact sequence and P is isomorphic to a direct summand of a free A-module. This proves that P is projective.

Example (7.3.5). — *Let A be a commutative ring, let S be a multiplicative subset of A and let P be a projective A-module. Then $S^{-1}P$ is a projective $S^{-1}A$-module.*

Let indeed L be a free A-module admitting submodules L' and L'' such that $L = L' \oplus L''$ and $L' \simeq P$. Then the modules $S^{-1}L'$ and $S^{-1}L''$ identify with submodules of the free $S^{-1}A$-module $S^{-1}L$ such that $S^{-1}L = S^{-1}L' \oplus S^{-1}L''$. In particular, $S^{-1}L'$ is a projective $S^{-1}A$-module. Moreover, $S^{-1}L'$ is isomorphic to $S^{-1}P$, so that $S^{-1}P$ is a projective $S^{-1}A$-module.

Remark (7.3.6). — Let P be a projective A-module and let $S \subset P$ be a generating set. The proof of implication (i)$\Rightarrow$(ii) in proposition 7.3.2 shows that P is (isomorphic to) a direct summand of the free A-module $A^{(S)}$. In particular, if P is a finitely generated projective A-module, then P is a direct summand of a free finitely generated A-module.

Example (7.3.7). — Let A be a principal ideal domain and let P be a finitely generated projective A-module. Then, there exists an integer n such that P is a direct summand of A^n. In particular, P is isomorphic to a submodule of A^n. By proposition 5.4.1, P is free.

Theorem (7.3.8) (Kaplansky). — *Let A be a local ring and let P be a projective A-module. Then P is free.*

Let A be a ring and let J be its Jacobson radical. We recall from definition 2.1.8 that A is local if and only if A/J is a division ring; then J is the unique maximal left ideal of A.

Remark (7.3.9). — Let us first prove this theorem under the additional assumption that P is finitely generated. Let J be the maximal left ideal of A. Let $(e_i)_{1 \le i \le n}$ be a family of minimal cardinality in P such that the e_i generate P/PJ; let us prove that it is a basis of P.

We first consider the morphism $\varphi\colon A^n \to P$ defined by $(a_1, \dots, a_n) \mapsto \sum e_i a_i$. Let $P' = \mathrm{Im}(\varphi)$; by assumption, one has $P = P' + P'J$. By Nakayama's lemma (corollary 6.1.2), one has $P = P'$ and φ is surjective.

Let us now prove that φ is an isomorphism. Let $Q = \mathrm{Ker}(\varphi)$. Since P is projective, the morphism φ has a right inverse ψ, and the image of ψ, which is isomorphic to Q, is a direct summand of Q. We get an isomorphism $P \oplus Q \simeq A^n$. Taking quotients modulo J, we also have an isomorphism $P/PJ \oplus Q/QJ \simeq (A/J)^n$. Since $n = \dim_{A/J}(P/PJ)$, we obtain the equality $Q = QJ$. Since Q is a direct summand of A^n, it is finitely generated. By Nakayama's lemma again (theorem 6.1.1), one has $Q = 0$. Consequently, φ is an isomorphism and P is free.

The remainder of this section is devoted to the proof of Kaplansky's theorem.

Lemma (7.3.10). — *Let A be a ring and let P be a projective A-module. There exists a family (P_i) of submodules of P, each of them generated by a countable family, such that $P = \bigoplus P_i$.*

Proof. — We may assume that there are a free A-module M containing P and a submodule Q of M such that $M = P \oplus Q$. Let $(e_i)_{i \in I}$ be a basis of M. For any subset J of I, let M_J be the submodule of M generated by the elements e_j, for $j \in J$. Observe that M_J has a direct summand in M, namely the submodule of M generated by the elements e_i, for $i \in I - J$; in particular, a submodule of some M_J which has a direct summand in M_J also has a direct summand in M.

Let S be a well-ordered set such that there is no injection from S to $\mathfrak{P}(I)$ (Hartogs's lemma, proposition A.2.9). Using the *transfinite construction principle*, we will construct an increasing family $(J_\alpha)_{\alpha \in S}$ of subsets of I satisfying the following properties, for every $\alpha \in S$:

(i) If $\alpha \in S$ is the successor of an element β, then the set $J_\alpha - J_\beta$ is at most countable, and is empty if and only if $J_\beta = I$;

(ii) If α is a limit element of S, then J_α is the union of the J_β for $\beta < \alpha$;

(iii) $M_{J_\alpha} = (P \cap M_{J_\alpha}) \oplus (Q \cap M_{J_\alpha})$.

Let $\alpha \in S$.

If α is a limit element of S, then the prescription (ii) defines a subset J_α of I. Let us show that condition (iii) is satisfied. Since $P \cap Q = 0$, it suffices to show that M_{J_α} is the sum of $P \cap L_{J_\alpha}$ and $Q \cap M_{J_\alpha}$. So let $m \in M_{J_\alpha}$. Only finitely many coordinates of m in the basis (e_i) are non-zero, so that there exists an element $\beta < \alpha$ such that $m \in M_{J_\beta}$; consequently, $m \in (P \cap L_{J_\beta}) + (Q \cap M_{J_\beta})$ and *a fortiori*, $m \in (P \cap L_{J_\alpha}) + (Q \cap M_{J_\alpha})$.

Let us now assume that α is the successor of $\beta \in S$. If $I = J_\beta$, we let $J_\alpha = J_\beta$. Otherwise, choose some element $j \in I - J_\beta$ and define a sequence $(K_n)_{n \in \mathbf{N}}$ of finite subsets of I as follows. Let $K_0 = \{j\}$. Assume K_n has been defined; for every element $k \in K_n$, write e_k as a sum $p_k + q_k$, where $p_k \in P$ and $q_k \in Q$. Then, let K_{n+1} be the union, for all $k \in K_n$, of the indices $i \in I$ such that the ith coordinate of p_k or q_k is non-zero. For every integer n, K_n is finite, so that the union $K = \bigcup K_n$ is either finite or countable. Then, set $J_\alpha = J_\beta \cup K$; condition (i) is satisfied. Let us prove that (iii) holds. Again, it suffices to show that any element $m \in M_{J_\alpha}$ can be written as the sum of an element of $P \cap M_{J_\alpha}$ and of an element of $Q \cap M_{J_\alpha}$. It suffices to prove this property for every element of the form e_k, for some $k \in J_\alpha$. It holds by induction if $k \in J_\beta$, hence we may assume that $k \in K$. By definition, there is an integer n such that $k \in K_n$. Then $e_k = p_k + q_k$ and by construction of K_{n+1}, $p_k \in P \cap M_{K_{n+1}}$ and $q_k \in Q \cap M_{K_{n+1}}$. In particular, $p_k \in P \cap M_{J_\alpha}$ and $q_k \in Q \cap M_{J_\alpha}$, as was to be shown.

To rigorously apply the construction principle, we need to use the axiom of choice, which implies the existence of a *function* that constructs J_α from J_β.

We then obtain a family $(J_\alpha)_{\alpha \in S}$ which satisfies the given requirements.

The map $\alpha \mapsto J_\alpha$ from S to $\mathfrak{P}(I)$ cannot be injective. Since it is increasing, there exists $\beta, \gamma \in S$ such that $\beta < \gamma$ and $J_\beta = J_\gamma$, which we may choose to be minimal. Since the family (J_α) is increasing, one has $J_\beta = J_\alpha$ for every $\alpha \in S$ such that $\alpha \geq \beta$, so that γ is the successor of β. By property (i) of the family (J_α), one has $J_\beta = I$, and $J_\alpha = I$ for every $\alpha \in S$ such that $\alpha \geq \beta$.

We then define a family $(P_\alpha)_{\alpha \in S}$ of submodules of P such that $P \cap M_{J_\alpha} = \bigoplus_{\beta \leq \alpha} P_\beta$ for every $\alpha \in S$. By the transfinite construction principle, we may assume that the family $(P_\beta)_{\beta < \alpha}$ is constructed so that $P \cap M_{J_\beta} = \bigoplus_{\gamma \leq \beta} P_\gamma$, and we have to construct P_α.

If α is a limit element of S, we set $P_\alpha = 0$. Otherwise, there exists an element $\beta \in S$ such that α is the successor of β. Then, $P \cap M_{J_\beta}$ is a direct summand of M_{J_β}, hence a direct summand of M. Restricting to $P \cap M_{J_\alpha}$ a left inverse of the injection $(P \cap M_{J_\beta}) \to M$, we see that $P \cap M_{J_\beta}$ has a direct summand P_α in $P \cap M_{J_\alpha}$. Moreover, P_α is countably generated.

By the transfinite construction principle, we thus obtain the desired family $(P_\alpha)_{\alpha \in S}$. Taking $\omega \in S$ large enough so that $J_\omega = I$, we have $P = P \cap M_{J_\omega} = \bigoplus_{\alpha \leq \omega} P_\alpha$.   □

Say that an A-module M has property (K) if for every element $m \in M$, there exists a direct summand N of M containing m which is free.

Lemma (7.3.11). — *Let A be a ring and let P be a countably generated A-module. Assume that any direct summand M of P satisfies property (K). Then P is a free A-module.*

Proof. — Let $(p_n)_{n \geq 1}$ be a generating family of P. We construct by induction families $(P_n)_{\geq 1}$ and $(Q_n)_{n \geq 0}$ of submodules of P such that $P_0 = P$, $P_{n-1} = P_n \oplus Q_n$ for every $n \geq 1$, and $p_1, \dots, p_n \in Q_1 \oplus \dots \oplus Q_n$, and such that for every integer n, Q_n is free. Assume $P_0, \dots, P_{n-1}, Q_1, \dots, Q_{n-1}$ have been defined. Observe that $P = P_{n-1} \oplus (Q_1 \oplus \dots \oplus Q_{n-1})$ and let $f : P \to P_{n-1}$ be the projection with kernel $Q_1 \oplus \dots \oplus Q_{n-1}$. Apply property (K) to the element $f(p_n)$ of P_{n-1}: there exists a decomposition of P_{n-1} as a direct sum $P_n \oplus Q_n$, where Q_n is free and contains $f(p_n)$. Consequently, $p_n \in Q_1 \oplus \dots \oplus Q_n$. By induction on n, this concludes the asserted construction.

Let us now show that P is the direct sum of the family $(Q_n)_{n \geq 1}$. By construction, this family is in direct sum in $\sum_{n \geq 1} Q_n$. Moreover, for every $m \geq 1$, one has $p_m \in Q_m \subset \sum_{n \geq 1} Q_n$. Since the family (p_m) generates P, this proves that $P = \sum_{n \geq 1} Q_n$ and concludes the proof of the lemma.   □

Lemma (7.3.12). — *Let A be a local ring. Any projective A-module has property (K).*

Proof. — Let P be a projective A-module. Let Q be an A-module such that $M = P \oplus Q$ is a free A-module. Let $p \in P$. Let $(e_i)_{i \in I}$ be a basis of M in which the coordinates (a_i) of p have the least possible non-zero entries. For simplicity, assume that these coordinates have indices $1, \dots, n$. For each $i \in \{1, \dots, n\}$, write $e_i = p_i + q_i$, where $p_i \in P$ and $q_i \in Q$. Then

$$p = \sum_{i=1}^{n} e_i a_i = \sum_{i=1}^{n} p_i a_i + \sum_{i=1}^{n} q_i a_i$$

is a decomposition of p as the sum of an element of P and an element of Q, so that

$$p = \sum_{i=1}^{n} e_i a_i = \sum_{i=1}^{n} p_i a_i.$$

We may then decompose each vector p_i along the basis (e_i): letting b_{ji} be the jth coordinate, we write

$$p_i = \sum_{j=1}^{n} e_j b_{ji} + p_i',$$

where $p_i' = \sum_{j \in J-\{1,\dots,n\}} e_j b_{ji}$ is the sum of terms corresponding to the coordinates other than $1, \dots, n$. Combining the last two equations, we obtain

$$\sum_{i=1}^{n} e_i a_i = \sum_{i=1}^{n} \sum_{j=1}^{n} e_j b_{ji} a_i + \sum_{i=1}^{n} \sum_{j \in J-\{1,\dots,n\}} e_j b_{ji} a_i.$$

Equating the coefficients of $e_1, \dots, e_n$, we get

$$a_j = \sum_{i=1}^{n} b_{ji} a_i, \qquad (1 \le j \le n).$$

If $j \in J - \{1, \dots, n\}$, we also get

$$\sum_{i=1}^{n} b_{ji} a_i = 0,$$

hence

$$p = \sum_{j=1}^{n} e_j a_j = \sum_{i=1}^{n} \sum_{j=1}^{n} e_j b_{ji} a_i.$$

Let us assume that there exists i, j such that $1 \le i, j \le n$ such that $1 - b_{ii}$ (if $j = i$) or b_{ji} (if $j \ne i$) is invertible in A. Then we can rewrite a_j as a left linear combination

$$a_j = \sum_{\substack{i=1 \\ i \ne j}}^{n} c_i a_i$$

of the a_i, for $i \ne j$. Let us then write

$$p = \sum_{i=1}^{n} e_i a_i = \sum_{\substack{i=1 \\ i \ne j}}^{n} (e_i a_i + e_j c_i a_i) = \sum_{\substack{i=1 \\ i \ne j}}^{n} (e_i + e_j c_i) a_i.$$

Let us then define a family (e'_i) by $e'_i = e_i$ if $i \notin \{1, \ldots, n\}$, $e'_j = e_j$, and $e'_i = e_i + e_j c_i$ if $i \in \{1, \ldots, n\}$ and $i \neq j$. Observe that it is a basis of M. However, the vector p has $n - 1$ non-zero coordinates in this new basis, which contradicts the assumption made at the beginning of the proof.

This shows that $1 - b_{ii}$ (for $1 \leq i \leq n$) and b_{ij} (for $1 \leq i \neq j \leq n$) belong to the Jacobson radical J of the local ring A. In other words, the image of the matrix $B = (b_{ji})$ in the quotient ring A/J is the identity matrix. By corollary 6.1.3, the matrix $B = (b_{ji})$ is invertible in $\mathrm{Mat}_n(A)$. Consequently, the family (f_i) defined by $f_j = \sum_{i=1}^{n} e_j b_{ji}$ (for $j \in \{1, \ldots, n\}$) and $f_j = e_j$ (otherwise) is a basis of M.

Since $p_i = f_i + p'_i$ and p'_i belongs to the submodule spanned by the e_j, for $j \notin \{1, \ldots, n\}$, we see that the family consisting of $p_1, \ldots, p_n$ and the e_j, for $j \in J - \{1, \ldots, n\}$ is a basis of M.

By construction, the submodule F of P generated by $p_1, \ldots, p_n$ contains p. It is also free and it admits a direct summand in M, because it is generated by a subset of a basis of M. Consequently, the injection $F \to M$ admits a left inverse, so that the injection $F \to P$ has a left inverse too; this implies that F has a direct summand in P (lemma 7.1.6). This concludes the proof of the lemma. $\square$

Proof (Proof of Kaplansky's theorem). — Let A be a local ring and let P be a projective A-module. By lemma 7.3.10, there exists a family (P_i) of countable generated submodules of P of which P is the direct sum. Since P_i is a direct summand of a projective A-module, it is itself projective. The ring A being local, P_i satisfies property (K), by lemma 7.3.12. Moreover, each direct summand of P_i is projective, hence satisfies property (K), so that P_i satisfies the hypotheses of lemma 7.3.11. Consequently, P_i is a free A-module. It follows that P is a direct sum of free A-modules, hence is free. $\square$

Proposition (7.3.13). — *Let A be a commutative ring and let M be a finitely presented A-module. The following properties are equivalent:*

(i) *The A-module M is projective;*

(ii) *For every prime ideal P of A, the A_P-module M_P is free;*

(iii) *For every maximal ideal P of A, the A_P-module M_P is free.*

Proof. — The implication (i)$\Rightarrow$(ii) follows from Kaplansky's theorem (theorem 7.3.8) while the implication (ii)$\Rightarrow$(iii) is obvious. Let us thus assume that the A_P-module M_P is free, for every maximal ideal P of A. Let $u: N \to N'$ be a surjective morphism of A-modules and let us show that the map $u_*: \mathrm{Hom}_A(M, N) \to \mathrm{Hom}_A(M, N')$ is surjective. It is a morphism of A-modules; let Q be its cokernel. Let P be a maximal ideal of A. By exactness of localization, the exact sequence

$$\mathrm{Hom}_A(M, N) \xrightarrow{u_*} \mathrm{Hom}_A(M, N') \to Q \to 0$$

gives rise to an exact sequence of A_P-modules

$$\text{Hom}_A(M, N)_P \xrightarrow{(u_*)_P} \text{Hom}_A(M, N')_P \to Q_P \to 0.$$

On the other hand, the hypothesis that M is a finitely presented A-module allows us to identify $\text{Hom}_A(M, N)_P$ with $\text{Hom}_{A_P}(M_P, N_P)$, and $\text{Hom}_A(M, N')_P$ with $\text{Hom}_{A_P}(M_P, N'_P)$ (proposition 7.2.7). The preceding exact sequence then becomes

$$\text{Hom}_{A_P}(M_P, N_P) \xrightarrow{(u_P)_*} \text{Hom}_{A_P}(M_P, N'_P) \to Q_P \to 0.$$

Since M_P is a projective A_P-module, the morphism $(u_P)_*$ is surjective, hence $Q_P = 0$.

This implies that $\text{Supp}(Q)$ contains no maximal ideal of A. On the other hand, if $\text{Supp}(Q)$ contains a prime ideal P, then it contains all maximal ideals that contain P (theorem 6.5.4, *b*) and Krull's theorem 2.1.3), so that $\text{Supp}(Q) = \emptyset$. By theorem 6.5.4, *a*), this shows that $Q = 0$, so that u_* is surjective. Consequently, M is a projective A-module. $\square$

7.4. Injective Modules

Definition (7.4.1). — Let A be a ring. One says that an A-module M is injective if every short exact sequence

$$0 \to M \to N \to P \to 0$$

is split.

The notion of an injective module is kind of dual of that of a projective module. However, the category of modules is quite different from its opposite category. As is pointed out by (MATSUMURA, 1986, p. 277), there does not exist a notion that would be the dual to that of a free module. Consequently, proposition 7.3.2 has no analogue for injective modules and the existence of injective modules is not obvious. We start with a partial counterpart to proposition 7.3.2.

Proposition (7.4.2). — *Let A be a ring and let M be a right A-module. The following conditions are equivalent.*

(i) *The module M is injective;*

(ii) *For all right A-modules N and P, every injective morphism $i: N \to P$ and any morphism $f: N \to M$, there exists a morphism $f: P \to M$ such that $f = g \circ i$;*

(iii) *For every right ideal I of A and any morphism $f: I \to M$, there exists a morphism $g: A \to M$ such that $f = g|_I$;*

(iv) *Every injective morphism from M to any A-module has a left inverse.*

The second condition is often taken as the definition of an injective module.

Proof. — The equivalence of (i) and (iv) follows from lemma 7.1.6.

(iv)$\Rightarrow$(ii). Let $i\colon N \to P$ be an injective morphism and let $f\colon N \to M$ be a morphism. We want to prove that there exists a morphism $g\colon P \to M$ such that $f = g \circ i$.

Let Q be the image of the morphism $x \mapsto (f(x), -i(x))$ from N to $M \times P$. Let $p\colon M \times P \to (M \times P)/Q$ be the canonical projection and let $j\colon M \to (M \times P)/Q$ be the map given by $j(x) = p(x, 0)$ for $x \in M$. Let us show that j is injective; let $x \in M$ be such that $(x, 0) \in Q$. Then there exists a $y \in N$ such that $f(y) = x$ and $-i(y) = 0$. Since i is injective, $y = 0$, hence $x = 0$, as was to be shown.

By (iv), there exists a morphism $\gamma\colon (M \times P)/Q \to M$ such that $\gamma \circ j = \mathrm{id}_M$, that is, $\gamma(p(x, 0)) = x$ for every $x \in M$. Let $g\colon P \to M$ be the morphism given by $y \mapsto \gamma(p(0, y))$. For every $x \in N$, one has

$$g(i(x)) = \gamma(p(0, i(x))) = \gamma(p(-f(x), i(x))) + \gamma(p(f(x), 0)) = f(x),$$

so that $f = g \circ i$, as required.

(ii)$\Rightarrow$(iii). It suffices to take for modules $N = I$ and $P = A$, the injective morphism i being the injection from I into A.

(iii)$\Rightarrow$(iv). Let $i\colon M \to N$ be an injective morphism. We want to prove that there exists a morphism $f\colon N \to M$ which is a left inverse of i, that is, $f \circ i = \mathrm{id}_M$.

Let $\mathscr{F}$ be the set of all pairs (N', f) where N' is a submodule of N containing $i(M)$ and $f\colon N' \to M$ is a morphism of A-modules such that $f \circ i = \mathrm{id}_M$.

Let us define an ordering $\prec$ on $\mathscr{F}$ by setting

$$(N'_1, f_1) \prec (N'_2, f_2) \qquad \Leftrightarrow \qquad N'_1 \subset N'_2 \quad \text{and} \quad f_2|_{N'_1} = f_1.$$

Let us show that the ordered set $\mathscr{F}$ is inductive. Indeed, let (N_α, f_α) be a totally ordered family of elements of $\mathscr{F}$. If it is empty, we take for its upper bound the pair $(N' = i(M), f)$, where $f\colon i(M) \to M$ is the inverse of the injective map i. Otherwise, since the non-empty family (N_α) of submodules of N is totally ordered, its union $N' = \bigcup N_\alpha$ is a submodule of N. We may then define a morphism $f\colon N' \to M$ by setting $f(y) = f_\alpha(y)$ for any α such that $y \in N_\alpha$. Indeed, it does not depend on the choice of α.

By Zorn's lemma (corollary A.2.13), the ordered set $\mathscr{F}$ has a maximal element (N', f). Let us prove by contradiction that $N' = N$. Otherwise, let $y \in N - N'$ and let $I = \{a \in A\,;\, ya \in N'\}$.

Let $\varphi\colon I \to N$ be the morphism given by $a \mapsto f(ya)$, for $a \in I$. By assumption, there exists a morphism $\psi\colon A \to N$ such that $\psi|_I = \varphi$. Set $N'_1 = N' + yA$ and define a morphism $f'_1\colon N'_1 \to M$ by setting $f_1(x + ya) = f(x) + \psi(a)$ for $x \in N'$ and $a \in A$. It is well-defined; indeed, if $x + ya = x' + ya'$, with $x, x' \in N'$ and $a, a' \in A$, then $x - x' = y(a' - a)$, so that $a' - a \in I$ and

$$\psi(a') - \psi(a) = \varphi(a' - a) = f(y(a' - a)) = f(x - x') = f(x) - f(x'),$$

so that $f(x') + \psi(a') = f(x) + \psi(a)$. It is easy to show that f_1 is a morphism of A-modules. Moreover, the construction of f_1 shows that $f_1|_{N'} = f$. In particular, (N'_1, f_1) is an element of $\mathscr{F}$, but this contradicts the hypothesis that (N', f) is a maximal element of $\mathscr{F}$, because $N'_1 \supsetneq N'$.

It follows that any maximal element of $\mathscr{F}$ is of the form (N, f), where $f: N \to M$ is a left inverse to i. This concludes the proof of (iv). $\qquad\square$

Corollary (7.4.3). — *Products of injective A-modules are injective.*

Proof. — Let $(M_s)_{s \in S}$ be a family of injective right A-modules and let $M = \prod_{s \in S} M_s$. For every $s \in S$, let p_s be the canonical projection from M to M_s.

Let I be a right ideal of A and let $f: I \to M$ be a morphism. For every $s \in S$, let $f_s = p_s \circ f$. Since M_s is injective, there exists a morphism $g_s: A \to M_s$ such that $g_s|_I = f_s$. Let $g = (g_s): A \to M$ be the unique morphism such that $p_s \circ g = g_s$ for every $s \in S$. One has $g|_I = (g_s|_I) = (f_s) = f$. This shows that M is an injective A-module. $\qquad\square$

Corollary (7.4.4). — *Let A be a principal ideal domain and let M be an A-module. Then M is an injective module if and only if for any non-zero $a \in A$, the morphism $\mu_a: M \to M$, $x \mapsto ax$, is surjective.*

The condition that the morphisms μ_a be bijective, for all $a \neq 0$, is rephrased as saying that M is a *divisible* A-module.

Proof. — Let us assume that M is an injective module; let $a \in A - \{0\}$. Let $x \in M$. Let I be the ideal (a) in A. Since $a \neq 0$ and A is a domain, the map from A to I given by $t \mapsto at$ is an isomorphism; consequently, there exists a unique morphism $f: I \to M$ such that $f(at) = tx$ for every $t \in A$. By property (iii) in proposition 7.4.2, there exists a morphism $g: A \to M$ such that $g(at) = f(at) = tx$ for any $t \in A$. Then $ag(1) = g(a) = x$. This implies that μ_a is surjective.

Conversely, let us assume that μ_a is surjective for every non-zero $a \in A$. Let I be an ideal of A and let $f: I \to M$ be a morphism of A-modules; we need to show that there exists a morphism $g: A \to M$ such that $g|_I = f$. If $I = 0$, we may take $g = 0$. Otherwise, since A is a principal ideal domain, there exists a non-zero element $a \in A$ such that $I = (a)$. By hypothesis, there exists an element $x \in M$ such that $ax = f(a)$. Let $g: A \to M$ be the morphism given by $g(t) = tx$. For any $t \in A$, one has $g(at) = atx = tf(a) = f(at)$, so that $g|_I = f$. By proposition 7.4.2, the A-module M is injective. $\qquad\square$

Example (7.4.5). — The **Z**-module **Q**/**Z** is injective.

Recall that the existence of projective A-modules is no great mystery, since free modules are projective. More generally, every A-module is the quotient of a free A-module, *a fortiori* a projective A-module. The dual property asserts that every A-module is a submodule of an injective A-module and is more difficult. It is the object of the following theorem.

Theorem (7.4.6) (Baer). — *Let A be a ring and let M be an A-module. There exists an injective A-module N and an injective morphism $i: M \to N$.*

Let A be a ring and let M be a right A-module (resp. a left A-module). Then $M^* = \text{Hom}_{\mathbf{Z}}(M, \mathbf{Q}/\mathbf{Z})$ is an abelian group. We endow it with the structure of a left A-module (resp. of a right A-module) given by $(a \cdot f)(x) = f(xa)$ (resp. $(f \cdot a)(x) = f(ax)$) for every $a \in A$, $f \in M^*$ and $x \in M$.

Lemma (7.4.7). — *Let A be a ring and let M be a right A-module. For any $x \in M$, let $i(x)$ be the map from M^* to $\mathbf{Q}/\mathbf{Z}$ given by $f \mapsto f(x)$. One has $i(x) \in M^{**}$ and the map $i: M \to M^{**}$ so defined is an injective morphism of A-modules.*

Proof. — It is clear that $i(x)$ is a morphism of abelian groups, hence $i(x) \in M^{**}$. Moreover, for every $x \in M$ and every $a \in A$, one has

$$i(xa)(f) = f(xa) = (a \cdot f)(x) = i(x)(a \cdot f) = (i(x) \cdot a)(f).$$

Finally,

$$i(x + y)(f) = f(x + y) = f(x) + f(y) = i(x)(f) + i(y)(f),$$

for every $x, y \in M$, which shows that $i: M \to M^{**}$ is a morphism of right A-modules.

Let us finally prove that i is injective. Let $x \in M$ be a non-zero element. We need to show that there exists a morphism of abelian groups, $g: M \to \mathbf{Q}/\mathbf{Z}$, such that $g(x) \neq 0$. The abelian group $x\mathbf{Z}$ generated by x in M is either isomorphic to $\mathbf{Z}$, or to $\mathbf{Z}/n\mathbf{Z}$, for some integer $n \geq 2$. We define $f: x\mathbf{Z} \to \mathbf{Q}/\mathbf{Z}$ by $f(xm) = m/2 \pmod{\mathbf{Z}}$ in the former case, and $f(xm) = m/n \pmod{\mathbf{Z}}$ in the latter. Observe that $f(x) \neq 0$.

Since $\mathbf{Q}/\mathbf{Z}$ is an injective $\mathbf{Z}$-module, there exists a morphism $g: M \to \mathbf{Q}/\mathbf{Z}$ of $\mathbf{Z}$-modules such that $g|_{x\mathbf{Z}} = f$ (proposition 7.4.2). The morphism g belongs to M^* and satisfies $g(x) \neq 0$.   □

Lemma (7.4.8). — *Let A be a ring.*

(i) *The right A-module $(A_s)^*$ is injective.*

(ii) *For every free left A-module M, the right A-module M^* is injective.*

Proof. — a) Let I be a right ideal of A and let $f: I \to (A_s)^*$ be a morphism of A-modules. Let $\varphi: I \to \mathbf{Q}/\mathbf{Z}$ be the map given by $\varphi(x) = f(x)(1)$, for $x \in I$. It is a morphism of abelian groups. Since $\mathbf{Q}/\mathbf{Z}$ is an injective $\mathbf{Z}$-module, there exists a morphism $\gamma: A \to \mathbf{Q}/\mathbf{Z}$ of abelian groups such that $\gamma|_I = \varphi$.

Let $g: A \to (A_s)^*$ be the map defined by $g(x)(y) = \gamma(xy)$, for $x, y \in A$. It is additive. Let $a, x \in A$; then $g(xa)$ is the element $y \mapsto \gamma(xay)$ of $(A_s)^*$, while $g(x) \cdot a$ is the map $y \mapsto g(x)(ay) = \gamma(xay)$. Consequently, $g(xa) = g(x) \cdot a$ and g is A-linear.

Let $x \in I$. For any $y \in A$, one has $g(x)(y) = \gamma(xy) = \varphi(xy)$ since $xy \in I$ and γ extends φ. Consequently, $g(x)(y) = f(xy)(1)$. Since f is A-linear, $f(xy) = f(x) \cdot y$, so that $f(xy)(1) = f(x)(y)$. This shows that $g(x)(y) = f(x)(y)$, hence $g(x) = f(x)$. We thus have shown that $g|_I = f$.

By proposition 7.4.2, the right A-module $(A_s)^*$ is injective.

b) Let $(e_i)_{i \in I}$ be a basis of M, so that M is identified with $(A_s)^{(I)}$. Then, the morphism $f \mapsto (f(e_i))$ is an isomorphism of right A-modules from M^* to $(A_s^*)^I$. Consequently, M^* is a product of injective A-modules. It then follows from corollary 7.4.3 that M^* is an injective A-module.   □

Proof (Proof of theorem 7.4.6). — We are now able to prove Baer's theorem that every A-module is a submodule of an injective A-module.

Let M be a right A-module. Let F be a free left A-module together with a surjective A-linear morphism $p: F \to M^*$. For example, one may take for F the free-A-module $A^{(M^*)}$ with basis indexed by M^*, and for p the map that sends the basis vector e_f indexed by $f \in M^*$ to f, for every $f \in M^*$. Since p is surjective and A-linear the map p^* from M^{**} to F^* given by $\varphi \mapsto \varphi \circ p$ is injective and A-linear. Let i be the morphism from M to M^{**} defined in lemma 7.4.7; it is injective. Then, the map $p^* \circ i: M \to F^*$ is A-linear and injective. By lemma 7.4.8, the right A-module F^* is injective. This concludes the proof.   □

7.5. Exactness Conditions for Functors

7.5.1. Definitions — Let A and B be two rings. Let us recall that a *functor* F from the category $\boldsymbol{Mod}_A$ of A-modules to the category $\boldsymbol{Mod}_B$ of B-modules associates to each A-module M a B-module F(M) and to each morphism $f: M \to N$ of A-modules a morphism $F(f): F(M) \to F(N)$ subject to the following constraints:

– For every A-module M, one has $F(\mathrm{id}_M) = \mathrm{id}_{F(M)}$;

– For all A-modules M, N, P and all morphisms $f: M \to N$ and $g: N \to$ P, one has $F(g \circ f) = F(g) \circ F(f)$.

Such a functor is said to be *covariant* by opposition to *contravariant functors* which reverse the direction of maps.

Indeed, a contravariant functor F from the category of A-modules to the category of B-modules associates to each A-module M a B-module F(M) and to each morphism $f: M \to N$ of A-modules a morphism $F(f): F(N) \to F(N)$ subject to the analogous requirements:

– For every A-module M, one has $F(\mathrm{id}_M) = \mathrm{id}_{F(M)}$;

– For all A-modules M, N, P and all morphisms $f: M \to N$ and $g: N \to$ P, one has $F(g \circ f) = F(f) \circ F(g)$.

Example (7.5.2) (Forgetful functors). —Let A and B be rings and let $u: B \to A$ be a morphism of rings. Then, every A-module M may be viewed as a B-module $u^*(M)$, with external multiplication given by the composition of the morphism u and of the external multiplication of M. Let $f: M \to N$ be a morphism of A-modules, the map f can be viewed as a map $u^*(f): u^*(M) \to$

$u^*(N)$ which is obviously B-linear. So $u^*(M)$ and $u^*(f)$ are just M and f with different names to indicate that we no longer consider them as A-modules but only as B-modules through u. In particular, $u^*(g \circ f) = u^*(g) \circ u^*(f)$ if $g: N \to P$ is a second morphism of A-modules.

We thus have defined a covariant functor from the category of A-modules to the category of B-modules. In particular, if $B = \mathbf{Z}$, we obtain a functor from the category $\mathbf{Mod}_A$ of A-modules to the category $\mathbf{AbGr} = \mathbf{Mod}_\mathbf{Z}$ of abelian groups. Since these functors *forget* the initial A-module structure (at least, part of it), they are called forgetful functors.

7.5.3. — One says that a functor F (covariant or contravariant) from the category $\mathbf{Mod}_A$ of A-modules to the category $\mathbf{Mod}_B$ of B-modules is *additive* if the maps $\varphi \mapsto F(\varphi)$ from $\mathrm{Hom}_A(M, N)$ to $\mathrm{Hom}_B(F(M), F(N))$ (in the covariant case, and to $\mathrm{Hom}_B(F(N), F(M))$ in the contravariant case) are morphisms of abelian groups: $F(0) = 0$ and $F(f + g) = F(f) + F(g)$ for every $f, g \in \mathrm{Hom}_A(M, N)$.

Remark (7.5.4). — Let $f: M \to N$ and $g: N \to P$ be two morphisms of A-modules such that $g \circ f = 0$, so that the diagram $M \xrightarrow{f} N \xrightarrow{g} P$ is a complex. Then, for any covariant additive functor F: $\mathbf{Mod}_A \to \mathbf{Mod}_B$, the diagram $F(M) \xrightarrow{F(f)} F(N) \xrightarrow{F(g)} F(P)$ is a complex of B-modules. Indeed, one has $F(g) \circ F(f) = F(g \circ f) = F(0) = 0$.

Of course, an analogous remark holds for contravariant functors.

Definition (7.5.5). — Let F be an additive (covariant) functor from the category of A-modules to the category of B-modules.

One says that F is *left exact* if for every exact sequence

$$0 \to M \to N \to P,$$

the complex

$$0 \to F(M) \to F(N) \to F(P)$$

is exact.

One says that F is *right exact* if for every exact sequence

$$M \to N \to P \to 0,$$

the complex

$$F(M) \to F(N) \to F(P) \to 0$$

is exact.

One says that F is *exact* if it is both left and right exact.

A left exact functor preserves injective morphisms; a right exact functor preserves surjective morphisms. Observe that the left-exactness of F means that the morphism $F(M) \to F(N)$ is an isomorphism from $F(M)$ to the kernel of the morphism $F(N) \to F(P)$. Similarly, the right-exactness of F means that the morphism $F(N) \to F(P)$ is surjective and its kernel is the image of the morphism $F(M) \to F(N)$. Saying that a functor F is exact means that for any module M and any submodule N of M, $F(N)$ is identified (via $F(i)$) with a submodule of $F(N)$ and $F(M)/F(N)$ is identified (via $F(p)$) with $F(M/N)$, where $i\colon N \to M$ and $p\colon M \to M/N$ are the canonical injection and surjection.

There are analogous definitions for contravariant additive functors. Such a functor F is *right exact* if for every exact sequence of the form $0 \to M \to N \to P$, the complex $F(P) \to F(N) \to F(M) \to 0$ is exact. One says that F is *left exact* if for every exact sequence $M \to N \to P \to 0$, the complex $0 \to F(P) \to F(N) \to F(M)$ is an exact sequence. One says that F is *exact* if it is both left and right exact.

Forgetful functors associated to a morphism of rings transform any diagram of modules into the same diagram, where only part of the linearity of the maps has been forgotten. In particular, they transform any exact sequence into an exact sequence: *these forgetful functors are exact.*

Remark (7.5.6). — (i) If a functor F is exact, then for every short exact sequence $0 \to M \xrightarrow{u} N \xrightarrow{v} P \to 0$, the complex $0 \to F(M) \xrightarrow{F(u)} F(N) \xrightarrow{F(v)} F(P) \to 0$ is exact, since the two complexes obtained by removing an extremity are exact, by assumption.

Conversely, let us show that the exactness of such complexes is sufficient. If $u\colon M \to N$ is an injective morphism, considering the exact sequence $0 \to M \xrightarrow{u} N \to N/u(M) \to 0$ and its image by the functor F shows that $F(u)$ is injective. Similarly, if $v\colon N \to P$ is a surjective morphism, we see, considering the exact sequence $0 \to \mathrm{Ker}(v) \to N \xrightarrow{v} P \to 0$ and its image by F, that $F(v)$ is surjective.

Let now $0 \to M \xrightarrow{u} N \xrightarrow{v} P$ be an exact sequence. Let $v'\colon N \to \mathrm{Im}(v)$ be the morphism deduced from v and let $j\colon \mathrm{Im}(v) \to P$ be the inclusion, so that $v = j \circ v'$. Then $0 \to M \xrightarrow{u} N \xrightarrow{v'} \mathrm{Im}(v) \to 0$ is a short exact sequence, so that its image by F, $0 \to F(M) \xrightarrow{F(u)} F(N) \xrightarrow{F(v')} F(\mathrm{Im}(v)) \to 0$, is an exact sequence too. In particular, $\mathrm{Ker}(F(v')) = \mathrm{Im}(F(u))$. On the other hand, j is injective, so that $F(j)$ is injective as well, and the relation $F(v) = F(j) \circ F(v')$ implies that $\mathrm{Ker}(F(v')) = \mathrm{Ker}(F(v))$. This proves that the complex $0 \to F(M) \xrightarrow{F(u)} F(N) \xrightarrow{F(v)} F(P)$ is exact. In particular, F is left exact.

We prove analogously that F is right exact. Indeed, let $M \xrightarrow{u} N \xrightarrow{v} P \to 0$ be an exact sequence. Let $u'\colon M/\mathrm{Ker}(u) \to N$ be the morphism deduced from u, and let $p\colon M \to M/\mathrm{Ker}(u)$ be the canonical projection so that $u = u' \circ p$. Then $0 \to M/\mathrm{Ker}(u) \xrightarrow{u'} N \xrightarrow{v} P \to 0$ is an exact sequence, hence $0 \to F(M/\mathrm{Ker}(u)) \xrightarrow{F(u')} F(N) \xrightarrow{F(v)} F(P) \to 0$ is exact as well, hence $\mathrm{Ker}(F(v)) = \mathrm{Im}(F(u'))$. Since p is surjective, $F(p)$ is surjective; the relation

$F(u) = F(u') \circ F(p)$ then implies that $\mathrm{Im}(F(u)) = \mathrm{Im}(F(u'))$. Consequently, $F(M) \xrightarrow{F(u)} F(N) \xrightarrow{F(v)} F(P) \to 0$ is an exact sequence, as was to be shown.

(ii) Assume that F is an exact functor. Let us prove that for every exact sequence $M \xrightarrow{u} N \xrightarrow{v} P$, the complex $F(M) \xrightarrow{F(u)} F(N) \xrightarrow{F(v)} F(P)$ is exact.

Let $p \colon M \to M/\mathrm{Ker}(u)$ be the projection and let $u' \colon M/\mathrm{Ker}(u) \to N$ be the unique morphism such that $u' \circ p = u$. Let $j \colon \mathrm{Im}(v) \to P$ and let $v' \colon N \to \mathrm{Im}(v)$ be the morphism deduced from v so that $v = j \circ v'$. The exactness of the initial diagram furnishes an exact sequence

$$0 \to M/\mathrm{Ker}(u) \xrightarrow{u'} N \xrightarrow{v'} \mathrm{Im}(v) \to 0.$$

Since F is exact, the diagram

$$0 \to F(M/\mathrm{Ker}(u)) \xrightarrow{F(u')} F(N) \xrightarrow{F(v')} F(\mathrm{Im}(v)) \to 0$$

is thus an exact sequence, which means that the image of $F(u')$ is the kernel of $F(v')$.

On the other hand, the relation $u' \circ p = u$ implies that $F(u') \circ F(p) = F(u)$; since F is right exact and p is surjective, $F(p)$ is surjective as well, so that $\mathrm{Im}(F(u')) = \mathrm{Im}(F(u))$. Similarly, the relation $v = j \circ v'$ implies that $F(v) = F(j) \circ F(v')$; since F is left exact and j is injective, $F(j)$ is injective and $\mathrm{Ker}(F(v)) = \mathrm{Ker}(F(v'))$.

This proves that $\mathrm{Ker}(F(v)) = \mathrm{Im}(F(u))$, as claimed.

7.5.7. The Hom functors — Let A be a ring and let Q be a fixed (A, B)-bimodule. For every left A-module M, let us consider the abelian group $\mathrm{Hom}_A(Q, M)$, endowed with its natural structure of a left B-module. (For $u \in \mathrm{Hom}_A(Q, M)$ and $b \in B$, let $bu \in \mathrm{Hom}_A(Q, M)$ be the map $x \mapsto u(xb)$. For $b, b' \in B$ and $x \in Q$, $b(b'u)$ is the morphism given by $x \mapsto (b'u)(xb) = u(xbb')$, so that $b(b'u) = (bb')u$.) If $f \colon M \to N$ is a morphism of left A-modules, let us consider the map $f_* \colon \mathrm{Hom}_A(Q, M) \to \mathrm{Hom}_A(Q, N)$ given by $\varphi \mapsto f \circ \varphi$. For $f \colon M \to N$ and $g \colon N \to P$, one has $(g \circ f)_*(\varphi) = g \circ f \circ \varphi = g \circ (f \circ \varphi) = g_*(f_*(\varphi)) = (g_* \circ f_*)(\varphi)$ for every $\varphi \in \mathrm{Hom}_A(Q, M)$. Consequently, we have defined a covariant functor $\mathrm{Hom}_A(Q, \bullet)$ from the category of left A-modules to the category of left B-modules.

We can also consider the abelian group $\mathrm{Hom}_A(M, Q)$ endowed with its natural structure of a right B-module defined by $(bu)(x) = (b(x))u$ for $u \in \mathrm{Hom}_A(M, Q)$, $b \in B$ and $x \in M$. For two left A-modules M and N, and any morphism $f \colon M \to N$, let us consider the map $f^* \colon \mathrm{Hom}_A(N, Q) \to \mathrm{Hom}_A(M, Q)$ defined by $\varphi \mapsto \varphi \circ f$. For $f \colon M \to N$ and $g \colon N \to P$, one has $(g \circ f)^*(\varphi) = \varphi \circ g \circ f = (\varphi \circ g) \circ f = f^*(g^*(\varphi))$, so that $(g \circ f)^* = f^* \circ g^*$. We thus have defined a contravariant functor $\mathrm{Hom}_A(\bullet, Q)$ from the category of left A-modules to the category of right B-modules.

Proposition (7.5.8). — *Let* A, B *be rings and* Q *be a* (A, B)-*bimodule.*

a) *The functor* $\mathrm{Hom}_A(Q, \bullet)$ *is left exact. It is exact if and only if the* A-*module* Q *is projective.*

b) *The functor* $\mathrm{Hom}_A(\bullet, Q)$ *is left exact. It is exact if and only if the* A-*module* Q *is injective.*

Proof. — a) Let us consider an exact sequence of A-modules

$$0 \to M \xrightarrow{f} N \xrightarrow{g} P$$

and let

$$0 \to \mathrm{Hom}_A(Q, M) \xrightarrow{f_*} \mathrm{Hom}_A(Q, N) \xrightarrow{g_*} \mathrm{Hom}_A(Q, P)$$

be the complex obtained by applying the functor $\mathrm{Hom}_A(Q, \bullet)$. We need to show that this complex is exact.

Exactness at $\mathrm{Hom}_A(Q, M)$. Let $\varphi \in \mathrm{Hom}_A(Q, M)$ be such that $f_*(\varphi) = f \circ \varphi = 0$. This means that for every $x \in Q$, $f(\varphi(x)) = 0$; since f is injective, $\varphi(x) = 0$. Consequently, $\varphi = 0$ and f_* is injective.

Exactness at $\mathrm{Hom}_A(L, N)$. We want to show that $\mathrm{Ker}(g_*) = \mathrm{Im}(f_*)$; the inclusion $\mathrm{Im}(f_*) \subset \mathrm{Ker}(g_*)$ holds since we have a complex. So let $\varphi \in \mathrm{Hom}_A(Q, N)$ be such that $g_*(\varphi) = 0$ and let us show that there exists a morphism $\psi \in \mathrm{Hom}_A(Q, M)$ such that $\varphi = f_*(\psi)$. Let $x \in Q$; then $g(\varphi(x)) = 0$, hence $\varphi(x) \in \mathrm{Ker}(g)$. Since $\mathrm{Ker}(g) = \mathrm{Im}(f)$ by assumption, $\mathrm{Im}(\varphi) \subset f(M)$. In other words, φ is really a morphism from Q to $f(M)$. Since f is injective, it induces an isomorphism from M to $f(M)$ and there exists a morphism $\psi \colon Q \to M$ such that $\varphi = f \circ \psi = f_*(\psi)$, as desired.

Exactness. The functor $\mathrm{Hom}_A(Q, \bullet)$ is right exact if and only if for every exact sequence

$$0 \to M \xrightarrow{f} N \xrightarrow{g} P \to 0,$$

the complex

$$0 \to \mathrm{Hom}_A(Q, M) \xrightarrow{f_*} \mathrm{Hom}_A(Q, N) \xrightarrow{g_*} \mathrm{Hom}_A(Q, P) \to 0$$

is exact. The exactness at $\mathrm{Hom}_A(Q, P)$ is the only missing point, that is, the surjectivity of g_* assuming that of g. Precisely, g_* is surjective if and only if, for every morphism $\varphi \colon Q \to P$, there exists a morphism $\psi \colon Q \to N$ such that $\varphi = g_*(\psi) = g \circ \psi$. Since g is surjective, proposition 7.3.2 asserts that g_* is surjective if and only if Q is a projective A-module.

b) Let us now consider an exact sequence

$$M \xrightarrow{f} N \xrightarrow{g} P \to 0$$

and let

$$0 \to \mathrm{Hom}_A(P, Q) \xrightarrow{g^*} \mathrm{Hom}_A(N, Q) \xrightarrow{f^*} \mathrm{Hom}_A(M, Q)$$

be the complex obtained by applying the contravariant functor $\mathrm{Hom}_A(\bullet, Q)$. We need to show that this complex is exact.

Exactness at $\mathrm{Hom}_A(P, Q)$. Let us show that g^* is injective. Let $\varphi \in \mathrm{Hom}_A(P, Q)$ be such that $g^*(\varphi) = \varphi \circ g = 0$. This means that $\mathrm{Ker}(\varphi)$ contains $g(N)$. Since g is surjective, $g(N) = P$ and $\varphi = 0$.

Exactness at $\mathrm{Hom}_A(N, Q)$. We have to show that $\mathrm{Ker}(f^*) = \mathrm{Im}(g^*)$. Let $\varphi \in \mathrm{Hom}_A(N, Q)$ be such that $\varphi \circ f = 0$ and let us show that there exists a morphism $\psi \in \mathrm{Hom}_A(P, Q)$ such that $\varphi = g^*(\psi) = \psi \circ g$. Since $\varphi \circ f = 0$, $\mathrm{Im}(f)$ is contained in $\mathrm{Ker}(\varphi)$. Passing to the quotient, we deduce a morphism $\psi_0 \colon N/\mathrm{Im}(f) \to Q$ such that $\varphi(x) = \psi_0(\mathrm{cl}(x))$ for every $x \in N$, where $\mathrm{cl} \colon N \to N/\mathrm{Im}(f)$ is the canonical projection. Since g is surjective, it induces an isomorphism g' from $N/\mathrm{Ker}(g)$ to P. By assumption, $\mathrm{Ker}(g) = \mathrm{Im}(f)$; let then $\psi \colon P \to Q$ be the composition $\psi = \psi_0 \circ (g')^{-1}$. By construction, for every $x \in P$, $g^*(\psi)(x) = \psi_0 \circ (g')^{-1} \circ g(x) = \psi_0(\mathrm{cl}(x)) = \varphi(x)$. In other words, $g^*(\psi) = \varphi$, as requested.

Exactness. Let us now consider an exact sequence

$$0 \to M \xrightarrow{f} N \xrightarrow{g} P \to 0$$

and let

$$0 \to \mathrm{Hom}_A(P, Q) \xrightarrow{g^*} \mathrm{Hom}_A(N, Q) \xrightarrow{f^*} \mathrm{Hom}_A(M, Q) \to 0$$

be the complex obtained by applying the contravariant functor $\mathrm{Hom}_A(\bullet, Q)$. We need to show that this complex is exact if and only if Q is an injective A-module.

Only the exactness at $\mathrm{Hom}_A(M, Q)$ has not been shown, that is, assuming the injectivity of f, the surjectivity of f^*. Surjectivity of f^* means that for every $\varphi \in \mathrm{Hom}_A(M, Q)$, there exists a morphism $\psi \in \mathrm{Hom}_A(N, Q)$ such that $\psi \circ g = \varphi$. By proposition 7.4.2, this condition holds if and only if Q is an injective A-module. $\square$

In fact, the Hom functor can be used to detect exactness of complexes.

Lemma (7.5.9). — *Let* A *be a ring. A complex*

$$M \xrightarrow{f} N \xrightarrow{g} P \to 0$$

of A-*modules is exact if and only if the complex*

$$0 \to \mathrm{Hom}_A(P, Q) \xrightarrow{g^*} \mathrm{Hom}_A(N, Q) \xrightarrow{f^*} \mathrm{Hom}_A(M, Q)$$

is exact for every A-*module* Q.

Proof. — One direction of the assertion is given by the left-exactness of the functor $\mathrm{Hom}(\bullet, Q)$. Conversely, let us assume that

$$0 \to \mathrm{Hom}_A(P, Q) \xrightarrow{g^*} \mathrm{Hom}_A(N, Q) \xrightarrow{f^*} \mathrm{Hom}_A(M, Q)$$

is exact for every A-module Q and let us show that the diagram

$$M \xrightarrow{f} N \xrightarrow{g} P \to 0$$

is exact.

We first prove that g is surjective. Let $Q = \mathrm{Coker}(g) = P/g(N)$ and let $\varphi \colon P \to Q$ be the canonical surjection. By construction, $g^*(\varphi) = \varphi \circ g = 0$. Since g^* is injective, $\varphi = 0$. This means that $Q = 0$, hence g is surjective.

By assumption, $g \circ f = 0$, hence $\mathrm{Im}(f) \subset \mathrm{Ker}(g)$. Let $Q = N/\mathrm{Im}(f)$ and let $\varphi \colon N \to Q$ be the canonical surjection. By construction, $f \circ \varphi = 0$, hence φ belongs to the kernel of the map f^* from $\mathrm{Hom}_A(N, Q)$ to $\mathrm{Hom}_A(M, Q)$. By assumption, there exists a morphism $\psi \in \mathrm{Hom}_A(P, Q)$ such that $\varphi = g^*(\psi) = \psi \circ g$. In particular, $\mathrm{Ker}(g)$ is contained in $\mathrm{Ker}(\varphi) = \mathrm{Im}(f)$, as was to be shown. □

Remark (7.5.10). — It is tempting to state a dual version that asserts that a complex

$$0 \to M \xrightarrow{f} N \xrightarrow{g} P$$

of A-modules is exact if and only if the complex

$$0 \to \mathrm{Hom}_A(Q, M) \xrightarrow{f_*} \mathrm{Hom}_A(Q, N) \xrightarrow{g_*} \mathrm{Hom}_A(Q, P)$$

is exact for every A-module Q. Of course, one direction of this statement follows from the left-exactness of the functor $\mathrm{Hom}_A(Q, \bullet)$. However, the other is trivial, since setting $Q = A$ recovers the initial complex, which is thus an exact sequence.

7.5.11. Localization functors — Let A be a commutative ring and let S be a multiplicative subset of A. For any A-module M, we have defined in section 3.6 an $S^{-1}A$-module $S^{-1}M$. Moreover, if $f \colon M \to N$ is any morphism of A-modules, we have constructed (*cf.* proposition 3.6.4) a morphism $S^{-1}f \colon S^{-1}M \to S^{-1}N$ characterized by the property that $(S^{-1}f)(m/s) = f(m)/s$ for every $m \in M$ and every $s \in S$. Moreover, $S^{-1}(f \circ g) = S^{-1}f \circ S^{-1}g$.

In other words, localization induces a functor from the category of A-modules to that of $S^{-1}A$-modules. It is crucial that these localization functors are exact; this is in fact a reformulation of proposition 3.6.6.

Proposition (7.5.12) (Exactness of localization). — *Let A be a ring and let S be a multiplicative subset of A. For any exact sequence $0 \to M \xrightarrow{f} N \xrightarrow{g} P \to 0$ of A-modules, the complex $0 \to S^{-1}M \xrightarrow{S^{-1}f} S^{-1}N \xrightarrow{S^{-1}g} S^{-1}P \to 0$ is an exact sequence. In other words, the localization functor from A-modules to $S^{-1}A$-modules is an exact functor.*

Proof. — We can use the injective morphism f to identify M with the submodule $f(M)$ of N, and the surjective morphism g to identify P with the quotient $N/f(M)$ of N. By proposition 3.6.6, the morphism $S^{-1}f$ is injective; moreover, the map $S^{-1}p\colon S^{-1}N \to S^{-1}P$ identifies the quotient $S^{-1}N/S^{-1}M$ with $S^{-1}(N/M) = S^{-1}P$. In other words, the morphism $S^{-1}g$ is surjective and its kernel is the image of $S^{-1}f$. This means that the diagram

$$0 \to S^{-1}M \xrightarrow{S^{-1}f} S^{-1}N \xrightarrow{S^{-1}g} S^{-1}P \to 0$$

is an exact sequence. $\square$

7.6. Adjoint Functors

Let A and B be rings. Let F be a functor from the category of A-modules to the category of B-modules, and let G be a functor from the category of B-modules to the category of A-modules.

Definition (7.6.1). — An *adjunction* for the pair (F, G) is the datum, for any A-module M and any B-module N, of a bijection

$$\Phi_{M,N}\colon \operatorname{Hom}_B(F(M), N) \to \operatorname{Hom}_A(M, G(N))$$

such that for any morphism $f\colon M \to M'$ of A-modules and any morphism $g\colon N \to N'$ of B-modules, one has

$$f^* \circ G(g)_* \circ \Phi_{M',N} = \Phi_{M,N'} \circ F(f)^* \circ g_*.$$

Explicitly, this condition means that for every $\varphi \in \operatorname{Hom}_B(F(M'), N)$, the two elements

$$G(g) \circ \Phi_{M',N}(\varphi) \circ f \quad \text{and} \quad \Phi_{M,N'}(g \circ \varphi \circ F(f))$$

of $\operatorname{Hom}_A(M, G(N'))$ are equal. It is often visually represented by saying that the diagram

$$
\begin{array}{ccc}
\operatorname{Hom}_B(F(M'), N) & \xrightarrow{\Phi_{M',N}} & \operatorname{Hom}_A(M', G(N)) \\
{\scriptstyle F(f)^* \circ g_*}\Big\downarrow & & \Big\downarrow{\scriptstyle f^* \circ G(g)_*} \\
\operatorname{Hom}_B(F(M), N') & \xrightarrow{\Phi_{M,N'}} & \operatorname{Hom}_A(M, G(N'))
\end{array}
$$

is commutative.

   If there exists such an adjunction, then one says that F and G *form a pair of adjoint functor, or, simply, are adjoint; one also says that* G *is a right adjoint of* F *and that* F *is a left adjoint of* G.

   Although we will not make use of this fact, let us mention the consequence of Yoneda's lemma (proposition A.3.14) that if two functors F and G are adjoint, then each of them essentially determines the other.

Lemma (7.6.2). — *Assume that* F *and* G *are adjoint functors and let* $\Phi = (\Phi_{M,N})$ *be an adjunction for the pair* (F, G).

(i) *For any* A-*module* M, *let* $\alpha_M = \Phi_{M,F(M)}(\mathrm{id}_{F(M)})\colon M \to G(F(M))$. *For any* B-*module* N *and any morphism* $g\colon F(M) \to N$, *one has* $\Phi_{M,N}(g) = G(g) \circ \alpha_M$.

(ii) *For any* B-*module* N, *let* $\beta_N\colon F(G(N)) \to N$ *be the unique morphism such that* $\Phi_{G(N),N}(\beta_N) = \mathrm{id}_{G(N)}$. *Then, for any* A-*module* M *and any morphism* $f\colon M \to G(N)$, *one has* $\Phi_{M,N}^{-1}(f) = \beta_N \circ F(f)$.

Proof. — For *a*), one applies the definition with $M' = M$, $f = \mathrm{id}_M$, and $\varphi = \mathrm{id}_{F(M)}$. This gives

$$\Phi_{M,N}(g) = \Phi_{M,N}(g \circ \mathrm{id}_{F(M)} \circ F(\mathrm{id}_M))$$
$$= G(g) \circ \Phi_{M,F(M)}(\mathrm{id}_{F(M)}) \circ \mathrm{id}_M$$
$$= G(g) \circ \alpha_M.$$

For *b*), one considers $M' = G(N)$, $N' = N$ and $g\ \mathrm{id}_N$. Then, one gets

$$\Phi_{M,N}(\beta_N \circ F(f)) = \Phi_{M,N'}(\mathrm{id}_N \circ \beta_N \circ F(f))$$
$$= G(\mathrm{id}_N) \circ \Phi_{G(N),N}(\beta_N) \circ f$$
$$= \mathrm{id}_{G(N)} \circ \mathrm{id}_{G(N)} \circ f$$
$$= f.$$

Example (7.6.3). — Let A and B be rings, let $f\colon B \to A$ be a morphism of rings and let F be the corresponding forgetful functor from right A-modules to right B-modules. Let P be the ring A viewed as an (A, B)-bimodule via the laws $a \cdot x \cdot b = axf(b)$ for any $a \in A$, $x \in P$ and $b \in B$. Let then G be the functor that associates to any right B-module N the right A-module $\mathrm{Hom}_B(P, N)$.

Let M be a right A-module, let N be an abelian group; let us define a bijection $\Phi_{M,N}$ from $\mathrm{Hom}_B(F(M), N)$ to $\mathrm{Hom}_A(M, G(N))$. Let $\varphi \in \mathrm{Hom}_B(F(M), N)$; in other words, φ is an additive map from M to N such that $\varphi(mf(b)) = \varphi(m)b$ for any $m \in M$ and any $b \in B$. For any $m \in M$, let $\psi_m\colon P \to N$ be the map given by $a \mapsto \varphi(ma)$; it is additive; for any $a \in A$ and any $b \in B$, one has $\psi_m(a)b = \varphi(ma)b = \varphi(maf(b)) = \psi_m(af(b))$. Consequently, ψ_m belongs to $\mathrm{Hom}_B(P, N) = G(N)$. The map $\psi\colon m \mapsto \psi_m$ from M to G(N) is additive. It is even A-linear. Indeed, let $a \in A$ and let $m \in M$; the map ψ_{ma} sends $a' \in A$ to $\varphi(maa')$; on the other hand, $\psi_m \cdot a$ is the map $a' \mapsto \psi_m(aa') = \varphi(maa')$. Consequently, $\psi_{ma} = \psi_m \cdot a$, as was to be shown. Let us then set $\Phi_{M,N}(\varphi) = \psi$.

The map $\Phi_{M,N}\colon \mathrm{Hom}_B(F(M), N) \to \mathrm{Hom}_A(M, G(N))$ is bijective. Indeed, if $\psi\colon M \to \mathrm{Hom}_B(P, N)$ is a morphism of A-modules, then the morphism $\varphi\colon F(M) \to N$ given by $\varphi(m) = \psi(m)(1)$ satisfies $\Phi_{M,N}(\varphi) = \psi$, since

$$\Phi_{M,N}(\varphi)(m)(a) = \varphi(ma) = \psi(ma)(1) = \psi(m)(a),$$

and it is the only one.

We now check that the maps $\Phi_{M,N}$ satisfy the condition given in the definition of adjoint functors. Let $f \colon M \to M'$ be a morphism of right A-modules, let $g \colon N \to N'$ be a morphism of right B-modules. Let $\varphi \in \mathrm{Hom}_A(F(M'), N)$ and let $m \in M$. Then,

$$f^* \circ G(g)_* \circ \Phi_{M',N}(\varphi)(m) \quad \text{and} \quad \Phi_{M,N'} \circ F(f)^* \circ g_*(\varphi)(m)$$

are both equal to the map $a \mapsto g(\varphi(f(m)a))$.

Consequently, F and G form a pair of adjoint functors.

Let us also describe the maps α_M and β_N. Let M be an A-module, so that $G(F(M))$ is the right A-module $\mathrm{Hom}_B(A, M)$. The morphism $\alpha_M \colon M \to G(F(M))$ is defined as $\Phi_{M,F(M)}(\mathrm{id}_{F(M)})$. Consequently, for any $m \in M$, $\alpha_M(m)$ is the element $a \mapsto ma$ of $\mathrm{Hom}_B(A, M)$. In particular, *the morphism α_M is injective.*

Let N be a B-module; $F(G(N))$ is the right B-module $\mathrm{Hom}_B(A, N)$. Moreover, $\beta_N \colon F(G(N)) \to N$ maps an element $u \in \mathrm{Hom}_B(A, N)$ to $u(1) \in N$. Indeed, let us write β'_N for this morphism and let us compute $\Phi_{G(N),N}(\beta'_N)$. With the previous notation, we set $M = G(N)$, $\varphi = \beta'_N$ and $\psi = \Phi_{G(N),N}(\beta'_N)$. For $u \in G(N)$, $\psi_u \colon P \to N$ is the map $a \mapsto \beta'_N(u \cdot a)$. By definition, $\beta'_N(u \cdot a) = (u \cdot a)(1) = u(a)$, so that $\psi_u = u$. Consequently, $\psi = \mathrm{id}_{G(N)}$ and $\beta'_N = \beta_N$, as was to be shown. We observe in particular that *the morphism β_N is surjective.*

In the preceding example, the functor G, being of the form $\mathrm{Hom}_B(P, \bullet)$, is left exact. The next proposition shows that this property is shared by all functors which are right adjoints.

Proposition (7.6.4). — *Assume that the functors F and G are adjoint. Then F is right exact and G is left exact.*

Proof. — *a*) Let $M \xrightarrow{f} N \xrightarrow{g} P \to 0$ be an exact sequence of A-modules and let us prove that the complex

$$F(M) \xrightarrow{F(f)} F(N) \xrightarrow{F(g)} F(P) \to 0$$

is exact. By lemma 7.5.9, it suffices to show that for every B-module Q, the complex

$$0 \to \mathrm{Hom}_A(F(P), Q) \xrightarrow{F(g)^*} \mathrm{Hom}_A(F(N), Q) \xrightarrow{F(f)^*} \mathrm{Hom}_A(F(M), Q)$$

is exact. Applying the adjunction property, this complex identifies with the complex

$$0 \to \mathrm{Hom}_A(P, G(Q)) \xrightarrow{g^*} \mathrm{Hom}_A(N, G(Q)) \xrightarrow{f^*} \mathrm{Hom}_A(M, G(Q))$$

which is exact, because the functor $\mathrm{Hom}_A(\bullet, G(Q))$ is left exact.

b) Let $0 \to M \xrightarrow{f} N \xrightarrow{g} P$ be an exact sequence of B-modules and let us prove that the complex

$$0 \to G(M) \xrightarrow{G(f)} G(N) \xrightarrow{G(g)} G(P)$$

of A-modules is exact. Let Q be an A-module and let us apply the functor $\mathrm{Hom}_A(Q, \bullet)$ to the preceding complex: we get

$$0 \to \mathrm{Hom}_A(Q, G(M)) \xrightarrow{G(f)_*} \mathrm{Hom}_A(Q, G(N)) \xrightarrow{G(g)_*} \mathrm{Hom}_A(Q, G(P)).$$

Using the adjunction property of the functors F and G, we can rewrite this complex as

$$0 \to \mathrm{Hom}_A(F(Q), M) \xrightarrow{f_*} \mathrm{Hom}_A(F(Q), N) \xrightarrow{g_*} \mathrm{Hom}_A(F(Q), P),$$

which is an exact sequence since the functor $\mathrm{Hom}_A(F(Q), \bullet)$ is left exact. $\square$

Proposition (7.6.5). — *Let A and B be rings. Let F be a functor from the category of A-modules to the category of B-modules and let G be a functor from the category of B-modules to the category of A-modules. Assume that F and G form an adjoint pair.*

a) If F is exact, then G(N) is an injective A-module for any injective B-module N.
b) If G is exact, then F(M) is a projective B-module for any projective A-module M.

Proof. — Let $(\Phi_{M,N})$ be an adjunction for the pair (F, G). For any A-module M, let $\alpha_M = \Phi_{M,F(M)}(\mathrm{id}_{F(M)})$. For any B-module N, let $\beta_N = \Phi^{-1}_{G(N),N}(\mathrm{id}_{G(N)})$.

a) Let M be an A-module and let $f : G(N) \to M$ be an injective morphism; let us show that f has a left inverse. Let $P = \mathrm{Coker}(f)$ so that we have an exact sequence

$$0 \to G(N) \xrightarrow{f} M \to P \to 0.$$

Applying the exact functor F, we obtain an exact sequence

$$0 \to F(G(N)) \xrightarrow{F(f)} F(M) \to F(P) \to 0.$$

In particular, $F(f)$ is injective. Since N is an injective B-module and $F(f)$ is injective, proposition 7.4.2 implies that there exists a morphism $v : F(M) \to N$ such that $v \circ F(f) = \beta_N$ (apply assertion (ii) with $i = F(f)$ and $f = \beta_N$). Set $g = \Phi_{M,N}(v)$. By the definition of adjoint functors, one has

$$g \circ f = \Phi_{M,N}(v) \circ f = \Phi_{G(N),N}(v \circ F(f)) = \Phi_{G(N),N}(\beta_N) = \mathrm{id}_{G(N)}.$$

This shows that g is a left inverse of f and concludes the proof that G(N) is an injective A-module.

b) This is analogous. Let N be a B-module and let $f : N \to F(M)$ be a surjective morphism of B-modules. Let $P = \mathrm{Ker}(f)$, so that we have an exact sequence of B-modules

$$0 \to P \to N \xrightarrow{f} F(M) \to 0.$$

Applying the exact functor G, we obtain an exact sequence

$$0 \to G(P) \to G(N) \xrightarrow{G(f)} G(F(M)) \to 0$$

of A-modules; in particular, the morphism $G(F(f))$ is surjective. Since M is a projective A-module, there exists a morphism $u \colon M \to G(N)$ such that $G(f) \circ u = \alpha_M$. Let $g = \Phi_{N,M}^{-1}(u)$. By definition of an adjoint pair, one has

$$\Phi_{M,F(M)}(f \circ g) = G(f) \circ \Phi_{N,M}(g) = G(f) \circ u = \alpha_M.$$

Consequently, $f \circ g = \mathrm{id}_{F(M)}$. In particular, f has a right inverse. This shows that $F(M)$ is a projective B-module.   □

Remark (7.6.6). — As an application of the fact that the right adjoint of an exact functor preserves injectives, let us give another (easier) proof that every module can be embedded in an injective module.

We first treat the case of the ring Z. So let M be a Z-module. Write M as the quotient of a free Z-module: let $(e_i)_{i \in I}$ be a generating family in M, let $f \colon \mathbf{Z}^{(I)} \to M$ be the unique morphism given by $(a_i)_i \mapsto \sum m_i a_i$ and let $N = \mathrm{Ker}(f)$, so that $M \simeq \mathbf{Z}^{(I)}/N$. View $\mathbf{Z}^{(I)}$ as a submodule of $\mathbf{Q}^{(I)}$ and define $M' = \mathbf{Q}^{(I)}/N$. The injection $\mathbf{Z}^{(I)} \to \mathbf{Q}^{(I)}$ induces an injection from M to M'. Since $\mathbf{Q}^{(I)}$ is divisible, so is M'. By corollary 7.4.4, M' is an injective Z-module.

Let now A be a ring and let F be the forgetful functor from right A-modules to Z-modules. By example 7.6.3, the functor $G = \mathrm{Hom}_{\mathbf{Z}}(A, \bullet)$ is a right adjoint to F. Moreover, F is exact and G is left exact (proposition 7.5.8), so that G preserves injective modules. Let M be a right A-module. Let $f \colon F(M) \to M'$ be an injection from the Z-module $F(M)$ to an injective Z-module M'. Then, $G(M')$ is an injective A-module and the morphism $G(f)$ is injective.

Moreover, we have seen that in this case, the morphism $\alpha_M \colon M \to G(F(M))$ is injective. The composition $G(f) \circ \alpha_M$ is thus an injective morphism from M to an injective A-module.

7.7. Differential Modules. Homology and Cohomology

Definition (7.7.1). — Let A be a ring. A *differential A-module* is a pair (M, d) where M is an A-module and d is an endomorphism of M such that $d^2 = 0$, called the *differential* of M.

Let (M, d_M) and (N, d_N) be differential A-modules. One says that a morphism $f \colon M \to N$ of A-modules is a *morphism of differential modules* if $d_N \circ f = f \circ d_M$.

Definition (7.7.2). — To any differential A-module (M, d), one associates the following A-modules:

  – the module $Z(M) = \operatorname{Ker} d$ of *cycles*;
  – the module $B(M) = \operatorname{Im} d$ of *boundaries*;
  – the module $H(M) = Z(M)/B(M) = \operatorname{Ker}(d)/\operatorname{Im}(d)$ of *homologies*.

Since $d^2 = 0$, observe that $\operatorname{Im}(d) \subset \operatorname{Ker}(d)$, so that the definition of the module of homologies makes sense.

Lemma (7.7.3). — *Let $f : (M, d_M) \to (N, d_N)$ be a morphism of differential modules. One has $f(Z(M)) \subset Z(N)$ and $f(B(M)) \subset B(N)$. Consequently, f induces a morphism $H(f) : H(M) \to H(N)$ of A-modules.*

Proof. — By definition $f \circ d_M = d_N \circ f$. Consequently, if $x \in M$ is such that $d_M(x) = 0$, one has $d_N(f(x)) = f(d_M(x)) = 0$, so that $d_M(x) \in Z(N)$. Similarly, if $x \in M$, one has $f(d_M(x)) = d_N(f(x))$ so that $f(d_M(x)) \in B(N)$ and $f(B(M)) \subset B(N)$. The definition of $H(f)$ follows by passing to the quotient.□

Remark (7.7.4). — Morphisms of differential A-modules can be composed, and the identity morphism is a morphism of differential A-modules, so that differential A-modules form a category. We have associated to every differential A-module (M, d_M) its homology module $H(M)$, which is an A-module, and to every morphism $f : (M, d_M) \to (N, d_N)$ of differential A-modules a morphism $H(f) : H(M) \to H(N)$ of A-modules.
  Observe that if $f = \operatorname{id}_M$, then $H(f) = \operatorname{id}_{H(M)}$. Moreover, if $f : (M, d_M) \to (N, d_N)$ and $g : (N, d_N) \to (P, d_P)$ are morphisms of differential modules, then $H(g \circ f) = H(g) \circ H(f)$, since both of these morphisms map the class of $x \in Z(M)$ modulo $B(M)$ to the class of $g(f(x)) \in Z(P)$ modulo $B(P)$. Moreover, if $f, g : (M, d_M) \to (N, d_N)$ are morphisms of differential modules, one has
$$H(f + g) = H(f) + H(g)$$
  In other words, H is an additive functor from the category of differential A-modules to the category of A-modules.

In many important applications, for example to topology, the differential A-modules that come in are in fact *graded* differential A-modules.

Definition (7.7.5). — A *graded differential A-module* is a differential module (M, d) such that

  – The module M is the direct sum of a family $(M_n)_{n \in \mathbf{Z}}$ (one says that M is **Z**-graded, or simply, *graded*);
  – There exists an integer r such that $d(M_n) \subset M_{n+r}$ for every $n \in \mathbf{Z}$ (one says that d has *degree r*).

In this case, one defines $Z_n(M) = Z(M) \cap M_n$, $B_n(M) = B(M) \cap M_n$ and $H_n(M) = Z_n(M)/B_n(M)$.

Lemma (7.7.6). — *The modules of cycles, boundaries and homologies of a graded differential A-module are graded. Explicitly, if (M, d) is a graded differential A-module, one has equalities $Z(M) = \bigoplus_n Z_n(M)$ and $B(M) = \bigoplus B_n(M)$; they induce an isomorphism $H(M) \simeq \bigoplus_{n \in \mathbf{Z}} H_n(M)$.*

Proof. — Observe that $Z_n(M)$ is a submodule of M_n; since the modules M_n are in direct sum, so are the $Z_n(M)$. Moreover, $Z_n(M) \subset Z(M)$ for every n. On the other hand, let $x \in Z(M)$; one can write x as a sum $x = \sum_{n \in \mathbf{Z}} x_n$ of an almost-null family, where $x_n \in M_n$ for every $n \in \mathbf{Z}$. Then $d(x) = \sum_{n \in \mathbf{Z}} d(x_n)$. For every n, $d(x_n) \in M_{n+r}$; since the modules M_n are in direct sum, $d(x_n) = 0$ for every n. Consequently, $x_n \in Z(M_n)$ for every n and $Z(M) = \bigoplus_n Z_n(M)$.

The modules $B_n(M)$ are in direct sum, and are submodules of $B(M)$. Conversely, let $x \in B(M)$; let $y \in M$ be such that $x = d(y)$. One may write $y = \sum_n y_n$, where $y_n \in M_n$ for every n. It follows that $x = \sum_n d(y_n)$. For every n, $d(y_n) \in B_{n+r}$, hence $x \in \sum_n B(M_n)$.

Finally, one has isomorphisms

$$H(M) = Z(M)/B(M) = \left(\bigoplus_n Z_n(M) \right) \Big/ \left(\bigoplus_n B_n(M) \right)$$

$$= \bigoplus_n (Z_n(M)/B_n(M)) = \bigoplus_n H_n(M).$$

This proves the lemma.   □

Example (7.7.7). — Let us consider a complex

$$\dots \xrightarrow{f_{n-1}} M_{n-1} \xrightarrow{f_n} M_n \xrightarrow{f_{n+1}} M_{n+1} \to \dots$$

of A-modules. Let us define $M = \bigoplus_{n \in \mathbf{Z}} M_n$ and let $d \in \mathrm{End}(M)$ be the unique endomorphism of M such that $d(x) = f_n(x)$ for every $x \in M_n$ and every $n \in \mathbf{Z}$.

Then (M, d) is a graded differential A-module whose differential has degree 1. Moreover, $H_n(M) = \mathrm{Ker}(f_{n+1})/\mathrm{Im}(f_n)$. In other words, the vanishing of $H_n(M)$ witnesses the exactness of the complex at the module M_n. Tradition denotes these modules by $H^n(M)$ and calls them *cohomology modules* of the complex.

Example (7.7.8) (De Rham complex of a manifold). —Let M be a manifold. For every $p \in \mathbf{N}$, let $\Omega^p(M)$ be the $\mathbf{R}$-vector space of differential forms of degree p on M. The exterior differential $d\colon \Omega^p(M) \to \Omega^{p+1}(M)$ satisfies $d \circ d = 0$. Consequently, the direct sum of the $\Omega^p(M)$ is a graded differential $\mathbf{R}$-module, the *de Rham complex* of M. Its cohomology modules are called the de Rham cohomology groups of M; they encode important information about the topology of M.

Let us describe them more explicitly in the particular case of an open subset U of $\mathbf{R}^2$. The spaces $\Omega^p(U)$ are zero unless $0 \leq p \leq 2$, and the spaces $\Omega^0(U), \Omega^1(U), \Omega^2(U)$ can be described explicitly as follows. For $p = 0$,

$\Omega^0(U) = \mathscr{C}^\infty(U)$ is the real vector space of $\mathscr{C}^\infty$-functions on U. For $p = 1$, $\Omega^1(U)$ is a free $\mathscr{C}^\infty(U)$-modules of rank 2, with basis (dx, dy). In other words, a differential form ω of degree 1 on U can be uniquely written as

$$\omega = A(x, y)\, dx + B(x, y)\, dy.$$

Finally, $\Omega^2(U)$ is a free $\mathscr{C}^\infty(U)$-module of rank 1, with basis denoted $dx \wedge dy$. The "exterior differential" consists of the maps

$$d: \Omega^0(U) \to \Omega^1(U)$$

$$f \mapsto \frac{\partial f}{\partial x}\, dx + \frac{\partial f}{\partial y}\, dy$$

$$d: \Omega^1(U) \to \Omega^2(U)$$

$$A(x, y)\, dx + B(x, y)\, dy \mapsto \left(\frac{\partial B(x, y)}{\partial x} - \frac{\partial A(x, y)}{\partial y} \right) dx \wedge dy$$

and d vanishes identically on $\Omega^2(U)$.

Observe that $d \circ d = 0$. The only necessary computation is that of $d^2(f)$ for $f \in \Omega^0(U)$. Then,

$$d^2(f) = d\left(\frac{\partial f}{\partial x}\, dx + \frac{\partial f}{\partial y}\, dy \right) = \left(\frac{\partial^2 f}{\partial x \partial y} - \frac{\partial^2 f}{\partial y \partial x} \right) dx \wedge dy,$$

hence $d^2(f) = 0$ by Schwarz's theorem.

Consequently, we have defined a complex $\Omega^\bullet$:

$$0 \to \Omega^0(U) \xrightarrow{d} \Omega^1(U) \xrightarrow{d} \Omega^2(U) \to 0,$$

the *de Rham complex of* U. Its cohomology groups $H^i_{dR}(U)$ are called the *de Rham cohomology groups of* U. These real vector spaces are of fundamental interest for topology.

Let us compute $H^0_{dR}(U)$. Let $f \in Z^0(\Omega^\bullet)$; this means that $f \in \mathscr{C}^\infty(U)$ and that $df = 0$, that is, $\frac{\partial f}{\partial x} = \frac{\partial f}{\partial y} = 0$. Consequently, f is constant on every connected component of U. Since $B^0(\Omega^\bullet) = 0$, we obtain an isomorphism $H^0_{dR}(U) = \mathbf{R}^{\pi_0(U)}$, where $\pi_0(U)$ is the set of connected components of U.

Let us assume that U is simply connected; for example, U could be $\mathbf{R}^2$, or contractible, or star-shaped. Then, Poincaré's lemma asserts that any differential form ω on U of degree > 0 which is a cycle (one says that ω is closed) is a boundary (one says that ω is exact). In Physics or Vector calculus, this lemma appears under a more elementary formulation: a vector field whose rotational is zero is a gradient.

Let us prove it when U is star-shaped with respect to the origin 0 of $\mathbf{R}^2$. Let $\omega = A(x, y)\, dx + B(x, y)\, dy$ be any closed differential form of degree 1. Since U is star-shaped with respect to the origin, for every point $(x, y) \in U$, and every $t \in [0, 1]$, one has $(tx, ty) \in U$. We may thus set

$$f(x, y) = \int_0^1 \left(xA(tx, ty) + yB(tx, ty) \right) \, dt.$$

Then f is $\mathscr{C}^\infty$ (see a course in calculus) and its partial derivatives can be computed by differentiating under the integral-sign. Thus one obtains

$$\frac{\partial f}{\partial x}(x, y) = \int_0^1 \left(A(tx, ty) + tx\frac{\partial A}{\partial x}(tx, ty) + ty\frac{\partial B}{\partial x}(tx, ty) \right) \, dt.$$

Since $d\omega = 0$, $\frac{\partial B}{\partial x} = \frac{\partial A}{\partial y}$ so that

$$\frac{\partial f}{\partial x}(x, y) = \int_0^1 \left(A(tx, ty) + tx\frac{\partial A}{\partial y}(tx, ty) + ty\frac{\partial B}{\partial x}(tx, ty) \right) \, dt$$

$$= \int_0^1 \left(A(tx, ty) + t\frac{d}{dt}(A(tx, ty)) \right) \, dt$$

$$= \int_0^1 \frac{d}{dt}(tA(tx, ty)) = \left[tA(tx, ty) \right]_0^1$$

$$= A(x, y).$$

One proves similarly that $\frac{\partial f}{\partial y}(x, y) = B(x, y)$, so that $df = \omega$.

On the other hand, if $U = \mathbf{R}^2 - \{0\}$, one can prove that $H_{dR}^1(U)$ has dimension 1 and is generated by the class of the differential form

$$-\frac{y}{x^2 + y^2}dx + \frac{x}{x^2 + y^2}dy.$$

Theorem (7.7.9). — *Let* A *be a ring, let* M, N, P *be differential modules and let* $f \colon M \to N$ *and* $g \colon N \to P$ *be morphisms of differential modules. Assume that one has an exact sequence* $0 \to M \xrightarrow{f} N \xrightarrow{g} P \to 0$. *(These conditions can be summed up by saying that we have an exact sequence of differential modules.) Then, there exists a morphism of* A-*modules* $\partial \colon H(P) \to H(M)$ *such that*

$$\mathrm{Ker}(\partial) = \mathrm{Im}(H(g)), \quad \mathrm{Ker}(H(g)) = \mathrm{Im}(H(f)), \quad \mathrm{Ker}(H(f)) = \mathrm{Im}(\partial).$$

Consequently, the morphism ∂ sits in an "exact triangle":

Proof. — a) Let us show that $\mathrm{Ker}(H(g)) = \mathrm{Im}(H(f))$. Since $g \circ f = 0$, one has $H(g) \circ H(f) = H(g \circ f) = 0$ and $\mathrm{Im}(H(f)) \subset \mathrm{Ker}(H(g))$. Conversely, let $\xi \in \mathrm{Ker}(H(g))$. By definition, ξ is the class of an element $x \in \mathrm{Ker}(d_N)$. Then,

$H(g)(\xi) = \mathrm{cl}(g(x)) = 0$, so that there exists a $y \in P$ such that $g(x) = d_P(y)$. Since the morphism g is surjective, there exists a $z \in N$ such that $y = g(z)$. Then $g(x) = d_P(y) = d_P(g(z)) = g(d_N(z))$ hence $x - d_N(z)$ belongs to $\mathrm{Ker}(g)$. Since $\mathrm{Ker}(g) = \mathrm{Im}(f)$, there exists a $t \in M$ such that $x - d_N(z) = f(t)$. It follows that $\xi = \mathrm{cl}(x) = \mathrm{cl}(d_N(z) + f(t)) = \mathrm{cl}(f(t)) = H(f)(\mathrm{cl}(t))$. Consequently, $\mathrm{Ker}(H(g)) \subset \mathrm{Im}(H(f))$, as was to be shown.

b) Let us now define a morphism $\partial \colon H(P) \to H(M)$. Let $\xi \in H(P)$. Write $\xi = \mathrm{cl}(x)$ for some $x \in \mathrm{Ker}\, d_P$. Since g is surjective, there exists a $y \in N$ such that $x = g(y)$. Then $0 = d_P(x) = d_P(g(y)) = g(d_N(y))$ hence $d_N(y) \in \mathrm{Ker}(g)$. Since $\mathrm{Ker}(g) = \mathrm{Im}(f)$, there exists a $z \in M$ such that $d_N(y) = f(z)$. One has $f(d_M(z)) = d_N(f(z)) = d_N \circ d_N(y) = 0$ since $d_N^2 = 0$. Since f is injective, $d_M(z) = 0$. We are going to set $\partial(\xi) = \mathrm{cl}(z) \in H(M)$.

Let us first check that $\mathrm{cl}(z)$ is independent of any choice. Recall that x, y, z have been chosen in such a way that one has $f(z) = d_N(y)$ and $x = g(y)$. Let x', y', z' be such that $f(z') = d_N(y')$ and $x' = g(y')$. Assume that $\mathrm{cl}(x) = \mathrm{cl}(x')$; then there exists an element $x'' \in P$ such that $x = x' + d_P(x'')$. Let us also choose $y'' \in N$ such that $x'' = g(y'')$. One has $g(y' - y) = x' - x = d_P(x'') = d_P(g(y'')) = g(d_N(y''))$, hence there exists a $z'' \in M$ such that $y' - y - d_N(y'') = f(z'')$. Consequently, $f(z' - z) = d_N(y') - d_N(y) = d_N(d_N(y'') + f(z'')) = d_N(f(z'')) = f(d_M(z''))$. Since f is injective, $z' - z = d_M(z'')$ and $\mathrm{cl}(z') = \mathrm{cl}(z)$ in $H(M)$.

Moreover, ∂ is a morphism of A-modules. Indeed, having chosen the triple (x_1, y_1, z_1) for ξ_1 and the triple (x_2, y_2, z_2) for ξ_2, we may choose the triple $(x_1 a_1 + x_2 a_2, y_1 a_1 + y_2 a_2, z_1 a_1 + z_2 a_2)$ for $\xi_1 a_1 + \xi_2 a_2$. Then $\partial(\xi_1 a_1 + \xi_2 a_2) = \partial(\xi_1) a_1 + \partial(\xi_2) a_2$.

c) Let us show that $\mathrm{Ker}(H(f)) = \mathrm{Im}(\partial)$. Let $\xi \in H(M)$ be such that $H(f)(\xi) = 0$ and let $x \in Z(M)$ be such that $\xi = \mathrm{cl}(x)$. Then $H(f)(\xi) = \mathrm{cl}(f(x))$, hence $f(x) \in \mathrm{Im}(d_N)$ and there exists a $y \in N$ such that $f(x) = d_N(y)$. By definition of the morphism ∂, one has $\partial(\mathrm{cl}(g(y))) = \mathrm{cl}(x) = \xi$, hence $\mathrm{Ker}(H(f)) \subset \mathrm{Im}(\partial)$.

Conversely, let $\eta \in \mathrm{Im}(\partial)$, let $\xi \in H(P)$ such that $\eta = \partial(\xi)$ and let $x \in Z(P)$ be such that $\xi = \mathrm{cl}(x)$. Let $y \in N$ be such that $x = g(y)$ and let $z \in M$ be such that $d_N(y) = f(z)$; then $\eta = \partial(\xi) = \mathrm{cl}(z)$, by the construction of the morphism ∂ in *b*). We then have $H(f)(\eta) = \mathrm{cl}(f(z)) = \mathrm{cl}(d_N(y)) = 0$, so that $\eta \in \mathrm{Ker}(H(f))$, as was to be shown.

d) Let us show that $\mathrm{Im}(H(g)) = \mathrm{Ker}(\partial)$.

Let $\xi \in H(P)$ be such that $\partial(\xi) = 0$; let $x \in Z(P)$ be such that $\xi = \mathrm{cl}(x)$, let $y \in N$ be such that $x = g(y)$ and let $z \in M$ be such that $d_N(y) = f(z)$, so that $\partial(\xi) = \mathrm{cl}(z)$. By assumption, $z \in \mathrm{Im}(d_M)$; let then $z' \in M$ be such that $z = d_M(z')$. We have $f(z) = f(d_M(z')) = d_N(f(z')) = d_N^2(y) = 0$. Since f is injective, $z = 0$ and $d_N(y) = f(z) = 0$. Moreover, the class of y in $H(N)$ satisfies $H(g)(\mathrm{cl}(y)) = \mathrm{cl}(g(y)) = \mathrm{cl}(x) = \xi$, which proves that $\xi \in \mathrm{Im}\, H(g)$.

Conversely, let $\xi \in \mathrm{Im}(H(g))$; there exists a $y \in Z(N)$ such that $\xi = H(g)(\mathrm{cl}(y))$, that is, $\xi = \mathrm{cl}(g(y))$. Set $x = g(y)$; one has $x \in Z(P)$; let then $z \in M$ be such that $d_N(y) = f(z)$. Then $f(z) = 0$, hence $z = 0$ since f is injective, and the definition of ∂ implies that $\partial(\xi) = \mathrm{cl}(z) = 0$. $\qquad\square$

Corollary (7.7.10). — *Let us assume moreover that* M, N, P *are graded differential modules and that the morphisms* f *and* g *are of degree 0 (meaning* $f(M_n) \subset N_n$ *and* $f(N_n) \subset P_n$ *for every* n). *Then, the morphism* ∂ *constructed in theorem 7.7.9 has degree 1 and its restriction* ∂^n *to* $H^n(P)$ *gives rise to an exact sequence*

$$\cdots \to H^n(M) \to H^n(N) \to H^n(P) \xrightarrow{\partial^n} H^{n+1}(M) \to \cdots$$

The conditions of the corollary are often used by saying that one has an exact sequence of complexes, namely a diagram

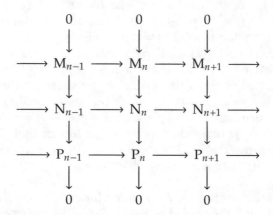

where the rows are complexes and the columns are exact sequences.

Proof. — Recall that for any $\xi \in H(P)$, $\partial(\xi)$ has been defined as $\mathrm{cl}(z)$, where $x \in Z(P)$, $y \in N$ and $z \in M$ are such that $\xi = \mathrm{cl}(x)$, $x = g(y)$ and $d_N(y) = f(z)$. Assume that $\xi \in H^n(P)$. Then, one may replace x, y, z by their components in P_n, N_n and M_{n+1} respectively, so that $\partial(\xi) \in H^{n+1}(M)$. The rest of the corollary follows from lemma 7.7.11. □

Lemma (7.7.11). — *Let* I *be a set, let* $(M_i)_{i\in I}$ *and* $(N_i)_{i\in I}$ *be families of A-module; for every* i, *let* $f_i \colon M_i \to N_i$ *be a morphism of A-modules. Set* $M = \bigoplus_{i\in I} M_i$, $N = \bigoplus_{i\in I} N_i$ *and let* $f \colon M \to N$ *be the morphism of A-modules given by* $f((m_i)) = (f_i(m_i))$ *for* $(m_i) \in M$. *Then for every* $i \in I$, *one has* $\mathrm{Ker}(f_i) = \mathrm{Ker}(f) \cap M_i$, $\mathrm{Im}(f_i) = \mathrm{Im}(f) \cap N_i$ *and*

$$\mathrm{Ker}(f) = \bigoplus_{i\in I} \mathrm{Ker}(f_i) \quad and \quad \mathrm{Im}(f) = \bigoplus_i \mathrm{Im}(f_i).$$

Proof. — For $m = (m_i) \in M$, one has $f(m) = (f_i(m_i)) = 0$, so $m \in \mathrm{Ker}(f)$ if and only if $m_i \in \mathrm{Ker}(f_i)$ for every i. This shows that $\mathrm{Ker}(f) \cap M_i = \mathrm{Ker}(f_i)$. Moreover, for $m \in \mathrm{Ker}(f)$, one has $m = \sum m_i$, hence $\mathrm{Ker}(f) = \bigoplus \mathrm{Ker}(f_i)$.

Let also $n = (n_i) \in N$. If $n = f(m)$, for $m \in M$, then $f(m_i) = n_i$ for every i, so that $n_i \in \mathrm{Im}(f_i)$. Conversely, assume that $n_i \in \mathrm{Im}(f_i)$ for every i; let J be the finite set of $i \in I$ such that $n_i \neq 0$. For $i \in J$, choose $m_i \in M_i$ such that $n_i = f_i(m_i)$; for $i \in I - J$, set $m_i = 0$. Then $f((m_i)) = n$, hence $n \in \mathrm{Im}(f)$. The relation $n = \sum n_i$ also shows that $n \in \bigoplus \mathrm{Im}(f_i)$. □

7.8. Exercises

Exercise (7.8.1). — Let A be a ring, let M_1, M_2, M_3, M_4 be A-modules and let $w: M_1 \to M_2$, $v: M_2 \to M_3$ and $u: M_3 \to M_4$ be morphisms. If f is a morphism of A-modules, we denote by $\rho(f)$ the length of $\mathrm{Im}(f)$, or $+\infty$ if this length is infinite.

a) Deduce from w an A-morphism

$$f: \mathrm{Ker}(u \circ v \circ w)/\mathrm{Ker}(u \circ v) \to \mathrm{Ker}(u \circ v)/\mathrm{Ker}(v).$$

b) Deduce from v an A-morphism

$$g: \mathrm{Ker}(u \circ v)/\mathrm{Ker}(v) \to \mathrm{Im}(v)/\mathrm{Im}(v \circ w).$$

c) Deduce from u an A-morphism

$$h: \mathrm{Im}(v)/\mathrm{Im}(v \circ w) \to \mathrm{Im}(u \circ v)/\mathrm{Im}(u \circ v \circ w).$$

d) Show that the diagram

$$0 \longrightarrow \mathrm{Ker}(u \circ v \circ w)/\mathrm{Ker}(u \circ v) \xrightarrow{f} \mathrm{Ker}(u \circ v)/\mathrm{Ker}(v)$$

$$\xrightarrow{g} \mathrm{Im}(v)/\mathrm{Im}(v \circ w) \xrightarrow{h} \mathrm{Im}(u \circ v)/\mathrm{Im}(u \circ v \circ w) \longrightarrow 0$$

is an exact sequence.

e) Prove the inequality (due to Frobenius):

$$\rho(v \circ w) + \rho(u \circ v) \le \rho(v) + \rho(u \circ v \circ w).$$

Exercise (7.8.2). — Let A be a ring.

a) Let $M_0 \xrightarrow{f_0} M_1 \xrightarrow{f_1} \dots \xrightarrow{f_{n-1}} M_n$ be a complex of A-modules. Show that this complex is an exact sequence if and only if, for every i, the diagram $(0) \to \mathrm{Ker}(f_i) \to M_i \xrightarrow{f_i} \mathrm{Ker}(f_{i+1}) \to (0)$ is a short exact sequence.

b) Let $0 \to M_1 \xrightarrow{f_1} M_2 \xrightarrow{f_2} \dots \xrightarrow{f_{n-1}} M_n \to 0$ be an exact sequence. Assuming that the modules M_i have finite length, show that $\sum_{i=1}^{n}(-1)^i \, \ell_A(M_i) = 0$.

c) More generally, let $0 \to M_1 \xrightarrow{f_1} M_2 \xrightarrow{f_2} \dots \xrightarrow{f_{n-1}} M_n \to 0$ be a complex. Assume that the modules M_i have finite length. For every i, let $H_i = \mathrm{Ker}(f_i)/\mathrm{Im}(f_{i-1})$; prove that $H_1, \dots, H_n$ have finite length and that

$$\sum_{i=1}^{n}(-1)^i \, \ell_A(M_i) = \sum_{i=1}^{n}(-1)^i \, \ell_A(H_i).$$

Exercise (7.8.3). — Let A be a commutative ring, let M be an A-module and let (a, b) be two elements of A.

a) Show that the map $d_1 \colon M \to M \times M$ and $d_2 \colon M \times M \to M$ defined by

$$d_1(x) = (ax, bx) \quad \text{and} \quad d_2(x, y) = bx - ay$$

give rise to a complex $M^{\bullet}$:

$$0 \to M \xrightarrow{d_1} M \times M \xrightarrow{d_2} 0.$$

b) Show that $H^0(M^{\bullet}) = \{x \in M ; ax = by = 0\}$ and that $H^2(M^{\bullet}) = M/(aM + bM)$.

c) Assume that the multiplication by a is injective in M. Then show that multiplication by b in M/aM is injective if and only if $H^1(M^{\bullet}) = 0$.

Exercise (7.8.4). — Let A be a commutative ring and let I be an ideal of A.

a) Assume that the canonical morphism cl: $A \to A/I$ admits a right inverse f. Show that there exists an $a \in I$ such that $a = a^2$ and $I = (a)$.

b) Assume that A is an integral domain. Then show that A/I is a projective A-module if and only if $I = 0$ or $I = A$.

Exercise (7.8.5). — Let k be a field and let A be the subring $k[X^2, XY, Y^2]$ of $k[X, Y]$. Show that the ideal $I = (X^2, XY)$ of A is not a projective A-module.

Exercise (7.8.6). — Let $A = \mathscr{C}([0, 1], \mathbf{R})$ be the ring of continuous functions on $[0, 1]$ with real values and let I be the set of functions $f \in A$ which are identically 0 in some neighborhood of 0.

a) Show that I is not finitely generated.

b) Show that $I = I^2$.

c) Show that there exists a family $(\theta_n)_{n \geq 1}$ of continuous functions on $[0, 1]$ such that $\theta_n(x) = 0$ if $x \notin [\frac{1}{n}, \frac{2}{n}]$, and that $\sum_{n \geq 1} \theta_n(x) > 0$ for every $x \in (0, 1]$. (*For example, set* $\theta_n(x) = (nx - 1)(2 - nx)$ *for* $x \in [\frac{1}{n}, \frac{2}{n}]$.)

d) Show that the ideal I is a projective A-module.

Exercise (7.8.7). — Let k be a ring and let $A = \mathrm{Mat}_n(k)$ be the ring of matrices of size n with entries in k.

a) Let $M = k^n$, viewed as a left A-module. Show that $A \simeq M^n$.

b) Assume that k is commutative or that it is a division ring. Show that M is not free.

Exercise (7.8.8). — Let A be a ring and let P be a finitely generated projective A-module. Show that $\mathrm{End}_A(P)$ is again a finitely generated and projective A-module.

Exercise (7.8.9). — Let A be a ring and let $0 \to M \xrightarrow{f} N \xrightarrow{g} P \to 0$ be an exact sequence of A-modules.

a) Assume that this exact sequence is split. Then show that for every A-module X, the diagrams

$$0 \to \mathrm{Hom}(X, M) \xrightarrow{f_*} \mathrm{Hom}(X, N) \xrightarrow{g_*} \mathrm{Hom}(X, P) \to 0$$

and

$$0 \to \mathrm{Hom}(P, X) \xrightarrow{g^*} \mathrm{Hom}(N, X) \xrightarrow{f^*} \mathrm{Hom}(M, X) \to 0$$

are split exact sequences.

b) Conversely, assume that for every A-module X, the diagram

$$0 \to \mathrm{Hom}(X, M) \xrightarrow{f_*} \mathrm{Hom}(X, N) \xrightarrow{g_*} \mathrm{Hom}(X, P) \to 0$$

is an exact sequence. Show that the initial exact sequence is split.

c) Solve the same question for the second sort of exact sequences.

Exercise (7.8.10). — a) Prove that $\mathbf{Z}/3\mathbf{Z}$ is a projective $(\mathbf{Z}/6\mathbf{Z})$-module which is not free.

b) Let A be a commutative ring. Let $e \in A - \{0, 1\}$ be a non-trivial idempotent ($e^2 = e$). Prove that eA is a projective submodule of A which is not free.

Exercise (7.8.11). — Let A be a ring and let

$$
\begin{array}{ccc}
M & \xrightarrow{\ f\ } & N \\
\downarrow{\scriptstyle u} & & \downarrow{\scriptstyle v} \\
M' & \xrightarrow{\ f'\ } & N'
\end{array}
$$

be a commutative diagram of A-modules.

a) Let $\psi \colon M \to M' \oplus N$ and $\varphi \colon M' \oplus N \to N'$ be the maps given by by $\psi(x) = (u(x), f(y))$ and $\varphi(x, y) = f'(x) - v(y)$. Prove that $\varphi \circ \psi = 0$.

b) Assume that ψ is injective and $\mathrm{Im}(\psi) = \mathrm{Ker}(\varphi)$. (One says that this diagram is a *pull-back*.) Prove that for every A-module P, every morphism $p \colon P \to N$ and every morphism $q \colon P \to M'$ such that $f' \circ q = v \circ p$, there exists a unique morphism $r \colon P \to M$ such that $p = f \circ r$ and $q = u \circ r$.

c) Assume that φ is surjective and $\mathrm{Im}(\psi) = \mathrm{Ker}(\varphi)$. (One says that this diagram is a *push-out*.) Prove that for every A-module Q, every morphism $p \colon N \to Q$ and every morphism $q \colon M' \to Q$ such that $p \circ f = q \circ u$, there exists a unique morphism $r \colon N' \to Q$ such that $p = r \circ v$ and $q = r \circ f'$.

d) Let A be a ring and let I, J be two-sided ideals of A. Prove that the diagram

$$A/(I \cap J) \longrightarrow A/I$$
$$\downarrow \qquad\qquad \downarrow$$
$$A/J \longrightarrow A/(I+J)$$

(whose maps are the obvious ones) is both a pull-back and a push-out.

Exercise (7.8.12). — Let A be a ring and let

$$\begin{array}{ccc} N & \xrightarrow{f} & P \\ \downarrow u & & \downarrow v \\ N' & \xrightarrow{f'} & P' \end{array}$$

be a commutative diagram of A-modules.

  a) Assume that this diagram is a pull-back (see exercise 7.8.11). Let $g' : M \to N'$ be a morphism of A-modules such that $\mathrm{Im}(g') = \mathrm{Ker}(f')$. Prove that there exists a unique morphism $g : P \to M$ such that $g' = u \circ g$ and the sequence

$$M \xrightarrow{g} N \xrightarrow{f} P$$

is exact.

  b) Assume that this diagram is a push-out (see exercise 7.8.11). Let $g : P \to Q'$ be a morphism of A-modules such that $\mathrm{Ker}(g) = \mathrm{Im}(f)$. Prove that there exists a unique morphism $g' : P' \to Q'$ such that $g = g' \circ v$ and the sequence

$$N' \xrightarrow{f'} P' \xrightarrow{g'} Q'$$

is exact.

Exercise (7.8.13). — Prove this generalization of the snake lemma (theorem 7.2.1). Let A be a ring and let

$$\begin{array}{ccccc} M & \xrightarrow{f} & N & \xrightarrow{g} & P \\ \downarrow u & & \downarrow v & & \downarrow w \\ M' & \xrightarrow{f'} & N' & \xrightarrow{g'} & P' \end{array}$$

be a commutative diagram of A-modules with exact rows, namely $\mathrm{Ker}(g) = \mathrm{Im}(f)$ and $\mathrm{Ker}(g') = \mathrm{Im}(f')$.

  a) Prove that the given morphisms induce exact sequences

$$0 \to \mathrm{Ker}(f) \xrightarrow{u} \mathrm{Ker}(f' \circ u) \xrightarrow{f} \mathrm{Ker}(v) \xrightarrow{g} \mathrm{Ker}(w) \cap \mathrm{Im}(g)$$

and

$$M'/(\mathrm{Im}(u) + \mathrm{Ker}(f')) \xrightarrow{f'} \mathrm{Coker}(v) \xrightarrow{g'} \mathrm{Coker}(w \circ g) \xrightarrow{\mathrm{id}_{P'}} \mathrm{Coker}(g') \to 0.$$

b) Construct a morphism of modules

$$\delta: \mathrm{Ker}(w) \cap \mathrm{Im}(g) \rightarrow M'/(\mathrm{Im}(u) + \mathrm{Ker}(f'))$$

that connects the preceding two exact sequences and gives rise to an exact sequence. (In other words, $\mathrm{Im}(g) = \mathrm{Ker}(\delta)$ and $\mathrm{Im}(\delta) = \mathrm{Ker}(f')$.)

c) Discuss the particular cases : (i) f' is injective; (ii) g is surjective; (iii) f' is injective and g is surjective.

d) Show that the given morphisms induce an exact sequence

$$0 \xrightarrow{\phantom{\mathrm{id}_M}} \mathrm{Ker}(f) \xrightarrow{\mathrm{id}_M} \mathrm{Ker}(g \circ f) \xrightarrow{f} \mathrm{Ker}(g)$$
$$\downarrow{\delta}$$
$$0 \xleftarrow{\phantom{\mathrm{id}_P}} \mathrm{Coker}(g) \xleftarrow{\mathrm{id}_P} \mathrm{Coker}(g \circ f) \xleftarrow{g} \mathrm{Coker}(f)$$

in which the morphism δ is defined as in b).

Exercise (7.8.14). — Give examples of commutative rings A, multiplicative subsets S of A, and A-modules M, N such that the natural map of proposition 7.2.7 $\delta_{M,N}: S^{-1}\mathrm{Hom}_A(M,N) \rightarrow \mathrm{Hom}_{S^{-1}A}(S^{-1}M, S^{-1}N)$ is not injective, resp. not surjective.

Exercise (7.8.15). — Let A be a ring, let M be a A-module and let N be a submodule of M.

a) If N is injective, then it is a direct summand.

b) If M/N is projective, then N is a direct summand.

Exercise (7.8.16). — Let A be a ring, let P be an A-module and let M be a direct summand of P. If P is an injective A-module, show that M is injective.

Exercise (7.8.17). — Let A be a ring.

a) Assume that A is right noetherian. Show that any direct sum of injective right A-modules is injective. (*Apply criterion (ii) of proposition 7.4.2.*)

b) Let $(I_n)_{n\in\mathbf{N}}$ be an increasing sequence of right ideals of A and let I be its union. For every n, let M_n be a right A-module containing A/I_n and let $f_n: A \rightarrow M_n$ be the canonical surjection from A to A/I_n composed with the injection to M_n. Let M be the direct sum of the modules M_n.

Show that there exists a unique map $f: I \rightarrow M$ such that $f(a) = (f_n(a))_n$ for every $a \in I$.

c) (*continued*) Assuming that M is injective, prove that there exists an integer $n \in \mathbf{N}$ such that $I_n = I$.

d) (*continued*) Assume that every direct sum of injective right A-modules is injective. Show that A is right noetherian (the Bass–Papp theorem).

Exercise (7.8.18). — Let A be a ring, let P be an A-module and let M be a submodule of P. One says that a submodule N of P *essential* for M if it contains M and if $N_1 \cap M \neq 0$ for every non-zero submodule N_1 of N.

a) Show that there exists a maximal submodule N of P containing M which is essential for M. (*Let $\mathscr{S}$ be the set of submodules of* P *containing* M *which are essential. Show that $\mathscr{S}$ is inductive and apply Zorn's lemma.*)

b) Show that there exists a maximal submodule N′ of P such that $M \cap N' = 0$. (*Show that the set of such submodules is inductive.*)

c) Assume that P is an injective A-module. Show that $P = N \oplus N'$. Using exercise 7.8.16, conclude that N is an injective A-module.

Exercise (7.8.19). — Let A be a ring and let M be a right A-module. Let $j: M \to N$ and $j': M \to N'$ be two morphisms of M into right A-modules N and N′.

a) Assume that N is essential with respect to $j(M)$ and that N′ is injective. Prove that there exists an injective morphism $f: N \to N'$ such that $j' = f \circ j$.

b) Assume furthermore that N is injective and that N′ is essential with respect to $j(M')$. Prove that f is an isomorphism.

If M *is an A-module, theorem 7.4.6 shows that there exists an injective A-module* N *containing* M. *By exercise 7.8.18, there exists such an injective module* N *which is essential with respect to* M. *The result of the present exercise shows that* N *is well-defined.*

Exercise (7.8.20). — Let A be a ring, let M, N be A-modules and let $f: M \to N$ be a morphism.

a) Prove that f is an epimorphism in the category of A-modules if and only f is surjective.

b) Prove that f is a monomorphism in the category of A-modules if and only if f is injective.

Exercise (7.8.21). — Let A, B be rings. Let F be a functor from the category of A-modules to that of B-modules and let G be a right adjoint to F. For any A-module M and any B-module N, let $\Phi_{M,N}: \mathrm{Hom}_B(F(M), N) \to \mathrm{Hom}_A(M, G(N))$ be the adjunction map.

a) Assume that the functor F is faithful. Show that for any injective morphism g of B-modules, the morphism $G(g)$ is injective. (*Use the characterization of exercise 7.8.20.*)

b) Assume that the functor G is faithful. Show that for any surjective morphism f of A-modules, the morphism $F(f)$ is surjective.

Exercise (7.8.22). — Let A and B be rings, let F be a functor from the category of A-modules to the category of B-modules and let G be a functor from the category of B-modules to the category of A-modules. Assume that these functors form a pair of adjoint functors and let $(\Phi_{M,N})$ be an adjunction.

a) If F and G are additive, prove that the maps $\Phi_{M,N}$ are additive.

b) Assume that $A = B$ is a commutative ring and that the functors F, G are A-linear, that is, they are additive and map a homothety of ratio a to a homothety of ratio a. Prove that the maps $\Phi_{M,N}$ are morphisms of A-modules.

Exercise (7.8.23). — Let A be a ring and let P be an A-module. Assume that for every injective A-module Q, every surjective morphism $p \colon Q \to Q''$ and every morphism $f \colon P \to Q''$, there exists a morphism $g \colon P \to Q$ such that $p \circ g = f$. The goal of the first two questions is to prove that P is projective.

a) Let $0 \to M' \xrightarrow{j} M \xrightarrow{p} M'' \to 0$ be an exact sequence of A-modules and let $f \colon P \to M''$ be a morphism of A-modules. Let Q be an injective A-module and $i \colon M \to Q$ an injective morphism (Baer). Let $Q' = i(j(M'))$, let $Q'' = Q/Q'$ and let $q \colon Q \to Q''$ be the canonical surjection Prove that there exists a morphism $k \colon M'' \to Q''$ such that $k \circ p = q \circ i$.

b) Let $g \colon P \to Q$ be a morphism such that $q \circ g = k \circ f$. Prove that $\mathrm{Im}(g) \subset \mathrm{Im}(i)$. Conclude.

c) Let Q be an A-module. Show that Q is injective if and only if, for every projective A-module P, every injective morphism $i \colon P' \to P$ and every morphism $f \colon P' \to Q$, there exists a morphism $g \colon P \to Q$ such that $g \circ i = f$.

Exercise (7.8.24). — One says that a ring A is right hereditary if every right ideal of A is a projective A-module.

a) Let A be a right hereditary ring and let n be an integer. Prove that every submodule M of A^n is a projective A-module. (*Argue by induction on n; consider the morphism from M to A given by the last coordinate.*)

b) More generally, if A is right hereditary, then every submodule of a free right A-module is a direct sum of ideals. (*Let L be a free A-module and let M be a submodule of L. Using Zermelo's theorem (corollary A.2.18), choose a well-ordering of a basis of L, and argue similarly.*)

c) Using exercise 7.8.24, prove that A is right hereditary if and only if every quotient of an injective A-module is injective.

(*Commutative*) integral domains which are right hereditary are called Dedekind rings. They will be studied in §9.6.

Exercise (7.8.25). — Let $A = \mathbf{Z}[T]$ and let $K = \mathbf{Q}(T)$ be its fraction field.

a) Let $I = (2, T)$. Prove that there exists a morphism $f \colon I \to K/A$ such that $f(2) = 0$ and $f(T)$ is the class of $1/2$. Prove that f does not extend to A.

b) Prove that K/A is divisible but is not an injective A-module.

Exercise (7.8.26). — Let A be a ring.

a) Let $0 \to P' \to P \to M \to 0$ and $0 \to Q' \to Q \to M \to 0$ be exact sequences of right A-modules.

If P and Q are projective, then $P \oplus Q' \simeq P' \oplus Q$.

b) Give an example that shows that the projectivity assumption is necessary.

c) Let $0 \to M \to P \to P' \to 0$ and $0 \to M \to Q \to Q' \to 0$ be exact sequences of right A-modules.

If P and Q are injective, then $P \oplus Q' \simeq P' \oplus Q$. (This result is called *Schanuel's lemma.*)

Chapter 8.
Tensor Products and Determinants

A companion chapter to linear algebra, multilinear *algebra studies situations where quantities depend on many variables and vary linearly in each of them. The* tensor product *construction takes two modules* M *and* N *and makes a module* M ⊗ N *out of it, which reflects bilinear maps from* M × N. *When* M *and* N *are free, then so is* M ⊗ N, *and the* tensors *of old-style differential/riemannian geometry, a bunch of real numbers with indices or exponents, are nothing but the coordinates of such algebraic objects. Tensor products are also important in other fields of mathematics, such as representation theory where they are used to define induced representations.*

This construction can also serve to associate various algebras with a module over a commutative ring. Despite its commutative character, the symmetric algebra *will not be studied thoroughly here; in the simplest case, when the given module is free, it furnishes another construction of rings of polynomials. However, because of its relation with determinants and, henceforth, with linear algebra, I give a more detailed presentation of the* exterior algebra.

In the next section, I prove that tensoring with a given module preserves some exact sequences. This is an extremely important property which I first prove by relatively explicit computations. I then give another, more abstract, proof, based on the fact that this functor is right exact, *a property that itself derives from the universal property of the tensor product construction.*

However, tensoring with a given module is not left exact in general, and modules which preserve all exact sequences are called flat. *Free modules, more generally projective modules, are flat, and it is useful to characterize flat modules. Over principal ideal domains, flat modules are just torsion-free modules, but this condition is not sufficient in general. I then prove a theorem of Lazard that shows that a finitely presented module is flat if and only if it is projective.*

Faithful flatness is a reinforcement of flatness. Its importance is revealed in the next section where I explain Grothendieck's idea of faithfully flat descent. *If* $f: A \to B$ *is a morphism of commutative rings, the tensor product construction furnishes a functor* M ↦ B ⊗ₐ M *from A-modules to B-modules. When the morphism* f *is faithfully flat, the module* B ⊗ₐ M *reflects many properties of* M, *and I will show some of them. Even more remarkably, Grothendieck proved that the module*

© Springer Nature Switzerland AG 2021
A. Chambert-Loir, *(Mostly) Commutative Algebra*, Universitext,
https://doi.org/10.1007/978-3-030-61595-6_8

$B \otimes_A M$ *carries an additional structure, called a* descent datum, *which allows one to recover the initial A-module M.*

In a final section, I show how when f is a finite Galois extension of fields, this general theorem translates to Galois descent *via a too lengthy computation.*

8.1. Tensor Products of Two Modules

Let A be a ring, let M be a right A-module and N be a left A-module. We are interested here in the *balanced bi-additive maps* from $M \times N$ into an abelian group P, that is to say, the maps $f : M \times N \to P$ which satisfy the following properties:

(i) For all $m, m' \in M$ and all $n \in N$, one has $f(m+m', n) = f(m, n)+f(m', n)$ (*left additivity*);

(ii) For all $m \in M$ and all $n, n' \in N$, one has $f(m, n+n') = f(m, n)+f(m, n')$ (*right additivity*);

(iii) For all $m \in M$, all $n \in N$ and all $a \in A$, one has $f(ma, n) = f(m, an)$ (*balancing condition*).

Our first goal is to *linearize* the map f, that is, to replace it by a linear map φ between appropriate modules so that the study of φ would be more or less equivalent to that of f.

8.1.1. Construction — Let $T_1 = \mathbf{Z}^{(M \times N)}$ be the free **Z**-module with basis $M \times N$; for $m \in M$ and $n \in N$, let $e_{m,n}$ be the element (m, n) of the basis of T_1. Let T_2 be the submodule of T_1 generated by the following elements:

$$e_{m+m',n} - e_{m,n} - e_{m',n}, \quad e_{m,n+n'} - e_{m,n} - e_{m,n'}, \quad e_{ma,n} - e_{m,an},$$

where m, m' run along M, n, n' run along N and a runs along A. Finally, let $T = T_1/T_2$ and let $m \otimes n$ denote the class of $e_{m,n}$ in the quotient T. Let θ be the map from $M \times N$ to $M \otimes N$ given by $\theta(m, n) = m \otimes n$. By construction, the map θ is balanced and bi-additive. Indeed, $\theta(m + m', n) - \theta(m, n) - \theta(m', n)$ is the class in T_1/T_2 of $e_{m+m',n} - e_{m,n} - e_{m',n}$, hence is 0; the two other axioms are verified in the same way.

By the universal property of free **Z**-modules, there is a unique morphism φ_1 from T_1 to P such that $f(m, n) = \varphi_1(e_{(m,n)})$ for every $(m, n) \in M \times N$. Then, the kernel of φ_1 contains T_2 if and only if φ_1 vanishes on each of the given generators of T_2, in other words, if and only if f is A-balanced and bi-additive. Consequently, there exists a morphism of abelian groups $\varphi : T \to P$ such that $\varphi(m \otimes n) = f(m, n)$.

Definition (8.1.2). — The abelian group T defined above is called the *tensor product* of the A-modules M and N; it is denoted $M \otimes_A N$. Its elements are called *tensors*. The map $\theta : M \times N \to M \otimes_A N$ defined by $\theta(m, n) = m \otimes n$ is called the *canonical A-balanced bi-additive map*. The elements of $M \otimes_A N$

of the form $m \otimes n$, for some $(m, n) \in M \times N$, (the image of θ) are called *split tensors*.

Since the split tensors are the image in $T = T_1/T_2$ of a basis of T_1, they form a generating family of T.

Proposition (8.1.3) (Universal property of tensor products). — *For any abelian group P and any A-balanced and bi-additive map* $f : M \times N \to P$, *there exists a unique morphism of abelian groups* $\varphi : M \otimes_A N \to P$ *such that* $\varphi \circ \theta = f$ *(explicitly:* $f(m, n) = \varphi(m \otimes n)$ *for every* $(m, n) \in M \times N$).

Proof. — We just showed the existence of a morphism φ such that $\varphi \circ \theta = f$. Let φ, φ' be two such morphisms, and let us prove that $\varphi = \varphi'$. Let $\delta = \varphi' - \varphi$; it is a morphism of an abelian groups from $M \otimes_A N$ to P. By assumption,

$$\delta(m \otimes n) = \varphi'(m \otimes n) - \varphi(m \otimes n) = f(m, n) - f(m, n) = 0$$

for every $(m, n) \in M \otimes N$. Consequently, the kernel of δ contains every split tensor. Since they generate $M \otimes_A N$, one has $\mathrm{Ker}(\delta) = M \otimes_A N$ and $\delta = 0$. $\square$

Remark (8.1.4). — It is probably useful to repeat things once more.

The property for the map $\theta : (m, n) \mapsto m \otimes n$ to be A-balanced and bi-additive is equivalent to the following relations

$$(m + m') \otimes n = m \otimes n + m' \otimes n,$$
$$m \otimes (n + n') = m \otimes n + m \otimes n',$$
$$ma \otimes n = m \otimes an$$

for $m, m' \in M$, $n, n' \in N$ and $a \in A$.

Moreover, both the universal property of the tensor product $M \otimes_A N$ expressed by proposition 8.1.3 and its construction above show that to define a morphism φ from the abelian group $M \otimes_A N$ to an abelian group P, we just need to construct an A-balanced and bi-additive map f from $M \times N$ to P, the relation between φ and f being the equality $f(m, n) = \varphi(m \otimes n)$.

Finally, to check that a morphism from a tensor product $M \otimes_A N$ to an abelian group P vanish, it suffices to check that it maps all split tensors to 0. Similarly, to check that two morphisms from a tensor product $M \otimes_A N$ to P coincide, it suffices to check that their difference vanishes, hence that these two morphisms coincide on split tensors.

8.1.5. Functoriality — Let M' be another right A-module and let N' be another left A-module. Let $u : M \to M'$ and $v : N \to N'$ be morphisms of A-modules. The map from $M \times N$ to $M' \otimes_A N'$ that sends (m, n) to $u(m) \otimes v(n)$ is A-balanced and bi-additive, as the following computations show:

$$u(m + m') \otimes v(n) = (u(m) + u(m')) \otimes v(n) = u(m) \otimes v(n) + u(m') \otimes v(n),$$
$$u(m) \otimes v(n + n') = u(m) \otimes v(n) + u(m) \otimes v(n'),$$
$$u(ma) \otimes v(n) = u(m)a \otimes v(n) = u(m) \otimes av(n) = u(m) \otimes v(an).$$

By the universal property of the tensor product $M \otimes_A N$, there exists a unique morphism of abelian groups from $M \otimes_A N$ to $M' \otimes_A N'$, written $u \otimes v$, such that $(u \otimes v)(m \otimes n) = u(m) \otimes v(n)$ for every $(m, n) \in M \times N$.

Let M'' and N'' be yet other respectively right and left A-modules and let $u' \colon M' \to M''$ and $v' \colon N' \to N''$ be morphisms of A-modules. Applied to a split tensor $m \otimes n$, the morphisms $(u' \circ u) \otimes (v' \circ v)$ and $(u' \otimes v') \circ (u \otimes v)$, both furnish the split tensor $(u' \circ u(m)) \otimes (v' \circ v(n))$. Consequently, these two morphisms are equal.

Let us now assume that $M = M'$, $N = N'$ and let us take for morphisms the identity maps of M and N. One has $\mathrm{id}_M \otimes \mathrm{id}_N = \mathrm{id}_{M \otimes N}$, since these two morphisms send a split tensor $m \otimes n$ to itself.

Let us still assume that $M' = M$ and $N' = N$, but let u, v be arbitrary endomorphisms of M and N, respectively. Then, $u \otimes v$ is an endomorphism of $M \otimes_A N$. Moreover, the maps $\mathrm{End}_A(M) \to \mathrm{End}(M \otimes_A N)$ and $\mathrm{End}_A(N) \to \mathrm{End}(M \otimes_A N)$ respectively given by $u \mapsto u \otimes \mathrm{id}_N$ and $v \mapsto \mathrm{id}_M \otimes v$ are morphisms of rings.

Remark (8.1.6). — Let A be a ring, let M be a right A-module and let N be a left A-module. Let π be an element of the tensor product $M \otimes_A N$.

By definition, it can be written $\sum_{i \in I} m_i \otimes n_i$ for a finite set I and finite families $(m_i)_{i \in I}$ in M and $(n_i)_{i \in I}$ in N. Let M_0 be the submodule of M generated by the family (m_i), and let N_0 be the submodule of N generated by the family (n_i); they are finitely generated, by construction. Moreover, the element π is the image of an element $\pi_0 \in M_0 \otimes_A N_0$ by the canonical morphism $f_0 \otimes g_0 \colon M_0 \otimes_A N_0 \to M \otimes_A N$ deduced from the inclusions $f_0 \colon M_0 \to M$ and $g_0 \colon N_0 \to N$: it suffices to set $\pi_0 = \sum_{i \in I} m_i \otimes n_i$, where these split tensors are viewed in $M_0 \otimes_A N_0$.

However, a representation of a tensor $\pi = \sum m_i \otimes n_i$ is by no means unique, and we may well have $\pi = 0$ and $\pi_0 \neq 0$. If $\pi = 0$, the construction of the tensor product module $M \otimes_A N$ shows that the element $\sum e_{m_i, n_i}$ of $T_1 = \mathbf{Z}^{(M \times N)}$, in the notation of 8.1.1, belongs to T_2, hence is itself some finite linear combination of basic elements of the form $e_{m+m', n} - e_{m, n} - e_{m', n}$, $e_{m, n+n'} - e_{m, n} - e_{m, n'}$ and $e_{ma, n} - e_{m, an}$. Let M_1 be the submodule of M generated by M_0 and by all the elements m, m' that appear; similarly, let N_1 be the submodule of N generated N_0 and by all the elements n, n' that appear in such a linear combination. Let $f_1 \colon M_1 \to M$ and $g_1 \colon N_1 \to N$ be the inclusions. Under the morphism $f_1 \otimes g_1$, the tensor $\pi_1 = \sum m_i \otimes n_i$ of $M_1 \otimes_A N_1$ maps to π. Moreover, one has $\pi_1 = 0$.

In other words, the nullity of an explicitly written tensor π of $M \otimes_A N$ can be witnessed in the tensor product $M_1 \otimes_A N_1$, where M_1 and N_1 are appropriate finitely generated submodules of M and N respectively.

8.1.7. Bimodules — Let us assume in particular that M be endowed with the structure of a (B, A)-bimodule and that N be endowed with the structure of an (A, C)-bimodule. If we have no such structure at our disposal, we can always take $B = \mathbf{Z}$ or $C = \mathbf{Z}$. If A is a commutative ring, we can also choose $B = C = A$.

Anyway, we have a morphism of rings $B \to \mathrm{End}_A(M)$ (which sends $b \in B$ to the left-multiplication λ_b by b in M) and a morphism of rings $C^o \to \mathrm{End}_B(N)$ (which sends $c \in C$ to the right-multiplication ρ_c by c in N). This shows that we would essentially consider the most general situation by assuming that $B = \mathrm{End}_A(M)$ and $C = \mathrm{End}_A(M)^o$, the above morphism of rings would then be identities.

By composition, we get two ring morphisms $B \to \mathrm{End}(M \otimes_A N)$ and $C^o \to \mathrm{End}(M \otimes_A N)$, given by $b \mapsto \lambda_b \otimes \mathrm{id}_N$ and $c \mapsto \mathrm{id}_M \otimes \rho_c$. They endow the abelian group $M \otimes_A N$ with a structure of a left B-module and of a right C-module. Moreover, for $b \in B$, $c \in C$, $m \in M$ and $n \in N$, we have

$$ b \cdot ((m \otimes n) \cdot c) = b \cdot (m \otimes nc) = bm \otimes nc = (b \cdot (m \otimes n)) \cdot c, $$

so that these two structures are compatible. This shows that $M \otimes_A N$ has a natural structure of a (B, C)-bimodule.

Let P be a (B, C)-bimodule. For a morphism of abelian groups $\varphi \colon M \otimes_A N \to P$ to be (B, C)-linear, it is necessary and sufficient that the A-balanced and bi-additive map $f = \varphi \circ \theta \colon M \times N \to P$ satisfies $f(bm, nc) = b f(m, n) c$ for every $m \in M$, every $n \in N$, every $b \in B$ and every $c \in C$. In particular, if $u \colon M \to M'$ is a morphism of (B, A)-bimodules and $v \colon N \to N'$ is a morphism of (A, C)-bimodules, then the morphism $u \otimes v$ from $M \otimes_A N$ to $M' \otimes_A N'$ is a morphism of (B, C)-modules.

If A is a commutative ring and $B = C = A$, we then have $a(m \otimes n) = (ma) \otimes n = m \otimes (an) = (m \otimes n)a$, so that the two given structures of an A-module on $M \otimes_A N$ coincide.

Proposition (8.1.8) (Compatibility with direct sums). —*Let $(M_i)_{i \in I}$ be a family of right A-modules and $(N_j)_{j \in J}$ be a family of left A-modules. For every $(i, j) \in I \times J$, let $P_{i,j} = M_i \otimes_A N_j$. Set $M = \bigoplus_{i \in I} M_i$, $N = \bigoplus_{j \in J} N_j$, $P = \bigoplus_{(i,j) \in I \times J} P_{i,j}$. For $i \in I$ and $j \in J$, let $\alpha_i \colon M_i \to M$, $\beta_j \colon N_j \to N$, $\gamma_{i,j} \colon P_{i,j} \to P$ be the canonical injections. There is a unique morphism $\lambda \colon P \to M \otimes_A N$ such that $\lambda \circ \gamma_{i,j} = \alpha_i \otimes \beta_j$ for every $(i, j) \in I \times J$. Moreover, λ is an isomorphism.*

Proof. — Existence and uniqueness of λ follows from the universal property of the direct sum. To prove that λ is an isomorphism, we shall construct its inverse explicitly. For $i \in I$ and $j \in J$, let $p_i \colon M \to M_i$, $q_j \colon N \to N_j$ and $r_{i,j} \colon P \to M_i \otimes N_j$ be the canonical projections.

By construction of the direct sums $M = \bigoplus M_i$ and $N = \bigoplus N_j$, the families $(p_i(m))_{i \in I}$ and $(q_j(n))_{j \in J}$ are almost null, for any $(m, n) \in M \times N$. Consequently, in the family $(p_i(m) \otimes q_j(n))_{(i,j) \in I \times J}$, only finitely many terms are non-zero, so that this family is not only an element of the product module $\prod_{(i,j) \in I \times J} P_{i,j}$ but one of the direct sum $P = \bigoplus P_{i,j}$. This defines a map $f \colon M \times N \to P$. Let us show that this map is A-balanced and bi-additive. Indeed, for any $m, m' \in M$, $n, n' \in N$ and any $a \in A$,

$$f(m + m', n) = \big(p_i(m + m') \otimes q_j(n)\big)_{i,j}$$
$$= \big(p_i(m) \otimes q_j(n) + p_i(m') \otimes q_j(n)\big)_{i,j}$$
$$= \big(p_i(m) \otimes q_j(n)\big) + \big(p_i(m') \otimes q_j(n)\big)_{i,j}$$
$$= f(m, n) + f(m', n)$$

and a similar proof shows that

$$f(m, n + n') = f(m, n) + f(m, n').$$

Moreover, for every $a \in A$, one has

$$f(ma, n) = \big(p_i(ma) \otimes q_j(n)\big)_{i,j}$$
$$= \big(p_i(m)a \otimes q_j(n)\big)_{i,j}$$
$$= \big(p_i(m) \otimes aq_j(n)\big)_{i,j}$$
$$= \big(p_i(m) \otimes q_j(an)\big)_{i,j}$$
$$= f(m, an),$$

since the maps p_i and q_j are A-linear. Consequently, there exists a unique morphism of abelian groups $\varphi \colon M \otimes_A N \to P$ such that $\varphi(m \otimes n) = f(m, n)$ for every $(m, n) \in M \times N$.

For every $(m, n) \in M \times N$, one has

$$\lambda(\varphi(m \otimes n))\lambda(f(m, n)) = \lambda\Big(\sum_{i,j} p_i(m) \otimes q_j(m) \Big)$$

$$= \sum_{i,j} \lambda(p_i(m) \otimes q_j(n)) = \sum_{i,j} \alpha_i(p_i(m)) \otimes \beta_j(q_j(n))$$

$$= \Big(\sum_i \alpha_i(p_i(m)) \Big) \otimes \Big(\sum_j \beta_j(q_j(n)) \Big) = m \otimes n.$$

Since the split tensors generate $M \otimes_A N$, one has $\lambda(\varphi(\xi)) = \xi$ for every $\xi \in M \otimes_A N$, so that $\lambda \circ \varphi$ is the identity on $M \otimes_A N$. On the other hand, for any $(u, v) \in I \times J$, any $m \in M_u$ and any $n \in N_v$, one has

$$\varphi(\lambda(\gamma_{u,v}(m \otimes n))) = \varphi(\alpha_u(m) \otimes \beta_v(n)) = \sum_{i,j} \big(p_i(\alpha_u(m)) \otimes q_j(\beta_v(n))\big).$$

By definition of the projections p_i, one has $p_i(\alpha_u(m)) = m$ if $i = u$, and $p_i(\alpha_u(m)) = 0$ if $i \neq u$; similarly, $q_j(\beta_v(n)) = n$ if $j = v$ and is zero otherwise. This implies that

$$\varphi(\lambda(\gamma_{u,v}(m \otimes n))) = \gamma_{u,v}(m \otimes n).$$

Since the split tensors $m \otimes n$, for $m \in M_u$ and $n \in N_v$, generate $M_u \otimes_A N_v = P_{u,v}$, one has $\varphi(\lambda(\gamma_{u,v}(\xi))) = \gamma_{u,v}(\xi)$ for every $\xi \in P_{u,v}$. Since the submodules $\gamma_{u,v}(P_{u,v})$ of P generate P, it follows that $\varphi \circ \lambda = \mathrm{id}_P$.

This implies that λ is an isomorphism and that φ is its inverse. □

Remark (8.1.9). — Let us keep the hypotheses of the proposition. Let B, C be rings and let us assume moreover that the modules M_i are (B, A)-bimodules, and the modules N_j are (A, C)-bimodules. Then, M is a (B, A)-bimodule, N is an (A, C)-bimodule and λ is an isomorphism of (B, C)-bimodules.

We need to show that $\lambda(btc) = b\lambda(t)c$ for every $b \in B$, every $c \in C$ and every $t \in M \otimes_A N$. It follows from the definition of λ that

$$\lambda(b(m \otimes n)c) = \lambda((bm) \otimes (nc)) = b\lambda(m \otimes n)c$$

for every $m \in M$, $n \in N$, $b \in B$ and $c \in C$. Since the split tensors $m \otimes n$ generate $M \otimes_A N$, the result follows.

Proposition (8.1.10). — *Let A be a ring, let M be a right A-module and let I be a left ideal of A. Then, the map f from M to $M \otimes_A (A/I)$ which maps m to the tensor $m \otimes \mathrm{cl}(1)$ is a surjective morphism of abelian groups. Its kernel is the subgroup MI of M, and f induces an isomorphism of abelian groups $\varphi \colon M/MI \to M \otimes_A (A/I)$.*

Moreover, if B is a ring and M is a (B, A)-bimodule, then f is a morphism of left B-modules.

Proof. — It is obvious that f is additive (and B-linear if M is a (B, A)-bimodule). Let $m \in M$ and let $a \in I$; we have

$$f(ma) = ma \otimes \mathrm{cl}(1) = m \otimes a\,\mathrm{cl}(1) = m \otimes 0 = 0,$$

so that the kernel of f contains the abelian subgroup MI of M. Consequently, f defines, by quotient, a morphism $\varphi \colon M/MI \to M \otimes_A (A/I)$. To prove that φ is an isomorphism, we will construct its inverse.

The map $g \colon M \times (A/I) \to M/MI$ sending $(m, \mathrm{cl}(a))$ to the class of ma is well-defined: if $\mathrm{cl}(a) = \mathrm{cl}(b)$, there exists an element $x \in I$ such that $b = a + x$; then, $mb = ma + mx = ma$ (mod MI). It is obviously bi-additive, as well as A-balanced (for $m \in M$ and $a, b \in A$, $g(mb, \mathrm{cl}(a)) = \mathrm{cl}(mba) = g(m, \mathrm{cl}(ba))$). Therefore, there exists a unique morphism of abelian groups $\gamma \colon M \otimes_A (A/I) \to M/MI$ such that $\gamma(m \otimes \mathrm{cl}(a)) = \mathrm{cl}(ma)$ for every $m \in M$ and every $a \in A$.

If $m \in M$, one has $\gamma(\varphi(\mathrm{cl}_{MI}(m))) = \gamma(m \otimes 1) = \mathrm{cl}_{MI}(m)$, hence $\gamma \circ \varphi$ is the identity of M/MI. On the other hand, for any $m \in M$ and any $a \in A$, one has

$$\varphi(\gamma(m \otimes \mathrm{cl}(a))) = \varphi(\mathrm{cl}(ma)) = ma \otimes \mathrm{cl}(1) = m \otimes a\,\mathrm{cl}(1) = m \otimes \mathrm{cl}(a).$$

Since the split tensors generate $M \otimes (A/I)$, $\varphi \circ \gamma$ is the identity morphism of $M \otimes_A (A/I)$. This concludes the proof of the proposition. □

Remark (8.1.11). — Let M be a right A-module, N be a left A-module, P be an abelian group, and let $f \colon M \otimes_A N \to P$ be a morphism of abelian groups.

Suppose we need to prove that f is an isomorphism. Surjectivity of f is often easy, because in most cases it is quite obvious to spot preimages of suitable generators of P. On the other hand, the injectivity is more difficult. A direct approach would begin by considering a tensor $\sum m_i \otimes n_i$ with image 0, and then trying to prove that this tensor is zero. By the given definition of the tensor product, this requires us to prove that the element $\sum e_{(m_i,n_i)}$ in the free abelian group $\mathbf{Z}^{(M\times N)}$ is a linear combination of the elementary relations. But how?

In all important cases, an efficient way of proving that f is an isomorphism consists in defining its inverse g. With that aim, the proof of the surjectivity of f is useful. Indeed, one has often written a formula for the inverse $g(z)$ of some elements z which generate P. Then try to prove that there exists a unique morphism g from P to $M \otimes_A N$; in general, this amounts to checking that some formula does not depend on choices needed to write it down. To conclude the proof, show that f and g are inverses of one another.

8.1.12. Change of base ring — Let A, C be rings and let $\alpha\colon A \to C$ be a morphism of rings. Let M be a right A-module. The ring C can be seen as a left A-module via the morphism α, and as a right C-module; consequently, the abelian group $M \otimes_A C$ has the natural structure of a right C-module.

In this setting, the universal property of the tensor product furnishes the following result: *For every right C-module P and any A-linear morphism $f\colon M \to P$, there exists a unique C-linear morphism $\varphi\colon M \otimes_A C \to P$ such that $\varphi(m \otimes 1) = f(m)$ for every $m \in M$.*

Indeed, let $f\colon M \to P$ be an A-linear map, that is, an additive map such that $f(ma) = f(m)\alpha(a)$ for every $a \in A$. Let us first show that there is at most one C-linear map φ as required. Indeed, if φ and φ' are two such maps, one has

$$\varphi(m \otimes c) = \varphi((m \otimes 1)c) = \varphi(m \otimes 1)c = f(m)c$$

for every pair $(m, c) \in M \times C$, and $\varphi'(m \otimes c) = f(m)c$ likewise. Consequently, the C-linear map $\varphi' - \varphi$ vanishes on all split tensors of $M \otimes_A C$, hence is zero.

To prove the existence of a map φ such that $\varphi(m \otimes 1) = f(m)$ for every $m \in M$, we first consider the map $f_1\colon (m, c) \mapsto f(m)c$ from $M \times C$ to P. It is obviously bi-additive, as well as A-balanced since $f_1(ma, c) = f(ma)c = f(m)ac = f_1(m, ac)$. Consequently, there exists a unique morphism of abelian groups $\varphi\colon M \otimes_A C \to P$ such that $\varphi(m \otimes c) = f(m)c$ for every $m \in M$ and every $c \in C$. In particuliar, $\varphi(m \otimes 1) = f(m)$.

Let us show that φ is C-linear. For any $c' \in C$, one has $\varphi((m \otimes c)c') = \varphi(m \otimes cc') = f(m)cc' = \varphi(m \otimes c)c'$. Since the split tensors generate $M \otimes_A C$, one has $\varphi(\xi c') = \varphi(\xi)c'$ for every $\xi \in M \otimes_A C$ and every $c' \in C$. In other words, φ is C-linear.

8.2. Tensor Products of Modules Over a Commutative Ring

This is probably the most important case and we first sum-up the situation.

Let A be a commutative ring and let M, N be A-modules. We constructed an abelian group $M \otimes_A N$ and a A-balanced bi-additive map $\theta: M \times N \rightarrow M \otimes_A N$. We also endowed the group $M \otimes_A N$ with the unique structure of an A-module for which

$$a(m \otimes n) = (am) \otimes n = m \otimes (an).$$

Consequently, the map θ satisfies the relations

(i) For all $m, m' \in M$ and all $n \in N$, one has $\theta(m + m', n) = \theta(m, n) + \theta(m', n)$,

(ii) For all $m \in M$ and all $n, n' \in N$, one has $\theta(m, n+n') = \theta(m, n) + \theta(m, n')$,

(iii) For all $m \in M$, all $n \in N$ and all $a \in A$, one has $\theta(am, n) = \theta(m, an) = a\theta(m, n)$,

for $m \in M$, $n \in N$ and $a \in A$. In other words, θ is A-*bilinear*.

Let f be an A-bilinear map from $M \times N$ to an A-module P. There exists a unique map which is bi-additive and A-balanced $\varphi: M \otimes_A N \rightarrow P$ such that $\varphi(m \otimes n) = f(m, n)$ for every $(m, n) \in M \times N$. Moreover, if $a \in A$, then

$$\varphi(a(m \otimes n)) = \varphi(am \otimes n) = f(am, n) = af(m, n) = a\varphi(m \otimes n).$$

Since the split tensors generate $M \otimes_A N$, we have $\varphi(a\xi) = a\varphi(\xi)$ for every $\xi \in M \otimes_A N$. This means that φ is A-linear and is the unique A-linear map from $M \otimes_A N$ to P such that $\varphi \circ \theta = f$.

Proposition (8.2.1). — *Let A be a commutative ring and let M and N be A-modules. Let $(e_i)_{i \in I}$ and $(f_j)_{j \in J}$ be families of elements respectively in M and N.*

If these families (e_i) and (f_j) are generating families (resp. are bases), then the family $(e_i \otimes f_j)_{(i,j) \in I \times J}$ generates (resp. is a basis of) $M \otimes_A N$.

Proof. — Let us assume that both families (e_i) and (f_j) are generating families and let us show that the family $(e_i \otimes f_j)$ generates $M \otimes_A N$. Since the split tensors of $M \otimes_A N$ generate $M \otimes_A N$, it suffices to show that every split tensor is a linear combination of elements of the form $e_i \otimes f_j$. Let thus $m \in M$ and $n \in N$. By assumption, there exists a family $(a_i) \in A^{(I)}$ such that $m = \sum a_i e_i$, as well as a family $(b_j) \in A^{(J)}$ such that $n = \sum b_j f_j$. Then,

$$m \otimes n = \left(\sum_i a_i e_i \right) \otimes \left(\sum_j b_j f_j \right) = \sum_{i,j} (a_i e_i) \otimes (b_j f_j) = \sum_{i,j} a_i b_j e_i \otimes f_j,$$

hence the claim.

Let us now assume that the families (e_i) and (f_j) are bases. Since we have already shown that the family $(e_i \otimes f_j)$ generates $M \otimes_A N$, we just need to show that it is free. With that aim, let $(a_{i,j})$ be an almost-null family of elements of A, indexed by $I \times J$, such that $\sum_{i,j} a_{i,j} e_i \otimes f_j = 0$. This implies

$$\sum_{i \in I} e_i \otimes \left(\sum_{j \in I} a_{i,j} f_j \right) = 0.$$

Let (e_i^*) be the "dual basis" of the basis (e_i) and let (f_j^*) be the "dual basis" of the basis (f_j) (they do not form bases unless I is finite and J is finite). Let $(p, q) \in I \times J$; the map from $M \times N$ to A given by $(m, n) \mapsto e_p^*(m) e_q^*(n)$ is A-bilinear, hence there exists a unique morphism of A-modules from $M \otimes_A N$ to A which maps $m \otimes n$ to $e_p^*(m) e_q^*(n)$ for every $(m, n) \in M \times N$. Let us apply this morphism to our linear combination $\sum_{i,j} a_{i,j} e_i \otimes f_j$. We get

$$0 = \sum_{\substack{i \in I \\ j \in J}} a_{i,j} e_p^*(e_i) e_q^*(e_j) = a_{p,q}.$$

Consequently, $a_{i,j} = 0$ for every $(i, j) \in I \times J$, so that the family $(e_i \otimes f_j)$ is free, as we needed to show.   □

Example (8.2.2). — Let A be a commutative ring, let S be a multiplicative subset of A and let M be an A-module.

The map $m \mapsto m \otimes 1$ from M to $M \otimes_A S^{-1}A$ is a morphism of A-modules. Since $M \otimes_A S^{-1}A$ is an $S^{-1}A$-module, it extends to a morphism of $S^{-1}A$-modules, $\varphi \colon S^{-1}M \to M \otimes_A S^{-1}A$ given, explicitly, by $m/s \mapsto m \otimes (1/s)$ for $m \in M$ and $s \in S$. Let us show that φ is an isomorphism.

Let $g \colon M \times_A S^{-1}A \to S^{-1}M$ be the map given by $g(m, a/s) = am/s$. One checks that is well defined, that is, if $a/s = b/t$, then $am/s = bm/t$, and A-bilinear. Consequently, there exists a unique morphism of A-modules, $\psi \colon M \otimes_A S^{-1}A \to S^{-1}M$, such that $\psi(m \otimes a/s) = am/s$ for all $a \in A$, $s \in S$ and $m \in M$.

One has

$$\varphi \circ \psi(m \otimes a/s) = \varphi(am/s) = am \otimes (1/s) = m \otimes a/s,$$

so that $\varphi \circ \psi$ is the identity morphism. Similarly,

$$\psi \circ \varphi(m/s) = \psi(m \otimes (1/s)) = m/s,$$

hence $\psi \circ \varphi$ is the identity morphism as well. Consequently, φ and ψ are isomorphisms of A-modules, inverse to one another.

8.2.3. — Let M be an A-module, let $M^\vee = \mathrm{Hom}_A(M, A)$ be its dual. Let N be an A-module.

Let $d \colon M^\vee \times N \to \mathrm{Hom}_A(M, N)$ be the map such for every $\varphi \in M^\vee$ and every $n \in N$, $d(\varphi, n)$ is the morphism $x \mapsto \varphi(x)n$ from M to N. It is A-bilinear. By the universal property of the tensor product, there exists a unique morphism of A-modules $\delta_{M,N} \colon M^\vee \otimes_A N \to \operatorname{Hom}_A(M, N)$ such that $\delta_{M,N}(\varphi \otimes n) = d(\varphi, n)$ for every $\varphi \in M^\vee$ and every $n \in N$.

Proposition (8.2.4). — *Let M be an A-module.*

a) *If M is finitely generated and projective, then the morphism $\delta_{M,N} \colon M^\vee \otimes_A N \to \operatorname{Hom}_A(M, N)$ is an isomorphism for every A-module N.*

b) *If the morphism $\delta_{M,M} \colon M^\vee \otimes_A M \to \operatorname{End}_A(M)$ is surjective, then M is finitely generated and projective.*

Proof. — a) Let $(e_i)_{1 \le i \le r}$ be a finite generating family of M, let $f \colon A^r \to M$ be the surjective morphism given by $f(a_1, \dots, a_n)) = \sum_{i=1}^r a_i e_i$. Since M is projective, there exists a morphism $g \colon M \to A^r$ such that $f \circ g = \operatorname{id}_M$. For $i \in \{1, \dots, r\}$, let $\varphi_i \in M^\vee$ be the composition of g with the ith projection $A^r \to A$. By construction, one has $m = \sum_{i=1}^r \varphi_i(m) e_i$ for every $m \in M$. In other words, $\operatorname{id}_M = \delta_{M,M}(\sum_{i=1}^r \varphi_i \otimes e_i)$.

Let N be an A-module, let $u \colon M \to N$ be a morphism of A-modules. Let $\xi = \sum_{i=1}^r \varphi_i \otimes u(e_i)$. By definition, the morphism $\delta_{M,N}(\xi)$ sends an element $m \in M$ to $\sum_{i=1}^r \varphi_i(m) u(e_i) = u(\sum_{i=1}^r \varphi_i(m) e_i) = u(m)$, hence $u = \delta_{M,N}(\xi)$. This proves that $\delta_{M,N}$ is surjective.

Let then $\lambda \colon \operatorname{Hom}_A(M, N) \to M^\vee \otimes N$ be the map given by $\lambda(u) = \sum_{i=1}^r \varphi_i \otimes u(e_i)$. It is A-linear and the previous computations show that $\delta_{M,N} \circ \lambda = \operatorname{id}_{\operatorname{Hom}_A(M,N)}$. Conversely, let $\varphi \in M^\vee$ and $n \in N$; since $\delta_{M,N}(\varphi \otimes n)(e_i) = \varphi(e_i) n$, we have

$$\lambda \circ \delta_{M,N}(\varphi \otimes n) = \sum_{i=1}^r \varphi_i \otimes \varphi(e_i) n = \sum_{i=1}^r \varphi(e_i) \varphi_i \otimes n = \varphi' \otimes n,$$

where $\varphi' = \sum_{i=1}^r \varphi(e_i) \varphi_i$. For every $i \in \{1, \dots, r\}$, one has

$$\varphi'(e_i) = \sum_{i=1}^r \varphi(e_i) \varphi_i(e_i) = \varphi(\sum_{i=1}^r \varphi_i(e_i) e_i) = \varphi(e_i),$$

so that $\varphi' = \varphi$. This proves that $\lambda \circ \delta_{M,N}(\varphi \otimes n) = \varphi \otimes n$. Consequently, $\lambda \circ \delta_{M,N}$ coincides with the identity morphism on split tensors, hence on the whole of $M^\vee \otimes_A N$. In particular, the morphism $\delta_{M,N}$ is injective. We thus have shown that $\delta_{M,N}$ is an isomorphism.

b) Assume that $\delta_{M,M}$ is surjective and let $\xi \in M^\vee \otimes M$ be such that $\delta_{M,M}(\xi) = \operatorname{id}_M$; write $\xi = \sum_{i=1}^r \varphi_i \otimes e_i$, with $\varphi_1, \dots, \varphi_r \in M^\vee$ and $e_1, \dots, e_r \in M$.

Let $m \in M$. The equality $m = \operatorname{id}_M(m) = \delta_{M,M}(\xi)$ rewrites as

$$m = \sum_{i=1}^r \varphi_i(m) e_i.$$

This proves that $(e_1, \dots, e_r)$ generates M, which is thus finitely generated.

Let $f \colon A^r \to M$ be the morphism given by $f(a_1, \dots, a_r) = \sum_{i=1}^r a_i e_i$. Let $g \colon M \to A^r$ be the morphism given by $g(m) = (\varphi_1(m), \dots, \varphi_r(m))$. One has

$f \circ g = \mathrm{id}_M$. This implies that M is isomorphic to the submodule $g(M)$ of A^r, and $g(M)$ has a direct summand. In particular, M is projective.   □

8.2.5. — Let M be an A-module and let $M^\vee = \mathrm{Hom}_A(M, A)$. Consider the morphism $\delta_M \colon M^\vee \otimes_A M \to \mathrm{End}_A(M)$ of §8.2.3.

Let $t \colon M^\vee \times M \to A$ be the map given by $t(\varphi, m) = \varphi(m)$ for $\varphi \in M^\vee$ and $m \in M$. Again, it is A-bilinear. Consequently, there exists a unique linear form τ_M on $M^\vee \otimes_A M$ such that $t(\varphi \otimes m) = \varphi(m)$ for every $\varphi \in M^\vee$ and every $m \in M$.

Example (8.2.6). — Let us assume that M is free and finitely generated. In this case, proposition 8.2.4 asserts that $\delta_{M,N} \colon M^\vee \otimes_A N \to \mathrm{Hom}_A(M, N)$ is an isomorphism for every A-module N.

Let $(e_1, \ldots, e_r)$ be a basis of M and let $(e_1^*, \ldots, e_r^*)$ be its dual basis. Take N to be free and finitely generated and let $(f_j)_{1 \leq j \leq s}$ be a basis of N. Then the family $(e_i^* \otimes f_j)_{\substack{1 \leq i \leq r \\ 1 \leq j \leq s}}$ is a basis if $M^\vee \otimes_A N$. Let $\xi = \sum_{i=1}^r \sum_{j=1}^s a_{i,j} e_i^* \otimes f_j$ be any element of $M^\vee \otimes_A N$ and let $u = \delta_M(\xi)$. By definition, $u(e_i) = \sum_{j=1}^s a_{i,j} f_j$, so that $U = (a_{i,j}) \in \mathrm{Mat}_{s,r}(A)$ is the matrix of u in the bases $(e_1, \ldots, e_r)$ of M and $(f_1, \ldots, f_s)$ of N.

Now assume that M = N. Then

$$\tau_M(u) = \sum_{i,j=1}^n a_{i,j} e_i^*(e_j) = \sum_{i=1}^n a_{i,i} = \mathrm{Tr}(U),$$

and the linear form $\tau_M \colon M^\vee \otimes_A M \to A$ corresponds to the *trace* of an endomorphism of M.

8.2.7. Tensor product of algebras — Let k be a commutative ring, let A and B be (not necessarily commutative) k-algebras. Let us show how to endow the k-module $A \otimes_k B$ with the structure of a k-algebra. For $x \in A$ (or B), write λ_x for the endomorphisms of A (or B) given by left-multiplication by x. Let $(a, b) \in A \times B$. The endomorphism $\lambda_a \otimes \lambda_b$ of $A \otimes_k B$ is k-linear. Moreover, the map $(a, b) \mapsto \mathrm{End}_k(A \otimes_k B)$ is k-bilinear since the images of k in A and B are central. By the universal property of the tensor product, this gives a canonical k-linear morphism from $A \otimes_k B$ to $\mathrm{End}_k(A \otimes_k B)$, denoted by $t \mapsto \lambda_t$, which maps a split tensor $a \otimes b$ to $\lambda_a \otimes_k \lambda_b$. This morphism then furnishes a k-bilinear morphism μ from $(A \otimes_k B) \times (A \otimes_k B)$ to $A \otimes_k B$ given by $(\xi, \eta) \mapsto \lambda_\xi(\eta)$. Observe that

$$M(a \otimes b, a' \otimes b') = \lambda_{a \otimes b}(a' \otimes b') = (\lambda_a \otimes \lambda_b)(a' \otimes b') = (aa') \otimes bb'.$$

From that point on, it is straightforward (but a bit tedious) to prove that the composition law defined by M is associative, that $1 \otimes 1$ is its unit element, and that it is commutative if A and B are commutative. This endows $A \otimes_k B$ with the structure of a k-algebra.

8.3. Tensor Algebras, Symmetric and Exterior Algebras

In this section, A is a commutative ring.

The tensor product of two modules represents bilinear maps; we first generalize the construction for multilinear maps on products of more than two modules.

Lemma (8.3.1). — *Let $(M_i)_{i \in I}$ be a family of A-modules. There exists an A-module T and a multilinear map $\theta \colon \prod_i M_i \to T$ that possesses the following universal property: for every A-module N and every multilinear map $f \colon \prod M_i \to P$, there exists a unique morphism of A-modules $\varphi \colon T \to P$ such that $\varphi \circ \theta = f$.*

Proof. — Let F be the free A-module with basis $\prod_i M_i$. For any element $m = (m_i)_{i \in I}$ of $\prod_i M_i$, one writes e_m for the corresponding basis element of F. One then defines T as the quotient of the free A-module F by the submodule R generated by the following elements:

– the elements $e_m - e_{m'} - e_{m''}$ whenever $m = (m_i)$, $m' = (m_i')$, $m'' = (m_i'')$ are three elements of $\prod_i M_i$ for which there exists a $j \in I$ such that $m_i = m_i' = m_i''$ if $i \neq j$, and $m_j = m_j' + m_j''$;

– the elements $ae_m - e_{m'}$ whenever $a \in A$ and $m = (m_i)$, $m' = (m_i')$ are two elements of $\prod_i M_i$ for which there exists a $j \in I$ such that $m_i = m_i'$ if $i \neq j$, and $m_j' = am_j$.

For $m \in \prod_i M_i$, let $\theta(m)$ be the class of e_m in T. The map θ is multilinear, because we precisely modded out by all the necessary elements. Observe that the elements of the form e_m generate F, hence the image of θ generates T.

Let $f \colon \prod_i M_i \to N$ be any multilinear map, let $f_0 \colon F \to N$ be the unique morphism such that $f_0(e_m) = f(m)$, for every $m \in \prod_i M_i$. Since f is multilinear, the submodule R is contained in $\mathrm{Ker}(f_0)$; consequently, there exists a morphism $\varphi \colon T \to N$ such that $\varphi(\theta(m)) = f_0(e_m) = f(m)$ for every $m \in \prod_i M_i$. In other words, $\varphi \circ \theta = f$.

If φ and φ' are two morphisms such that $f = \varphi \circ \theta = \varphi' \circ \theta$, then $\varphi - \varphi'$ vanishes on the image of θ, hence on the submodule generated by θ which is T. Consequently, $\varphi = \varphi'$. □

The A-module T is called the *tensor product of the family* (M_i) *of A-modules*; it is denoted $\bigotimes_{i \in I} M_i$. When $I = \{1, \ldots, n\}$, one writes it rather $M_1 \otimes \cdots \otimes M_n$. For any element $m = (m_i) \in \prod M_i$, the element $\theta(m)$ of T is written $\otimes_i m_i$, or $m_1 \otimes \cdots \otimes m_n$ if $I = \{1, \ldots, n\}$.

Remark (8.3.2). — As for any solution of a universal problem, the pair (θ, T) is uniquely characterized by this universal property. Indeed, let (θ', T') be a pair consisting of an A-module T' and of a multilinear map $\theta' \colon \prod M_i \to T'$ satisfying the universal property. First of all, by the universal property applied to (θ, T) and to the multilinear map θ', there exists a morphism of A-modules $f \colon T \to T'$ such that $\theta' = f \circ \theta$. Similarly, there exists a morphism of A-modules $f' \colon T' \to T$ such that $\theta = f' \circ \theta'$. Then, one has $\theta = (f' \circ f) \circ \theta$;

by the universal property of (θ, T), id_T is the unique morphism $g\colon T \to T$ such that $\theta = g \circ \theta$; consequently, $f' \circ f = \mathrm{id}_T$. Reversing the roles of T and T′, we also have $f \circ f' = \mathrm{id}_{T'}$. This implies that f is an isomorphism.

Remark (8.3.3). — When $I = \{1, \ldots, n\}$, one can also construct the tensor product $M_1 \otimes \cdots \otimes M_n$ by induction. More precisely, let $M_1, \ldots, M_n$ be A-modules. Let P be a way of parenthesizing the product $x_1 \ldots x_n$. They are defined by induction using the following rules:

  – If $n = 1$, then x_1 can be uniquely parenthesized, as x_1;

  – If $n \geq 2$, one parenthesizes $x_1 \ldots x_n$ by choosing $j \in \{1, \ldots, n-1\}$, and $P = (P')(P'')$ where P′ is a way of parenthesizing $x_1 \ldots x_j$ and P″ is a way of parenthesizing $x_{j+1} \ldots x_n$.

For example, $x_1 x_2$ can be uniquely parenthesized, namely as $(x_1)(x_2)$; for $n = 5$, $((x_1)(x_2))(((x_3)(x_4))(x_5))$ is one of the 14 possible parenthesizations of $x_1 x_2 x_3 x_4 x_5$. (In general, the number of all possible parenthesizations of $x_1 \ldots x_n$ is known as the Catalan number $C_{n-1} = \dfrac{1}{n}\binom{2(n-1)}{n-1}$.)

To any such parenthesization, one can define by induction a module M_P and a multilinear map $\theta_P\colon \prod M_i \to M_P$. When $n = 1$, one sets $M_P = M_1$ and $\theta_P = \mathrm{id}$. When $n \geq 2$ and $P = (P')(P'')$, one defines $M_P = M_{P'} \otimes_A M_{P''}$ and $\theta_P = \theta_{P'} \otimes \theta_{P''}$. For example, the previous parenthesizing P of $x_1 x_2 x_3 x_4 x_5$ gives rise to

$$M_P = (M_1 \otimes M_2) \otimes ((M_3 \otimes M_4) \otimes M_5)$$

and

$$\theta_P(m_1, m_2, m_3, m_4, m_5) = (m_1 \otimes m_2) \otimes ((m_3 \otimes m_4) \otimes m_5).$$

However, a map on $M_1 \times \cdots \times M_n$ is multilinear if and only if it is multilinear with respect to both the first group of variables $(m_1, \ldots, m_j)$ and the second group $(m_{j+1}, \ldots, m_n)$. By induction, we see that the map θ_P is multilinear, hence that the module M_P is also a solution of the universal property. Consequently, there exists a unique morphism of A-modules from $\bigotimes_{i=1}^{n} M_i$ to M_P which maps $\theta(m_1, \ldots, m_n)$ to $\theta_P(m_1, \ldots, m_n)$.

In the sequel, we shall neglect these various "associativity" isomorphisms.

Proposition (8.3.4). — *Let $M_1, \ldots, M_n$ be A-modules. For every $i \in \{1, \ldots, n\}$, let $(e^i_j)_{j \in J_i}$ be a family of elements of M_i. Let $J = J_1 \times \cdots \times J_n$; for every $j = (j_1, \ldots, j_n) \in J$, let $e_j = e^1_{j_1} \otimes \cdots \otimes e^n_{j_n}$.*

  a) *If for each i, $(e^i_j)_{j \in J_i}$ generates M_i, then the family $(e_j)_{j \in J}$ generates the tensor product $M_1 \otimes \cdots \otimes M_n$.*

  b) *If for each i, $(e^i_j)_{j \in J_i}$ is a basis of M_i, then the family $(e_j)_{j \in J}$ is a basis of the tensor product $M_1 \otimes \cdots \otimes M_n$.*

Proof. — The case $n = 1$ is trivial, the case $n = 2$ is proposition 8.2.1. The general case follows by induction thanks to the step by step construction of $M_1 \otimes \cdots \otimes M_n$.   □

8.3.5. The tensor algebra — Let M be an A-module. Set $M_0 = A$ and, for any positive integer n, let M_n be the tensor product of n copies of the A-module M. Modulo the associativity isomorphisms, we have $M_n = M_p \otimes_A M_q$ whenever p, q, n are non-negative integers such that $p + q = n$. The map $(m_1, \ldots, m_n) \mapsto m_1 \otimes \cdots \otimes m_n$ from M^n to M_n is n-linear and the module M_n satisfies the following universal property: *for every n-linear map f from M^n to an A-module P, there exists a unique morphism of A-modules, $\varphi \colon M_n \to P$, such that $\varphi(m_1 \otimes \cdots \otimes m_n) = f(m_1, \ldots, m_n)$ for every $(m_1, \ldots, m_n) \in M^n$.*

If M is free and $(e_i)_{i \in I}$ is a basis of M, then for every $n \in \mathbf{N}$, the A-module M_n is free with basis the family $(e_{i_1} \otimes \cdots \otimes e_{i_n})$ indexed by $(i_1, \ldots, i_n) \in I^n$. Indeed, this follows from proposition 8.3.4.

Let $T(M)$ be the direct sum of all A-modules M_n, for $n \in \mathbf{N}$. The submodule M_n of $T(M)$ will be written $T^n(M)$. An element of $T^n(M)$ will be said to be of degree n.

The associativity isomorphisms $M_p \otimes M_q \simeq M_{p+q}$ furnish bilinear maps $T^p(M) \times T^q(M) \to T^{p+q}(M)$, for $p, q \geq 0$. There is a unique structure of an A-algebra on the A-module $T(M)$ of which multiplication law is given by these maps. The resulting algebra is called the *tensor algebra of the A-module* M. It contains $M = T^1(M)$ as a submodule. The elements m, for $m \in M = T^1(M)$, generate $T(M)$ as an A-algebra.

This algebra satisfies the following universal property: *For every A-algebra (associative, with unit) B, and every morphism of A-modules $f \colon M \to B$, there exists a unique morphism of A-algebras, $\varphi \colon T(M) \to B$, such that $\varphi(m) = f(m)$ for every $m \in M$.*

If M is free with basis $(e_i)_{i \in I}$, then $T(M)$ is a free A-module with basis the disjoint union of the bases $1 \in T^0(M)$ and $e_{i_1} \otimes \cdots \otimes e_{i_n}$ for $(i_1, \ldots, i_n) \in I^n$ and $n \geq 1$.

8.3.6. The symmetric algebra — The tensor algebra $T(M)$ is not commutative in general. By definition, the *symmetric algebra* of the A-module M is the quotient of $T(M)$ by the two-sided ideal I generated by elements of the form $m_1 \otimes m_2 - m_2 \otimes m_1$, for $m_1, m_2 \in M$. It is denoted $S(M)$.

For $n \in \mathbf{N}$, let $I_n = I \cap T^n(M)$; this is a subbmodule of $T^n(M)$. The ideal I is homogeneous in the sense that one has $I = \bigoplus_n I_n$. The inclusion $\bigoplus_n I_n \subset I$ is obvious. Conversely, let $m_1, m_2 \in M$ and write $m_1 = \sum m_1^{(n)}$ and $m_2 = \sum m_2^{(n)}$, where $m_i^{(n)} \in T^n(M)$ for all n. Then

$$m_1 \otimes m_2 - m_2 \otimes m_1 = \sum_{n \in \mathbf{N}} \sum_{p+q=n} \left(m_1^{(p)} \otimes m_2^{(q)} - m_1^{(q)} \otimes m_2^{(p)} \right)$$

belongs to $\bigoplus_n I_n$; this implies that the generators of I are contained in the two-sided ideal $\bigoplus I_n$, hence that $I = \bigoplus I_n$.

These computations also show that $I_0 = 0$ and $I_1 = 0$.

Write $S^n(M)$ for the image of $T^n(M)$ in $S(M)$; one has $S^n(M) = T^n(M)/I_n$ and $S(M) = \bigoplus S^n(M)$; moreover, $S^0(M) = A$ and $S^1(M) = M$.

Observe that, by construction, all elements of $S^1(M)$ pairwise commute, and that $S(M)$ is generated by $S^1(M)$. Consequently, the algebra $S(M)$ is commutative.

The algebra $S(M)$ is a commutative A-algebra together with a morphism $M \to S(M)$ which satisfies the following universal property: *For every commutative A-algebra B and every morphism $f\colon M \to B$ of A-modules, there exists a unique morphism of A-algebras, $\varphi\colon S(M) \to B$, such that $\varphi(m) = f(m)$ for every $m \in M = S^1(M)$.*

For every integer n, the A-module $S^n(M)$ is endowed with an n-linear symmetric map $M^n \to S^n(M)$. It satisfies the following universal property: *For every A-module P and any n-linear symmetric map $f\colon M^n \to P$, there exists a unique morphism of A-modules $\varphi\colon S^n(M) \to P$ such that $\varphi(m_1 \ldots m_n) = f(m_1, \ldots, m_n)$ for every $m_1, \ldots, m_n \in M$.*

8.3.7. The exterior algebra — As we have observed in section 3.8, alternating multilinear maps are a very important object. Let us understand them with the tool of the tensor product, analogously to the description of symmetric multilinear maps from the the symmetric algebra.

Let J be the two-sided ideal of $T(M)$ generated by all elements of the form $m \otimes m$, for $m \in M$. It is a "homogeneous ideal" of $T(M)$, that is to say, $J = \bigoplus_{n=0}^{\infty} J_n$, where $J_n = J \cap T^n(M)$. Indeed, $\bigoplus_n J_n$ is a two-sided ideal which contains all elements of the form $m \otimes m$, for $m \in M$, hence is equal to J. Moreover, $J_0 = 0$ and $J_1 = 0$.

Let $\Lambda(M)$ be the quotient of the algebra $T(M)$ by this two-sided ideal J. As an A-module, one has $\Lambda(M) = \bigoplus_{n=0}^{\infty} \Lambda^n(M)$, where $\Lambda^n(M) = T^n(M)/J_n$. In particular, $\Lambda^0(M) = A$ and $\Lambda^1(M) = M$. The multiplication of $\Lambda(M)$ is denoted by the symbol $\wedge$. For example, one has $m \wedge m = 0$ for every $m \in M$.

For any $m_1, m_2 \in M$, one has

$$m_1 \otimes m_2 + m_2 \otimes m_1 = (m_1 + m_2) \otimes (m_1 + m_2) - m_1 \otimes m_1 - m_2 \otimes m_2 \in J,$$

so that $m_1 \wedge m_2 = -m_2 \wedge m_1$. More generally, let $m_1, \ldots, m_n \in M$ and let $\sigma \in \mathfrak{S}_n$ be any permutation of $\{1, \ldots, n\}$. Then,

$$m_{\sigma(1)} \wedge \cdots \wedge m_{\sigma(n)} = \varepsilon(\sigma) m_1 \wedge \cdots \wedge m_n,$$

where $\varepsilon(\sigma)$ is the signature of σ.

The A-algebra $\Lambda(M)$ is called the *exterior algebra* on the module M. It satisfies the following universal property: *For every A-algebra B and any morphism $f\colon M \to B$ such that $f(m) \cdot f(m) = 0$ for every $m \in M$, there exists a unique morphism of A-algebras, $\varphi\colon \Lambda(M) \to B$, such that $\varphi(m) = f(m)$ for every $m \in M = \Lambda^1(M)$.*

An element of $\Lambda^p(M)$ is called a p-vector. Any p-vector can be written as a sum of split p-vectors, namely p-vectors of the form $m_1 \wedge \cdots \wedge m_p$, for $m_1, \ldots, m_p \in M$. By p-linearity, we see that we can restrict ourselves to vectors m_i in some generating family of M.

More precisely, let us assume that $(m_i)_{1 \leq i \leq n}$ generates M. Then any p-vector in $\Lambda^p(M)$ is a linear combination of vectors of the form $m_{i_1} \wedge \cdots \wedge m_{i_p}$, for $i_1, \ldots, i_p \in \{1, \ldots, n\}$. By the alternating property, we can even assume that $i_1 \leq \cdots \leq i_p$. Then, if $p > n$, we see that two consecutive indices are equal hence $m_{i_1} \wedge \cdots \wedge m_{i_p} = 0$ by the alternating property. This shows that $\Lambda^p(M) = 0$ for $p > n$.

Proposition (8.3.8). — *Let $n \in \mathbf{N}$. The A-module $\Lambda^n(M)$ is endowed with an alternating n-linear map $M^n \to \Lambda^n(M)$, $(m_1, \ldots, m_n) \mapsto m_1 \wedge \cdots \wedge m_n$. It satisfied the following universal property: For every A-module P and any n-linear alternating map $f \colon M^n \to P$, there exists a unique morphism $\varphi \colon \Lambda^n(M) \to P$ of A-modules such that $\varphi(m_1 \wedge \cdots \wedge m_n) = f(m_1, \ldots, m_n)$.*

Proof. — The uniqueness of the morphism φ follows from the fact that the products $m_1 \wedge \cdots \wedge m_n$, for $m_1, \ldots, m_n \in M$, generate $\Lambda^n(M)$. Let us show its existence. Let $\varphi_1 \colon T^n(M) \to P$ the canonical morphism of A-modules which is deduced from the multilinear map f; one has $\varphi_1(m_1 \otimes \cdots \otimes m_n) = f(m_1, \ldots, m_n)$ for every $(m_1, \ldots, m_n) \in M^n$. Let us show that φ_1 vanishes on the kernel J_n of the canonical morphism from $T^n(M)$ to $\Lambda^n(M)$.

Let $m_1, \ldots, m_n \in M$. If there exists an integer $i \in \{1, \ldots, n-1\}$ such that $m_i = m_{i+1}$, then the tensor

$$m_1 \otimes \cdots \otimes m_n = (m_1 \otimes \cdots \otimes m_{i-1}) \otimes (m_i \otimes m_i) \otimes (m_{i+2} \otimes \ldots m_n)$$

belongs to the ideal J, the kernel of the canonical morphism of algebras from $T(M)$ to $\Lambda(M)$. Moreover, when n varies, as well as $m_1, \ldots, m_n \in M$, we see that these tensors generate a two-sided ideal of $T(M)$ containing all elements of the form $m \otimes m$, hence is equal to J. In particular, any element of $J_n = J \cap T^n(M)$ is a linear combination of such tensors.

Since f is alternating, one has

$$\varphi_1(m_1 \otimes \cdots \otimes m_n) = f(m_1, \ldots, m_n) = 0$$

as soon as there exists an integer $i \in \{1, \ldots, n-1\}$ such that $m_i = m_{i+1}$. Consequently, φ_1 vanishes on J_n. This implies that there exists a morphism of A-modules $\varphi \colon \Lambda^n(M) \to P$ such that $\varphi(m_1 \wedge \cdots \wedge m_n) = \varphi_1(m_1 \otimes \ldots m_n)$ for every $(m_1, \ldots, m_n) \in M^n$, that is, $\varphi(m_1 \wedge \cdots \wedge m_n) = f(m_1, \ldots, m_n)$. This concludes the proof. $\qquad\square$

Proposition (8.3.9). — *Let M and N be A-modules and let $u \colon M \to N$ be a morphism. There exists a unique morphism of A-algebras, $\Lambda(u) \colon \Lambda(M) \to \Lambda(N)$, which coincides with u on $\Lambda^1(M)$. One has $\Lambda(\mathrm{id}_M) = \mathrm{id}_{\Lambda(M)}$. If P is another A-module and $v \colon N \to P$ is a morphism, then one has $\Lambda(v \circ u) = \Lambda(v) \circ \Lambda(u)$.*

Proof. — This follows at once from the universal property. Indeed, let $\theta \colon N \to \Lambda(N)$ be the canonical injection; then, the map $\theta \circ u$ maps every element m of M to the element $u(m)$ of $\Lambda^1(N)$, hence its square is zero. Consequently, there exists a morphism of algebras from $\Lambda(M)$ to $\Lambda(N)$ which coincides with u on $M \simeq \Lambda^1(M)$.

Moreover, $\Lambda(v) \circ \Lambda(u)$ is a morphism of algebras from $\Lambda(M)$ to $\Lambda(P)$ which coincides with $v \circ u$ on M, hence is equal to $\Lambda(v \circ u)$. $\square$

Let $n \in \mathbf{N}$. By restriction to $\Lambda^n(M)$, the morphism of algebras $\Lambda(u)$ defines a morphism $\Lambda^n(u) \colon \Lambda^n(M) \to \Lambda^n(M)$ of A-modules. It is characterized by the formula

$$\Lambda^n(u)(m_1 \wedge \ldots m_n) = u(m_1) \wedge \cdots \wedge u(m_n)$$

for every $(m_1, \ldots, m_n) \in M^n$.

8.4. The Exterior Algebra and Determinants

Proposition (8.4.1). — *Let M be a free A-module, let $(e_i)_{i \in I}$ be a basis of M and let $\leq$ be a total ordering on I. Then, the family of p-vectors $e_{i_1} \wedge \ldots e_{i_p}$, where $(i_1, \ldots, i_p)$ runs along the set of strictly increasing sequences of indices, is a basis of $\Lambda^p(M)$. In particular, if M is finitely generated and n is its rank, then $\Lambda^p(M)$ is free of rank $\binom{n}{p}$.*

Let J be a subset of I with p elements, and let $(i_1, \ldots, i_p)$ be the unique strictly increasing tuple such that $J = \{i_1, \ldots, i_p\}$; we set $e_J = e_{i_1} \wedge \ldots e_{i_p}$.

Proof. — Any p-vector in M is a linear combination of p-vectors of the form $e_{i_1} \wedge \cdots \wedge e_{i_p}$, where $i_1, \ldots, i_p$ belong to I. Using the relations $e_i \wedge e_j = -e_j \wedge e_i$ for $i \neq j$, as well as $e_i \wedge e_i = 0$, we may only consider families $(i_1, \ldots, i_p)$ such that $i_1 < \cdots < i_p$. This shows that the elements e_J, for all p-subsets J of I, generate $\Lambda^p(M)$.

Let us show that these elements actually form a basis of $\Lambda^p(M)$ and let us consider a linear dependence relation $\sum_J a_J e_J = 0$ among them. The A-module $T^p(M)$ is free and the family $(e_{i_1} \otimes \cdots \otimes e_{i_p})$, for $(i_1, \ldots, i_p) \in I^p$, is a basis. Let us fix a subset $J \subset \{1, \ldots, n\}$ with cardinality p; let us write $J = \{j_1, \ldots, j_p\}$ where $j_1 < \cdots < j_p$. Let $f_J \colon M^n$ be the unique p-linear form on M which maps $(e_{i_1}, \ldots, e_{i_p})$ to 0 if $\{i_1, \ldots, i_p\} \neq J$ and $(e_{i_1}, \ldots, e_{i_p})$ to the signature of the permutation which maps i_k to j_k, for any $k \in \{1, \ldots, p\}$. It is alternating, hence there exists a unique morphism of A-modules $\varphi_J \colon \Lambda^p(M) \to A$ which maps e_I to 0 if $I \neq J$ and e_J to 1. Let us apply φ_J to the dependence relation $\sum_J a_J e_J = 0$; we get $a_J = 0$. Consequently, the given linear dependence relation is trivial, hence the family (e_J) is free. It is thus a basis of $\Lambda^p(M)$. $\square$

Lemma (8.4.2). — *Let M be an A-module and let $f \colon M^{n-1} \to A$ be an alternating $(n-1)$-linear form. The map $f' \colon M^n \to M$ given by*

$$f'(x_1, \ldots, x_n) = \sum_{p=1}^{n} (-1)^p f(x_1, \ldots, \widehat{x_p}, \ldots, x_n) x_p,$$

where the $\widehat{x_p}$ means that this element is omitted from the tuple, is n-linear and alternating.

Proof. — It is obvious that f' is linear with respect to each variable. Moreover, if $x_i = x_{i+1}$, the terms $f(x_1, \ldots, \widehat{x_p}, \ldots, x_n) x_p$ are 0 whenever $p \neq i$ and $p \neq i+1$, and the terms with indices i and $i+1$ are negatives of one another. This proves the lemma.   □

Proposition (8.4.3). — *Let M be a free finitely generated A-module and let $x_1, \ldots, x_n$ be elements of M. For the elements $x_1, \ldots, x_n$ to be linearly dependent, it is necessary and sufficient that there exists an element $\lambda \in A - \{0\}$ such that*

$$\lambda x_1 \wedge \cdots \wedge x_n = 0.$$

Proof. — Assume that $x_1, \ldots, x_n$ are linearly dependent and let $a_1 x_1 + \cdots + a_n x_n = 0$ be a nontrivial linear dependence relation. To fix the ideas, assume that $a_1 \neq 0$. Then, $a_1 x_1$ is a linear combination of $x_2, \ldots, x_n$, so that $(a_1 x_1) \wedge x_2 \wedge \cdots \wedge x_n = 0$; this implies $a_1 x_1 \wedge \cdots \wedge x_n = 0$.

Conversely, let us show by induction on n that if $\lambda x_1 \wedge \cdots \wedge x_n = 0$, for some non-zero $\lambda \in A$, then $x_1, \ldots, x_n$ are linearly dependent.

This holds when $n = 0$ (there is nothing to prove) and when $n = 1$ (by definition). So, let us assume that this assertion holds up to $n - 1$ and assume that $\lambda x_1 \wedge \cdots \wedge x_n = 0$, where $\lambda \neq 0$. If $x_2, \ldots, x_n$ are linearly dependent, then $x_1, \ldots, x_n$ are also linearly dependent too, so that we may assume that $x_2, \ldots, x_n$ are linearly independent and that $x_2 \wedge \cdots \wedge x_n \neq 0$. Since $\Lambda^{n-1}(M)$ is a free A-module, there exists a linear form φ on $\Lambda^{n-1}(M)$ such that $\mu = \varphi(x_2 \wedge \cdots \wedge x_n) \neq 0$. In other words, the map $f \colon M^{n-1} \to A$ given by $(v_2, \ldots, v_n) \mapsto \varphi(v_2 \wedge \cdots \wedge v_n)$ is an alternating $(n-1)$-linear form on M and $f(x_2, \ldots, x_n) = \mu \neq 0$. Let $f' \colon M^n \to M$ be the alternating n-linear map defined in lemma 8.4.2 above. Since $\lambda x_1 \wedge \cdots \wedge x_n = 0$, one has $f'(\lambda x_1, x_2, \ldots, x_n) = 0$, hence a relation

$$-\lambda f(x_2, \ldots, x_n) x_1 + \lambda \sum_{p=2}^{n} f(x_1, \ldots, \widehat{x_p}, \ldots, x_n) x_p = 0.$$

Since $\lambda f(x_2, \ldots, x_n) \neq 0$, the vectors $x_1, \ldots, x_n$ are linearly dependent, hence the proposition.   □

Corollary (8.4.4). — *Let A be a non-zero commutative ring, let M be a free A-module, let $n \geq 1$ be an integer, and let $e_1, \ldots, e_n$ be elements of M. Then $(e_1, \ldots, e_n)$ is a basis of M if and only if $e_1 \wedge \cdots \wedge e_n$ is a basis of $\Lambda^n(M)$.*

Proof. — If $e_1, \ldots, e_n$ is a basis of M, we proved in proposition 8.4.1 that $e_1 \wedge \cdots \wedge e_n$ is a basis of $\Lambda^n(M)$.

Conversely, let us assume that $e_1 \wedge \cdots \wedge e_n$ be a basis of $\Lambda^n(M)$. By the preceding proposition, the family $(e_1, \ldots, e_n)$ is free. It then follows from proposition 8.4.1 that M is finitely generated, that its rank is equal to n, that $\Lambda^{n-1}(M) \neq 0$ and that $\Lambda^{n+1}(M) = 0$. Let f be the map from M^n to A such that $f(x_1, \ldots, x_n)$ is the unique element $\lambda \in A$ such that $x_1 \wedge \cdots \wedge x_n = \lambda e_1 \wedge \cdots \wedge e_n$, for any $x_1, \ldots, x_n \in M$. The map f is n-linear and alternating (it is the composition of the n-linear alternating map $M^n \to \Lambda^n(M)$ given by $(x_1, \ldots, x_n) \mapsto x_1 \wedge \cdots \wedge x_n$, and of the inverse of the isomorphism $A \to \Lambda^n(M)$, given by $a \mapsto ae_1 \wedge \cdots \wedge e_n$). Since $\Lambda^{n+1}(M) = 0$, the map f' from M^{n+1} to M defined in lemma 8.4.2 is zero, because it is $(n+1)$-linear and alternating. In particular, for every $x \in M$,

$$f'(x, e_1, \ldots, e_n) = -f(e_1, \ldots, e_n)x + \sum_{p=1}^{n} (-1)^{p+1} f(x, e_1, \ldots, \widehat{e_p}, \ldots, e_n)e_p = 0.$$

Since $f(e_1, \ldots, e_n) = 1$, one gets

$$x = \sum_{p=1}^{n} (-1)^{p+1} f(x, e_1, \ldots, \widehat{e_p}, \ldots, e_n)e_p = 0,$$

which proves that the family $(e_1, \ldots, e_n)$ generates M. It is thus a basis of M.$\square$

Corollary (8.4.5). — *Let A be a non-zero commutative ring and let M be an A-module which possesses a basis of cardinality n.*

a) *Every free family in M has at most n elements.*

b) *Every generating family in M has at least n elements.*

Proof. — Let $x_1, \ldots, x_p$ be elements of M. If $(x_1, \ldots, x_p)$ is free, then $x_1 \wedge \cdots \wedge x_p \neq 0$, hence $\Lambda^p(M) \neq 0$ and $p \leq n$. If $(x_1, \ldots, x_p)$ is generating, then $\Lambda^k(M) = 0$ for $k > p$, hence $n \leq p$. $\square$

Consequently, when A is a non-zero commutative ring, it is unambiguous to define the rank of a free finitely generated A-module as the cardinality of any basis of M.

Definition (8.4.6). — Let M be a free A-module of rank n and let u be an endomorphism of M. The endomorphism $\Lambda^n(u)$ of $\Lambda^n(M)$ is a homothety. Its ratio is called the *determinant* of u and written $\det(u)$.

By the preceding proposition, $\Lambda^n(M)$ is isomorphic to A. Let ε be a basis of $\Lambda^n(M)$ and let $\delta \in A$ by such that $\Lambda^n(u)(\varepsilon) = \delta\varepsilon$. For any $x \in \Lambda^n(M)$, let $a \in A$ be such that $x = a\varepsilon$; then, $\Lambda^n(x) = \Lambda^n(a\varepsilon) = a\Lambda^n(\varepsilon) = a\delta\varepsilon = \delta x$. This shows that $\Lambda^n(u)$ is a homothety.

Let u and v be endomorphism of a free A-module of rank n. One has $\Lambda^n(v \circ u) = \Lambda^n(v) \circ \Lambda^n(u)$, hence $\det(v \circ u) = \det(v)\det(u)$. One also has $\Lambda^n(\mathrm{id}_M) = \mathrm{id}$, so that $\det(\mathrm{id}_M) = 1$. If u is invertible, with inverse v, then $\Lambda^n(u)$ is invertible too, with inverse $\Lambda^n(v)$. It follows that $\det(u)$ is an invertible element of A, with inverse $\det(v)$. In other words,

$$\det(u^{-1}) = \det(u)^{-1}.$$

Proposition (8.4.7). — *Let* A *be a non-zero commutative ring, let* M *be a free* A*-module of rank* n *and let* u *be an endomorphism of* M.

a) *The morphism* u *is an isomorphism if and only if* u *is surjective, if and only if* $\det(u)$ *is invertible.*

b) *The morphism* u *is injective if and only if* $\det(u)$ *is regular in* A.

Proof. — Let $(e_1, \ldots, e_n)$ be a basis of M.

If u is an automorphism of M, we just proved that $\det(u)$ is invertible in A. More generally, let us assume that u is surjective. Since M is free, the morphism u has a right inverse v. (For every $i \in \{1, \ldots, n\}$, let $f_i \in M$ be any element such that $u(f_i) = e_i$, and let v be the unique endomorphism of M such that $v(e_i) = f_i$ for every i.) One has $u \circ v = \mathrm{id}_M$, hence $\det(u)\det(v) = \det(\mathrm{id}_M) = 1$. This shows that $\det(u)$ is invertible.

Let us now prove that if $\det(u)$ is invertible, then u is an automorphism of M. By definition, one has $u(e_1) \wedge \cdots \wedge u(e_n) = \det(u)e_1 \wedge \cdots \wedge e_n$. If $\det(u)$ is invertible, then $u(e_1) \wedge \cdots \wedge u(e_n)$ is a basis of $\Lambda^n(M)$. By corollary 8.4.4, $(u(e_1), \ldots, u(e_n))$ is a basis of M. Let v be the unique endomorphism of M such that $v(u(e_i)) = e_i$ for every $i \in \{1, \ldots, n\}$. One has $v \circ u(e_i) = e_i$ for every i, hence $v \circ u = \mathrm{id}$. Consequently, $u \circ v(u(e_i)) = u(e_i)$ for every i; since $(u(e_1), \ldots, u(e_n))$ is a basis of M, this implies that $u \circ v = \mathrm{id}$. In other words, u is invertible, and v is its inverse.

For the morphism u to be injective, it is necessary and sufficient that the family $(u(e_1), \ldots, u(e_n))$ be free. If this holds, proposition 8.4.3 asserts that for every $\lambda \in A - \{0\}$, $\lambda u(e_1) \wedge \cdots \wedge u(e_n) \neq 0$. Since $e_1 \wedge \cdots \wedge e_n$ is a basis of $\Lambda^n(M)$, we see that u is injective if and only if $\lambda \det(u) \neq 0$ for $\lambda \neq 0$, in other words, if and only if $\det(u)$ is regular in A. $\qquad\square$

8.4.8. Determinant of a matrix — One defines the determinant of a matrix $U \in \mathrm{Mat}_n(A)$ as the determinant of the endomorphism u of A^n it represents in the canonical basis $(e_1, \ldots, e_n)$ of A^n. One has

$$\Lambda^n(u)(e_1 \wedge \cdots \wedge e_n) = u(e_1) \wedge \cdots \wedge u(e_n)$$

$$= \sum_{i_1=1}^n \cdots \sum_{i_n=1}^n U_{i_1,1} \ldots U_{i_n,n} e_{i_1} \wedge e_{i_2} \wedge \cdots \wedge e_{i_n}$$

$$= \sum_{\sigma \in \mathfrak{S}_n} U_{\sigma(1),1} \ldots U_{\sigma(n),n} \varepsilon(\sigma) e_1 \wedge \cdots \wedge e_n.$$

Consequently,

$$\det(U) = \sum_{\sigma \in \mathfrak{S}_n} \varepsilon(\sigma) U_{\sigma(1),1} \ldots U_{\sigma(n),n}.$$

Since a permutation and its inverse have the same signature, we also obtain

$$\det(U) = \sum_{\sigma \in \mathfrak{S}_n} \varepsilon(\sigma) U_{1,\sigma(1)} \ldots U_{n,\sigma(n)}.$$

In particular, a matrix and its transpose have the same determinant.

8.4.9. Laplace expansion of a determinant — Let n be an integer and let U be an $n \times n$-matrix with coefficients in a commutative ring A. Let u be the endomorphism of A^n whose matrix in the canonical basis $(e_1, \ldots, e_n)$ is equal to U. For any subset I of $\{1, \ldots, n\}$, let $\bar{I}$ be the complementary subset. If $I = \{i_1, \ldots, i_p\}$, with $i_1 < \cdots < i_p$, we define $e_I = e_{i_1} \wedge \cdots \wedge e_{i_p}$. If I and J are any two subsets of $\{1, \ldots, n\}$, one has $e_I \wedge e_J = 0$ if $I \cap J \neq \emptyset$. Assume that $I \cap J = \emptyset$, and let m be the number of pairs $(i, j) \in I \times J$ such that $i > j$ and set $\varepsilon_{I,J} = (-1)^m$; then, $e_I \wedge e_J = \varepsilon_{I,J} e_{I \cup J}$.

Let I and J be subsets of $\{1, \ldots, n\}$; let $U_{I,J}$ be the submatrix $(U_{i,j})_{\substack{i \in I \\ j \in J}}$ of U obtained by selecting the rows with indices in I and the columns with indices in J. When I and J have the same cardinality, the determinant of $U_{I,J}$ is called the (I, J)-*minor* of U. With these notations, we can state the following proposition.

Proposition (8.4.10). — a) *Let* I *be any subset of* $\{1, \ldots, n\}$ *and let* $p = \text{Card}(I)$. *Then,* $\Lambda^p(u)(e_I) = \sum_J \det(U_{J,I}) e_J$, *where* J *runs among all subsets of cardinality* p *of* $\{1, \ldots, n\}$.

b) *Let* J *and* K *be subsets of* $\{1, \ldots, n\}$ *with the same cardinality. Then*

$$\sum_I \varepsilon_{J,\bar{J}} \varepsilon_{K,\bar{K}} \det(U_{I,J}) \det(U_{\bar{I},\bar{K}}) = \begin{cases} \det(U) & \text{if } J = K; \\ 0 & \text{otherwise}, \end{cases}$$

the sum being over all subsets I *of* $\{1, \ldots, n\}$ *with cardinality* p.

Proof. — a) Let $i_1, \ldots, i_p$ be the elements of I, written in increasing order. One has

$$\Lambda^p(u)(e_I) = u(e_{i_1}) \wedge \cdots \wedge u(e_{i_p})$$

$$= \sum_{j_1 = 1}^{n} \cdots \sum_{j_p = 1}^{n} U_{j_1, i_1} \ldots U_{j_p, i_p} e_{j_1} \wedge \cdots \wedge e_{j_p}$$

$$= \sum_{j_1 < \cdots < j_p} \left(\sum_{\sigma \in \mathfrak{S}_p} \varepsilon(\sigma) U_{j_{\sigma(1)}, i_1} \ldots U_{j_{\sigma(p)}, i_p} \right) e_{j_1} \wedge \cdots \wedge e_{j_p}$$

$$= \sum_J \det(U_{J,I}) e_J,$$

as was to be shown.

b) By definition,

$$\Lambda^p(u)(e_I) \wedge \Lambda^{n-p}(u)(e_{\bar{K}}) = \Lambda^n(u)(e_I \wedge e_{\bar{K}})$$

vanishes if I and $\bar{K}$ have a common element, that is if $I \neq K$, since they have the same cardinality. Otherwise, if $I = K$, it is equal to

$$\Lambda^n(u)(\varepsilon_{I,\bar{I}} e_1 \wedge \cdots \wedge e_n) = \varepsilon_{I,\bar{I}} \det(U) e_1 \wedge \cdots \wedge e_n.$$

On the other hand, it follows from a) that

$$\Lambda^p(u)(e_I) \wedge \Lambda^{n-p}(u)(e_{\bar{K}}) = \sum_{\text{Card}(J)=p} \det(U_{J,I})e_J \sum_{\text{Card}(L)=p} \det(U_{\bar{L},\bar{K}})e_{\bar{L}}$$

$$= \sum_{\text{Card}(J)=p} \sum_{\text{Card}(L)=p} \det(U_{J,I})\det(U_{\bar{L},\bar{K}})e_J \wedge e_{\bar{L}}$$

$$= \sum_{\substack{\text{Card}(J)=p \\ L=J}} \det(U_{J,I})\det(U_{\bar{J},\bar{K}})e_J \wedge e_{\bar{J}}$$

$$= \left(\sum_{\text{Card}(J)=p} \det(U_{J,I})\det(U_{\bar{J},\bar{K}})\varepsilon_{J,\bar{J}} \right) e_1 \wedge \cdots \wedge e_n.$$

Comparing these two formulas, we obtain that

$$\varepsilon_{I,\bar{I}} \sum_{\text{Card}(J)=p} \varepsilon_{J,\bar{K}} \det(U_{J,I})\det(U_{\bar{J},\bar{K}})$$

equals 0 if $I \neq K$, and $\det(U)$ otherwise. This concludes the proof of the proposition. □

Applied with $p = 1$, the formula of b) recovers the well-known formula for the expansion of a determinant along a column.

Corollary (8.4.11). — *Let U be any $n \times n$ matrix with coefficients in A. Let C be the* adjugate *matrix of U: for every pair (i, j) the coefficient $C_{i,j}$ is equal to $(-1)^{i+j}$ times the determinant of the matrix obtained from U by deleting line i and column j. One has $C^t \cdot U = U \cdot C^t = \det(U)I_n$.*

Proof. — The formula $C^t \cdot U = \det(U)I_n$ is nothing but the Laplace expansion formula for $p = 1$. The other then follows by transposition. Indeed, the adjugate matrix of U^t is the transpose of C. Consequently, $C^t \cdot U = \det(U^t)I_n$, hence $U^t \cdot C = \det(U)I_n$. □

8.5. Adjunction and Exactness

Theorem (8.5.1). — *Let A be a ring, let M be a right A-module and let $N_1 \overset{f}{\to} N_2 \overset{g}{\to} N_3 \to 0$ be an exact sequence of left A-modules. Then, the following diagram*

$$M \otimes_A N_1 \xrightarrow{\text{id}_M \otimes f} M \otimes_A N_2 \xrightarrow{\text{id}_M \otimes g} M \otimes_A N_3 \to 0$$

is exact.

Proof. — *Exactness at* $M \otimes_A N_3$. We need to prove that $\text{id}_M \otimes g$ is surjective. Since every element of $M \otimes_A N_3$ is a sum of split tensors, it suffices to show that for any $m \in M$ and any $z \in N_3$, the tensor $m \otimes z$ has a preimage in $M \otimes_A N_2$. Since g is surjective, there exists an element $y \in N_2$ such that $g(y) = z$. Then $m \otimes z = (\mathrm{id}_M \otimes g)(m \otimes y)$, hence the claim.

Exactness at $M \otimes_A N_2$. We need to prove that $\mathrm{Ker}(\mathrm{id}_M \otimes g) = \mathrm{Im}(\mathrm{id}_M \otimes f)$. The inclusion $\mathrm{Im}(\mathrm{id}_M \otimes f) \subset \mathrm{Ker}(\mathrm{id}_M \otimes g)$ is easy, since $(\mathrm{id}_M \otimes g) \circ (\mathrm{id}_M \otimes f) = \mathrm{id}_M \otimes(g \circ f) = 0$. The other inclusion is the difficult one. Indeed, it is quite illusory to prove it directly, each element at a time. The incredulous reader should try by himself, and, after some effort, read again remark 8.1.11.

Since $\mathrm{Im}(\mathrm{id}_M \otimes f) \subset \mathrm{Ker}(\mathrm{id}_M \otimes g)$, the morphism $\mathrm{id}_M \otimes g$ induces a morphism of abelian groups $g' \colon (M \otimes_A N_2)/\mathrm{Im}(\mathrm{id}_M \otimes f) \to M \otimes_A N_3$. Since $\mathrm{id}_M \otimes g$ is surjective, g' is surjective. In fact, we need to prove that g' is an isomorphism. Let us construct an inverse h' of g'. To shorten the notation, let $P = (M \otimes_A N_2)/\mathrm{Im}(\mathrm{id}_M \otimes f)$ and let $p \colon M \otimes_A N_2 \to P$ be the canonical surjection. The sought-for morphism h' goes from a tensor product $M \otimes_A N_3$ to P. By the universal property of the tensor product, its definition amounts to the definition of a biadditive and A-balanced map $h \colon M \times N_3 \to P$ such that $h(m, z) = h'(m \otimes z)$. Moreover, $h'(m \otimes z)$ is supposed to be a preimage of $m \otimes z$ by $\mathrm{id}_M \otimes g$. After this paragraph of explanation, let us pass to the construction.

Let $m \in M$ and let $z \in N_3$. For any y and $y' \in N_2$ such that $g(y) = g(y')$, one has $y' - y \in \mathrm{Ker}(g) = \mathrm{Im}(f)$, so there exists an element $x \in N_1$ such that $y' - y = f(x)$; consequently, $m \otimes y' - m \otimes y$ belongs to $\mathrm{Im}(\mathrm{id}_M \otimes f)$ and $p(m \otimes y') = p(m \otimes y)$. Therefore, one defines a map $h \colon M \times N_3 \to P$ by setting $h(m, z) = p(m \otimes y)$ where y is any element of N_2 such that $g(y) = z$.

The map h is biadditive. Indeed, for $z, z' \in N_3$, $y, y' \in N_2$ such that $g(y) = z$ and $g(y') = z'$, one has $g(y + y') = z + z'$, so that

$$h(m, z + z') = p(m \otimes (y + y')) = p(m \otimes y) + p(m \otimes y') = h(m, z) + h(m, z').$$

Similarly, for $m, m' \in M$, $z \in N_3$ and $y \in N_2$ such that $g(y) = z$, one has

$$h(m + m', z) = p((m + m') \otimes y) = p(m \otimes y) + p(m \otimes y') = h(m, z) + h(m', z).$$

Finally, the map h is A-balanced: let $m \in M$, $z \in N_3$, $y \in N_2$ such that $g(y) = z$, and $a \in A$. Then, $g(ay) = ag(y) = az$, since g is A-linear, so that

$$h(ma, z) = p(ma \otimes y) = p(m \otimes ay) = h(m, az).$$

Consequently, there exists a unique morphism of abelian groups $h' \colon M \otimes_A N_3 \to P$ such that $h'(m \otimes z) = h(m, z)$ for any $(m, z) \in M \times N_3$.

It remains to show that h' is the inverse of g'. Let $m \in M$ and $y \in N_2$, let $t = p(m \otimes y)$; one has $g'(t) = g'(p(m \otimes y)) = (\mathrm{id}_M \otimes g)(m \otimes y) = m \otimes g(y)$, so that $h'(g'(t)) = p(m \otimes y) = t$. Since the tensors of the form $m \otimes y$ generate $M \otimes_A N_2$ and p is surjective, this implies that $h' \circ g' = \mathrm{id}_P$. In particular, g' is injective. Since it is surjective, this is an isomorphism. Had we wished so, we could also have computed $g' \circ h'$: let $m \in M$ and let $z \in N_3$, let $y \in N_2$ be such that $g(y) = z$; then $g'(h'(m \otimes z)) = g'(h(m, z)) = g'(p(m \otimes y)) = m \otimes g(y) = m \otimes z$.

Since the tensors of the form $m \otimes z$ in $M \otimes N_3$ generate this module, it follows that $g'(h'(t)) = t$ for every $t \in M \otimes N_3$, hence $g' \circ h' = \mathrm{id}_{M \otimes N_3}$. □

8.5.2. — This important result gains in being rephrased in the categorical language. Let A be a ring and let P be a right A-module. Observe that one defines a functor T_P by associating to any left A-module M the abelian group $T_P(M) = P \otimes_A M$, and to any morphism $f \colon M \to N$ of left A-modules the morphism $T_P(f) = \mathrm{id}_M \otimes_A f$. Indeed, $T_P(f)$ goes from $P \otimes_A M = T_P(M)$ to $P \otimes_A N = T_P(N)$, $T_P(\mathrm{id}_M) = \mathrm{id}_P \otimes_A \mathrm{id}_M = \mathrm{id}_{P \otimes_A M} = \mathrm{id}_{T_P(M)}$, and

$$T_P(g \circ f) = \mathrm{id}_P \otimes (g \circ f) = (\mathrm{id}_P \otimes g) \otimes (\mathrm{id}_P \otimes f) = T_P(g) \circ T_P(f)$$

for all morphisms $f \colon M_1 \to M_2$ and $g \colon M_2 \to M_3$ of left A-modules. This "tensor product by P" functor is additive and we may rephrase theorem 8.5.1 as follows:

Corollary (8.5.3). — *Let A be a ring and let M be a right A-module. The functor "tensor product by M" is right exact.*

In fact, once stated in this language, theorem 8.5.1 may be given another proof, more categorical. Thanks to the following proposition, it will be a consequence of the general result that left adjoints are right exact (proposition 7.6.5).

Let us generalize this functor a little bit. Let B be a second ring and let us assume that P is a (B, A)-bimodule. Then, for any left A-module M, $P \otimes_A M$ is naturally a left B-module and for any morphism $f \colon M \to N$ of left A-modules, $\mathrm{id}_P \otimes f$ is B-linear. Consequently, the functor T_P goes from the category of left A-modules to the category of left B-modules.

Proposition (8.5.4). — *Let A and B be rings and let P be a (B, A)-bimodule. The "tensor product by P" functor T_P defined above is a left adjoint of the functor $\mathrm{Hom}_A(P, \bullet)$.*

Proof. — For any left A-module M and any left B-module N, we need to define a bijection $\Phi_{M,N} \colon \mathrm{Hom}_B(P \otimes_A M, N) \to \mathrm{Hom}_A(M, \mathrm{Hom}_B(P, N))$ satisfying the relations required by the definition of a pair of adjoint functors.

Before we give such a formula, let us prove it with words. By the universal property of the tensor product, $\mathrm{Hom}_B(P \otimes_A M, N)$ is the set of biadditive, A-balanced maps u from $P \times M$ to N which are also B-linear with respect to P. Such a map is additive in each of its variables, B-linear in the P-variable, and A-balanced. In particular, it induces, for every $m \in M$, a B-linear map $u(\cdot, m)$ from P to N; moreover, $u(pa, m) = u(p, am)$ for every $a \in A$, $p \in P$ and $m \in M$. Conversely, a family $(u_m)_{m \in M}$ of B-linear maps from P to N such that the map $m \mapsto u_m$ is A-linear gives exactly a map u as required.

For the sake of the reader, let us put this into a formula. Let $u \colon P \otimes_A M \to N$ be a morphism of B-modules. Let $m \in M$. Then the map $p \mapsto u(p \otimes m)$ is B-linear, hence is an element $\Phi_{M,N}(u)(m)$. For $a, a' \in A$, $m, m' \in M$, and $p \in P$, one has

$$\Phi_{M,N}(u)(am + a'm')(p) = u(p \otimes (am + a'm')) = u(pa \otimes m) + u(pa' \otimes m')$$
$$= \Phi_{M,N}(u)(m)(pa) + \Phi_{M,N}(u)(m)(pa'),$$

so that

$$\Phi_{M,N}(u)(am + a'm') = a \cdot \Phi_{M,N}(u)(m) + a' \cdot \Phi_{M,N}(u)(m').$$

Consequently, the map $m \mapsto \Phi_{M,N}(u)(m)$ is A-linear, so that $\varphi_{M,N}(u)$ belongs to $\mathrm{Hom}_A(M, \mathrm{Hom}_B(P, N))$. We thus have defined a map $\Phi_{M,N}$.

Let us prove that $\Phi_{M,N}$ is bijective. Let $v \in \mathrm{Hom}_A(M, \mathrm{Hom}_B(P, N))$. For $u \in \mathrm{Hom}_B(P \otimes_A M, N)$, the relations $\Phi_{M,N}(u) = v$ and $u(p \otimes m) = v(m)(p)$ are equivalent. Since the tensors of the form $p \otimes m$ generate $P \otimes_A M$, this implies that $\Phi_{M,N}$ is injective. Moreover, the map from $P \times M$ to N given by $h(p,m) = v(m)(p)$ is biadditive. Since $m \mapsto v(m)$ is A-linear, one has $v(am)(p) = (a \cdot v(m))(p) = v(m)(pa)$, so that h is A-balanced. Consequently, there exists a unique morphism of abelian groups $u: P \otimes_A M \to N$ such that $u(p \otimes m) = h(p,m)$ for $p \in P$ and $m \in M$. Since $v(m)$ is B-linear for every $m \in M$, one has

$$u(b \cdot (p \otimes m)) = u(bp \otimes m) = v(m)(bp) = b(v(m)(p)) = bu(p \otimes m)$$

and the map u is B-linear. It is the unique preimage of v in $\mathrm{Hom}_B(P \otimes_A M, N)$. This shows that $\Phi_{M,N}$ is bijective.

Let M, M' be left A-modules, N, N' be left B-modules, and let $f: M \to M'$ and $g: N \to N'$ be morphisms of modules. By construction, for any morphism $u: P \otimes_A M' \to N$ of B-modules, any $p \in P$ and any $m \in M$, one has

$$(g \circ \Phi_{M',N}(u) \circ f)(m)(p) = g(\Phi_{M',N}(u)(f(m))(p)) = g(u(p \otimes f(m))),$$

while

$$\Phi_{M,N'}(g \circ u \circ (\mathrm{id}_P \times f))(m)(p) = (g \circ u \circ \mathrm{id}_P \times f)(p \times m)$$
$$= (g \circ u)(p \otimes f(m))$$
$$= g(u(p \otimes f(m))).$$

This shows that the maps $\Phi_{M,N}$ satisfy the adjunction relations and concludes the proof of the proposition. $\qquad\square$

8.6. Flat Modules

Definition (8.6.1). — Let A be a ring and let M be a right A-module. One says that M is *flat* if the functor "tensor product by M", T_M, $M \otimes_A \bullet$, is exact.

Since this functor is always right exact (corollary 8.5.3), a module M is flat if and only if, for any injective morphism $f : N \to N'$ of left A-modules, $\mathrm{id}_M \otimes f$ is injective.

One defines similarly the notion of a flat left A-module M, by the exactness of the right exact functor $\bullet \otimes_A M$.

Example (8.6.2). — The map $N \to A_d \otimes_A N$ given by $n \mapsto 1 \otimes n$ is an isomorphism of A-modules, and a morphism u is changed into $\mathrm{id}_A \otimes u$ under this isomorphism. Consequently, A_d is a flat A-module.

Example (8.6.3). — Let A be a commutative ring and let S be a multiplicative subset of A. According to example 8.2.2, tensoring a module M with $S^{-1}A$ corresponds to taking the fraction module $S^{-1}M$. If $u : M \to N$ is a morphism of A-modules, then under the isomorphisms of that example, the morphisms $S^{-1}u : S^{-1}M \to S^{-1}N$ and $u \otimes \mathrm{id} : M \otimes_A S^{-1}A \to N \otimes_A S^{-1}A$ coincide. This identifies the functor $\bullet \otimes_A S^{-1}A$ with the functor $M \mapsto S^{-1}A$ which is exact, by proposition 7.5.12.

Consequently, $S^{-1}A$ is a flat A-module.

Remark (8.6.4). — It is possible that the module M is non-flat, but that the functor T_M preserves the exactness of some exact sequences $0 \to N \to N' \to N'' \to 0$.

For example, if $f : N \to N'$ is a split injection of A-modules, with left inverse $g : N' \to N$, then $\mathrm{id}_M \otimes g$ is a left inverse of $\mathrm{id}_M \otimes f$, so that $\mathrm{id}_M \otimes f$ is again a split injection.

A more advanced discussion of homological algebra will construct A-modules $\mathrm{Tor}_p^A(M, N)$, for $p \geq 1$, and a long exact sequence,

$$\cdots \to \mathrm{Tor}_2^A(M, N'') \to \mathrm{Tor}_1^A(M, N) \to \mathrm{Tor}_1^A(M, N') \to$$
$$\to \mathrm{Tor}_1^A(M, N'') \to M \otimes N \to M \otimes N' \to M \otimes N'' \to 0.$$

Proposition (8.6.5). — *Let A be a ring and let M be a direct sum of a family $(M_i)_{i \in I}$ of A-modules. Then M is flat if and only if M_i is flat for every i.*

Proof. — Let $M = \bigoplus_i M_i$ be a direct sum of submodules; for $i \in I$, let j_i be the injection of M_i into M and let $p_i : M \to M_i$ be the projection. Then, for any left A-module N, one has an isomorphism $\theta = \sum p_i \otimes \mathrm{id}_N : M \otimes_A N \simeq \bigoplus_i M_i \otimes_A N$ whose inverse is induced by the natural maps $j_i \otimes \mathrm{id}_N$ from $M_i \otimes_A N$ to $M \otimes_A N$. Let $u : N \to N'$ be a morphism of left A-modules. In the isomorphism θ, the morphism $\mathrm{id}_M \otimes u$ is transformed into the direct sum of the morphisms $\mathrm{id}_{M_i} \otimes u$. In particular, $\mathrm{id}_M \otimes u$ is injective if and only if $\mathrm{id}_{M_i} \otimes u$ is injective for every i.

Let $u : N \to N'$ be an injective morphism of A-modules. If all A-modules M_i are flat, then $\mathrm{id}_{M_i} \otimes u$ is injective for every i, hence $\mathrm{id}_M \otimes u$ is injective. This proves that M is flat.

Conversely, if M is flat, then $\mathrm{id}_{M_i} \otimes u$ is injective for every injective morphism $u : N \to N'$ and every i, so that M_i is flat for every i.   □

Corollary (8.6.6). — *Any projective A-module, in particular any free A-module, is flat.*

Proof. — Since the right A-module A_d is flat, every free right A-module is flat. Any projective module, being a direct summand of a free A-module, is flat.   □

Example (8.6.7). — Any module over a division ring is flat.

Theorem (8.6.8). — *Let A be a ring and let M be a right A-module. Then M is flat if and only if for every left ideal J of A, the canonical morphism from $M \otimes_A J$ to M (sending $m \otimes a$ to ma, for every $m \in M$ and every $a \in J$) is injective.*

Proof. — Assume that M is flat. Let J be a left ideal of A and let us consider the exact sequence $0 \to J \to A \to A/J \to 0$ of left A-modules. Since M is flat, tensoring it with M, we obtain an exact sequence $0 \to M \otimes J \to M \to M \otimes A/J \to 0$ of abelian groups, where the injection $M \otimes J \to M$ is the one of the lemma.

Conversely, let us now assume that the natural map $M \otimes_A J \to M$ is injective, for every left ideal J, and let us prove that M is a flat A-module.

Let $f: N \to N'$ be an injective morphism of left A-modules and let us show that $\mathrm{id}_M \otimes f$ is injective.

Step 1. We first prove this in the case $N' = A^n$, by induction on n, f being the inclusion of the submodule N. The case $n = 0$ is obvious ($N = N' = 0$), and the case $n = 1$ is exactly the assumption of the lemma (identifying N' with A, then N is a left ideal). Let $q': N' \to A^{n-1} = N'_1$ be the projection to the $(n-1)$ first coordinates and let $N'_2 = \{0\}^{n-1} \times A$ be its kernel; let $j': N'_2 \to N'$ be the canonical inclusion. Let $N_1 = q(N)$ and $N_2 = N \cap N'_2$; this gives rise to a commutative diagram with exact rows

$$
\begin{array}{ccccccccc}
0 & \longrightarrow & N_2 & \stackrel{j}{\longrightarrow} & N & \stackrel{q}{\longrightarrow} & N_1 & \longrightarrow & 0 \\
  &   & \downarrow{\scriptstyle f_2} & & \downarrow{\scriptstyle f} & & \downarrow{\scriptstyle f_1} & & \\
0 & \longrightarrow & N'_2 & \stackrel{j'}{\longrightarrow} & N' & \stackrel{q'}{\longrightarrow} & N'_1 & \longrightarrow & 0,
\end{array}
$$

where j and q are the restrictions of j' and q' to N_2 and N_1, and the morphisms $f_1: N_1 \to N'_1$ and $f_2: N_2 \to N'_2$ are deduced from f. Let us apply the functor "tensor product with M" to this diagram. We obtain a commutative diagram

$$
\begin{array}{ccccccccc}
 & & M \otimes_A N_2 & \stackrel{\mathrm{id}_M \otimes j}{\longrightarrow} & M \otimes_A N & \stackrel{\mathrm{id}_M \otimes q}{\longrightarrow} & M \otimes_A N_1 & \longrightarrow & 0 \\
 & & \downarrow{\scriptstyle \mathrm{id}_M \otimes f_2} & & \downarrow{\scriptstyle \mathrm{id}_M \otimes f} & & \downarrow{\scriptstyle \mathrm{id}_M \otimes f_1} & & \\
0 & \longrightarrow & M \otimes_A N'_2 & \stackrel{\mathrm{id}_M \otimes j'}{\longrightarrow} & M \otimes_A N' & \stackrel{\mathrm{id}_M \otimes q'}{\longrightarrow} & M \otimes_A N'_1 & \longrightarrow & 0.
\end{array}
$$

The upper row is an exact sequence, because the functor T_M is right exact. The lower row is exact as well, because the initial sequence is split (remark 8.6.4).

By the snake lemma (theorem 7.2.1), the morphism $\mathrm{id}_M \otimes f$ is injective, as was to be shown.

Step 2. Let us now prove the result under the assumption that N' is finitely generated. Let $s' \colon P' = A^n \to N'$ be a surjective morphism, let $K = \mathrm{Ker}(s')$ and let $P = (s')^{-1}(N)$. Write $j \colon P \to P'$ and $k \colon K \to P$ for the canonical inclusions; let also $k' = j \circ k \colon K \to P'$ and $s = s'|_P$, so that $K = \mathrm{Ker}(s)$. We sum up these constructions by the commutative diagram with exact rows:

$$
\begin{array}{ccccccccc}
0 & \longrightarrow & K & \overset{k}{\longrightarrow} & P & \overset{s}{\longrightarrow} & N & \longrightarrow & 0 \\
 & & \Big\| & & \Big\downarrow{\scriptstyle j} & & \Big\downarrow{\scriptstyle f} & & \\
0 & \longrightarrow & K & \overset{k'}{\longrightarrow} & P' & \overset{s'}{\longrightarrow} & N' & \longrightarrow & 0.
\end{array}
$$

Let us apply the functor "tensor product with M". By right-exactness of the tensor product, we obtain a commutative diagram with exact rows

$$
\begin{array}{ccccccc}
M \otimes_A K & \overset{\mathrm{id}_M \otimes k}{\longrightarrow} & M \otimes_A P & \overset{\mathrm{id}_M \otimes s}{\longrightarrow} & M \otimes_A N & \longrightarrow & 0 \\
\Big\| & & \Big\downarrow{\scriptstyle \mathrm{id}_M \otimes j} & & \Big\downarrow{\scriptstyle \mathrm{id}_M \otimes f} & & \\
M \otimes_A K & \overset{\mathrm{id}_M \otimes k'}{\longrightarrow} & M \otimes_A P' & \overset{\mathrm{id}_M \otimes s'}{\longrightarrow} & M \otimes_A N' & \longrightarrow & 0.
\end{array}
$$

By the first step, we also know that the morphisms $\mathrm{id}_M \otimes j$ and $\mathrm{id}_M \otimes k'$ are injective, so that $\mathrm{id}_M \otimes k$ is injective as well. Let us prove that $\mathrm{id}_M \otimes f$ is injective; the argument is analogous to that of theorem 7.2.1. Let then $\mu \in M \otimes_A N$ be an element of $\mathrm{Ker}(\mathrm{id}_M \otimes f)$. Let $\pi \in M \otimes_A P$ be an element such that $\mathrm{id}_M \otimes s(\pi) = \mu$. One has

$$
(\mathrm{id}_M \otimes s') \circ (\mathrm{id}_M \otimes j)(\pi) = (\mathrm{id}_M \otimes f) \circ (\mathrm{id}_M \otimes s)(\pi) = (\mathrm{id}_M \otimes f)(\mu) = 0.
$$

Consequently, there exists an element $\pi' \in M \otimes_A K$ such that $(\mathrm{id}_M \otimes j)(\pi) = (\mathrm{id}_M \otimes k')(\pi')$, and $\pi = (\mathrm{id}_M \otimes k)(\pi')$. Then

$$
\mu = (\mathrm{id}_M \otimes s)(\pi) = (\mathrm{id}_M \otimes s) \circ (\mathrm{id}_M \otimes k)(\pi') = 0,
$$

as was to be shown.

Step 3. We finally prove the assertion in general. Let $\mu \in \mathrm{Ker}(\mathrm{id}_M \otimes f)$ be given as a finite linear combination of split tensors $\mu = \sum m_i \otimes n_i$, with $m_i \in M$ and $n_i \in N$. By assumption, one has $\sum m_i \otimes f(n_i) = 0$ in $M \otimes_A N'$.

By remark 8.1.6, there exists a finitely generated submodule N'_0 of N, containing the elements $f(n_i)$, such that $\sum m_i \otimes f(n_i) = 0$ in $M \otimes_A N'_0$. Let N_0 be the submodule of N generated by the n_i, let $j \colon N_0 \to N$ be the inclusion, and let $f_0 \colon N_0 \to N'_0$ be the map induced by f; let $\mu_0 = \sum m_i \otimes n_i$ viewed in $M \otimes_A N_0$. By construction, one has $(\mathrm{id}_M \otimes f_0)(\mu_0) = 0$, that is, $\mu_0 \in \mathrm{Ker}(\mathrm{id}_M \otimes f_0)$. Since N'_0 is finitely generated, step 2 proves that $\mu_0 = 0$. Then $\mu = (\mathrm{id}_M \otimes j)(\mu_0) = 0$. This proves that $\mathrm{id}_M \otimes f$ is injective. □

Corollary (8.6.9). — *Let A be an integral domain and let M be an A-module. If M is flat, then M is torsion-free. The converse holds if A is a principal ideal domain.*

Proof. — Let a be a non-zero element of A. The ideal (a) is a free A-module, with basis a, so that the morphism $M \to (a) \otimes_A M$ given by $m \mapsto a \otimes m$ is an isomorphism. Under this identification, the morphism $(a) \otimes_A M \to M$ given by $b \otimes m \mapsto bm$ identifies with the morphism $m \mapsto am$. If M is flat, then this morphism is injective (theorem 8.6.8) so that no non-zero element $m \in M$ satisfies $am = 0$. This proves that a flat A-module is torsion-free. Conversely, let us assume that A is a principal ideal domain and M is torsion-free. In this case, this argument shows that the canonical morphism $J \otimes_A M \to M$ given by $b \otimes m \mapsto bm$ is injective for every non-zero ideal J of A. On the other hand, if $J = 0$, one has $J \otimes_A M = 0$ and this morphism is injective as well. By theorem 8.6.8, this shows that M is flat. □

8.6.10. — Let M and P be right A-modules. Let $M^\vee = \mathrm{Hom}_A(M, A)$ be the dual of M, viewed as a left A-module. For $\varphi \in M^\vee$ and $p \in P$, the map $p\varphi \colon x \mapsto p\varphi(x)$ from M to P is a morphism of A-modules, and the map $(p, \varphi) \mapsto p\varphi$ is biadditive and A-balanced, since $(pa)\varphi(x) = pa\varphi(x) = p(a\varphi)(x)$ for every $x \in M$, every $p \in P$ and every $\varphi \in M^\vee$. Consequently, there exists a unique morphism

$$\delta_M \colon P \otimes_A M^\vee \to \mathrm{Hom}_A(M, P)$$

such that $\delta_M(p \otimes \varphi) = p\varphi$ for every $\varphi \in M^\vee$ and every $p \in P$. (See also proposition 8.2.4.)

Proposition (8.6.11). — *Let A be a ring and let P be a flat right A-module. Then, for every finitely presented right A-module M, the morphism $\delta_M \colon P \otimes_A M^\vee \to \mathrm{Hom}_A(M, P)$ is an isomorphism.*

Proof. — Since M is finitely presented, there exists an exact sequence of A-modules
$$L' \xrightarrow{u} L \xrightarrow{v} M \to 0,$$
where L and L′ are free and finitely generated. Let us apply the right exact functor $\bullet^\vee = \mathrm{Hom}_A(\bullet, A)$ to this exact sequence; this furnishes an exact sequence
$$0 \to M^\vee \xrightarrow{v^*} L^\vee \xrightarrow{u^*} (L')^\vee.$$
We now tensor this exact sequence with P and obtain, since P is flat, an exact sequence
$$0 \to P \otimes_A M^\vee \xrightarrow{\mathrm{id}_P \otimes v^*} P \otimes L^\vee \xrightarrow{\mathrm{id}_P \otimes u^*} P \otimes (L')^\vee.$$
Similarly, we apply the right exact functor $\mathrm{Hom}_A(\bullet, P)$ and obtain an exact sequence
$$0 \to \mathrm{Hom}_A(M, P) \xrightarrow{v^*} \mathrm{Hom}_A(L, P) \xrightarrow{u^*} \mathrm{Hom}_A(L', P).$$
Via the various morphisms δ, these last two exact sequences can be combined into the following diagram

$$0 \longrightarrow P \otimes_A M^\vee \xrightarrow{\ \mathrm{id}_P \otimes v^*\ } P \otimes_A L^\vee \xrightarrow{\ \mathrm{id}_P \otimes u^*\ } P \otimes_A (L')^\vee$$

with vertical maps δ_M, δ_L, $\delta_{L'}$ and

$$0 \longrightarrow \mathrm{Hom}_A(M, P) \xrightarrow{\ v^*\ } \mathrm{Hom}_A(L, P) \xrightarrow{\ u^*\ } \mathrm{Hom}_A(L', P).$$

This left square is commutative: if $\varphi \in M^\vee$ and $p \in P$, since both $\delta_L \circ (\mathrm{id}_P \otimes v^*)(p \otimes \varphi)$ and $(v^* \circ \delta_M)(p \otimes \varphi)$ are the morphism $y \mapsto \varphi(v(y))p$ from L to P. Consequently, the morphisms $\delta_L \circ (\mathrm{id}_P \otimes v^*)$ and $v^* \circ \delta_M$ coincide on split tensors, hence are equal. This proves that the left-hand square is commutative. The commutativity of the right-hand one is analogous: the maps $\delta_{L'} \circ (\mathrm{id}_P \otimes u^*)$ and $u^* \circ \delta_L$ both map a split tensor $p \otimes \varphi$ (with $p \in P$ and $\varphi \in L^\vee$) to the morphism $y \mapsto p\varphi(u(y))$, hence they coincide.

The module L is free and finitely generated. As in the proof of proposition 8.2.4 (which concerns the case where A is commutative), we prove that the morphism δ_L is an isomorphism. Indeed, if $(e_i)_{i \in I}$ is a finite basis of L, with dual basis (e_i^*), and if $u \in \mathrm{Hom}_A(L, P)$, then $\mu = \sum_i u(e_i) \otimes e_i^*$ is the unique element of $P \otimes_A M^\vee$ such that $\delta_L(\mu) = u$. Similarly, $\delta_{L'}$ is an isomorphism.

Let us deduce that δ_M is an isomorphism. One has $\delta_L \circ (\mathrm{id}_P \otimes v^*) = v^* \circ \delta_M$, and $\delta_L \circ (\mathrm{id}_P \otimes v^*)$ is injective, so that δ_M is injective. Let $f \in \mathrm{Hom}_A(M, P)$. Consider the morphism $v^*(f) = f \circ v \in \mathrm{Hom}_A(L, P)$. Since δ_L is an isomorphism, there exists a unique element $\mu \in P \otimes_A L^\vee$ such that $\delta_L(\mu) = f \circ v$. Then

$$\delta_{L'} \circ (\mathrm{id}_P \otimes u^*)(\mu) = u^* \circ (\mathrm{id}_P \otimes \delta_L)(\mu) = u^*(v^*(f)) = f \circ v \circ u = 0.$$

Since $\delta_{L'}$ is an isomorphism, one has $\mathrm{id}_P \otimes u^*(\mu) = 0$. By the exactness of the first row of the diagram, there exists an element $\nu \in P \otimes_A M^\vee$ such that $\mu = (\mathrm{id}_P \otimes v^*)(\nu)$. Then,

$$\delta_M(\nu) \circ v = \delta_L \circ (\mathrm{id}_P \otimes v^*)(\nu) = \delta_L(\mu) = f \circ v.$$

Since v is surjective, $f = \delta_M(\nu)$. This concludes the proof that δ_M is an isomorphism. $\square$

Theorem (8.6.12) (Lazard). — *Let A be a ring and let P be a right A-module. The module P is flat if and only if, for every finitely presented right A-module M and every morphism $f: M \to P$, there exists a free finitely generated right A-module L and morphisms $g: M \to L$ and $h: L \to P$ such that $f = h \circ g$.*

Proof. — Assume that P is flat, let M be a finitely presented right A-module and let $f: M \to P$ be a morphism of A-modules. Let us choose a tensor $\mu \in P \otimes_A M^\vee$ such that $\delta_M(\mu) = f$ and let us write $\mu = \sum_{i=1}^n p_i \otimes \varphi_i$, with $\varphi_1, \dots, \varphi_n \in M^\vee$ and $p_1, \dots, p_n \in P$. Let $\varphi = (\varphi_1, \dots, \varphi_n): M \to A^n$ and $p: A^n \to P$ be given by $p(a_1, \dots, a_n) = p_1 a_1 + \dots + p_n a_n$; these are morphisms of A-modules. Moreover, for every $m \in M$, one has

$$p \circ \varphi(m) = \sum_{i=1}^{n} p_i \varphi_i(m) = \delta_M(\mu)(m) = f(m),$$

so that $m = p \circ \varphi$. This proves the desired property.

Conversely, let us assume that this property holds and let us prove that P is flat. Let J be a left ideal of A and let us show that the canonical morphism $f : P \otimes_A J \to P$ is injective. Let x be any element of $\operatorname{Ker}(f)$, written as $x = \sum_{i=1}^{n} x_i \otimes a_i$, where $a_1, \ldots, a_n \in A$ and $x_1, \ldots, x_n \in P$. Let M be the quotient of the free A-module A_d^n, with canonical basis $(e_1, \ldots, e_n)$, by its submodule generated by $e_1 a_1 + \cdots + e_n a_n$; let $p : A_d^n \to M$ be the canonical surjection. The module M is finitely presented. By construction, the morphism from A_d^n to P such that $e_i \mapsto x_i$ for each i vanishes on $(a_1, \ldots, a_n)A$. Consequently, there exists a unique morphism $f : M \to P$ such that $f(p(e_i)) = x_i$ for every i. Let m be an integer, let $g : M \to A_d^m$ and $h : A_d^m \to P$ be morphisms such that $f = h \circ g$. Let $(f_1, \ldots, f_m)$ be the canonical basis of A_d^m; For every $j \in \{1, \ldots, m\}$, let $y_j = h(f_j)$. For each $i \in \{1, \ldots, n\}$, let $(b_{i,j})$ be the coordinates of $g(p(e_i))$ in the basis $(f_1, \ldots, f_m)$; one has $g(p(e_i)) = \sum_{j=1}^{m} f_j b_{i,j}$. Since $\sum_i p(e_i) a_i = p(\sum e_i a_i) = 0$, one has

$$0 = \sum_{i=1}^{n} g(p(e_i) a_i) = \sum_{j=1}^{m} f_j \left(\sum_{i=1}^{n} b_{i,j} a_i \right),$$

hence $\sum_{i=1}^{n} b_{i,j} a_i = 0$ for every j. Moreover,

$$x_i = f(p(e_i)) = h(g(p(e_i))) = \sum_{j=1}^{m} y_j b_{i,j}$$

for every i. Then one has, in $P \otimes_A J$, the relations

$$\sum_{i=1}^{n} x_i \otimes a_i = \sum_{i=1}^{n} \left(\sum_{j=1}^{m} y_j b_{i,j} \right) \otimes a_i$$

$$= \sum_{j=1}^{m} \left(\sum_{i=1}^{n} y_j b_{i,j} \otimes a_i \right)$$

$$= \sum_{j=1}^{m} y_j \otimes \left(\sum_{i=1}^{n} b_{i,j} a_i \right)$$

$$= 0.$$

This proves that the canonical morphism $P \otimes_A J \to P$ is injective and concludes the proof that P is a flat A-module. $\square$

Corollary (8.6.13). — *Let A be a commutative ring and let M be a finitely presented A-module. Then M is flat if and only if M is projective.*

Proof. — A projective module is flat (corollary 8.6.6), hence it suffices to prove that M is projective if it is flat. By the preceding proposition applied to the identity morphism $\mathrm{id}_M \colon M \to M$, there exist a free finitely generated A-module L and morphisms $f \colon M \to L$ and $g \colon L \to M$ such that $\mathrm{id}_M = g \circ f$. The morphism f is injective and g is a left inverse to f. By lemma 7.1.6, applied to the exact sequence $0 \to M \xrightarrow{f} L \to L/M \to 0$, the submodule $f(M)$ of L has a direct summand, hence is projective (proposition 7.3.2). Since f induces an isomorphism from M to its image $f(M)$, the module M is projective, as was to be shown. $\qquad\square$

Remark (8.6.14). — By definition, a finitely presented right A-module M is the cokernel of an A-morphism $u \colon A_d^n \to A_d^m$. Such a morphism can be represented by a matrix $U \in \mathrm{Mat}_{m,n}(A)$. Moreover, a morphism from M to a free A-module A_s^p corresponds to a morphism $v \colon A_d^n \to A_d^p$ that vanishes on the image of u. The morphism v is represented by a matrix $V \in \mathrm{Mat}_{p,m}(A)$, and the condition that $v(\mathrm{Im}(u)) = 0$ translates as the relation $V \cdot U = 0$.

Thanks to these remarks, theorem 8.6.12 can be rewritten as the following *equational criterion of flatness*. Namely, the following properties are equivalent:

(i) The A-module M is flat;

(ii) For every family $(x_1, \ldots, x_m)$ of elements of M, and every family $(a_1, \ldots, a_m)$ in A such that $\sum_{i=1}^m x_i a_i = 0$, there exist an integer p, a family $(y_1, \ldots, y_p)$ in M and a matrix $(b_{k,i}) \in \mathrm{Mat}_{p,m}(A)$ such that $x_i = \sum_{k=1}^p y_k b_{k,i}$ for every $i \in \{1, \ldots, m\}$, and $\sum_{i=1}^m b_{k,i} a_i = 0$ for every k;

(iii) For every family $(x_1, \ldots, x_m)$ of elements of M, and every matrix $(a_{i,j}) \in \mathrm{Mat}_{m,n}(A)$ such that $\sum_{i=1}^m x_i a_{i,j} = 0$ for every $j \in \{1, \ldots, n\}$, there exist an integer p, a family $(y_1, \ldots, y_p)$ in M and a matrix $(b_{k,i}) \in \mathrm{Mat}_{p,m}(A)$ such that

$$x_i = \sum_{k=1}^p y_k b_{k,i}$$

for every $i \in \{1, \ldots, m\}$, and

$$\sum_{i=1}^m b_{k,i} a_{i,j} = 0$$

for every k and j.

Assertion (ii) is the particular case of (iii) with $n = 1$. The equivalence of (i) and (iii) is the content of theorem 8.6.12, and in the proof, we have seen that it is sufficient to consider A-modules M which are cokernels of a morphism $u \colon A_d \to A_d^m$, i.e., with $n = 1$, that is, to prove (ii).

Proposition (8.6.15). — *Let* A *be a commutative ring and let* M *be an* A-*module.*

a) *Let* S *be a multiplicative subset of* A. *If* M *is flat, then the* $S^{-1}A$-*module* $S^{-1}M$ *is flat.*

b) *If M_P is a flat A_P-module for every maximal ideal P of A, then M is flat.*

Proof. — a) For any $S^{-1}A$-module N, there is a morphism of $S^{-1}A$-modules $M \otimes_A N \simeq S^{-1}M \otimes_{S^{-1}A} N$ sending $m \otimes n$ to $(m/1) \otimes n$. This morphism is an isomorphism. Moreover, if $u : N \to N'$ is a morphism of $S^{-1}A$-modules, this morphism transforms $\mathrm{id}_M \otimes u$ into $\mathrm{id}_{S^{-1}M} \otimes u$.

If u is injective and M is flat, it follows that $\mathrm{id}_{S^{-1}M} \otimes u$ is injective. Since u is arbitrary, $S^{-1}M$ is a flat $S^{-1}A$-module.

b) Let $u : N \to N'$ be an injective morphism of A-modules. The image of $\mathrm{id}_M \otimes u$ is $M \otimes u(N)$; let Q be the kernel of $\mathrm{id}_M \otimes u$ so that we have an exact sequence

$$0 \to Q \to M \otimes_A N \xrightarrow{\mathrm{id}_M \otimes u} M \otimes_A u(N) \to 0.$$

Let P be a maximal ideal of A and let us tensor this exact sequence of A-modules by the flat A-module A_P; we obtain an exact sequence

$$0 \to Q_P \to M_P \otimes_A N_P \xrightarrow{\mathrm{id}_{M_P} \otimes u_P} M_P \otimes_{A_P} u(N)_P \to 0.$$

Since A_P is a flat A-module (example 8.6.3), the morphism u_P is injective. Since M_P is assumed to be flat, the morphism $\mathrm{id}_{M_P} \otimes u_P$ is injective. Consequently, $Q_P = 0$ for every $P \in \mathrm{Spec}(A)$. One thus has $\mathrm{Supp}(Q) = 0$, hence $Q = 0$ by theorem 6.5.4.

This implies that the map $\mathrm{id}_M \otimes u$ is injective. Since u is arbitrary, M is a flat A-module. □

8.7. Faithful Flatness

Definition (8.7.1). — Let A be a commutative ring and let M be an A-module. One says that M is *faithfully flat* if a complex $N_1 \xrightarrow{u} N_2 \xrightarrow{v} N_3$ of A-modules is exact if and only if the complex $M \otimes_A N_1 \xrightarrow{\mathrm{id}_M \otimes u} M \otimes_A N_2 \xrightarrow{\mathrm{id}_M \otimes v} M \otimes_A N_3$ is exact.

The "only if" direction of the definition shows that a faithfully flat A-module is flat.

Proposition (8.7.2). — *Let M be a flat A-module. The following properties are equivalent:*

(i) *The A-module M is faithfully flat;*

(ii) *For every morphism $u : N \to N'$ of A-modules such that $\mathrm{id}_M \otimes u = 0$, one has $u = 0$;*

(iii) *For every non-zero A-module N, one has $M \otimes_A N \neq 0$;*

(iv) *For every prime ideal P of A, the A-module $M \otimes_A K_P$ is non-zero, where K_P is the field of fractions of the integral domain A/P;*

(v) *For every maximal ideal* P *of* A, *one has* M $\neq$ PM.

Proof. — (i)$\Rightarrow$(ii). Assume that $\mathrm{id}_M \otimes u = 0$. Let us consider the complex N $\xrightarrow{u}$ N' $\xrightarrow{\mathrm{id}_{N'}}$ N' and let us tensor it with M. We obtain the complex

$$M \otimes_A N \xrightarrow{\mathrm{id}_M \otimes u} M \otimes_A N' \xrightarrow{\mathrm{id}} M \otimes_A N',$$

which is exact since $\mathrm{id}_M \otimes u = 0$, so that its image, 0, equals the kernel of the identity isomorphism. Since M is faithfully flat the initial complex was exact and $\mathrm{Im}(u) = \mathrm{Ker}(\mathrm{id}_{N'}) = 0$, hence $u = 0$.

(ii)$\Rightarrow$(iii). One has $M \otimes N = 0$ if and only if $\mathrm{id}_M \otimes \mathrm{id}_N = 0$; if (ii) holds, then $\mathrm{id}_N = 0$, hence $N = 0$.

The implications (iii)$\Rightarrow$(iv) and (iv)$\Rightarrow$(v) are obvious.

Let us prove the implication (iii)$\Rightarrow$(i). Let $N_1 \xrightarrow{u} N_2 \xrightarrow{v} N_3$ be a complex of A-modules which becomes exact after tensorization by M. Let $K = \mathrm{Ker}(v)/\mathrm{Im}(u)$.

Let $j \colon u(N_1) \to N_2$ be the inclusion; since M is flat, the morphism $\mathrm{id}_M \otimes j$ is injective and its image is a submodule of $M \otimes_A N_2$ isomorphic to $M \otimes_A \mathrm{Im}(u)$. By right exactness, this submodule is also $\mathrm{Im}(\mathrm{id}_M \otimes u)$. Then consider the exact sequence $0 \to \mathrm{Ker}(v) \xrightarrow{k} N_2 \xrightarrow{v} N_2$, where k is the inclusion morphism, and apply the tensor product by M; since M is flat, this gives an exact sequence

$$0 \to M \otimes_A \mathrm{Ker}(v) \xrightarrow{\mathrm{id}_M \otimes k} M \otimes_A N_2 \xrightarrow{\mathrm{id}_M \otimes v},$$

so that $\mathrm{id}_M \otimes k$ is an isomorphism from $M \otimes_A \mathrm{Ker}(v)$ to $\mathrm{Ker}(\mathrm{id}_M \otimes v)$.

By right exactness of the tensor product, one has

$$M \otimes_A K = M \otimes_A (\mathrm{Ker}(v)/\mathrm{Im}(u)) \simeq (M \otimes_A \mathrm{Ker}(v))/(M \otimes_A \mathrm{Im}(u)).$$

Consequently, $M \otimes_A K = 0$, by assumption. Since M is faithfully flat, this shows that $K = 0$ and the initial complex is exact.

We finally prove that (v)$\Rightarrow$(iii). Let N be an A-module such that $M \otimes_A N = 0$ and let us prove that $N = 0$. Let $x \in N$ and let $I = \{a \in A \,;\, ax = 0\}$, so that the morphism $a \mapsto ax$ induces an injection $A/I \hookrightarrow N$. By flatness of M, we thus have an injection $M \otimes_A (A/I) \hookrightarrow M \otimes_A N$, so that $M \otimes_A (A/I) = 0$ and $M = IM$. If $I \neq A$, there exists a maximal ideal P of A that contains I; then $M = IM \subset PM$, which contradicts (v). Hence $I = A$ and $x = 0$, so that $N = 0$.□

Definition (8.7.3). — One says that a morphism $f \colon A \to B$ of commutative rings is *flat* (resp. *faithfully flat*) if B is a flat (resp. faithfully flat) A-module.

Proposition (8.7.4). — *Let* $f \colon A \to B$ *be a* flat *morphism of commutative rings. The following properties are equivalent:*

(i) *The morphism* f *is faithfully flat;*

(ii) *The continuous map* $f^* \colon \mathrm{Spec}(B) \to \mathrm{Spec}(A)$ *is surjective;*

(iii) *For every maximal ideal* M *of* A, *there exists a prime ideal* P *of* B *such that* M = f^{-1}(P).

Proof. — Let P be a prime ideal of A and let K_P be the fraction field of the residue ring A/P. Note that K_P = A_P/PA_P. Let T = $f(A - P)$; it is a multiplicative subset of B; by abuse of language, we write B_P for $T^{-1}B$.

Let us consider the following commutative diagram of ring morphisms

$$
\begin{array}{ccccc}
A & \xrightarrow{\ f\ } & B & \longrightarrow & B/f(P)B \\
\downarrow & & \downarrow & & \downarrow \\
A_P & \xrightarrow{\ f_P\ } & B_{f(P)} & \longrightarrow & B_P/f(P)B_P,
\end{array}
$$

where the unlabeled vertical arrows represent the obvious localization morphisms, and the unlabeled horizontal arrows represent the obvious quotient maps. Moreover, one has an isomorphism $B_P/f(P)B_P \simeq T^{-1}(B/f(P)B) \simeq B \otimes_A K_P$.

Under these morphisms, a prime ideal Q of $B_P/f(P)B_P$ corresponds to a prime ideal of B_P that contains $f(P)B_P$, or to a prime ideal of B containing $f(P)$ and disjoint from T, and its preimage $f^{-1}(Q)$ is a prime ideal of A containing P and disjoint from A−P, hence $f^{-1}(Q)$ = P. Conversely, such an ideal furnishes a prime ideal of $B_P/f(P)B_P \simeq B \otimes_A K_P$.

If f is faithfully flat, then $B \otimes_A K_P$ is non-zero, so that it admits a prime ideal Q, and $f^{-1}(Q)$ = P belongs to the image of f^*. This proves the implication (i)⟹(ii).

If M is a maximal ideal, then K_M = A/M and the above argument shows that (iii) implies that $B \otimes_A (A/M) \neq 0$, that is, $B \neq MB$. Consequently, B is faithfully flat.

Finally, the implication (ii)⟹(iii) is obvious.   □

Examples (8.7.5). — Let A be a commutative ring.

a) For every multiplicative subset S of A, the morphism $f \colon A \to S^{-1}A$ given by $f(a) = a/1$ is flat (example 8.6.3). It is faithfully flat if and only if S consists of invertible elements, in which case it is an isomorphism. Indeed, a prime ideal P of A is of the form $f^{-1}(Q)$ for a prime ideal Q of $S^{-1}A$ if and only if $P \cap S = \emptyset$. If S contains a noninvertible element a, then there is a prime ideal of A containing a, and this prime ideal does not belong to the image of f^*.

b) Let $a_1, \ldots, a_n$ be a family of elements of A, let $B = \prod_{i=1}^n A_{a_i}$ and let $f \colon A \to B$ be the morphism given by $f(a) = (a/1, \ldots, a/1)$ for every $a \in A$. The morphism f is flat.

It is faithfully flat if and only if, for every prime ideal P of A there exists an integer $i \in \{1, \ldots, n\}$ such that $a_i \notin P$, in other words, if and only if $\bigcap_{i=1}^n V((a_i)) = \emptyset$ in Spec(A). Indeed, prime ideals of B are of the form $Q_1 \times \cdots \times Q_n$, where, for some $i \in \{1, \ldots, n\}$, Q_i is a prime ideal of A_{a_i}, and $Q_j = A_{a_j}$ otherwise. Its preimage by f is then $f^{-1}(Q_i)$, a prime ideal that does not contain a_i, and every such prime ideal of A can be obtained.

Example (8.7.6). — Let $f\colon A \to B$ be a flat morphism of commutative rings. Assume that f has a left inverse, that is, that there exists a morphism $f'\colon B \to A$ such that $f' \circ f = \mathrm{id}_A$. Then f is faithfully flat. Indeed, one has $f^* \circ (f')^* = \mathrm{id}_{\mathrm{Spec}(A)}$, hence f^* is surjective.

8.7.7. — Let $f\colon A \to B$ be a morphism of commutative rings. We consider the diagram $\mathscr{D}_f$ of A-modules:

$$0 \to A \xrightarrow{f} B \xrightarrow{g} B \otimes_A B,$$

where $g(b) = 1 \otimes b - b \otimes 1$, for every $b \in B$. For $a \in A$, one has $1 \otimes f(a) = f(a) \otimes 1 = a \cdot (1 \otimes 1)$ in $B \otimes_A B$, so that $g \circ f = 0$: the diagram $\mathscr{D}_f$ is a complex.

More generally, for every A-module M, let $f_M = f \otimes \mathrm{id}_M\colon M \to B \otimes_A M$ be the map given by $m \mapsto 1 \otimes m$, and let $g_M = g \otimes \mathrm{id}_M\colon B \otimes_A M \to B \otimes_A B \otimes_A M$ be given by $g_M(b \otimes m) = 1 \otimes b \otimes m - b \otimes 1 \otimes m$, and consider the diagram $\mathscr{D}_f^M$:

$$0 \to M \xrightarrow{f_M} B \otimes_A M \xrightarrow{g_M} B \otimes_A B \otimes_A M.$$

Again, $g_M \circ f_M = 0$, so that $\mathscr{D}_f^M$ is a complex of A-modules.

Proposition (8.7.8). — *Assume that there exists a morphism of rings $f'\colon B \to A$ such that $f' \circ f = \mathrm{id}_A$. Then the diagram $\mathscr{D}_f$ is an exact sequence. More generally, for every A-module M, the complex $\mathscr{D}_f^M$ is exact.*

We will generalize this proposition below, in theorem 8.7.10, and prove that its conclusion holds as soon as f is faithfully flat.

Proof. — Since $f' \circ f = \mathrm{id}_A$, the morphism f is injective.

One has $g \circ f = 0$, hence $\mathrm{Im}(f) \subset \mathrm{Ker}(g)$. Conversely, the map from $B \times B$ to B given by $(b, b') \mapsto f'(b)b'$ is A-bilinear, hence there exists a unique morphism of A-modules $g'\colon B \otimes_A B \to B$ such that $g'(b \otimes b') = f'(b)b'$. One has $g' \circ g(b) = g'(1 \otimes b) - g'(b \otimes 1) = b - f(f'(b))$, hence $g' \circ g = \mathrm{id}_B - f \circ f'$. Assume that $g(b) = 0$. Then $g'(g(b)) = 0$, hence $b = f(f'(b))$ and $b \in \mathrm{Im}(f)$, hence $\mathrm{Ker}(g) \subset \mathrm{Im}(f)$. This proves that $\mathrm{Ker}(g) = \mathrm{Im}(f)$.

Let us prove the more general result. Since the map from $B \times M$ to M given by $(b, m) \mapsto f'(b)m$ is A-bilinear, there exists a unique morphism of A-modules $f'_M\colon B \otimes_A M \to M$ such that $f'_M(b \otimes m) = f'(b)m$. One has $f'_M \circ f_M(m) = f'_M(1 \otimes m) = m$ for every $m \in M$, hence $f'_M \circ f_M = \mathrm{id}_M$. This proves that f_M is injective.

Similarly, there exists a unique morphism $g'_M\colon B \otimes_A B \otimes_A M \to B \otimes_A M$ such that $g'_M(b \otimes b' \otimes m) = g'(b \otimes b') \otimes m = f'(b)b' \otimes m$. One has

$$
\begin{aligned}
g'_M \circ g_M(b \otimes m) &= g'_M(1 \otimes b \otimes m - b \otimes 1 \otimes m) \\
&= b \otimes m - f'(b)1_{B'} \otimes m \\
&= (b - f(f'(b)) \otimes m,
\end{aligned}
$$

for every $b, b' \in B$ and every $m \in M$. Consequently, $g'_M \circ g_M = (\mathrm{id}_B - f \circ f')_M$. Let us now prove that $\mathrm{Ker}(g_M) = \mathrm{Im}(f_M)$. Since $g_M \circ f_M = 0$, one has $\mathrm{Im}(f_M) \subset \mathrm{Ker}(g_M)$.

Conversely, let $\xi \in \mathrm{Ker}(g_M)$. Then $g'_M(g_M(\xi)) = 0$, that is $\xi = f_M \circ f'_M(\xi)$: we have proved that $\xi \in \mathrm{Im}(f_M)$. □

Corollary (8.7.9). — *For every* A*-module* M, *the diagram of* B*-modules*

$$0 \to B \otimes_A M \xrightarrow{\mathrm{id}_B \otimes f_M} B \otimes_A M \xrightarrow{\mathrm{id}_B \otimes g_M} B \otimes_A B \otimes_A M$$

is exact.

Proof. — The map $f_B \colon B \to B \otimes_A B$ given by $f_B(b) = 1 \otimes b$ is a morphism of rings. There is a unique morphism of A-modules $g \colon B \otimes_A B \to B$ such that $g(b \otimes b') = bb'$ for every $b, b' \in A$; this is even a morphism of rings, and one has $g \circ f_B = \mathrm{id}_B$.

Moreover, the morphism $g_B \colon B \otimes_A B \to B \otimes_A B \otimes_A B$ is given by

$$g_B(b \otimes b') = 1 \otimes b \otimes b' - b \otimes 1 \otimes b'.$$

Under the isomorphism

$$(B \otimes_A B) \otimes_B (B \otimes_A B) \xrightarrow{\sim} B \otimes_A B \otimes_A B, (b \otimes b') \otimes (c \otimes c') \mapsto b \otimes b'c \otimes c,$$

it identifies with the morphism g associated with the morphism f_B. Consequently, the diagram of the question is an exact sequence. □

Theorem (8.7.10). — *Let* $f \colon A \to B$ *be a morphism of commutative rings. If* f *is faithfully flat, then for every* A*-module* M, *the complex* $\mathscr{D}_f^M$ *is exact.*

Proof. — Let M be an A-module. The preceding corollary shows that the complex $\mathscr{D}_f^N$ is exact, where $N = B \otimes_A M$. On the other hand, the complex $\mathscr{D}_f^N$ identifies with the complex deduced from $\mathscr{D}_f^M$ by tensoring with B. Since B is a faithfully flat A-module, the complex $\mathscr{D}_f^M$ is exact. □

8.8. Faithfully Flat Descent

8.8.1. — Let $f \colon A \to B$ be a faithfully flat morphism of commutative rings. Set $C = B \otimes_A B$; there are two ring morphisms $g_1, g_2 \colon B \to C$, respectively given by $g_1(b) = b \otimes 1$ and $g_2(b) = 1 \otimes b$, so that $g = g_2 - g_1$. Consequently, the exactness of the complex $\mathscr{D}_f$ asserted by theorem 8.7.10 says that the ring A can be recovered from B as the subring where these two ring morphisms coincide.

8.8.2. — Let M be an A-module, write $M_B = B \otimes_A M$ and $M_C = C \otimes_A M = B \otimes_A M_B$. By base change, these are a B-module and a C-module respectively, and we may view M_C as a B-module in two ways, via g_1 or g_2. The map $g_1^M = g_1 \otimes \mathrm{id}_M \colon M_B \to M_C$ maps $b \otimes b$ to $b \otimes 1 \otimes m = g_1(b) \otimes m$, hence is a morphism of B-modules if C is viewed as a B-module via g_1; similarly, the map $g_2^M = g_2 \otimes \mathrm{id}_M$ is a morphism of B-modules if C is viewed as a B-module via g_2. Moreover, theorem 8.7.10 asserts that M can be recovered from M_B as the A-submodule where these two morphisms coincide.

It is therefore natural to ask which properties of the A-module M can be witnessed on the B-module M_B. Similarly, a morphism $u \colon M \to M'$ of A-modules gives rise to a morphism of B-modules $u_B = \mathrm{id}_B \otimes u$, and one may wonder what properties of u are witnessed by u_B.

Proposition (8.8.3). — *Let $f \colon A \to B$ be a faithfully flat morphism of commutative rings. Let $u \colon M \to N$ be a morphism of A-modules.*

   a) *If $u_B = 0$, then $u = 0$;*

   b) *If u_B is surjective, then u is surjective;*

   c) *If u_B is injective, then u is injective.*

Proof. — Assertion a) holds, by proposition 8.7.2.

Let us prove b) and c). Let $K = \mathrm{Ker}(u)$ and let $j \colon K \to M$ be the inclusion; let $P = \mathrm{Coker}(u)$ and let $v \colon N \to P$ be the canonical projection. By construction, the diagram $0 \to K \xrightarrow{j} M \xrightarrow{u} N \xrightarrow{v} P \to 0$ is exact. Since f is flat, the diagram $0 \to K_B \xrightarrow{j_B} M_B \xrightarrow{u_B} N_B \xrightarrow{v_B} P_B \to 0$ is exact as well.

If u_B is surjective, then $P_B = 0$, hence $P = 0$ since f is faithfully flat; this shows that u is surjective.

If u_B is injective, then $K_B = 0$, hence $K = 0$ since f is faithfully flat, and u is injective. □

Proposition (8.8.4). — *Let $f \colon A \to B$ be a faithfully flat morphism of commutative rings and let M be an A-module.*

   a) *If M_B is a finitely generated B-module, then M is a finitely generated A-module;*

   b) *If M_B is a finitely presented B-module, then M is a finitely presented A-module;*

   c) *If M_B is a flat (resp. faithfully flat) B-module, then M is a flat (resp. faithfully flat) A-module;*

   d) *If M_B is a finitely generated projective B-module, then B is a finitely generated projective A-module;*

   e) *If M_B is a noetherian (resp. artinian) B-module, then M is a noetherian (resp. artinian) A-module.*

Proof. — a) Let $(m_i)_{i \in I}$ be a generating family of M. The family $(1 \otimes m_i)_{i \in I}$ generates M_B. Since M_B is finitely generated, there exists a finite subset J of I such that the family $(1 \otimes m_i)_{i \in J}$ generates M_B. Let $u \colon A^J \to M$ be the morphism that maps $(a_j)_{j \in J}$ to $\sum a_j m_j$. Then $u_B \colon B^J \to M_B$ is surjective, hence u is surjective. This proves that M is finitely generated.

b) By *a*), M is finitely generated. Let $u: A^n \to M$ be a surjective morphism and let $K = \text{Ker}(u)$; we want to show that $\text{Ker}(u)$ is finitely generated. The morphism $u_B: B^n \to M_B$ is surjective, and its kernel is finitely generated, because M_B is finitely presented. Since f is flat, one has $K_B = \text{Ker}(u_B)$. By *a*), K is finitely generated, as was to be shown.

c) We first show that M is flat if M_B is flat. Let us consider a complex $\mathscr{C}$, $N_1 \xrightarrow{u} N_2 \xrightarrow{v} N_3$, of A-modules, and let us consider the complex $\mathscr{C}_M$:

$$M \otimes_A N_1 \xrightarrow{\text{id}_M \otimes u} M \otimes_A N_2 \xrightarrow{\text{id}_M \otimes v} M \otimes_A N_3.$$

If we tensor it by B, we obtain the complex $\mathscr{C}_{M,B}$

$$B \otimes_A M \otimes_A N_1 \xrightarrow{\text{id}_B \otimes \text{id}_M \otimes u} B \otimes_A M \otimes_A N_2 \xrightarrow{\text{id}_B \otimes \text{id}_M \otimes v} B \otimes M \otimes_A N_3,$$

which identifies with the complex

$$M_B \otimes_B N_{1,B} \xrightarrow{\text{id}_{M_B} \otimes u_B} M_B \otimes_B N_{2,B} \xrightarrow{\text{id}_{M_B} \otimes v_B} M_B \otimes_B N_{3,B},$$

which is deduced from the complex $\mathscr{C}_B$ of B-modules

$$N_{1,B} \xrightarrow{u_B} N_{2,B} \xrightarrow{v_B} N_{3,B}$$

by tensor product with M_B. Assume that the complex $\mathscr{C}$ is exact. Since B is a flat A-module, the complex $\mathscr{C}_B$ is exact; since M_B is a flat B-module, the complex $\mathscr{C}_{M,B}$ is exact as well. Since B is a faithfully flat A-module, the complex $\mathscr{C}_M$ is exact. This shows that M is a flat A-module.

Let us now assume that M_B is faithfully flat. By what precedes, M is flat. Assume that the complex $\mathscr{C}_M$ is exact. Since B is flat, the complex $\mathscr{C}_{M,B}$ is flat. Since M_B is faithfully flat, the complex $\mathscr{C}$ is exact. This shows that M is faithfully flat.

d) Since M_B is finitely generated and projective, it is a finitely presented and flat B-module. By *b*) and *c*), M is a finitely presented and flat A-module. It then follows from corollary 8.6.13 that M is a finitely generated projective A-module.

e) Let us assume that M_B is noetherian. Let (M_n) be an increasing sequence of submodules of M. Since B is flat, the modules $M_{n,B}$ identify as submodules of M_B and the sequence $(M_{n,B})$ is increasing. Consequently, it is stationary and there exists an integer $m \in \mathbf{N}$ such that $M_{n,B} = M_{m,B}$ for $n \geq m$. Since $(M_n/M_m)_B \simeq M_{n,B}/M_{m,B}$, one has $(M_n/M_m)_B = 0$ for $n \geq m$; since B is a faithfully flat A-module, one has $M_n/M_m = 0$, hence (M_n) is stationary. This shows that M is a noetherian A-module.

The proof that M is artinian if M_B is artinian is analogous; just replace "increasing" by "decreasing" and the module M_n/M_m by the module M_m/M_n in the preceding proof.   □

Proposition (8.8.5). — *Let $f : A \to B$ be a faithfully flat morphism of commutative rings. Then B/A is a flat A-module.*

Proof. — Since f is faithfully flat, it is injective. Let us consider the exact sequence $0 \to A \xrightarrow{f} B \xrightarrow{g} B/A \to 0$, where g is the canonical projection. By tensor product by B, we obtain the exact sequence $0 \to B \xrightarrow{f_B} B \otimes_A B \xrightarrow{g_B} B \otimes_A (B/A) \to 0$, where $f_B(b) = b \otimes_A 1$ and $g_B(b) = b \otimes g(b)$. On the other hand, the morphism of rings $h : B \otimes_A B \to B$ such that $h(b \otimes b') = bb'$ satisfies $h \circ f_B = \mathrm{id}_B$ and is a morphism of B-modules. Consequently, the latter exact sequence is split and $B \otimes_A (B/A)$ is a direct summand of $B \otimes_A B$. Since B is a flat A-module, $B \otimes_A B$ is a flat B-module, and $B \otimes_A (B/A)$ is a flat B-module. By proposition 8.8.4, B/A is a flat A-module. $\square$

Corollary (8.8.6). — *Let $f : A \to B$ be a morphism of integral domains that induces an isomorphism on their fields of fractions. If f is faithfully flat, then f is an isomorphism.*

Proof. — Since f is faithfully flat, it is injective and we consider A as a subring of B. Every element $b \in B$ can be written as a fraction a'/a'', for $a', a'' \in A$ and $a'' \neq 0$, so that the image of b in B/A is torsion. Since B/A is a flat A-module, it is torsion-free (corollary 8.6.9), so that $b \in A$. This proves that $B = A$. $\square$

8.8.7. — Let $f : A \to B$ be a faithfully flat morphism of commutative rings. Given a morphism u of A-modules, proposition 8.8.3 asserts properties of u from similar properties of the morphism u_B. Similarly, given an A-module M, proposition 8.8.4 asserts properties of M from similar properties of the B-module M_B. One says that faithfully flat morphisms *descend* these properties. Following the impetus of GROTHENDIECK (1960), we can be willing to go further and descend morphisms and modules, that is:

  a) Recover a morphism $u : M \to M'$ of A-modules from the B-morphism $u \otimes \mathrm{id}_B : M \otimes_A B \to M' \otimes_A B$, in particular, describe which morphisms $v : M \otimes_A B \to M' \otimes_A B$ are of the form $u \otimes \mathrm{id}_B$;

  b) Recover an A-module M from the B-module $M \otimes_A B$ and an additional datum.

This process has been coined *faithfully flat descent* by Grothendieck.

8.8.8. — The map $B \times B \to B \otimes_A B$ given by $(b, b') \mapsto b' \otimes b$ is A-bilinear, hence there exists a morphism $\tau : B \otimes_A B \to B \otimes_A B$ of A-modules such that $\tau(b \otimes b') = b' \otimes b$ for all $b, b' \in B$. This is a morphism of A-algebras. For $b \in B$, notice the relation

$$g(b) = 1 \otimes b - b \otimes 1 = (\mathrm{id} - \tau)(1 \otimes b).$$

Let M be an A-module. We let $\tau_M = \tau \otimes \mathrm{id}_M$ be the automorphism of $B \otimes_A B \otimes_A M$. Observe that although the A-module $B \otimes_A B \otimes_A M$ can be viewed as a $B \otimes_A B$-module, the morphism τ_M is not $B \otimes_A B$-linear, but satisfies the relation

$$\tau_M(\beta \cdot \mu) = \tau(\beta) \cdot \tau_M(\mu)$$

for $\beta \in B \otimes_A B$ and $\mu \in B \otimes_A B \otimes_A M$. We say that τ_M is τ-linear. Indeed, if $\beta = b_1 \otimes b_1'$ and $\mu = b \otimes b' \otimes m$ are split tensors, this just follows from the equalities

$$\begin{aligned}
\tau_M(\beta \cdot \mu) &= \tau_M((b_1 \otimes b_1') \cdot (b \otimes b' \otimes m)) = \tau_M(b_1 b \otimes b_1' b' \otimes m) \\
&= b_1' b' \otimes b_1 b \otimes m = (b_1' \otimes b_1) \cdot (b' \otimes b \otimes m) \\
&= \tau(b_1 \otimes b_1') \cdot \tau_M(b \otimes b' \otimes m) = \tau(\beta) \cdot \tau_M(\mu).
\end{aligned}$$

For $b \in B$ and $m \in M$, notice the relation

$$g_M(b \otimes m) = 1 \otimes b \otimes m - b \otimes 1 \otimes m = (\mathrm{id} - \tau_M)(1 \otimes b \otimes m).$$

Consequently, for $\mu \in B \otimes_A M$, one has $g_M(\mu) = (\mathrm{id} - \tau_M)(1 \otimes \mu)$.

Proposition (8.8.9). — *Let $f: A \to B$ be a faithfully flat morphism of commutative rings and let M, M' be A-modules. The map $u \mapsto \mathrm{id}_B \otimes u$ from $\mathrm{Hom}_A(M, M')$ to $\mathrm{Hom}_B(M \otimes_A B, M' \otimes_A B)$ is injective; its image is the set of all B-morphisms $v: B \otimes_A M \to B \otimes_A M'$ such that $(\mathrm{id}_B \otimes v) \circ \tau_M = \tau_{M'} \circ (\mathrm{id}_B \otimes v)$.*

Proof. — The injectivity of this map follows from proposition 8.8.3, $a)$. Let $v \in \mathrm{Hom}_B(B \otimes_A M, B \otimes_A M')$. If the morphism v is of the form $\mathrm{id}_B \otimes u$, for $u \in \mathrm{Hom}_A(M, M')$, then for all $b, b' \in B$ and $m \in M$, one has

$$\begin{aligned}
\tau_{M'} \circ (\mathrm{id}_B \otimes v)(b \otimes b' \otimes m) &= \tau_{M'}(b \otimes b' \otimes u(m)) \\
&= b' \otimes b \otimes u(m) \\
&= (\mathrm{id}_B \otimes v)(b' \otimes b \otimes m) \\
&= (\mathrm{id}_B \otimes v) \circ \tau_M(b \otimes b' \otimes m).
\end{aligned}$$

This proves that the two morphisms $(\mathrm{id}_B \otimes v) \circ \tau_M$ and $\tau_{M'} \circ (\mathrm{id}_B \otimes v)$ coincide on split tensors, hence are equal, as was to be shown.

Conversely, let us assume that $(\mathrm{id}_B \otimes v) \circ \tau_M = \tau_{M'} \circ (\mathrm{id}_B \otimes v)$. Then, for every $m \in M$, one has

$$\begin{aligned}
\tau_{M'}(1 \otimes v(f_M(m))) &= \tau_{M'} \circ (\mathrm{id}_B \otimes v)(1 \otimes 1 \otimes m) \\
&= (\mathrm{id}_B \otimes v) \circ \tau_M(1 \otimes 1 \otimes m) \\
&= 1 \otimes v(1 \otimes m) = 1 \otimes v(f_M(m)),
\end{aligned}$$

and $g_{M'}(v(f(m))) = 0$. Consequently, the image of the morphism $v \circ f_M: M \to B \otimes_A M'$ is contained in $\mathrm{Ker}(g_{M'})$. Since $f_{M'}$ is an isomorphism from M' to $\mathrm{Ker}(g_{M'})$, by theorem 8.7.10, this implies that there exists a morphism $u: M \to M'$ such that $v \circ f_M = f_{M'} \circ u$. Then, for $b \in B$ and $m \in M$, one has

$$v(b \otimes m) = v(b \cdot (1 \otimes m)) = b \cdot v(1 \otimes m) = b \cdot (1 \otimes u(m)) = b \otimes u(m),$$

so that v and $\mathrm{id}_B \otimes u$ coincide on split tensors, hence are equal. $\square$

8.8.10. — If a B-module N is of the form $B \otimes_A M$, then τ_M is a τ-linear automorphism of the $B \otimes_A$ B-module $B \otimes_A N = B \otimes_A B \otimes_A M$.

In fact, this automorphism τ_M satisfies an additional relation that we have to make explicit.

To every τ-linear endomorphism θ of a $B \otimes_A$ B-module $B \otimes_A N$, we associate two endomorphisms of the A-module $B \otimes_A B \otimes_A N$, respectively given by

$$\theta_1 = \mathrm{id}_B \otimes \theta : b \otimes b' \otimes n \mapsto b \otimes \theta(b' \otimes n)$$

and

$$\tau_1 = \tau \otimes \mathrm{id}_N : b \otimes b' \otimes n \mapsto b' \otimes b \otimes n.$$

The maps $\sigma = \mathrm{id}_B \otimes \tau$ and $\sigma' = \tau \otimes \mathrm{id}_B$ are automorphisms of the A-algebra $B \otimes_A B \otimes_A B$. For b, b' and $b'' \in B$, one has $\sigma(b \otimes b' \otimes b'') = b \otimes b'' \otimes b'$ and $\sigma'(b \otimes b' \otimes b'') = b' \otimes b \otimes b''$: the automorphism σ swaps the last two components, while the automorphism σ' swaps the first two. A reflection of the braid identity

$$(2\,3)(1\,2)(2\,3) = (1\,3) = (1\,2)(2\,3)(1\,2)$$

in the symmetric group $\mathfrak{S}_3$, one has

$$\sigma \circ \sigma' \circ \sigma = \sigma' \circ \sigma \circ \sigma'.$$

If $N = B \otimes_A M$ and $\theta = \tau_M = \tau \otimes \mathrm{id}_M$, then $\theta_1 = \sigma \otimes \mathrm{id}_M$ and $\tau_1 = \sigma' \otimes \mathrm{id}_M$, so that

$$\theta_1 \circ \tau_1 \circ \theta_1 = \tau_1 \circ \theta_1 \circ \tau_1.$$

Definition (8.8.11). — Let $f : A \to B$ be a morphism of rings and let $\tau : B \otimes_A B \to B \otimes_A B$ be the automorphism of A-algebras such that $\tau(b \otimes b') = b' \otimes b$ for $b, b' \in B$. A *descent datum* of a B-module N with respect to the morphism $f : A \to B$ is a τ-linear automorphism θ of $B \otimes_A N$ such that

$$\theta_1 \circ \tau_1 \circ \theta_1 = \tau_1 \circ \theta_1 \circ \tau_1.$$

If N, N' are B-modules endowed with descent data θ, θ' with respect to f, a B-morphism $u : N \to N'$ is compatible with the given descent data if one has $(\mathrm{id}_B \otimes u) \circ \theta = \theta' \circ (\mathrm{id}_B \otimes u)$.

With this definition, we can reformulate the preceding results as follows, for a ring morphism $f : A \to B$:

a) If M is an A-module, the map $\tau_M = \tau \otimes \mathrm{id}_M$ is a descent datum of the B-module $B \otimes_A M$ with respect to f;

b) If M, M' are A-modules, and if f is faithfully flat, then the map $u \mapsto \mathrm{id}_B \otimes u$ induces a bijection from $\mathrm{Hom}_A(M, M')$ to the set of all B-morphisms $\mathrm{Hom}_B(B \otimes_A M, B \otimes_A M')$ which are compatible with the descent data τ_M and $\tau_{M'}$.

One says that τ_M *is the canonical descent datum of the B-module* $B \otimes_A M$.

On Alexander Grothendieck

Alexander GROTHENDIECK (1928–2014) was a stateless mathematician whose life and mathematical achievements made of him an extraordinary figure of mathematics in the twentieth century. Born in Germany, his childhood has been marked by his family escaping the Nazis in 1933, his and his mother's internment in French camps, and his father's arrest and deportation to Auschwitz.

The early works of GROTHENDIECK were in functional analysis. In general, the algebraic tensor product of normed vector spaces admits many natural norms serving different purposes, and for which it is rarely complete. The fundamental *Grothendieck inequality* relates some of these norms, and it was later seen to have applications in computer science, graph theory and quantum mechanics.

His immense ability to build new mathematical fields served him to prove extraordinary results such as the general Riemann–Roch theorem in algebraic geometry. In the 1960s, he entirely rebuilt that theory upon the concept of the spectrum of a ring, simultaneously enlarging its scope to arbitrary rings, in particular, allowing nilpotent elements or not necessarily finitely generated algebras. In retrospect, these new foundations give a geometric content to results such as the primary decomposition. The development of *étale cohomology of algebraic varieties* that he pursued with his mathematical school led to the proof by DELIGNE, in 1974, of the analogue of the Riemann hypothesis

Alexander Grothendieck (June 1950), during a bicycle trip to Pont-à-Mousson

Photographer: Paulo Ribenboim
Source: Philippe Douroux
Courtesy of Paulo Ribenboim

Photograph of the Grothendieck seminar (1967) at Institut des Hautes Études Scientifiques (Bures-sur-Yvette)   Unknown photographer. Source: Wikipedia. copyright: unknown.

for algebraic varieties (conjecture of WEIL, see the notice about ARTIN). To provide a deeper understanding of the cohomology of algebraic varieties, GROTHENDIECK envisioned the theory of *motives*, which structured a lot of research in algebraic geometry since 1980, leading in particular to VOEVODSKY's work and its applications to deep questions of Galois cohomology.

Homological algebra had been founded by EILENBERG, STEENROD and MAC LANE in the 1940s, and the 1956 book of CARTAN and EILENBERG promoted it to a mathematical field. Soon after, the *Tohôku* paper of GROTHENDIECK recast this theory into the framework of "abelian categories", providing these methods with an even larger field of applications. A few years later, he would again revisit these foundations with VERDIER's derived categories or in his even later works on homotopical algebra.

At first, the work of GROTHENDIECK certainly strikes the reader by the degree of generality and the innovative concepts it brings in. Remarkably, this generality always illuminates the initial concepts so much that it becomes almost unavoidable. Other examples of such innovations is his revisiting of Galois theory (*axiomatic conditions for a Galois theory*, later followed by the theory of tannakian categories), or the notion of *topos*, which he introduced to define the above-mentioned étale cohomology.

His political views, marked by pacifism and ecology, led him to resign from his institution in 1970 after he found that it was partly funded by the military, and to gradually retire from the mathematical world. In the 1980s, he wrote very long texts on mathematics, on his approach to mathematics, as well as on mystical matters. He spent the last twenty years of his life in almost total seclusion.

Observe that B-modules N endowed with a descent datum θ form a category: objects are pairs (N, θ), and a morphism from (N, θ) to (N', θ') is a B-morphism $f: N \to N'$ which is compatible with the descent datas. Let us denote this category by $Mod_{B/A}$. By what precedes, the tensor product functor $M \mapsto (B \otimes_A M, \tau_M)$ is a functor from Mod_A to $Mod_{B/A}$. In this language, proposition 8.8.9 says that this functor is fully faithful; the following theorem of Grothendieck implies that it is an equivalence of categories.

Theorem (8.8.12) (Grothendieck's "Faithfully flat descent"). — *Let $f: A \to B$ be a faithfully flat morphism of rings. Let N be a B-module, let θ be a descent datum of N with respect to f and let M be the subset of all $x \in N$ such that $\theta(1 \otimes x) = 1 \otimes x$. Then the following properties hold:*

a) *M is an A-submodule of N;*

b) *There exists a unique morphism $\psi: B \otimes_A M \to N$ such that $\psi(b \otimes x) = bx$ for all $b \in B$ and all $x \in M$;*

c) *The morphism ψ is compatible with the descent datum τ_M of $B \otimes_A M$ and the given descent datum θ of N;*

d) *The morphism ψ is an isomorphism of B-modules.*

Proof. — a) Since the morphism $\mathrm{id}_B \otimes \theta$ is A-linear, the set M is an A-submodule of N.

b) The map from $B \times_A M$ to N given by $(b, x) \mapsto bx$ is A-multilinear, hence there exists a unique morphism $\psi: B \otimes_A M \to N$ such that $\psi(b \otimes m) = bm$ for all $b \in B$ and all $m \in M$.

c) Let us show that ψ is compatible with the given descent data. Let $b, b' \in B$ and $m \in M$; then

$$\theta \circ (\mathrm{id}_B \otimes \psi)(b \otimes b' \otimes m) = \theta(b \otimes \psi(b' \otimes m)) = \theta(b \otimes b'm)$$
$$= \theta((b \otimes b')(1 \otimes m)) = \tau(b \otimes b') \cdot \theta(1 \otimes m)$$
$$= (b' \otimes b) \cdot (1 \otimes m) = b' \otimes bm.$$

On the other hand,

$$(\mathrm{id}_B \otimes \psi) \circ \tau_M(b \otimes b' \otimes m) = (\mathrm{id}_B \otimes \psi)(b' \otimes b \otimes m) = b' \otimes bm,$$

so that $\theta \circ (\mathrm{id}_B \otimes \psi) = (\mathrm{id}_B \otimes \psi) \circ \tau_M$, since these two A-morphisms coincide on split tensors. This proves that ψ is compatible with the given descent data.

d) By definition, M is the kernel of the morphism $f_N - \theta \circ f_N$, so that we have an exact sequence

$$0 \longrightarrow M \xrightarrow{\ j\ } N \xrightarrow[-\theta \circ f_N]{f_N} B \otimes_A N$$

of A-modules, where j is the inclusion of M into N. Since B is a faithfully flat A-module, tensoring this exact sequence by B furnishes again an exact sequence, which is the upper row of the following diagram, the second row being the exact sequence $\mathscr{D}_f^N$:

$$0 \longrightarrow B \otimes_A M \xrightarrow{\mathrm{id}_B \otimes j} B \otimes_A N \xrightarrow[{-\,\mathrm{id}_B \otimes \theta \circ f_N}]{\mathrm{id}_B \otimes f_N} B \otimes_A B \otimes_A N$$

with vertical maps ψ, θ, θ_1:

$$0 \longrightarrow N \xrightarrow{f_N} B \otimes_A N \xrightarrow{g_N} B \otimes_A B \otimes_A N.$$

Let us check that the diagram is commutative. For $b \in B$ and $m \in M$, one has

$$\theta \circ (\mathrm{id}_B \otimes j)(b \otimes m) = \theta(b \otimes m) = \theta\big((b \otimes 1) \cdot (1 \otimes m)\big)$$
$$= (1 \otimes b)\theta(1 \otimes m) = (1 \otimes b) \cdot (1 \otimes m)$$
$$= 1 \otimes bm = f_N \circ \psi(b \otimes m),$$

so that the morphisms $\theta \circ (\mathrm{id}_B \otimes j)$ and $f_N \circ \psi$ coincide on split tensors; they are thus equal, hence the commutativity of the left square.

Recall that one has $g_N = g_N^1 - g_N^2$, where $g_N^1(b \otimes n) = 1 \otimes b \otimes n$ and $g_N^2(b \otimes n) = b \otimes 1 \otimes n$, for $b \in B$ and $n \in N$. We first prove that $\theta_1 \circ \mathrm{id}_B \otimes f_N = g_B^1 \circ \theta$. Indeed, for $b \in B$ and $n \in N$, one has

$$\theta_1 \circ \tau_1 \circ (\mathrm{id}_B \otimes f_N)(b \otimes n) = \theta_1 \circ \tau_1(b \otimes 1 \otimes n)$$
$$= \theta_1(1 \otimes b \otimes n)$$
$$= 1 \otimes \theta(b \otimes n)$$
$$= g_N^1 \circ \theta(b \otimes n).$$

This proves that $\theta_1 \circ \tau_1 \circ (\mathrm{id}_B \otimes f_N) = g_N^1 \circ \theta$.

Let $b \in B$ and $n \in N$; let (b_i) and (n_i) be finite families such that $\theta(1 \otimes n) = \sum b_i \otimes n_i$; then

$$\theta(b \otimes n) = \theta\big((b \otimes 1) \cdot (1 \otimes n)\big) = (1 \otimes b) \cdot \theta(1 \otimes n) = \sum b_i \otimes bn_i.$$

Moreover,

$$\tau_1 \circ \theta_1 \circ \tau_1(b \otimes 1 \otimes n) = \tau_1 \circ \theta_1(1 \otimes b \otimes n)$$
$$= \tau_1(1 \otimes \theta(b \otimes n))$$
$$= \tau_1\big(\sum 1 \otimes b_i \otimes bn_i\big)$$
$$= \sum b_i \otimes 1 \otimes bn_i$$
$$= g_N^2\big(\sum b_i \otimes bn_i\big)$$
$$= g_N^2 \circ \theta(b \otimes n).$$

In other words,

$$g_N^2 \circ \theta = \tau_1 \circ \theta_1 \circ \tau_1 \circ g_N^2.$$

On the other hand,

$$(\mathrm{id}_B \otimes \theta \circ f_N)(b \otimes n) = b \otimes \theta(1 \otimes n)$$
$$= \theta_1(b \otimes 1 \otimes n)$$
$$= \theta_1 \circ g_N^2(b \otimes n),$$

so that

$$\theta_1 \circ \tau_1 \circ (\mathrm{id}_B \otimes \theta \circ f_N) = \theta_1 \circ \tau_1 \circ \theta_1 \circ g_N^2.$$

Thus the relation

$$\theta_1 \circ \tau_1 \circ (\mathrm{id}_B \otimes (\theta \circ f_N)) = g_N^2 \circ \theta$$

follows from the definition of a descent datum, and the right-hand square is commutative as well.

We now deduce from the snake lemma (theorem 7.2.1) that ψ is an isomorphism. In the above diagram, replace the source of the morphism θ_1 with the image of the morphism $\mathrm{id}_B \otimes (f_N - \theta \circ f_N)$, and the morphism θ_1 with its restriction θ'_1, which is then injective. Since $\mathrm{id}_B \otimes j$ is injective, the exact sequence of theorem 7.2.1 then reads as

$$0 \to \mathrm{Ker}(\psi) \to \mathrm{Ker}(\theta) \to \mathrm{Ker}(\theta'_1) \to \mathrm{Coker}(\psi) \to \mathrm{Coker}(\theta).$$

Since $\mathrm{Ker}(\theta) = \mathrm{Ker}(\theta'_1) = \mathrm{Coker}(\theta) = 0$, we conclude that $\mathrm{Ker}(\psi) = \mathrm{Coker}(\psi) = 0$, hence ψ is an isomorphism, as was to be shown.   □

8.9. Galois Descent

Proposition (8.9.1). — *Let* $K \to L$ *be a finite extension of fields. The following properties are equivalent:*

(i) *The extension is Galois;*

(ii) *There exists a finite set* I *and an isomorphism of* L-*algebras* $L \otimes_K L \simeq L^I$.

Assume that they hold and let $G = \mathrm{Gal}(L/K)$. *There exists a unique morphism of* K-*algebras from* $L \otimes_K L$ *to* L^G *such that* $a \otimes b \mapsto (a\sigma(b))_{\sigma \in G}$, *and this morphism is an isomorphism of* L-*algebras.*

Proof. — Assume that the extension $K \to L$ is Galois and let $G = \mathrm{Gal}(L/K)$. The map from $L \times L$ to L^G given by $(a, b) \mapsto (a\sigma(b))_{\sigma \in G}$ is K-bilinear, hence there exists a unique morphism of K-algebras $\varphi \colon L \otimes_K L \to L^G$ such that $\varphi(a \otimes b) = (a\sigma(b))_{\sigma \in G}$ for all $a, b \in L$. It is also L-linear when $L \otimes_K L$ is viewed as an L-algebra via the morphism $a \mapsto a \otimes 1$.

Let us prove that φ is injective. Let $(b_1, \ldots, b_d)$ be a basis of L as a K-vector space; then $(1 \otimes b_1, \ldots, 1 \otimes b_d)$ is a basis of $L \otimes_K L$ as an L-vector space. Let $(a_1, \ldots, a_d) \in L^d$ be such that $\varphi(\sum a_i(1 \otimes b_i)) = 0$. By assumption, this implies that $\sum_{i=1}^d a_i \sigma(b_i) = 0$ for every $\sigma \in G$. As a consequence, $\sum_{i=1}^d a_i \sigma(y) = 0$ for every $y \in L$, so that $\sum_{i=1}^d a_i \sigma = 0$ in the set of K-linear morphisms from L to L. By linear independence of characters, $a_1 = \cdots = a_d = 0$.

Since the extension $K \to L$ is Galois, one has $\mathrm{Card}(G) = [L : K] = d$, hence $\dim_L(L \otimes_K L) = \dim_L(L^G)$ so that the injective L-linear map φ is an isomorphism.

Let us now assume that there exists a finite set I and an isomorphism of L-algebras $\varphi \colon L \otimes_K L \simeq L^I$. Let Ω be an algebraic closure of L. Using the associativity isomorphism $\Omega \otimes_L L \otimes_K L \simeq \Omega \otimes_K L$, the morphism $\mathrm{id}_\Omega \otimes \varphi$ induces an isomorphism of Ω-algebras $\Omega \otimes_K L \simeq \Omega^I$. Let $b \in L$ and let $P \in K[T]$ be its minimal polynomial; then $K[b]$ is a sub-algebra of L isomorphic to $K[T]/(P)$, and $\Omega[T]/(P)$ identifies with a sub-algebra of Ω^I. Since Ω^I is reduced, this implies that P has only simple roots in Ω: it is a separable polynomial, and b is separable. Consequently, L is a separable extension of K. Let us now use the primitive element and choose $b \in L$ such that $L = K[b]$; let P be its minimal polynomial. Then $L^I \simeq L \otimes_K L \simeq L[T]/(P)$. On the other hand, if $P = \prod_{i=1}^m P_i$ is the decomposition of P as a product of irreducible polynomials in $L[T]$, we get an isomorphism $L^I \simeq \prod_{i=1}^m L[T]/(P_i)$. The maximal ideals of the left-hand side are of the kernels of the projections from L^I to L and their residue fields are L; the maximal ideals of the right-hand side are the kernels of the projections to $L[T]/(P_i)$, for $i \in \{1, \dots, m\}$, and their residue fields are $L[T]/(P_i)$. This implies that $\deg(P_i) = 1$ for every $i \in \{1, \dots, m\}$: the polynomial P is split in $L[T]$. Let us write $P = \prod_{i=1}^m (T - b_i)$. For every i, there exists a unique morphism of K-algebras σ_i from $L = K[b]$ to itself such that $\sigma_i(b) = b_i$. Since the extension $K \to L$ is finite, these are isomorphisms, hence are elements of $\mathrm{Gal}(L/K)$. This shows that $\mathrm{Card}(\mathrm{Gal}(L/K)) = m = [L : K]$; the extension $K \to L$ is then Galois. $\square$

8.9.2. — Let $f \colon K \to L$ be a Galois extension and let G be its Galois group. Since every module over a field is flat, this morphism f is flat, and it is even faithfully flat since $\mathrm{Spec}(K)$ and $\mathrm{Spec}(K)$ are both reduced to one element. Faithfully flat descent asserts that the tensor product functor from K-vector spaces to L-vector spaces, $V \mapsto L \otimes_K V$, induces an equivalence of categories from $\mathbf{Mod}_K$ to the category $\mathbf{Mod}_{L/K}$ of L-vector spaces W endowed with descent data θ.

The definition of a descent datum involves $L \otimes_K L$-modules and $L \otimes_K L \otimes_K L$-modules, and the goal here is to reformulate it in terms of L-vector spaces *plus* various maps indexed by G, using proposition 8.9.1 to replace the algebra $L \otimes_K L$ with the isomorphic algebra L^G. The result of this translation will be called *Galois descent*.

In fact, while the proposition describes an isomorphism $L \otimes_K L \simeq L^G$ of L-algebras, when $L \otimes_K L$ is viewed as an L-algebra by the morphism $a \mapsto a \otimes 1$, it will be more convenient in what follows to view $L \otimes_K L$ as an L-algebra by the morphism $b \mapsto 1 \otimes b$. It is also isomorphic to L^G, but via the morphism $\varphi \colon L \otimes_K L \to L^G$ such that $\varphi(a \otimes b) = (\sigma(a)b)$ for every $a, b \in L$.

8.9.3. — For every $\sigma \in G$, we let $\delta_\sigma = (\delta_{\sigma,\tau})_{\tau \in G}$ be the element of L^G such that $\delta_{\sigma,\tau} = 1$ if $\tau = \sigma$ and $\delta_{\sigma,\tau} = 0$ otherwise. One has $\delta_\sigma \delta_\tau = 0$ if $\sigma \neq \tau$, and $\delta_\sigma^2 = \delta_\sigma$; moreover, $1 = \sum_{\sigma \in G} \delta_\sigma$.

Let $\varepsilon_\sigma \in L \otimes_K L$ be the unique element such that $\varphi(\varepsilon_\sigma) = \delta_\sigma$. In particular, for all $\sigma, \tau \in G$, one has $\varepsilon_\sigma \varepsilon_\tau = 0$ if $\sigma \neq \tau$, and $\varepsilon_\sigma^2 = \varepsilon_\sigma$; moreover, $1 = \sum_{\sigma \in G} \varepsilon_\sigma$.

Lemma (8.9.4). — *Let $\theta \colon L \otimes_K L \to L \otimes_K L$ be the automorphism of K-algebras such that $\theta(a \otimes b) = b \otimes a$ for every $a, b \in L$. Then $\varphi \circ \theta \circ \varphi^{-1}$ is the automorphism $(a_\sigma) \mapsto (\sigma(a_{\sigma^{-1}}))$.*

Proof. — Let $(x_\sigma) \in L^G$. Then $(x_\sigma) = \sum_{\sigma \in G} \delta_\sigma x_\sigma$, so that $(x_\sigma) = \varphi(x)$, where $x = \sum_{\sigma \in G} \varepsilon_\sigma (1 \otimes x_\sigma)$. For every $\sigma \in G$, fix a representation $\varepsilon_\sigma = \sum a_i^\sigma \otimes b_i^\sigma$. Then $x = \sum_{\sigma \in G} \sum_i a_i^\sigma \otimes b_i^\sigma x_\sigma$, so that $\theta(x) = \sum_{\sigma \in G} \sum_i b_i^\sigma x_\sigma \otimes a_i^\sigma$. Let $y = (y_\tau)_{\tau \in G} = \varphi(\theta(x)) = \varphi(\theta(\varphi^{-1}(x_\sigma)))$. By definition, one has

$$
y_\tau = \sum_{\sigma \in G} \sum_i \tau(b_i^\sigma x_\sigma) a_i^\sigma = \tau \Big(\sum_{\sigma \in G} \sum_i \tau^{-1}(a_i^\sigma) b_i^\sigma x_\sigma \Big)
$$

$$
= \tau \Big(\sum_{\sigma \in G} \varphi(\varepsilon_\sigma)_{\tau^{-1}} x_\sigma \Big) = \tau(x_{\tau^{-1}}),
$$

as was to be shown.   □

Proposition (8.9.5). — *Let W be an L-vector space.*

a) *There exists a unique morphism of K-vector spaces $\varphi_W \colon L \otimes_K W \to W^G$ such that $\varphi_W(a \otimes v) = (\sigma(a)v)_{\sigma \in G}$ for all $a \in L$ and all $v \in W$. It is bijective.*

b) *A map $\lambda \colon L \otimes_K W \to L \otimes_K W$ is a descent datum if and only if there exists a family $(\lambda_\sigma)_{\sigma \in G}$ satisfying the properties:*

   (i) *For every $\sigma \in G$, λ_σ is a bijective K-linear map $W \to W$ such that $\lambda_\sigma(av) = \sigma(a)\lambda_\sigma(v)$ for every $a \in L$ and every $v \in W$;*
   (ii) *$\varphi_W \circ \lambda \circ \varphi_W^{-1}((v_\sigma)) = (\lambda_\sigma(v_{\sigma^{-1}}))$ for every $(v_\sigma) \in W^G$;*
   (iii) *For every $\sigma, \tau \in G$, one has $\lambda_{\sigma\tau} = \lambda_\sigma \circ \lambda_\tau$.*

Proof. — a) The map $L \times W \to W^G$ given by $(a, v) \mapsto (\sigma(a)v)_{\sigma \in G}$ is K-bilinear; consequently, there exists a unique morphism $\varphi_W \colon L \otimes_K W \to W^G$ such that $\varphi_W(a \otimes v) = (\sigma(a)v)$.

By definition, one has $\varphi_W((a \otimes b) \cdot (c \otimes v)) = \varphi_W(ac \otimes bv) = (\sigma(ac)bv)_\sigma = \varphi(a \otimes b)\varphi_W(c \otimes v)$. This proves that $\varphi_W(\alpha \cdot \xi) = \varphi(\alpha)\varphi_W(\xi)$ for every $\alpha \in L \otimes_K L$ and every $\xi \in L \otimes_K W$.

In particular, $\varphi_W(\varepsilon_\sigma(1 \otimes v)) = \delta_\sigma(v, \dots, v)$. Since $(v_\sigma) = \sum_{\sigma \in G} \delta_\sigma(v_\sigma, \dots)$, this implies that $(v_\sigma) = \sum_{\sigma \in G} \varphi_W(\varepsilon_\sigma(1 \otimes v_\sigma))$. Therefore, the K-linear map $\psi_W \colon W^G \to L \otimes_K W$ given by $\psi_W((v_\sigma)) = \sum_{\sigma \in G} \varepsilon_\sigma(1 \otimes v_\sigma)$ satisfies $\varphi_W \circ \psi_W = \mathrm{id}$; in particular, φ_W is surjective. If $\dim_K(W)$ is finite, then $\dim_K(L \otimes_K W) = [L : K]\dim_K(W) = \dim_K(W^G)$ so that φ_W is an isomorphism. Let ξ be an element of $\mathrm{Ker}(\varphi_W)$. Let us write $\xi = \sum a_i \otimes v_i$ and let W_1 be the subspace of W generated by the family (v_i); it is finite-dimensional. One has $\xi \in L \otimes_K W_1$, hence $\varphi_{W_1}(\xi) = 0$, hence $\xi = 0$. We thus have proved that φ_W is bijective.

b) Let $\lambda \colon L \otimes_K W \to L \otimes_K W$ be a K-linear map. We express in terms of the K-linear map $\lambda' = \varphi_W \circ \lambda \circ \varphi_W^{-1}$ the conditions for λ to be a descent datum.

Let $\theta' = \varphi \circ \theta \circ \varphi^{-1}$; by lemma 8.9.4, one has $\theta'((a_\sigma)) = (\sigma(a_{\sigma^{-1}}))$ for every $(a_\sigma) \in L^G$. In particular, $\theta'(\delta_\sigma) = \delta_{\sigma^{-1}}$.

First of all, λ has to be bijective and θ-linear, that is, $\lambda(\alpha\xi) = \tau(\alpha)\lambda(\xi)$ for every $\alpha \in L \otimes_K L$ and every $\xi \in L \otimes_K W$. Since φ_W is an isomorphism, this means $\lambda'(av) = \theta'(a)\lambda'(v)$ for every $a \in L^G$ and every $v \in W^G$.

Let us assume that this relation holds. Let $v \in W^G$; since $1 = \sum_{\sigma \in G} \delta_\sigma$, one has $v = \sum_{\sigma \in G} \delta_\sigma v = \sum_{\sigma \in G} \delta_\sigma v_\sigma$, where, by abuse, we write $\delta_\sigma v_\sigma$ for the element of W^G whose σ-coordinate is equal to v_σ, and all other coordinates are 0. We obtain

$$\lambda'(v) = \sum_{\sigma \in G} \lambda'(\delta_\sigma v) = \sum_{\sigma \in G} \theta'(\delta_\sigma)\lambda'(v) = \sum_{\sigma \in G} \delta_{\sigma^{-1}}\lambda'(v).$$

Moreover, using that $\delta_\sigma^2 = 1$, we have

$$\delta_{\sigma^{-1}}\lambda'(v) = \lambda'(\delta_\sigma v) = \lambda'(\delta_\sigma^2 v) = \delta_{\sigma^{-1}}\lambda(\delta_\sigma v).$$

In other words, the σ^{-1}-component of $\lambda'(v)$, being equal to $\lambda(\delta_\sigma v)$, only depends of v_σ. This proves the existence of maps $\lambda_\sigma \colon W \to W$ such that $\lambda'((v_\sigma)) = (\lambda_\sigma(v_{\sigma^{-1}}))$; these maps λ_σ are necessarily K-linear. Let $v = (v_\sigma) \in W^G$ and $a = (a_\sigma) \in L^G$; one has

$$\lambda'(av) = \lambda'((a_\sigma v_\sigma)) = (\lambda_\sigma(a_{\sigma^{-1}}v_{\sigma^{-1}}))$$

and

$$\theta'(a)\lambda'(v) = (\sigma(a_{\sigma^{-1}})(\lambda_\sigma(v_{\sigma^{-1}}) = (\sigma(a_{\sigma^{-1}})\lambda_\sigma(v_{\sigma^{-1}})),$$

so that the relation $\lambda'(av) = \theta'(a)\lambda'(v)$ is equivalent to the relations $\lambda_\sigma(av) = \sigma(a)\lambda_\sigma(v)$ for all $a \in L$, $v \in W$ and $\sigma \in G$.

The map λ is bijective if and only if λ' is bijective, which is equivalent to the maps λ_σ being bijective. We thus see that this first condition for a descent datum is equivalent to properties (i) and (ii).

Property (iii) will be a reformulation of the equality "$\theta_1 \circ \tau_1 \circ \theta_1 = \tau_1 \circ \theta_1 \circ \tau_1$" of maps from $L \otimes_K L \otimes_K W$ to itself.

With the present notation, "τ_1" is the map $\tilde{\theta} = \theta \otimes \mathrm{id}_W$ and "θ_1" is the map $\tilde{\lambda} = \mathrm{id}_L \otimes \lambda$. The map

$$\tilde{\varphi}_W = \varphi_{W^G} \circ (\mathrm{id}_L \otimes \varphi_W) \colon L \otimes_K \otimes_K W \to W^{G \times G}$$

satisfies

$$\tilde{\varphi}_W(a \otimes b \otimes v) = \varphi_{W^G}(a \otimes (\sigma(b)v)) = (\tau(a)\sigma(b)v)_{\sigma,\tau \in G},$$

for every $a, b \in L$ and every $v \in W$; it is a K-linear isomorphism.

Let $\tilde{\theta}' = \tilde{\varphi}_W \circ \tilde{\theta} \circ \tilde{\varphi}_W^{-1}$ and $\tilde{\lambda}' = \tilde{\varphi}_W \circ \tilde{\lambda} \circ \tilde{\varphi}_W^{-1}$. The second condition for λ to be a descent datum reads as $\tilde{\lambda}' \circ \tilde{\theta}' \circ \tilde{\lambda}' = \tilde{\theta}' \circ \tilde{\lambda}' \circ \tilde{\theta}'$.

Let us first compute $\tilde{\theta}'$. For $a, b \in L$ and $v \in W$, one has $\tilde{\theta}(a \otimes b \otimes v) = b \otimes a \otimes v$, so that $\tilde{\theta}'((\tau(a)\sigma(b)v)) = (\sigma(a)\tau(b)v)$. Consequently, $\tilde{\theta}'((v_{\sigma,\tau})) = (v_{\tau,\sigma})$ for every element $(v_{\sigma,\tau})$ of $W^{G \times G}$ of the form $\tilde{\varphi}_W(a \otimes b \otimes v)$, with $a, b \in L$ and $v \in W$. Since they generate $W^{G \times G}$ as a K-vector space, this proves that

$$\tilde{\theta}'((v_{\sigma,\tau})) = (v_{\tau,\sigma})_{\sigma,\tau\in G}$$

for every $(v_{\sigma,\tau}) \in W^{G\times G}$.

Let us now compute $\tilde{\lambda}'$. With that aim, we will decompose an element $(v_{\sigma,\tau})$ as a sum

$$\sum_{\sigma,\tau\in G} \delta_{(\sigma,\tau)}\delta(v_{\sigma,\tau}),$$

where $\delta_{(\sigma,\tau)}$ is the element of $L^{G\times G}$ all of whose coordinates are zero, but for the coordinate (σ,τ) which is equal to 1, and $v_{\sigma,\tau}$ is an abuse of language for the element of $W^{G\times G}$ all of whose coordinates are equal to $v_{\sigma,\tau}$.

Fix $\sigma_0, \tau_0 \in G$, as well as decompositions $\varepsilon_{\sigma_0} = \sum a_i \otimes b_i$ and $\varepsilon_{\tau_0} = \sum c_j \otimes d_j$ in $L \otimes_K L$ — in other words, $\sum \sigma(a_i)b_i = 1$ if $\sigma = \sigma_0$, and 0 otherwise, and $\sum \tau(c_j)d_j = 1$ if $\tau = \tau_0$ and 0 otherwise. Let $v \in W$ and let

$$\xi = \sum_{i,j}(c_j \otimes 1 \otimes d_j)(1 \otimes a_i \otimes b_i)(1 \otimes 1 \otimes v) = \sum_{i,j} c_j \otimes a_i \otimes b_i d_j v.$$

For $\sigma, \tau \in G$, the (σ, τ)-component of $\tilde{\varphi}_W(\xi)$ is equal to

$$\sum_{i,j}\tau(c_j)\sigma(a_i)(b_i d_j v) = \sum_j \tau(c_j)\sigma(d_j)\sum_i \sigma(a_i)b_i v$$

$$= \sum_j \tau(c_j)\sigma(d_j)\delta_{\sigma_0}v$$

$$= \begin{cases} v & \text{if } \tau = \tau_0 \text{ and } \sigma = \sigma_0 \\ 0 & \text{otherwise.} \end{cases}$$

Moreover,

$$\xi = \sum_{i,j}(c_j \otimes a_i \otimes b_i d_j)(1 \otimes 1 \otimes v),$$

so that

$$\mathrm{id}_L \otimes\lambda(\xi) = \sum_{i,j}(c_j \otimes b_i d_j \otimes a_i)(1 \otimes \lambda(1 \otimes v)),$$

since λ is θ-linear. Let us then choose a decomposition $\lambda(1 \otimes v) = \sum u_k \otimes v_k$, for some elements $u_k \in L$ and $v_k \in W$. One thus has

$$\mathrm{id}_L \otimes\lambda(\xi) = \sum_{i,j,k}(c_j \otimes b_i d_j \otimes a_i)(1 \otimes u_k \otimes v_k) = \sum_{i,j,k} c_j \otimes b_i d_j u_k \otimes a_i v_k.$$

It thus follows that for every $\sigma, \tau \in G$, one has

$$\lambda'(\delta_{(\sigma_0,\tau_0)}v) = \tilde{\varphi}_W(\mathrm{id} \otimes \lambda(\xi))_{\sigma,\tau}$$

$$= \sum_{i,j,k} \tau(c_j)\sigma(b_i d_j u_k) a_i v_k$$

$$= \sum_i \sigma(b_i) a_i \sum_j \tau(c_j)\sigma(d_j) \sum_k \sigma(u_k) v_k.$$

One has

$$\sum_i \sigma(b_i) a_i = \sigma\left(\sum_i \sigma^{-1}(a_i) b_i\right) = \sigma(\delta_{\sigma,\sigma_0^{-1}}) = \delta_{\sigma,\sigma_0^{-1}}$$

and

$$\sum_j \tau(c_j)\sigma(d_j) = \sigma\left(\sum_j \sigma^{-1}\tau(c_j) d_j\right) = \sigma(\delta_{\sigma^{-1}\tau,\tau_0}) = \delta_{\sigma^{-1}\tau,\tau_0}.$$

Moreover,

$$\left(\sum \sigma(u_k) v_k\right)_\sigma = \varphi_W(\lambda(1 \otimes v)) == \lambda'(\varphi_W(1 \otimes v)) = \lambda'((v)_\sigma) = (\lambda_\sigma(v)).$$

Consequently,

$$v_{\sigma,\tau} = \delta_{\sigma,\sigma_0^{-1}} \delta_{\sigma^{-1}\tau,\tau_0} \lambda_\sigma(v).$$

This is 0 unless $\sigma = \sigma_0^{-1}$ and $\sigma^{-1}\tau = \tau_0$, that is $\tau = \sigma_0^{-1}\tau_0$; in this case, one has

$$v_{\sigma_0^{-1},\sigma_0^{-1}\tau_0} = \lambda_{\sigma_0^{-1}}(v).$$

We thus obtain that

$$\lambda'(\delta_{(\sigma_0,\tau_0)}v) = \delta_{(\sigma_0^{-1},\sigma_0^{-1}\tau_0)}\lambda_{\sigma_0^{-1}}(v).$$

The relation $(\sigma,\tau) = (\sigma_0^{-1},\sigma_0^{-1}\tau_0)$ is equivalent to $(\sigma_0,\tau_0) = (\sigma^{-1},\sigma_0\tau) = (\sigma^{-1},\sigma^{-1}\tau)$, so that, for every element $v = (v_{\sigma,\tau}) \in W^{G\times G}$, we have

$$\lambda'(v) = (\lambda_\sigma(v_{\sigma^{-1},\sigma^{-1}\tau})).$$

At this point, we can finally express the relation $\lambda' \circ \theta' \circ \lambda' = \theta' \circ \lambda' \circ \theta'$. For $v = (v_{\sigma,\tau}) \in W^{G\times G}$, one has

$$(\lambda' \circ \theta' \circ \lambda'(v))_{\sigma,\tau} = \lambda_\sigma((\theta' \circ \lambda'(v))_{\sigma^{-1},\sigma^{-1}\tau})$$

$$= \lambda_\sigma((\lambda'(v))_{\sigma^{-1}\tau,\sigma^{-1}})$$

$$= \lambda_\sigma(\lambda_{\sigma^{-1}\tau}(v_{\tau^{-1}\sigma,\tau^{-1}})).$$

On the other hand, one has

$$(\theta' \circ \lambda' \circ \theta'(v))_{\sigma,\tau} = (\lambda' \circ \theta'(v))_{\tau,\sigma}$$

$$= \lambda_\tau((\theta'(v))_{\tau^{-1},\tau^{-1}\sigma}) \qquad = \lambda_\tau(v_{\tau^{-1}\sigma,\tau^{-1}}).$$

Consequently, the relation $\lambda' \circ \theta' \circ \lambda' = \theta' \circ \lambda' \circ \theta'$ is equivalent to the equalities $\lambda_\sigma \circ \lambda_{\sigma^{-1}\tau} = \lambda_\tau$ for all $\sigma, \tau \in G$. This concludes the proof of the proposition. $\qquad\qquad\square$

Theorem (8.9.6) (Galois descent). — *Let $K \subset L$ be a finite Galois extension. Let W be an L-vector space and let (λ_σ) be a family of K-linear maps, $\lambda_\sigma : W \to W$, satisfying the properties:*

(i) *$\lambda_\sigma(av) = \sigma(a)\lambda_\sigma(v)$ for all $a \in L$ and all $v \in W$;*

(ii) *$\lambda_{\sigma\tau} = \lambda_\sigma \circ \lambda_\tau$ for all $\sigma, \tau \in G$.*

Let V be the set of all elements $v \in W$ such that $\lambda_\sigma(v) = v$ for all $\sigma \in G$. Then the following properties hold:

a) *V is a K-vector subspace of W;*

b) *There exists a unique morphism $\psi : L \otimes_K V \to W$ such that $\psi(a \otimes v) = av$ for all $a \in L$ and all $v \in V$;*

c) *The morphism ψ is an isomorphism.*

Proof. — This is just a reformulation of theorem 8.8.12 in the case of the Galois extension $K \to L$. Let $\varphi_W : L \otimes_K W \to W^G$ be the K-linear isomorphism such that $\varphi_W(a \otimes v) = (\sigma(a)v)_\sigma$ for all $a \in L$ and all $v \in W$. The given conditions on the family (λ_σ) furnish a descent datum λ on W, given by $\varphi_W \circ \lambda \circ \varphi_W^{-1}((v_\sigma)) = (\lambda_\sigma(v_{\sigma^{-1}}))$ for all $(v_\sigma) \in W^G$. The condition $\lambda(1 \otimes v) = 1 \otimes v$ defining the submodule denoted by M in theorem 8.8.12 rewrites as $\lambda_\sigma(v) = v$ for all $\sigma \in G$, hence is equal to V. This concludes the proof. $\qquad\square$

8.10. Exercises

Exercise (8.10.1). — Let m and n be two coprime integers. Show that

$$(\mathbf{Z}/m\mathbf{Z}) \otimes_\mathbf{Z} (\mathbf{Z}/n\mathbf{Z}) = 0.$$

Exercise (8.10.2). — Let I and J be ideals of a commutative ring A. Construct an isomorphism of A-algebras

$$(A/I) \otimes_A (A/J) \simeq A/(I + J).$$

Exercise (8.10.3). — Let X be a topological space. Show that

$$C(X, \mathbf{R}) \otimes_\mathbf{R} \mathbf{C} \simeq C(X, \mathbf{C}).$$

Exercise (8.10.4). — Let M be a Z-module.

a) Show that $M \otimes_\mathbf{Z} \mathbf{Q}$ is torsion-free.

b) Let $S = \mathbf{Z} - \{0\}$. Construct an isomorphism between $M \otimes_\mathbf{Z} \mathbf{Q}$ and $S^{-1}M$.

c) Show that M_{tor} is the kernel of the canonical morphism from M to $M \otimes_Z Q$.

Exercise (8.10.5). — Let A, B, C, D be rings. Let M be an (A, B)-bimodule, N be a (B, C)-bimodule and P be a (C, D)-bimodule. Show that there is a unique morphism of abelian groups

$$M \otimes_B (N \otimes_C P) \to (M \otimes_B N) \otimes_C P$$

which maps $m \otimes (n \otimes p)$ to $m \otimes (n \otimes p)$. Show that it is an isomorphism of (A, D)-bimodules.

Exercise (8.10.6). — Let A be a ring, let M be a right A-module and let N be a left A-module. Let $(x_i)_{i \in I}$ be a generating family in M, let $(y_j)_{j \in J}$ be a generating family in N and let $(m_j)_{j \in J}$ be an almost-null family of elements of M; let $\mu = \sum m_j \otimes y_j \in M \otimes_A N$.

a) Assume that there exists an almost-null family $(a_{i,j})$ in A, indexed by $I \times J$, such that $m_j = \sum_{i \in I} x_i a_{i,j}$ for all $j \in J$, and $0 = \sum_{j \in J} a_{i,j} y_j$ for all $i \in I$. Prove that $\mu = 0$.

b) Conversely, assume that $\mu = 0$. Introducing the kernel K of the unique morphism $u : A^{(J)} \to N$ that maps the basis element of index j to y_j, for every $j \in J$, prove that there exists an almost-null family $(a_{i,j})$ as above.

Exercise (8.10.7). — Let M be a right A-module and let N be a submodule of M; let us denote by $j : N \to M$ the inclusion morphism. One says that j is *pure*, or that N is a pure submodule of M if, for every left A-module X, the morphism $j \otimes id_X : N \otimes_A X \to M \otimes_A X$ is injective.

a) Assume that N is pure. Let m, n be positive integers, let $U = (a_{i,j}) \in Mat_{m,n}(A)$ and let $(y_1, \ldots, y_n) \in N^n$. Assume that there exists $(x_1, \ldots, x_m) \in M^m$ such that $y_j = \sum_{i=1}^m x_j a_{j,i}$ for every j. Prove that there exists such a family in N^m.

b) Conversely, prove that this property implies that for every finitely presented left A-module X, the map $j \otimes id_X$ is injective.

c) Under this assumption, prove that N is a pure submodule of M.

Exercise (8.10.8). — Let k be a field, let A be the polynomial ring $k[X, Y]$ and let $M = (X, Y)$ be the ideal of A consisting of polynomials without constant term. Let $g : A \to A \oplus A$ be the morphism given by $g(P) = (YP, -XP)$ and $f : A \oplus A \to A$ be the morphism defined by $f(P, Q) = XP + YQ$.

a) Show that one has an exact sequence

$$0 \to A \xrightarrow{g} A \oplus A \xrightarrow{f} M \to 0.$$

b) By restriction and tensor product, deduce the following exact sequences:

$$0 \to A \xrightarrow{g} M \oplus M \xrightarrow{f} M^2 \to 0,$$

and

$$0 \to M \xrightarrow{g \otimes \mathrm{id}_M} M \oplus M \xrightarrow{f \otimes \mathrm{id}_M} M \otimes_A M \to 0.$$

(One writes M^2 for the square of the ideal M.)

c) Define a commutative diagram of A-modules with exact rows:

d) Let $t = X \otimes Y - Y \otimes Y \in M \otimes_A M$. Prove that $t \neq 0$ and that it generates the torsion submodule of $M \otimes_A M$.

e) Prove that the submodule of $M \otimes_A M$ generated by $X \otimes X$, $X \otimes Y$ and $Y \otimes Y$ is isomorphic to M^2.

f) Construct an isomorphism $M \otimes_A M \simeq M^2 \oplus k$.

Exercise (8.10.9). — Let A be a ring, let M be a right A-module and let N be a left A-module. Let $(e_i)_{i \in I}$ be generating family of M, let $(f_j)_{j \in J}$ be a generating family of N and let $(y_i)_{i \in I}$ be a family with finite support in N. Let $t = \sum_{i \in I} e_i \otimes y_i$.

a) Assume that there exists a family $(a_{i,j})$ with finite support in A, indexed by $(i, j) \in I \times J$, such that $y_i = \sum_{j \in J} a_{i,j} f_j$ for every $i \in I$ and $\sum_{i \in I} e_i a_{i,j} = 0$ for every $j \in J$. Prove that $t = 0$.

b) Let $f \colon A_d^{(I)} \to M$ be the morphism of A-modules given by $f((a_i)) = \sum e_i a_i$. Prove that $\mathrm{Ker}(f \otimes \mathrm{id}_N)$ is generated by the image of $\mathrm{Ker}(f) \otimes_A N$.

c) Assuming that $t = 0$, prove that there exists a family $(a_{i,j})$ as in question a).

Exercise (8.10.10). — Let k be a field and let $P \in k[X]$ be an irreducible polynomial of degree > 1. Let $K = k[X]/(P)$.

a) Show that K is a field.

b) Show that there exists a morphism φ of k-algebras from $K \otimes_k K$ to K which maps $a \otimes b$ to ab for every $a, b \in K$.

c) Show that φ is not injective and conclude that $K \otimes_k K$ is not a field.

Exercise (8.10.11). — Let A be a local ring, let J be its maximal ideal and let $K = A/J$ be its residue field.

Let M and N be two finitely generated A-modules, and also assume that N is free. Let $f \colon M \to N$ be a morphism of A-modules such that the morphism

$$f \otimes \mathrm{id}_K \colon M \otimes_A K \to N \otimes_A K$$

is an isomorphism.

a) Using Nakayama's lemma, show that f is surjective.

b) Show that f has right inverse g, that is, show that there exists a morphism $g: N \to M$ such that $f \circ g = \mathrm{id}_N$.

c) Show that g is surjective and conclude that f is an isomorphism.

Exercise (8.10.12). — Let A be a commutative ring and let M and N be A-modules.

a) Assume that M and N are finitely generated; show that $M \otimes_A N$ is finitely generated.

b) Assume that M and N are simple. Show that $M \otimes_A N$ is isomorphic to M if M and N are isomorphic, and is 0 otherwise.

c) Assume that M and N are A-modules of finite length. Show that $M \otimes_A N$ has finite length, and that

$$\ell_A(M \otimes_A N) \leq \ell_A(M)\, \ell_A(N).$$

Exercise (8.10.13). — Let A be a local commutative ring, let J be its maximal ideal and let $K = A/J$ be its residue field. Let M and N be two A-modules.

a) Recall the structures of K-modules on M/JM and N/JN. Construct a surjective homomorphism from $M \otimes_A N$ to $(M/JM) \otimes_K (N/JN)$.

b) Assume that $M \otimes_A N = 0$. Show that $M/JM = 0$ or $N/JN = 0$.

c) If, moreover, M and N are finitely generated, show using Nakayama's lemma that $M = 0$ or $N = 0$.

d) Give an example of a commutative local ring A and of non-zero A-modules M and N such that $M \otimes_A N = 0$.

Exercise (8.10.14). — Let A be a commutative ring and let M and N be finitely generated A-modules. Using exercise 8.10.13, prove that $\mathrm{Supp}(M \otimes_A N) = \mathrm{Supp}(M) \cap \mathrm{Supp}(N)$.

Exercise (8.10.15). — Let A be a local commutative ring and let J be its maximal ideal. Assume that A is an integral domain and let K be its field of fractions.

Let M be a finitely generated A-module such that

$$\dim_{A/J} M/JM = \dim_K M \otimes_A K.$$

Prove that M is a free A-module.

Exercise (8.10.16). — Let k be a field and let K and L be k-algebras which are (commutative) fields.

a) Observe that $K \otimes_k L$ is a commutative k-algebra.

b) Show that there are canonical morphisms from K to $K \otimes_k L$ and from L to $K \otimes_k L$ such that $a \mapsto a \otimes 1$ and $b \mapsto 1 \otimes b$ respectively. Observe that they are injective.

c) Show that there exists a field E together with field morphisms $i\colon K \to$ E and $j\colon L \to$ E. (In other words, there exists a field E which contains simultaneously isomorphic copies of K and L.)

Exercise (8.10.17). — Let $f\colon A \to B$ be a morphism of rings.

a) Prove that there exists a unique morphism of rings $f\colon B \otimes_A B \to B$ such that $f(b \otimes b') = bb'$ for every $b, b' \in B$.

b) Show that f is an epimorphism in the category of rings if and only if f is an isomorphism.

c) Assume that f is an epimorphism. Prove that $f(Z(A)) \subset Z(B)$.

d) Assume that f is an epimorphism and that B is commutative. Prove that A is commutative.

Exercise (8.10.18). — Let $f\colon A \to B$ be a morphism of commutative rings.

a) Let M be a finitely generated non-zero A-module. Using a surjective morphism of the form $g\colon M \to A/I$, prove that $M \otimes_A M \neq 0$.

b) Assume that f is an epimorphism and that B is finitely generated as an A-module; prove that f is surjective. (*First prove that* $(B/A) \otimes_A B = 0$ *and that* $(B/A) \otimes_A (B/A) = 0$.)

Exercise (8.10.19). — Localization and quotients of commutative rings produce epimorphisms of rings (see exercise 1.8.51); the former add inverses and the latter are not injective. This exercise furnishes an example (due to GERASCHENKO (2009)) of an epimorphism which is of a different nature.

Let K be a field, let $A = K[X, Y]$ be the ring of polynomials in two indeterminates with coefficients in K and let B be the subalgebra generated by $X, XY, XY^2 - Y$.

a) Prove that the inclusion morphism $j\colon B \to A$ is an epimorphism of rings.

b) Prove that $B^\times = A^\times \cap B$.

c) Prove that j is not surjective.

Exercise (8.10.20). — Let A be a ring and let B be a subring of A. The *dominion* of B in A is the set B' of all $a \in A$ such that, for every two morphisms $f, f'\colon A \to C$ to a ring such that $f|_B = f'|_B$, one has $f(c) = f'(c)$.

a) Prove that B' is a subring of A that contains B.

b) Prove that $B' = A$ if and only if the inclusion morphism $j\colon B \to A$ is an epimorphism.

c) Let B'' be the set of all elements $a \in A$ for which there exist matrices $X \in \mathrm{Mat}_{1,m}(A), P \in \mathrm{Mat}_{m,n}(B), Y \in \mathrm{Mat}_{n,1}(A)$ such that $(a) = XPY$ and such that the matrices XP, P and PY have coefficients in B. Prove that B'' is a subring of B and that $B \subset B'' \subset B'$.

d) Let $a \in B'$; prove that $1 \otimes a = a \otimes 1$ in $A \otimes_A B$. Using exercise 8.10.9, prove that $a \in B''$. (This result is due to MAZET (1967).)

Exercise (8.10.21). — Let $f: A \to B$ be a morphism of commutative rings such that for every A-module M, the map $f_M: M \to M \otimes_A B$ (given by $m \mapsto m \otimes 1$) is injective. (One says that f is *pure*.)

a) Prove that f is injective.

b) Prove that for every ideal I of A, one has $I = f^{-1}(f(I) \cdot B)$.

c) Assume that A is a domain. Prove that B is a domain.

d) Assume that A is a domain and is integrally closed; prove that the same holds for B.

Exercise (8.10.22). — Let $f: A \to B$ be a morphism of commutative rings such that $I = f^{-1}(f(I) \cdot B)$ for every ideal I of A. For every A-module M, one denotes by f_M the canonical morphism $M \to M \otimes_A B$.

a) Let I be an ideal of A. Prove that $f_{A/I}$ is injective.

b) Let M be a A-module and let N be a submodule of M. Assume that f_N and $f_{M/N}$ are injective; prove that f_M is injective.

c) Let M be a finitely generated A-module. Prove that f_M is injective.

d) Prove that f is pure, that is, f_M is injective for every A-module M.

Exercise (8.10.23). — Let M be a free A-module of rank n. Show that the symmetric algebra of M is isomorphic to the ring $A[T_1, \ldots, T_n]$ of polynomials in n indeterminates.

Exercise (8.10.24). — Let M be an A-module. Let n be a non-negative integer.

a) Define an action of the symmetric group $\mathfrak{S}_n$ on $T^n(M)$ in such a way that $\sigma(m_1 \otimes \cdots \otimes m_n) = m_{\sigma^{-1}(1)} \otimes \cdots \otimes m_{\sigma^{-1}(n)}$ for every $\sigma \in \mathfrak{S}_n$ and any $m_1, \ldots, m_n \in M$.

b) One says that a tensor x in $T^n(M)$ is symmetric $\sigma(x) = x$ for every $\sigma \in \mathfrak{S}_n$. Let $T^n(M)^{\text{sym}}$ be the set of all symmetric tensors in $T^n(M)$. Show that it is a submodule of $T^n(M)$.

c) Let then $T(M)^{\text{sym}}$ be the direct sum of the submodules $T^n(M)^{\text{sym}}$ in $T(M)$. Show that it is a commutative sub-algebra of $T(M)$.

d) Assume that $n!$ is invertible in A. Show that one defines an endomorphism s of $T^n(M)$ by defining $s(x) = \frac{1}{n!} \sum_{\sigma \in \mathfrak{S}_n} \sigma(x)$, for $x \in T^n(M)$ (symmetrization of the tensor x). Show that $s(x) \in T^n(M)^{\text{sym}}$ for every $x \in T^n(M)$. Deduce that the canonical map from $T^n(M)$ to $S^n(M)$ induces an isomorphism from $T^n(M)^{\text{sym}}$ to $S^n(M)$.

e) In particular, if A contains $\mathbf{Q}$, the A-module $T(M)^{\text{sym}}$ is canonically isomorphic to $S(M)$. This gives another structure of an A-algebra on $T(M)^{\text{sym}}$, which is the one deduced from that of $S(M)$ by this canonical isomorphism. Explicitly, if x and y are symmetric tensors, show that their new product is the symmetrization of the tensor $x \otimes y$.

f) Assume that $A = \mathbf{Z}$ and that $M = \mathbf{Z}^2$. Let (e, f) be the canonical basis of M. Show that $(e \otimes e, e \otimes f, f \otimes e, f \otimes f)$ is a basis of $T^2(M)$. Show that $(e \otimes e, e \otimes f + f \otimes e, f \otimes f)$ is a basis of $T^2(M)^{\text{sym}}$. Show that the canonical morphism from $T^2(M)^{sym}$ to $S^2(M)$ is not surjective. More precisely, show that the quotient of $S^2(M)$ by the submodule generated by $T^2(M)$ is isomorphic to $\mathbf{Z}/2\mathbf{Z}$.

Exercise (8.10.25). — Let M and P be A-modules and let $f: M^n \to P$ be a symmetric n-linear map from M^n to P. Show that there exists a unique morphism of A-modules, $\varphi: S^n(M) \to P$, such that $\varphi(m_1 \ldots m_n) = f(m_1, \ldots, m_n)$ for every $(m_1, \ldots, m_n) \in M^n$.

Exercise (8.10.26). — Let M be a free A-module of rank n and let u be an endomorphism of M. Let M[X] be the A[X]-module $M \otimes_A A[X]$.

  a) Identify M[X] with the abelian group $\bigoplus_{n \in \mathbf{N}} M$ endowed with the external multiplication $(\sum a_n X^n) \cdot (m_n) = (\sum_k a_k m_{n-k})_{n \in \mathbf{N}}$.

  b) Show that $(m_n) \mapsto (u(m_n))$ is an endomorphism of M[X], still denoted u.

  c) Show that the determinant of $X \operatorname{id}_{M[X]} - u$ is the characteristic polynomial P_u of u.

  d) Identify the quotient of M[X] by the image of the endomorphism $X - u: m \mapsto Xm - u(m)$ to the A[X]-module M_u. (Recall that M_u is the abelian group M with external multiplication given by the morphism $P \mapsto P(u)$, from A[X] to End(M)).

  e) Using exercise 3.10.33, show that $P_u(u) = 0$ (*the Cayley–Hamilton theorem*).

Exercise (8.10.27). — Let M be a finitely generated A-module and let u be an endomorphism of M. Let $f: A^n \to M$ be any surjective morphism of A-modules.

  a) Show that there exists an endomorphism v of A^n such that $f(v(x)) = u(f(x))$ for every $x \in A^n$.

  b) Show that there exist elements $a_1, \ldots, a_n \in A$ such that

$$u^n(x) + a_1 u^{n-1}(x) + \cdots + a_{n-1} u(x) + a_n x = 0$$

for every $x \in M$.

Exercise (8.10.28). — Let K be a field and let V be a K-vector space. Let $V' = \operatorname{Span}(x_1, \ldots, x_n)$ and $V'' = \operatorname{Span}(y_1, \ldots, y_n)$ be subspaces of V with the same dimension n.

  a) If $V' = V''$, show that the vectors $x_1 \wedge \cdots \wedge x_n$ and $y_1 \wedge \cdots \wedge y_n$ in $\Lambda^n(V)$ are collinear.

  b) Let $\varphi_1, \ldots, \varphi_n$ be linear forms on V. Show that there exists a unique linear form Φ on $\Lambda^n(V)$ such that $\Phi(e_1 \wedge \cdots \wedge e_n) = \det(\varphi_i(e_j))$ for every $e_1, \ldots, e_n \in V$.

  c) Show that two vectors of a vector space are collinear if and only if any linear form which vanishes on one vanishes on the other.

  d) If $x_1 \wedge \cdots \wedge x_n$ and $y_1 \wedge \cdots \wedge y_n$ are collinear, show that $V' = V''$.

Exercise (8.10.29). — Let M be an A-module and let M' and M'' be submodules of M such that $M = M' \oplus M''$.

a) Show that there exists a unique morphism of A-modules $\theta \colon \Lambda(M') \otimes \Lambda(M'') \to \Lambda(M)$ which maps $(e_1 \wedge \cdots \wedge e_p) \otimes (f_1 \wedge \cdots \wedge f_q)$ to $e_1 \wedge \cdots \wedge e_p \wedge f_1 \wedge \cdots \wedge f_q$ for every $e_1, \ldots, e_p \in M'$ and $f_1, \ldots, f_q \in M$.

b) If M' and M'' are free, show that θ is an isomorphism.

c) Show that θ is *not* a morphism of algebras (unless $M' = 0$ or $M'' = 0$).

d) Show that there exists a unique structure of an A-algebra on $\Lambda(M') \otimes \Lambda(M'')$ such that $(\xi' \otimes \xi'')(\eta' \otimes \eta'') = (-1)^{q'p''}(\xi' \wedge \eta') \otimes (\xi'' \wedge \eta'')$, if $\xi' \in \Lambda^{p'}(M')$, $\xi'' \in \Lambda^{p''}(M'')$, $\eta' \in \Lambda^{q'}(M')$, $\eta'' \in \Lambda^{q''}(M'')$.

e) Show that θ is an isomorphism for this structure.

Exercise (8.10.30). — Let A and B be noetherian local rings, let M_A and M_B denote their maximal ideals and let $f \colon A \to B$ be a morphism of rings such that $M_A = f^{-1}(M_B)$.

a) Prove that $\dim(B) \leq \dim(A) + \dim(B/M_A B)$. (*Use theorem 9.4.3.*)

b) Assume, moreover, that f is flat. Let $n = \dim(A)$ and let $(P_0, \ldots, P_n = M_A)$ be a chain of prime ideals of A. Prove that there exists a chain $(Q_0, \ldots, Q_n)$ of prime ideals of B such that $P_j = f^{-1}(Q_j)$ for all j. (This is a *going down theorem for flat morphisms.*)

c) Still assuming that f is flat, prove that $\dim(B) = \dim(A) + \dim(B/M_A B)$.

Exercise (8.10.31). — Let A be a principal ideal domain and let $a \in A$. Let M be an $A/(a)$-module.

a) Let $b, c \in A$ be such that $a = bc$ and let $I = bA/aA$. Show that the morphism $x \mapsto bx$ induces an isomorphism from A/cA to I, and an isomorphism from M/cM to $I \otimes_{A/(a)} M$.

b) Prove that M is flat if and only if the following condition holds: for every pair (b, c) in A such that $a = bc$, and for every $x \in M$ such that $bx = 0$, there exists a $y \in M$ such that $x = cy$.

Exercise (8.10.32). — Let $f \colon A \to B$ be a flat morphism of commutative rings. Let $a \in A$ be a regular element. Prove that $f(a)$ is a regular element of B.

Exercise (8.10.33). — Let A be a commutative ring, let $a_1, \ldots, a_n \in A$ and let $B = A[T_1, \ldots, T_n]/(a_0 + a_1 T_1 + \cdots + a_n T_n)$.

a) Assume that there exists an $i \in \{1, \ldots, n\}$ such that a_i is a unit. Prove that R is a free A-module.

b) Assume that $(a_1, \ldots, a_n)$ generate the unit ideal. Prove that B is a faithfully flat A-module.

c) Assume that $(a_0, \ldots, a_n)$ generate the unit ideal. Prove that B is a flat A-module, and that it is faithfully flat if and only if $(a_1, \ldots, a_n)$ generate the unit ideal. (*Prove that a prime ideal $P \in \mathrm{Spec}(A)$ is of the form $A \cap Q$, for $Q \in \mathrm{Spec}(B)$ if and only if $P \notin V((a_1, \ldots, a_n))$.*)

(The more general exercise 6.4 of EISENBUD (1995) treats the case where $a_0 + a_1 T_1 + \cdots + a_n T_n$ is replaced by any polynomial $f \in A[T_1, \ldots, T_n]$, but its proof goes beyond the techniques of this book.)

Exercise (8.10.34). — Let $f : A \to B$ be a flat morphism of commutative local rings. Let J_A be the maximal ideal of A and let J_B be the maximal ideal of B; assume that $f^{-1}(J_B) = J_A$. Prove that for every A-module M, one has

$$\ell_B(M \otimes_A B) = \ell_A(M)\, \ell_B(B/J_A B).$$

Exercise (8.10.35). — Let A be a commutative ring.

a) Let $a \in A$ be such that the A-module $A/(a)$ is flat. Prove that there exists a $b \in A$ such that $a = a^2 b$.

b) If every A-module is flat, then A is a von Neumann ring.

c) Conversely, if A is a von Neumann ring, then every A-module is flat.

Exercise (8.10.36). — Let A be a (possibly noncommutative) local ring and let J be its maximal ideal. Let M be a finitely presented right A-module.

a) Construct a morphism $f : A^n \to M$ that induces an isomorphism $(A/J)^n \to M/MJ$.

b) Let $P = \mathrm{Ker}(f)$. Construct a morphism $g : A^m \to P$ that induces an isomorphism $(A/J)^m \to P/PJ$.

c) Prove that the canonical morphism $P \otimes_A J \to P$ is an isomorphism.

d) Conclude that $P = 0$ and that M is a free A-module.

e) Let A be any commutative ring (not necessary local), and let M be a finitely presented flat A-module. Prove that for every prime ideal P of A, the A_P-module M_P is free. Conclude that M is projective (cf. corollary 8.6.13).

Chapter 9.
The Normalization Theorem, Dimension Theory and Dedekind Rings

In this final chapter, where all rings are assumed to be commutative, I initiate the study of general noetherian rings, with a special emphasis on finitely generated algebras over a field. Once translated into geometric properties of algebraic subsets, these results form the foundations of algebraic geometry, but this interpretation will only be slightly evoked here.

Noether's normalization theorem is a tool to reduce the study of a general finitely generated algebra over a field to the study of a polynomial ring. The proof I explain is valid over any field, and the classical proof over infinite fields is given as an exercise. I then explore a few of its consequences: a general proof of Hilbert's Nullstellensatz and Zariski's version (which, however, can be proved more directly, see the exercises), the fact that every prime ideal of such an algebra is an intersection of maximal ideals, a property which gives rise to the notion of a Jacobson ring, etc.

I give another application of Noether's normalization theorem, that if an integral domain A is a finitely generated algebra over a field, then the integral closure of A is a finitely generated A-module. (To avoid technicalities, its proof is restricted to the case of fields of characteristic zero.)

The next three sections are devoted to the notions of dimension and codimension of rings. Based on lengths of chains of prime ideals, they embody the idea that a point is contained in a curve, which is contained in a surface, etc. For integral finitely generated algebras over a field, I show that the dimension equals the transcendence degree of the field of fractions. The theory of codimension requires Krull's Hauptidealsatz that embodies another intuition, that p equations, if well chosen, define a subset of codimension p. I then establish the expected relation between dimension and codimension.

A last section studies Dedekind rings, the algebraic analogues of curves. Their importance had first been revealed in algebraic number theory, for rings of integers in number fields furnish fundamental examples. Although Dedekind rings are not unique factorization domain in general, their (non-zero) ideals have a similar factorization as a product of maximal ideals. Their defect of unique factorization is encoded in their class groups; in the case of rings of integers of number fields, I prove Minkowski's theorem that this class group is finite.

© Springer Nature Switzerland AG 2021
A. Chambert-Loir, *(Mostly) Commutative Algebra*, Universitext,
https://doi.org/10.1007/978-3-030-61595-6_9

9.1. Noether's Normalization Theorem

Combined with the properties of integral morphisms, Noether's normalization theorem is a powerful tool for the study of finitely generated algebras
over a field. This section and the following illustrate this fact.

Theorem (9.1.1) (Normalization theorem). — *Let* K *be a field and let* A *be
a finitely generated* K*-algebra. Then there exist an integer* $n \geq 0$ *and elements*
$a_1, \ldots, a_n \in A$ *such that the unique morphism of* K*-algebras from* $K[X_1, \ldots, X_n]$
to A *such that* $X_i \mapsto a_i$ *is injective and integral.*

Proof. — Let $(x_1, \ldots, x_m)$ be a family of elements of A such that $A = K[x_1, \ldots, x_m]$. Let us prove the theorem by induction on m. If $m = 0$, then
$A = K$ and the result holds with $n = 0$. We thus assume that $m \geq 1$ and
that the result holds for any K-algebra which is generated by at most $m - 1$
elements.

Let $\varphi \colon K[X_1, \ldots, X_m] \to A$ be the unique morphism of K-algebras such
that $\varphi(X_i) = x_i$. If φ is injective, the result holds, taking $n = m$ and $a_i = x_i$
for every i.

We may thus assume that there is a non-zero polynomial $P \in K[X_1, \ldots, X_m]$
such that $P(x_1, \ldots, x_m) = 0$. Let $(c_{\mathbf{n}})$ be the coefficients of P, so that

$$P = \sum_{\mathbf{n} \in \mathbf{N}^m} c_{\mathbf{n}} \prod_{i=1}^{m} X_i^{n_i}.$$

Let r be an integer strictly greater than the degree of P in each variable; in
other words, for every $\in \mathbf{N}^m$, one has $c_{\mathbf{n}} = 0$ unless $n_i < r$ for all i; then set
$y_i = x_i - x_1^{r^{i-1}}$ for $i \in \{2, \ldots, m\}$. Let $B = K[y_2, \ldots, y_m]$ be the subalgebra of A
generated by $y_2, \ldots, y_m$; we are going to show that A is integral over B.

We define a polynomial $Q \in B[T]$ by

$$Q(T) = P(T, y_2 + T^r, \ldots, y_m + T^{r^{m-1}})$$

$$= \sum_{\mathbf{n} \in \mathbf{N}^m} c_{\mathbf{n}} T^{n_1} (y_2 + T^r)^{n_2} \ldots (y_m + T^{r^{m-1}})^{n_m}$$

$$= \sum_{\mathbf{n} \in \mathbf{N}^m} \sum_{j_2=0}^{n_2} \cdots \sum_{j_m=0}^{n_m} \binom{n_2}{j_2} \cdots \binom{n_m}{j_m} c_{\mathbf{n}} y_2^{n_2-j_2} \cdots y_m^{n_m-j_m} T^{n_1 + \sum_{i=2}^{m} j_i r^{i-1}}$$

and observe that

$$Q(x_1) = P(x_1, y_2 + x_1^r, \ldots, y_m + x_1^{r^{m-1}}) = P(x_1, x_2, \ldots, x_m) = 0.$$

Order $\mathbf{N}^m$ with the "reverse lexicographic order": first in the order of the
last coordinate then, in case of equality, in the order of the penultimate, etc.
In formulas, we say that $(n_1', \ldots, n_m') < (n_1, \ldots, n_m)$ if and only if there exists
an integer $k \in \{1, \ldots, m\}$ such that $n_k' < n_k$ and $n_j = n_j'$ for every integer j
such that $k < j \leq m$. If one only considers sequences $\mathbf{n}$ such that $\sup(n_j) < r$,

in particular for elements $\mathbf{n} \in \mathbf{N}^m$ such that $c_{\mathbf{n}} \neq 0$, this ordering corresponds with the one given by the expansion in base r (written in the reverse order): $(n'_1, \ldots, n'_m) < (n_1, \ldots, n_m)$ if and only if

$$n'_m r^{m-1} + n'_{m-1} r^{m-2} + \cdots + n'_2 r + n'_1 < n_m r^{m-1} + n_{m-1} r^{m-2} + \cdots + n_2 r + n_1.$$

Let $\mathbf{n}$ be the largest multi-index in $\mathbf{N}^m$, for that ordering, such that $c_{\mathbf{n}} \neq 0$. For any $\mathbf{n}' \in \mathbf{N}^m$ such that $c_{\mathbf{n}'} \neq 0$, one has $n'_i < r$ for every i, by definition of r, so that for any $j_2 \in \{0, \ldots, n'_2\}, \ldots, j_m \in \{0, \ldots, n'_m\}$,

$$n'_1 + j_2 r + \cdots + j_m r^{m-1} \leq n'_1 + n'_2 r + \cdots + n'_m r^{m-1} \leq n_1 + n_2 r + \cdots + n_m r^{m-1},$$

and equalities are only possible if $\mathbf{n}' = \mathbf{n}$ and $j_2 = n'_2, \ldots, j_m = n'_m$. This implies that the degree of Q is equal to $n_1 + n_2 r + \cdots + n_m r^{m-1}$ and that only the term with $j_k = n_k$ for $k \in \{2, \ldots, m\}$ contributes the leading coefficient, which thus equals $c_{\mathbf{n}}$, a non-zero element of K. In particular, Q is a polynomial in B[T] whose leading coefficient is a unit, so that x_1 is integral over B.

Consequently, $B[x_1]$ is integral over B. For every $i \in \{2, \ldots, m\}$, one has $x_i = y_i - x_1^{r^{i-1}} \in B[x_1]$. Since $A = K[x_1, \ldots, x_m]$, we conclude that $A = B[x_1]$ and A is integral over B.

By induction, there exist an integer $n \leq m - 1$ and elements $a_1, \ldots, a_n \in B$ such that the unique morphism $f : K[T_1, \ldots, T_n] \to B$ of K-algebras such that $f(T_i) = a_i$ for all i is injective and such B is integral over $K[a_1, \ldots, a_n]$. Then A is integral over $K[a_1, \ldots, a_n]$ as well, and this concludes the proof of the theorem. □

As a first application, let us prove the general case of Hilbert's Nullstellensatz (theorem 2.3.1), which we had only proved under the assumption that the field was uncountable.

The most fundamental form of this theorem is due to Zariski and claims:

Theorem (9.1.2) (Zariski). — *Let K be a field and let A be a finitely generated K-algebra. Assume that A is field. Then A is a finite extension of K.*

Proof. — By Noether's normalization theorem (theorem 9.1.1), there exist an integer $n \geq 0$ and an injective morphism of K-algebras $f : K[X_1, \ldots, X_n] \to A$ such that A is integral over the image B of f. Since A is a field, proposition 4.2.5 implies that B is a field as well. However, B, being the image of the injective morphism f, is isomorphic to the ring $K[X_1, \ldots, X_n]$ of polynomials in n variables with coefficients in K. For $n \geq 1$, this ring is not a field. Consequently $n = 0$ and A is algebraic over K. Being moreover finitely generated as an algebra, corollary 4.2.2 implies that it is a finite extension of K. □

As a corollary, we can also give a complete proof of theorem 2.3.1. Let us first recall its statement.

Corollary (9.1.3). — *Let n be a positive integer, let K be an algebraically closed field and let M be a maximal ideal of the ring $K[X_1, \ldots, X_n]$. There exists a unique element $(a_1, \ldots, a_n) \in K^n$ such that $M = (X_1 - a_1, \ldots, X_n - a_n)$.*

Proof. — Let L be the residue field of the maximal ideal M, that is, L = $K[X_1, \ldots, X_n]/M$ and let $\theta \colon K[X_1, \ldots, X_n] \to L$ be the canonical surjection. By construction, L is a finitely generated K-algebra, and it is a field. By Zariski's theorem (theorem 9.1.2), L is an algebraic extension of K. Since K is algebraically closed, L = K. For $i \in \{1, \ldots, n\}$, let $a_i = \theta(X_i)$; by definition of the quotient ring, it is the only element of K such that $X_i - a_i \in M$. Then M contains the ideal $(X_1 - a_1, \ldots, X_n - a_n)$. On the other hand, the same proof as for theorem 2.3.1 implies that this ideal is maximal, so that $M = (X_1 - a_1, \ldots, X_n - a_n)$. □

At this point, we advise the reader to read again about the correspondence between ideals and algebraic sets discussed in section 2.3.

We now pass to other important consequences of theorem 9.1.2.

Corollary (9.1.4). — *Let K be a field, let A be a finitely generated K-algebra and let M be a prime ideal of A. Then M is maximal if and only if A/M is finitely dimensional over K; it is then a finite extension of K.*

Proof. — Assume that M is maximal; then A/M is a finitely generated K-algebra which is a field. By theorem 9.1.2, it is a finite extension of K and, in particular, its dimension as a K-vector space is finite.

Conversely, assume that the quotient K-algebra L = A/M is finite-dimensional. It is also an integral domain. For every non-zero $a \in L$, the map $b \mapsto ab$ from L to itself is K-linear, and injective, hence it is bijective; in particular, every non-zero element of L is invertible and L is a field. Consequently, M is a maximal ideal of A. □

Corollary (9.1.5). — *Let K be a field and let $f \colon A \to B$ be a morphism of finitely generated K-algebras. For every maximal ideal M of B, $f^{-1}(M)$ is a maximal ideal of A.*

Proof. — We know that the operation f^{-1} induces a map from Spec(B) to Spec(A), so that $f^{-1}(M)$ is a prime ideal; we need to prove that it is even maximal. Passing to the quotient, f induces an injective morphism $\varphi \colon A/f^{-1}(M) \to B/M$ of finitely generated K-algebras. Moreover, B/M is a field. By theorem 9.1.2, it is thus a finite extension of K, hence $f^{-1}(M)$ is a prime ideal of A such that $A/f^{-1}(M)$ is finite-dimensional. By the preceding corollary, $f^{-1}(M)$ is a maximal ideal. □

The following corollary strengthens proposition 2.2.11.

Corollary (9.1.6). — *Let K be a field and let A be a finitely generated K-algebra.*

(i) *The nilpotent radical and the Jacobson radical of A coincide.*

(ii) *For every ideal I of A, $\sqrt{I}$ is the intersection of all maximal ideals of A which contain I.*

(iii) *In particular, every prime ideal P of A is the intersection of the maximal ideals of P which contain P.*

One sums up assertions *b*) and *c*) by saying that A is a *Jacobson ring*.

Proof. — *a*) We need to prove that an element $a \in A$ is nilpotent if and only if it belongs to every maximal ideal of I. One direction is clear: if a is nilpotent, it belongs to every prime ideal of I, hence to every maximal ideal of I. Conversely, let us assume that a is not nilpotent and let us show that there exists a maximal ideal M of A such that $a \notin M$. Since a is not nilpotent, the K-algebra $A_a = A[1/a]$ is not null; it is also a finitely generated K-algebra. By corollary 9.1.5 the inverse image in A of a maximal ideal of A_a is a maximal ideal of A which does not contain a. This concludes the proof of assertion *a*).

b) Let $B = A/I$; it is a finitely generated K-algebra and its maximal ideals are of the form M/I, where M is a maximal ideal of A containing I. By part *a*), the nilpotent radical of B is the intersection of the maximal ideals of B. Since the class in B of an element $a \in A$ is nilpotent if and only if $a \in \sqrt{I}$, this implies that $\sqrt{I}$ is the intersection of all maximal ideals of A which contain I.

c) follows from *b*) applied to P = I.   □

Let us give another important application.

Theorem (9.1.7). — *Let K be an algebraically closed field and let A, B be two finitely generated K-algebras.*

(i) *If A and B are integral domains, then $A \otimes_K B$ is also an integral domain.*

(ii) *If A and B are reduced, then $A \otimes_K B$ is also reduced.*

Proof. — We first prove the theorem under the assumption that A and B are finitely generated.

(i) Assume that A and B are integral domains.

The tensor product of two non-zero K-vector spaces is a non-zero K-vector space; consequently, $A \otimes_K B \neq 0$.

Let then f and g be two elements of $A \otimes_K B$ such that $fg = 0$. We may decompose f as a sum $\sum_{i=1}^{r} a_i \otimes b_i$ of split tensors, where $b_1, \ldots, b_r$ are linearly independent over K. Similarly, we write $g = \sum_{j=1}^{s} a'_j \otimes b'_j$, where $b'_1, \ldots, b'_s$ are linearly independent over K.

Let M be a maximal ideal of A. The quotient ring A/M is a finitely generated K-algebra, and is a field; consequently, it is an algebraic extension of K, hence is isomorphic to K since K is algebraically closed. Let $cl_M : A \to K$ be the corresponding morphism of K-algebras with kernel M. Let also $\theta_M : A \otimes_K B \to B$ be the morphism $cl_M \otimes id_B$; it is a morphism of K-algebras. Since

$$\theta_M(f)\theta_M(g) = \theta_M(fg) = 0$$

and B is an integral domain, either $\theta_M(f) = 0$ or $\theta_M(g) = 0$. Moreover, one has

$$\theta_M(f) = \sum_{i=1}^{r} cl_M(a_i)b_i \quad \text{and} \quad \theta_M(g) = \sum_{j=1}^{s} cl_M(a'_j)b'_j.$$

Assume that $\theta_M(g) = 0$. Since $b_1, \ldots, b_r$ are linearly independent over K, we conclude that $\text{cl}_M(a_i) = 0$ for every $i \in \{1, \ldots, r\}$; in other words, the ideal $I = (a_1, \ldots, a_r)$ is contained in M.

Similarly, if $\theta_M(f) = 0$, we obtain that the ideal $J = (a'_1, \ldots, a'_s)$ is contained in M.

In any case, $I \cap J \subset M$.

This is valid for any maximal ideal M of A. By corollary 9.1.6, every element of $I \cap J$ is nilpotent. Since A is an integral domain, $I \cap J = 0$.

Assume that $f \neq 0$. Then $I \neq 0$; let thus x be a non-zero element of I. For every $y \in J$, one has $xy \in I \cap J$, hence $xy = 0$. Since A is an integral domain, this implies $y = 0$, hence $J = 0$, hence $a'_1 = \cdots = a'_s = 0$ and $g = 0$. This concludes the proof that $A \otimes_K B$ is an integral domain.

(ii) The proof is analogous. Let $f \in A \otimes_K B$ be a nilpotent element. We write $f = \sum_{i=1}^{r} a_i \otimes b_i$, where $a_1, \ldots, a_r \in A$ and $b_1, \ldots, b_r \in B$, chosen such that $b_1, \ldots, b_r$ are linearly independent over K.

Let M be a maximal ideal of A, let $\text{cl}_M \colon A \to A/M \simeq K$ be the quotient morphism and let $\theta_M = \text{cl}_M \otimes \text{id}_B \colon A \otimes_K B \to B$. Since f is nilpotent, $\theta_M(f)$ is nilpotent in B, hence $\theta_M(f) = 0$, because B is reduced. Since $\theta_M(f) = \sum_{i=1}^{r} \text{cl}_M(a_i)b_i$, this implies that the ideal $I = (a_1, \ldots, a_r)$ is contained in M.

By corollary 9.1.6, every element of I is nilpotent. Since A is reduced, $I = (0)$ and $a_1 = \cdots = a_r = 0$. This proves $f = 0$.

We now remove the assumption that A and B are finitely generated.

For $a)$, we consider $f, g \in A \otimes_K B$ such that $fg = 0$; for $b)$, we consider $f \in A \otimes_K B$ such that f is nilpotent. In both cases, there exist finitely generated subalgebras $A' \subset A$ and $B' \subset B$ such that $f \in A'$ and $g \in B'$ (with the notation introduced above, just take $A' = K[a_1, \ldots, a_r, a'_1, \ldots, a'_s]$ and $B' = K[b_1, \ldots, b_r, b'_1, \ldots, b'_s]$ in case $a)$, and $A' = K[a_1, \ldots, a_r]$ and $B' = K[b_1, \ldots, b_r]$ in case $b)$). These algebras A' and B' are integral domains (resp. reduced), and $A' \otimes_K B'$ is a subalgebra of $A \otimes_K B$. One thus has $fg = 0$ in $A' \otimes_K B'$ (resp. f is nilpotent in $A' \otimes_K B'$). By the finitely generated case, $f = 0$ or $g = 0$ (resp. $f = 0$), as was to be shown. $\qquad\square$

9.2. Finiteness of Integral Closure

Proposition (9.2.1). — *Let A be a noetherian integral domain and let E be its field of fractions. Let F be a* separable *finite algebraic extension of F and let B be the integral closure of A in F. If A is integrally closed, then B is a finitely generated A-module, in particular, a noetherian ring.*

Proof. — Let $(e_1, \ldots, e_n)$ be a basis of F as an E-vector space. Up to multiplying them by a non-zero element of A, we may assume that they belong to B. Since the extension $E \subset F$ is separable, the symmetric bilinear form defined by the trace is non-degenerate (theorem 4.7.7). Consequently, there exists a basis $(f_1, \ldots, f_n)$ of F as an E-vector space such that for every i and $j \in \{1; \ldots; n\}$,

$\mathrm{Tr}_{E/F}(e_i f_j) = 0$ if $i \neq j$, and $\mathrm{Tr}_{E/F}(e_i f_i) = 1$. Let D be a non-zero element of A such that $Df_i \in B$ for every i.

Let then x be an element of B, and let $x_1, \ldots, x_n \in E$ be such that $x = \sum_{i=1}^{n} x_i e_i$. For every $i \in \{1; \ldots; n\}$, one has $(Df_i)x \in B$, hence $\mathrm{Tr}_{F/E}(Df_i x) \in A$, by corollary 4.7.6, hence $Dx_i \in A$. Consequently, $B \subset D^{-1} \sum_{i=1}^{n} Ae_i$. In other words, B is an A-submodule of a free A-module of rank n. Since A is a noetherian ring, this implies that B is a finitely generated A-module.

Since ideals of B are A-submodules, this also implies that B is noetherian.□

Theorem (9.2.2). — *Let k be a field and let A be a finitely generated k-algebra which is an integral domain. Let E be the field of fractions of A and let $E \subset F$ be a finite algebraic extension. Finally, let B be the integral closure of A in F. Then, B is a finitely generated A-module.*

To simplify the proof, *we assume that the characteristic of k is zero.* The general case can be proved along the same lines but requires an additional study of inseparable extensions in the style of remark 4.4.17.

Proof. — Let us apply Noether's normalization theorem (theorem 9.1.1) to A. Let $x_1, \ldots, x_n$ be elements of A, algebraically independent over k, such that A is integral over $A_0 = k[x_1, \ldots, x_n]$.

Let us remark that an element of F is integral over A if and only if it is integral over A_0. Consequently, B is the integral closure of A_0 in F. Since the extension $E_0 = k(x_1, \ldots, x_n) \subset E$ is finite, the extension $E_0 \subset F$ is finite too.

Let us also remark that A_0 is a unique factorization domain (theorem 6.3.15) hence is integrally closed (theorem 4.1.9). Since we assumed that the characteristic of k is zero, the extension $E_0 \subset F$ is separable. Consequently, proposition 9.2.1 implies that B is a finitely generated A_0-module, hence a finitely generated A-module. □

9.3. Dimension and Transcendence Degree

9.3.1. — Let A be a ring. A *chain of prime ideals* of A is a set of prime ideals of A which is totally ordered by inclusion.

As in general ordered sets (see §A.1.4.4), the elements of a non-empty finite chain of prime ideals of A can be uniquely enumerated as a finite sequence $(P_0, \ldots, P_n)$ of prime ideals of A such that $P_0 \subsetneq P_1 \cdots \subsetneq P_n$. The length of such a chain is then defined as n.

In the sequel, we will not distinguish between the set $\{P_0, \ldots, P_n\}$ and the sequence $(P_0, \ldots, P_n)$. With this notation, a chain $(P_0, \ldots, P_n)$ is *maximal* if it cannot be extended by inserting another prime ideal, either before P_0, between P_{k-1} and P_k for any $k \in \{1, \ldots, n\}$, or after P_n.

Definition (9.3.2). — Let A be a ring.

(i) The *dimension* of A, denoted $\dim(A)$, is the supremum of the lengths of finite chains of prime ideals of A.

(ii) Let P be a prime ideal of A. The *height* of P is the supremum, denoted by $ht_A(P)$, of the lengths of finite chains of prime ideals of A which are contained in P.

Let us assume that $A \neq 0$. Since every non-empty finite chain has a largest element, which is a prime ideal of A, one has

$$\dim(A) = \sup_{P \in \mathrm{Spec}(A)} ht_A(P).$$

The zero ring has no prime ideal, hence the only chain of prime ideals is empty, with length -1, which gives $-1 = \dim(0)$. On the other hand, the right-hand side of the preceding formula is either 0 (if the supremum is taken in $\mathbf{N} \cup \{+\infty\}$), or $-\infty$ (if it is taken in $\mathbf{Z} \cup \{\pm\infty\}$).

Remark (9.3.3). — Without any hypotheses, the height of a prime ideal, or the dimension of a ring, may be infinite. However, the main results of dimension theory assert that the dimension of a finitely generated algebra over a field is finite (theorem 9.3.11) and that the dimension of a local noetherian ring is finite (theorem 9.4.3). However, there exist noetherian rings of infinite dimension (exercise 9.7.14): while the noetherian hypothesis forbids the existence of infinite strictly increasing sequences of prime ideals, it does not prevent the existence of arbitrary long such sequences.

Example (9.3.4). — One has $ht_A(P) = 0$ if and only if P is a minimal prime ideal of A. If A is a field, then $\dim(A) = 0$. More generally, Akizuki's theorem (theorem 6.4.14) implies that the dimension of an artinian ring is equal to 0.

Example (9.3.5) (Principal ideal domains). — *If A is a principal ideal domain which is not a field, then* $\dim(A) = 1$. *In particular,* $\dim(\mathbf{Z}) = 1$ *and* $\dim(K[X]) = 1$ *for any field K.*
    The prime ideals of A are (0), and the maximal ideals (p), where p ranges among all irreducible elements of A. One has $ht_A((0)) = 0$. Moreover, the chains of prime ideals beginning with (0) are (0) and $((0), (p))$, where p is an irreducible element of A, so that $ht_A((p)) = 1$. This shows that $\dim(A) = 1$.

Example (9.3.6). — Recall that the map $Q \mapsto QA_P$ induces a bijection between the prime ideals of A contained in P and the prime ideals of A_P, and this bijection respects the inclusion relation. Consequently, $ht_A(P) = \dim(A_P)$.

Example (9.3.7). — *Let A be a unique factorization domain. A prime ideal of A has height 1 if and only if it is generated by an irreducible element of A.*
    Let P be a prime ideal of A such that $ht_A(P) = 1$. In particular, $P \neq 0$; let thus f be any non-zero element of P. Since P is prime, f is not a unit in A and one of its irreducible factors, say a, must belong to P. Since A is a unique factorization domain, the ideal (a) is prime and one has inclusions $(0) \subsetneq (a) \subset P$. The hypothesis that $ht(P) = 1$ implies the equality $P = (a)$.
    Conversely, let a be an irreducible element of A, so that (a) is a prime ideal of A. The strict inclusion $(0) \subsetneq (a)$ implies that $ht((a)) \geq 1$. Let P be a

non-zero ideal of A contained in (a). Let $x \in P$ be a non-zero element whose number of irreducible factors is minimal. Since $P \subset (a)$, there exists a $y \in A$ such that $x = ay$; then the number of irreducible factors of y is one less than the number of irreducible factors of x, so that $y \notin P$. Since P is prime, this implies that $a \in P$, hence $P = (a)$. This shows that $\mathrm{ht}((a)) = 1$.

Remark (9.3.8) (Geometrical interpretations). — *a*) Let K be an algebraically closed field, let I be an ideal of $K[X_1, \ldots, X_n]$ and let $A = K[X_1, \ldots, X_n]/I$. Any prime ideal of A defines an irreducible algebraic set contained in $\mathcal{Z}(I)$, and conversely (see exercise 2.8.10). Consequently, the dimension of A is the supremum of the lengths of chains of irreducible algebraic sets contained in $\mathcal{Z}(I)$.

b) Let A be an arbitrary ring. Irreducible closed subsets of $\mathrm{Spec}(A)$ are subsets of the form $V(P)$, for some prime ideal P of A, and conversely (see exercise 2.8.8). Consequently, $\dim(A)$ is the supremum of the lengths of chains of irreducible closed subsets of $\mathrm{Spec}(A)$. For any prime ideal P of A, $\mathrm{ht}(P)$ is the supremum of the lengths of those chains of closed subsets which contain $V(P)$ or, equivalently, which contain the point $P \in \mathrm{Spec}(A)$. This is interpreted as the *codimension* of $V(P)$ in $\mathrm{Spec}(A)$.

c) Let K be a field. Generalizing the equality $\dim(K[X]) = 1$ from example 9.3.5, we will prove in theorem 9.3.11 below that $\dim(K[X_1, \ldots, X_n]) = n$. The chain of prime ideals

$$(0) \subsetneq (X_1) \subsetneq (X_1, X_2) \subsetneq \cdots \subsetneq (X_1, \ldots, X_n)$$

shows that $\dim(K[X_1, \ldots, X_n]) \geq n$, but the opposite inequality is more difficult.

In order to relate dimension and transcendence degrees of K-algebras, we first study the behavior of dimension with respect to integral extensions of rings.

Theorem (9.3.9) (Going up theorem of Cohen–Seidenberg). — *Let B be a ring and let A be a subring of B. Assume that B is integral over A.*

(i) *Let Q be a prime ideal of B and let $P = Q \cap A$. Then P is a maximal ideal of A if and only if Q is a maximal ideal of B.*

(ii) *Let $Q \subset Q'$ be prime ideals of B such that $Q \cap A = Q' \cap A$. Then $Q = Q'$.*

(iii) *The canonical map from $\mathrm{Spec}(B)$ to $\mathrm{Spec}(A)$ is surjective: for every prime ideal P of A, there exists a prime ideal Q of B such that $Q \cap A = P$.*

Proof. — *a*) Passing to the quotients, one gets an integral extension of integral domains $A/P \subset B/Q$. By proposition 4.2.5, A/P is a field if and only if B/Q is a field; in other words, P is maximal in A if and only if Q is maximal in B.

b) Let $P = Q \cap A$ and let us consider the integral extension of rings $A_P \subset B_P$ induced by localization by the multiplicative subset $A - P$ (it is indeed injective by proposition 3.6.6, and integral by lemma 4.2.4). The ideal $A_P \cap QB_P$ of A_P contains the maximal ideal PA_P of A_P, but does not contain 1, hence it is equal to PA_P. Then, *a*) implies that QB_P is a maximal ideal of B_P.

Since $Q \cap A = P = Q' \cap A$, the same arguments imply that $Q'B_P$ is a maximal ideal of B_P. However, the inclusion $Q \subset Q'$ implies an inclusion $QB_P \subset Q'B_P$. One thus has $QB_P = Q'B_P$.

Since localization induces a bijection from the set of prime ideals of B disjoint from $A - P$ to the set of prime ideals of B_P, one gets $Q = Q'$.

c) Let P be a prime ideal of A and let us consider the integral extension $A_P \subset B_P$ of localized rings. Let M be a maximal ideal of B_P. By a), $M \cap A_P$ is a maximal ideal of A_P, hence $M \cap A_P = PA_P$, and $P \subset M \cap A$. There exists a unique prime ideal Q of B such that $Q \cap (A - P) = \emptyset$ and $M = QB_P$. The intersection $Q \cap A$ is a prime ideal of A, contained in P by construction; by what precedes, it contains P, hence $Q \cap A = P$. This concludes the proof of the theorem of Cohen–Seidenberg. □

Corollary (9.3.10). — *Let* B *be a ring and let* A *be a subring of* B. *If* B *is integral over* A, *then* $\dim(A) = \dim(B)$.

Proof. — Let $(Q_0, \ldots, Q_n)$ be a chain of prime ideals of B. Let us intersect these ideals with A; this gives an increasing family $(Q_0 \cap A, \ldots, Q_n \cap A)$ of prime ideals of A. By part b) of theorem 9.3.9, this is even a chain of prime ideals, so that $\dim(A) \geq \dim(B)$.

Conversely, let $P_0 \subsetneq \cdots \subsetneq P_n$ be a chain of prime ideals of A. For each $m \in \{0, \ldots, n\}$, let us construct by induction a prime ideal Q_m of B such that $Q_m \cap A = P_m$ and such that $Q_0 \subset \cdots \subset Q_n$. This will imply that $\dim(B) \geq \dim(A)$, hence the corollary.

By part c) of theorem 9.3.9, there exists a prime ideal Q_0 of B such that $Q_0 \cap A = Q_0$. Assume $Q_0, \ldots, Q_m$ are defined. Let us consider the integral extension $A/P_m \subset B/Q_m$ of integral domains. By theorem 9.3.9, c), applied to the prime ideal P_{m+1}/P_m of A/P_m, there exists a prime ideal Q of the ring B/Q_m such that $Q \cap (A/P_m) = P_{m+1}/P_m$. Then, there exists a prime ideal Q_{m+1} containing Q_m such that $Q = Q_{m+1}/Q_m$. Moreover, $Q_{m+1} \cap A = P_{m+1}$. □

The next theorem is at the basis of a dimension theory in algebraic geometry.

Theorem (9.3.11). — *Let* K *be a field and let* A *be a finitely generated* K-*algebra. Assume that* A *is an integral domain and let* F *be its field of fractions. Then* $\dim(A) = \operatorname{tr} \deg_K(F)$.

Proof. — We prove the theorem by induction on the transcendence degree of F.

If $\operatorname{tr} \deg_K(F) = 0$, then A is integral over K; being a finitely generated K-algebra, it is finite-dimensional, hence a field. One thus has $\dim(A) = 0$.

Now assume that the theorem holds for finitely generated K-algebras whose field of fractions has transcendence degree strictly less than $\operatorname{tr} \deg_K(F)$.

By Noether's normalization theorem (theorem 9.1.1), there exists an integer $n \geq 0$ and elements $a_1, \ldots, a_n$ of A such that the unique morphism of algebras f from $K[X_1, \ldots, X_n]$ to A such that $f(X_i) = a_i$ is injective, and such that A is integral over the subring $B = K[a_1, \ldots, a_n] = f(K[X_1, \ldots, X_n])$. Moreover, $n = \operatorname{tr} \deg_K(F)$. By corollary 9.3.10, it suffices to prove that the dimension of the polynomial ring $K[X_1, \ldots, X_n]$ is equal to n.

We already stated in remark 9.3.8 that $\dim(K[X_1, \ldots, X_n]) \geq n$, in view of the chain $((0), (X_1), (X_1, X_2), \ldots, (X_1, \ldots, X_n))$ of prime ideals of $K[X_1, \ldots, X_n]$.

Conversely, let $((0), P_1, \ldots, P_m)$ be a chain of prime ideals of $K[X_1, \ldots, X_n]$ and let us set $A' = K[X_1, \ldots, X_n]/P_1$. Since P_1 is a prime ideal, the ring A' is an integral domain; let F' be its field of fractions. Moreover, A' is a finitely generated K-algebra and $\dim(A') \geq m - 1$, because $((0), P_2/P_1, \ldots, P_m/P_1)$ is a chain of prime ideals of A' of length $m - 1$. On the other hand, any non-zero polynomial $f \in P_1$ furnishes a non-trivial algebraic dependence relation between the classes $x_1, \ldots, x_n$ of $X_1, \ldots, X_n$ in A'. Consequently, $\operatorname{tr} \deg_K(F') \leq n - 1$. (See also example 4.8.11.) By induction, one has $\operatorname{tr} \deg_K(F') = \dim(A')$, hence $m - 1 \leq n - 1$, and $m \leq n$. This concludes the proof. $\square$

In the course of the proof of the theorem, we established the following particular case.

Corollary (9.3.12). — *For any field K, one has* $\dim(K[X_1, \ldots, X_n]) = n$.

9.4. Krull's Hauptidealsatz and Applications

By theorem 9.3.11, the dimension of a finitely generated algebra over a field is finite. One might think that the same property holds for noetherian rings: after all, the dimension of a ring involves strictly increasing sequences of prime ideals, and the noetherian property tells us that every such sequence is finite. However, it does not: although there are no *infinite* strictly increasing sequences of prime ideals, there may be such sequences of arbitrarily large length.

One of the main consequences of this section is that *local* noetherian rings are finite-dimensional. Stated differently, the height of any prime ideal of a noetherian ring is finite.

Since any such prime ideal is generated by a finite set (this is the definition of a noetherian ring!), it is natural to investigate the behavior of dimension when one quotients a ring by a principal ideal. Geometrically, this will amount to understanding the dimension of a hypersurface.

Theorem (9.4.1) (Krull's Hauptidealsatz). — *Let A be a noetherian ring, let $a \in A$ and let P be a prime ideal of A, minimal among those containing a.*

(i) $\operatorname{ht}(P) \leq 1$.

(ii) *If, moreover, a is regular, then* $\operatorname{ht}(P) = 1$.

Proof. — *a)* We need to prove that that there is no chain (Q', Q, P) of prime ideals in A. So let Q', Q, P be prime ideals of A such that $Q' \subset Q \subsetneq P$ and let us prove that $Q = Q'$.

On Wolfgang Krull

Wolfgang KRULL (1899–1971) was a German mathematician who brought many important contributions to commutative algebra.

After STEINITZ's 1910 paper on abstract fields and NOETHER's 1921 paper on the primary decomposition in abstract "noetherian" rings, the development of abstract commutative algebra could start, and KRULL places himself explicitly in this lineage.

The "generalized abelian groups" that KRULL introduced in 1925 are essentially modules; generalizing results of REMAK and SCHMIDT, he proved a unique decomposition theorem under the assumption that they have finite length (see exercise 6.7.10).

While they are necessary for results such as primary decomposition, a general ring theory needs to be freed of chain conditions. In 1929, KRULL proved that every (non-trivial) ideal is contained in a *maximal* ideal ("highest prime ideal", in his terminology).

In 1928, he generalized Galois theory to an infinite algebraic extension $K \subset L$, assumed to be normal and separable. The Galois group is still the automorphism group of L as a K-algebra but it is now endowed with a topology, the Galois correspondence being between sub-extensions of L and *closed* subgroups of the Galois group.

In 1928 again, KRULL invented the dimension theory of rings (definition 9.3.2), and proved the major results about it: *Hauptidealsatz* (theo-

Photograph of Wolfgang Krull (1969), on the occasion of the 7th Brazilian Mathematics Colloquium in Poços de Caldas (Brazil).

Photographer: Paul Halmos
Source: *Who's That Mathematician?*
Paul R. Halmos Collection,
Mathematical Association of America.

ERGEBNISSE DER MATHEMATIK
UND IHRER GRENZGEBIETE
HERAUSGEGEBEN VON DER SCHRIFTLEITUNG
DES
„ZENTRALBLATT FÜR MATHEMATIK"
VIERTER BAND

———— 3 ————

IDEALTHEORIE

VON

W. KRULL

BERLIN
VERLAG VON JULIUS SPRINGER
1935

rem 9.4.1) and the characterization of the height of a prime ideal by "systems of parameters" (theorem 9.4.3). His main tool was the primary decomposition. These results, accomplishing for dimension theory what NOETHER had done for primary decomposition, were at the heart of his 1935 monograph *Idealtheorie*.

He completed these results in his 1938 paper *Dimensionstheorie in Stellenringen* ("Dimension theory in local rings"), where he defines local rings (assumed to be noetherian) and regular rings (exercise 9.7.15) and where he proves the first important results about them, notably the *intersection theorem* (6.6.10). In that proof, KRULL also introduces a particular case of what is usually called "Nakayama's lemma".

Further results would need to wait the end of the 1950s, with the introduction by SERRE of homological methods in commutative algebra.

Among other important concepts invented by KRULL are the notions of a general valuation (1932), which is however not addressed in this book. He began the study of *étale* extensions of rings, of which ABHYANKAR and GROTHENDIECK would show the relevance for algebraic geometry. He also introduced the notion of a *Jacobson ring* (1951).

We first pass to the quotient by Q': in the noetherian ring A/Q', we have the prime ideals $0 \subset Q/Q' \subsetneq P/Q'$. Moreover, the prime ideal P/Q' of A/Q' is minimal among the prime ideals of A/Q' containing the class of a. This allows us to assume that A is an integral domain and $Q' = 0$.

We then replace A by its localization at the prime ideal P. The ring A_P is a noetherian local ring, with inclusions $0 \subset QA_P \subsetneq PA_P$ of prime ideals, and the maximal ideal PA_P is minimal among the prime ideals of A_P containing $a/1$.

We are thus reduced to the following particular case: the ring A is an integral domain, local, noetherian, its maximal ideal P is the only prime ideal containing the element a, the ideal Q of A is prime and distinct from P, and we need to prove that $Q = 0$.

For every integer $n \geq 1$, let $Q_n = Q^n A_Q \cap A$. By definition of localization, Q_n is the set of elements $x \in A$ such that there exists an element $y \in A - Q$ such that $xy \in Q^n$.

The sequence $(Q_n)_n$ is decreasing; let us prove that it is stationary.

The ring A/aA is noetherian, and since its prime ideals are in bijection with the prime ideals of A containing a, $P(A/aA)$ is its only prime ideal. Consequently (see theorem 6.4.14), A/aA is an artinian ring. In particular, the decreasing sequence $(Q_n(A/aA))_n$ of ideals of A/aA is stationary and there exists an integer n_0 such that

$$Q_n + aA = Q_{n+1} + aA$$

for every integer $n \geq n_0$.

Let n be an integer $\geq n_0$ and let $x \in Q_n$. There exists an element $y \in A$ such that $x + ay \in Q_{n+1}$; since $Q_{n+1} \subset Q_n$, we have $ay \in Q_n$, hence there exists an element $z \notin Q$ such that $ayz \in Q^n$. Since $a \notin Q$, the product az does not belong to Q, so that y belongs to Q_n. (We could have used the result of exercise 6.7.37 that Q_n is a Q-primary ideal.) Consequently, $x \in aQ_n + Q_{n+1}$, hence $x \in Q_{n+1} + PQ_n$. This implies the equality of ideals

$$Q_n = Q_{n+1} + PQ_n.$$

Since the ring A is noetherian, Q_n is finitely generated and Nakayama's lemma (corollary 6.1.2, observe that P is the Jacobson radical of A) implies that $Q_n = Q_{n+1}$ for all $n \geq n_0$.

Then,

$$Q^n A_Q = Q_n A_Q = Q_{n+1} A_Q = Q^{n+1} A_Q = Q \cdot Q^n A_Q.$$

Again, A_Q is noetherian and $Q^n A_Q$ is finitely generated; consequently, Nakayama's lemma (theorem 6.1.1) implies that $Q^n A_Q = 0$. Since A is an integral domain, this implies that $Q = 0$, as was to be shown.

b) If a is regular, then it does not belong to any associated prime ideal of A (theorem 6.5.8, a)), hence no minimal prime ideal of A contains a. Since $a \in P$, the prime ideal P is not minimal and $\mathrm{ht}_A(P) > 0$. By part a), the height of P is equal to 1. $\qquad\square$

Corollary (9.4.2). — *Let* A *be a noetherian integral domain. Then* A *is a unique factorization domain if and only if every prime ideal of height 1 is a principal ideal.*

Proof. — We have already proved in example 9.3.7 that if A is a unique factorization domain, then its prime ideals of height 1 are principal ideals (generated by irreducible elements).

Let us now assume that every prime ideal of A which is of height 1 is a principal ideal and let us prove that A is a unique factorization domain. Since A is noetherian, every increasing sequence of principal ideals of A is stationary, so that it suffices to prove that irreducible elements generate prime ideals. Let $p \in A$ be an irreducible element and let P be a prime ideal of A, minimal among those containing p. Since A is a domain and $p \neq 0$, theorem 9.4.1 implies that $\mathrm{ht}_A(P) = 1$. By assumption, the ideal P is principal, hence there exists an $a \in A$ such that $P = (a)$. The inclusion $(p) \subset P = (a)$ implies that a divides p; let $b \in A$ be such that $p = ab$. Since P is prime, a is not a unit. By definition of an irreducible element, b is a unit and $P = (a) = (p)$. This concludes the proof of the corollary.   □

Theorem (9.4.3). — *Let* A *be a noetherian ring and let* P *be a prime ideal of* A.

(i) *The height* $\mathrm{ht}_A(P)$ *of* P *is finite.*

(ii) *More precisely,* $\mathrm{ht}_A(P)$ *is the least integer n such that there exist* $a_1, \ldots, a_n \in P$ *such that the prime ideal* P *is minimal among those containing* $a_1, \ldots, a_n$.

Proof. — a) We first prove the following statement by induction on n: Let A be a noetherian ring, let $a_1, \ldots, a_n \in A$ and let P be a prime ideal which is minimal among those containing $a_1, \ldots, a_n$, then $\mathrm{ht}(P) \leq n$. (Observe that for $n = 1$, this is exactly Krull's Hauptidealsatz, theorem 9.4.1.)

The assertion holds if $\mathrm{ht}(P) = 0$ (trivially), or if $n = 0$ (because P is then a minimal prime ideal of A, hence $\mathrm{ht}(P) = 0$). Let us thus assume that $\mathrm{ht}(P) > 0$ and $n > 0$. In the fraction ring A_P, the ideal PA_P is the maximal ideal, and is minimal among those containing $a_1/1, \ldots, a_n/1$; moreover, $\mathrm{ht}_A(P) = \mathrm{ht}_{A_P}(PA_P)$. We may thus assume that A is a local ring with maximal ideal P. By the definition of the height of a prime ideal, to establish that $\mathrm{ht}_A(P) \leq n$, we need to prove that for any prime ideal Q of A such that $Q \subsetneq P$, one has $\mathrm{ht}_A(Q) \leq n - 1$. Since A is noetherian, there exists a prime ideal Q' such that $Q \subset Q' \subsetneq P$, and which is maximal among those ideals. Since one has $\mathrm{ht}_A(Q) \leq \mathrm{ht}_A(Q')$, it suffices to prove that $\mathrm{ht}_A(Q') \leq n - 1$. We may thus assume that there is no prime ideal Q' in A such that $Q \subsetneq Q' \subsetneq P$.

Since $Q \subsetneq P$, the definition of P implies that there exists an integer i such that $a_i \notin Q$. To fix the notation, let us assume that $a_1 \notin Q$. Then $Q \subsetneq Q + (a_1) \subset P$, so that P is a minimal prime ideal which contains $Q + (a_1)$, hence is the unique prime ideal containing $Q + (a_1)$.

Consequently, every element of P is nilpotent modulo $Q + (a_1)$. In particular, there exist an integer $m \geq 1$ and elements $x_2, \ldots, x_n \in A$, $y_2, \ldots, y_n \in Q$, such that

$$a_2^m = a_1 x_2 + y_2, \ldots, a_n^m = a_1 x_n + y_n.$$

These relations show that any prime ideal of A which contains a_1 and $y_2, \ldots, y_n$ also contains $a_2, \ldots, a_n$, hence contains P. In other words, the quotient $P/(y_2, \ldots, y_n)$ is a prime ideal of the quotient ring $A/(y_2, \ldots, y_n)$ which is minimal among those containing the class of a_1. By Krull's Hauptidealsatz (theorem 9.4.1), the height of $P/(y_2, \ldots, y_n)$ in $A/(y_2, \ldots, y_n)$ is ≤ 1. Since $(y_2, \ldots, y_n) \subset Q \subsetneq P$, the ideal $Q/(y_2, \ldots, y_n)$ is then a minimal prime ideal of $A/(y_2, \ldots, y_n)$. Equivalently, Q is a minimal prime ideal of A among those containing $(y_2, \ldots, y_n)$. By induction, $\mathrm{ht}_A(Q) \leq n - 1$, as was to be shown.

b) It follows that the height of any prime ideal P of A is finite. Indeed, since A is noetherian, the ideal P is finitely generated and there exists $a_1, \ldots, a_n \in A$ such that $P = (a_1, \ldots, a_n)$. By the part of the proof already shown, one has $\mathrm{ht}_A(P) \leq n$.

c) We now prove by induction on n the following statement: *Let A be noetherian ring, let P be a prime ideal of A and let $n = \mathrm{ht}_A(P)$. There exist elements $a_1, \ldots, a_n \in P$ such that P is the minimal prime ideal containing $(a_1, \ldots, a_n)$.*

This holds if $n = 0$, for P is then a minimal prime ideal.

Assume that $n = \mathrm{ht}_A(P) > 0$ and that the result holds for every prime ideal of A of height $< n$. Since $n > 0$, P is not contained in any minimal prime ideal of A. Since the set of minimal prime ideals of A is finite, the Prime avoidance lemma 2.2.12 implies that there exists an element $a_1 \in P$ which does not contain any minimal prime ideal of A. Consequently, the height of any prime ideal of A which is minimal among those containing a_1 is strictly positive. Thus, any chain of prime ideals of A containing a_1 and contained in P can be extended by one of the minimal prime ideals of A, so that the height of the prime ideal $P/(a_1)$ of the ring $A/(a_1)$ is at most $\mathrm{ht}_A(P) - 1 = n - 1$. By induction, there exist $a_2, \ldots, a_n \in P$ such that the height of the prime ideal $P/(a_1, \ldots, a_n)$ of the ring $A/(a_1, \ldots, a_n)$ is zero. This proves that P is a minimal prime ideal among those containing $(a_1, \ldots, a_n)$. This concludes the proof of the theorem.   □

Corollary (9.4.4). — *Let A be a noetherian local ring. Then* $\dim(A)$ *is finite. More precisely,* $\dim(A)$ *is the least integer n such that there exist $a_1, \ldots, a_n$ such that $\sqrt{(a_1, \ldots, a_n)}$ is the maximal ideal of A.*

Proof. — If A is local, every maximal chain of prime ideals ends at its maximal ideal M, hence $\dim(A) = \mathrm{ht}_A(M)$. The corollary thus follows from theorem 9.4.3 applied to M.   □

9.5. Heights and Dimension

Proposition (9.5.1). — *Let A be an integral domain, let E be its field of fractions and let $E \to F$ be a finite normal extension of fields. Assume that A is integrally closed in E and let B the integral closure of A in F.*

For every prime ideal P of A, the group $\mathrm{Aut}(F/E)$ *acts transitively on the set of prime ideals Q of B such that $Q \cap A = P$.*

Proof. — Let G = Aut(F/E). It is a finite group of automorphisms of F; the subfield F^G of F is a radicial extension of E (corollary 4.6.9) and $F^G \subset F$ is a Galois extension of group G (Artin's lemma 4.6.4).

First of all, according to the going up theorem (theorem 9.3.9), the set of prime ideals Q of B such that $Q \cap A = P$ is non-empty.

Let Q and Q′ be two prime ideals of B such that $Q \cap A = Q' \cap A = P$.

Let $x \in Q'$ and let $y = \prod_{\sigma \in G} \sigma(x)$; since $y \in F^G$ and since the extension $E \to F^G$ is radicial, there exists an integer $q \geq 1$ such that $y^q \in E$ (lemma 4.4.16; one has $q = 1$ if the characteristic of E is zero).

By definition of B, x is integral over A, as well as its images $\sigma(x)$, for $\sigma \in G$. Consequently, y is integral over A, hence y^q is integral over A. Since $y^q \in E$ and A is integrally closed in E, one has $y^q \in A$. Since $x \in Q'$, it follows that $y^q \in Q' \cap A = P$. Using that $P = Q \cap A$, we deduce that $y^q \in Q$. Finally, since Q is a prime ideal, we conclude that $y \in Q$.

Moreover, since $y = \prod_{\sigma \in G} \sigma(x)$, there exists an element $\sigma \in G$ such that $\sigma(x) \in Q$. Since G is a group, this gives the inclusion $Q' \subset \bigcup_{\sigma \in G} \sigma(Q)$.

Let $\sigma \in G$. Let us show that $\sigma(Q)$ is a prime ideal of B. Indeed, if $b \in F$, then b and $\sigma(b)$ satisfy the same polynomial relations with coefficients in E. In particular, b is integral over A if and only if $\sigma(b)$ is integral over A. This implies that $\sigma(B) = B$, hence $\sigma(Q)$ is a prime ideal of B.

By the prime avoidance lemma (lemma 2.2.12), there exists a $\sigma \in G$ such that $Q' \subset \sigma(Q)$. Then $P = Q' \cap A \subset \sigma(Q) \cap A = Q \cap A = P$, hence $Q' \cap A = \sigma(Q) \cap A$. By the Cohen–Seidenberg going up theorem (theorem 9.3.9), we thus have $Q' = \sigma(Q)$. □

Theorem (9.5.2) (Going down theorem of Cohen–Seidenberg). — *Let B be an integral domain and let A be a subring of B. Assume that A is integrally closed in its field of fractions and that B is integral over A.*

Let $P_0, \ldots, P_n$ be prime ideals of A such that $P_0 \subset \ldots P_n$ and let Q_n be a prime ideal of B such that $Q_n \cap A = P_n$. Then, there exist prime ideals $Q_0, \ldots, Q_{n-1}$ in B such that $Q_0 \subset \cdots \subset Q_n$ and $Q_m \cap A = P_m$ for every $m \in \{0, \ldots, n\}$.

Proof. — Let E be the field of fractions of A, let F be the field of fractions of B and let F′ be a finite normal extension of E containing F. Let B′ be the integral closure of A in F′. According to theorem 9.3.9, there exist prime ideals Q'_m in B′ (for $0 \leq m \leq n$), such that $Q'_0 \subset \cdots \subset Q'_n$ and $P_m = Q'_m \cap A$ for every $m \in \{0, \ldots, n\}$.

Let Q′ be a prime ideal of B′ such that $Q' \cap B = Q_n$. Since A is integrally closed in its field of fractions, proposition 9.5.1 asserts the existence of an automorphism $\sigma \in \mathrm{Aut}(F'/E)$ such that $\sigma(Q'_n) = Q'$. For every integer $m \in \{0, \ldots, n-1\}$, let $Q_m = \sigma(Q'_m) \cap B$. By construction, $(Q_0, \ldots, Q_{n-1})$ is an increasing family of prime ideals of B contained in Q_n. Moreover, for every $m \in \{0, \ldots, n-1\}$,

$$Q_m \cap A = \sigma(Q'_m) \cap B \cap A = \sigma(Q'_m \cap A) = \sigma(P_m) = P_m.$$

The theorem is proved. □

Corollary (9.5.3). — *Let* B *be an integral domain and let* A *be a subring of* B. *Assume that* A *is integrally closed in its field of fractions and that* B *is integral over* A. *Then for every prime ideal* Q *of* B, $\mathrm{ht}_B(Q) = \mathrm{ht}_A(Q \cap A)$.

Proof. — Let $P = Q \cap A$. Let $(Q_0, \ldots, Q_n)$ be a chain of prime ideals of B, contained in Q. By theorem 9.3.9, their intersections with A, $(Q_0 \cap A, \ldots, Q_n \cap A)$, form a chain of prime ideals of A contained in P. This implies that $\mathrm{ht}_A(P) \geq \mathrm{ht}_B(Q)$.

Conversely, let $(P_0, \ldots, P_n)$ be a chain of prime ideals of A contained in P. By theorem 9.5.2, there exists a family $(Q_0, \ldots, Q_n)$ of prime ideals of B such that $Q_0 \subset \cdots \subset Q_n \subset Q$ and $Q_m \cap A = P_m$ for every $m \in \{0, \ldots, n\}$. By theorem 9.3.9, $(Q_0, \ldots, Q_n)$ is a *chain* of prime ideals in B. Consequently, $\mathrm{ht}_B(Q) \geq \mathrm{ht}_A(P)$.

This concludes the proof of the corollary.   □

Theorem (9.5.4). — *Let* K *be a field and let* A *be a finitely generated* K-*algebra. Assume that* A *is an integral domain. Then, for every prime ideal* P *of* A,

$$\mathrm{ht}_A(P) = \dim(A) - \dim(A/P).$$

In particular, $\mathrm{ht}_A(M) = \dim(A)$ *for every maximal ideal* M *of* A.

Proof. — Let $(P_0, \ldots, P_n)$ be a chain of prime ideals of A contained in P. Let also $(Q_0, \ldots, Q_m)$ be a chain of prime ideals of A/P. For every integer $i \in \{0, \ldots, m\}$, let Q'_i be the preimage of Q_i in A by the canonical surjection from A to A/P. Then $(Q'_0, \ldots, Q'_m)$ is a chain of prime ideals of A containing P. Now observe that $(P_0, \ldots, P_n, Q'_1, \ldots, Q'_m)$ is chain of prime ideals of A — the inclusions $P_n \subset P \subset Q'_0 \subsetneq Q'_1$ establish the strict inclusion $P_n \subsetneq Q'_1$, and the others are obvious. We thus have shown the inequality $\dim(A) \geq \mathrm{ht}_A(P) + \dim(A/P)$.

The converse inequality is more delicate and we prove it by induction on $\dim(A)$. It clearly holds if $\dim(A) = 0$, since then A is a field and $P = (0)$. Let us assume that it holds for all prime ideals of a finitely generated K-algebra of dimension $< \dim(A)$ which is an integral domain.

Let us first apply Noether's normalization theorem: there exist an integer n and elements $a_1, \ldots, a_n \in A$, algebraically independent over K, such that A is integral over the subring $B = K[a_1, \ldots, a_n]$, and $\dim(A) = \dim(B) = n$. Let $Q = P \cap B$. Since the natural morphism from B/Q to A/P is injective and integral, one also has $\dim(A/P) = \dim(B/Q)$. Moreover, theorem 9.5.2 asserts that $\mathrm{ht}_A(P) = \mathrm{ht}_B(Q)$. Consequently, we may assume that $A = B$, hence that $A = K[X_1, \ldots, X_n]$.

If $P = (0)$, then $\mathrm{ht}_A(P) = 0$, $A \simeq A/P$, so that the equality $\mathrm{ht}_A(P) + \dim_A(A/P) = \dim(A)$ holds.

Let us assume that $P \neq (0)$. Then, since it is a prime ideal, P contains an irreducible element, say f. By example 9.3.7, the ideal (f) has height 1. Then set $A' = A/(f)$ and $P' = P/(f)$. The K-algebra A' is finitely generated; it is an integral domain and, by example 4.8.11, its field of fractions has transcendence degree $n - 1$ over K. By theorem 9.3.11,

$$dim(A') = n - 1 = dim(A) - 1.$$

Moreover, A'/P' is isomorphic to A/P, so that

$$dim(A/P) = dim(A'/P').$$

On the other hand, a chain of length m of prime ideals of A' contained in P' corresponds to a chain of prime ideals of A contained in P and containing (f); adjoining (0), this gives a chain of length $m+1$ of prime ideals of A contained in P. Consequently,

$$ht_{A'}(P') + 1 \le ht_A(P).$$

Finally, the induction hypothesis asserts that

$$ht_{A'}(P') + dim(A'/P') = dim(A').$$

Combining these four relations, we obtain

$$ht_A(P) + dim(A/P) \ge ht_{A'}(P') + 1 + dim(A'/P')$$
$$\ge dim(A') + 1 = dim(A).$$

This concludes the proof of the theorem. $\qquad\square$

Corollary (9.5.5). — *Let* K *be a field and let* A *be a finitely generated* K-*algebra which is an integral domain. Every maximal chain of prime ideals of* A *has length* dim(A).

Such rings, in which all maximal chains of prime ideals have the same length, are called *catenary*.

Proof. — Let $(P_0, P_1, \ldots, P_n)$ be a maximal chain of prime ideals of A. This means that this chain cannot be extended by inserting prime ideals (which is a different assertion than asserting that this sequence has maximal length). In particular, one has $P_0 = (0)$ because A is an integral domain, and P_n is a maximal ideal.

Let us now argue by induction on n. If $n = 0$, then P_0 is a maximal ideal, hence A is a field and $dim(A) = 0$.

Let us now assume that $n \ge 1$. Since the chain (P_0, P_1) is maximal among all chains terminating at P_1, we have $ht_A(P_1) = 1$. The K-algebra A/P_1 is finitely generated and an integral domain, and $(P_1/P_1, \ldots, P_n \subset P_1)$ is a maximal chain of prime ideals of A/P_1. By induction, one has $n - 1 = dim(A/P_1)$. Consequently,

$$n = (n - 1) + 1 = dim(A/P_1) + ht_A(P_1) = dim(A),$$

by theorem 9.5.4. $\qquad\square$

9.6. Dedekind Rings

Definition (9.6.1). — Let A be an integral domain. One says that A is a *Dedekind ring* if every ideal of A is a projective A-module.

Example (9.6.2). —   (i) A field is a Dedekind ring.

(ii) Since non-zero principal ideals of an integral domain A are free A-modules of rank 1, *any principal ideal domain is a Dedekind ring.*

(iii) Let A be a Dedekind ring and let S be a multiplicative subset of A. Then $S^{-1}A$ is a Dedekind ring.

Indeed, let J be an ideal of $S^{-1}A$. There exists an ideal I of A such that $J = S^{-1}I$. Since A is a Dedekind ring, the A-module I is projective. Consequently, the $S^{-1}A$-module $S^{-1}I$ is projective (example 7.3.5).

9.6.3. — Let A be an integral domain and let K be its field of fractions.

The set $\mathscr{I}(A)$ of non-zero ideals of A is a commutative monoid with respect to multiplication of ideals. The subset $\mathscr{P}(A)$ of $\mathscr{I}(A)$ consisting of principal ideals is a submonoid. The quotient monoid $\mathscr{C}(A) = \mathscr{I}(A)/\mathscr{P}(A)$ is called the *class monoid* of A. This monoid is trivial if and only if every ideal of A is principal. As we will show, Dedekind rings are actually characterized by the property that $\mathscr{C}(A)$ is a group.

To study $\mathscr{C}(A)$, it is convenient to consider the more general setup of fractional ideals. By definition, a *fractional ideal* of A is a non-zero submodule of K of the form aI, where I is an ideal of A and $a \in K^\times$. Fractional ideals can be multiplied as ideals can be, and the set $\mathscr{I}_0(A)$ of fractional ideals is a monoid, with the set $\mathscr{P}_0(A)$ of principal fractional ideals as a submonoid.

The inclusion $\mathscr{I}(A) \to \mathscr{I}_0(A)$ induces a morphism of monoids from $\mathscr{C}(A)$ to $\mathscr{I}_0(A)/\mathscr{P}_0(A)$. Since the class of a fractional ideal aI, where I is an ideal of A, is equal to the class of I, this morphism is surjective. Let then I and J be non-zero ideals of A which have the same class in $\mathscr{I}_0(A)/\mathscr{P}_0(A)$; let $a, b \in K^\times$ be such that $aI = bJ$; write $a = a_1/a_2$ and $b = b_1/b_2$, with $a_1, a_2, b_1, b_2 \in A$; then $a_1 b_2 I = b_1 a_2 J$ and $[I] = [a_1 b_2 I] = [b_1 a_2 J] = [J]$ in $\mathscr{C}(A)$.

Finally, a fractional ideal of A is isomorphic, as an A-module, to an ideal of A, so that a ring A is a Dedekind ring if and only if every fractional ideal of A is projective.

9.6.4. — Let A be an integral domain and let K be its field of fractions. Let I be a fractional ideal of A. Writing I^{-1} for the set of elements $a \in K$ such that $aI \in A$, one has $I \cdot I^{-1} \subset A$.

The map $I^{-1} \to I^\vee$ given by $a \mapsto (x \mapsto ax)$ is a morphism of A-modules. It is injective, because $I \neq 0$. Let us show that it is surjective; let $\varphi \in I^\vee$. Let x be a non-zero element of I and let $a = \varphi(x)x^{-1}$. Let y be a non-zero element of I; there exist $u, v \in A - \{0\}$ such that $ux = vy$; then $u\varphi(x) = v\varphi(y)$, so that $\varphi(y) = uv^{-1}\varphi(x) = ay$. In particular, $ay \in A$ for every $y \in I$, and φ is the image of a.

Lemma (9.6.5). — *Let I be a fractional ideal of A. The following properties are equivalent:*

(i) $I \cdot I^{-1} = A$;

(ii) *The A-module I is projective;*

(iii) *The fractional ideal I is invertible in the monoid $\mathscr{I}_0(A)$;*

(iv) *The class of I is invertible in the class monoid $\mathscr{C}(A)$.*

If they hold, then the fractional ideal I is finitely generated.

If these properties hold, one simply says that the fractional ideal I is invertible.

Proof. — (i)⟹(ii). Since $I \cdot I^{-1} = A$, there exist elements $a_1, \ldots, a_n \in I$ and $b_1, \ldots, b_n \in I^{-1}$ such that $1 = \sum_{i=1}^n a_i b_i$. Then the morphism $\delta_{I,I} \colon I^\vee \otimes I \to \mathrm{End}_A(I)$ maps $\sum b_i \otimes a_i$ to id_I, so that I is finitely generated and projective.

(ii)⟹(i). Assume that I is projective. Let $(\varphi_i)_i$ and (a_i) be families in $I^\vee$ and I respectively such that $x = \sum \varphi_i(x) a_i$ for every $x \in I$ (remark 7.3.4). For every i, let $b_i \in I^{-1}$ be such that $\varphi_i(x) = b_i x$ for every $x \in I$. Since $I \neq 0$, this implies $\sum a_i b_i = 1$. Then $1 \in I \cdot I^{-1}$, so that $I \cdot I^{-1} = A$.

(i)⟹(iii). Assume that $I \cdot I^{-1} = A$. By definition, this says that I is invertible in $\mathscr{I}_0(A)$, with inverse I^{-1}.

(iii)⟹(iv). If I is invertible in the monoid $\mathscr{I}_0(A)$, then its class $[I]$ is invertible in the class monoid $\mathscr{C}(A)$.

(iv)⟹(i). Finally, assume that the class $[I]$ of I is invertible in $\mathscr{C}(A)$ and let J be a fractional ideal of A such that $I \cdot J$ is a principal fractional ideal aA. Replacing J by $a^{-1}J$, we may assume that $I \cdot J = A$. Then, $J \subset I^{-1}$, hence $A = I \cdot J \subset I \cdot I^{-1} \subset A$, so that $I \cdot I^{-1} = A$. This proves (i). □

Corollary (9.6.6). — *An integral domain A is a Dedekind ring if and only if the class monoid $\mathscr{C}(A)$ is a group.*

When I is an invertible ideal, one can define I^n for every integer n (positive or negative) by the formula $I^n = (I^{-1})^{-n}$ if $n < 0$. From the relation $I \cdot I^{-1} = A$, one deduces that $I^{n+m} = I^n \cdot I^m$ for every $m, n \in \mathbf{Z}$.

Proposition (9.6.7). — *Let A be a local integral domain. Then A is a Dedekind ring if and only if it is a principal ideal domain.*

Proof. — By example 9.6.2, principal ideal domains are Dedekind rings; it thus suffices to prove that a Dedekind ring A which is a local ring is principal. Let P be the maximal ideal of A.

Let I be an ideal of A; it is finitely generated because A is noetherian. If $I = PI$, then Nakayama's lemma (theorem 6.1.1) implies that $I = 0$. In particular, I is principal.

Otherwise, let $a \in I - PI$ and let us prove that $I = (a)$. Since $aA \subset I$, one has $aI^{-1} \subset A$. If $aI^{-1} \neq A$, then aI^{-1} is contained in the maximal ideal P of A, that is, $aI^{-1} \subset P$; this implies $aA = aI^{-1} \cdot I \subset PI$, a contradiction. Consequently, the ideal I is principal. □

On Richard Dedekind

Richard DEDEKIND (1831–1916) was a German mathematician. He was the last student of GAUSS, but it is DIRICHLET, who had been appointed at Göttingen at the death of GAUSS, who would have a strong influence on him. DEDEKIND already understood the importance of abstract concepts, and he has been the first at Göttingen to lecture on Galois theory.

While he had to teach differential and integral calculus, DEDEKIND felt the need to revisit the fundamental concepts of calculus, as he was unsatisfied with the presentation of the theorem that a bounded increasing sequence has a limit. In 1858, he thus defind a real number in terms of *Dedekind cuts* — representing a real number by the set of all rational numbers which are smaller than it. This was a bold step since "actual" infinite sets were not yet accepted at that time; GAUSS, for example, refused them and the papers of CANTOR only appeared at the beginning of the 1880s. He would publish these thoughts in 1872 in a small book *Continuity and irrational numbers*.

DEDEKIND already based all mathematical notions upon the concept of a set (the word he used was *system*). He understood the idea of a bijection and was the first to propose a definition of an infinite set: a set which is in

Portrait of Richard Dedekind (before 1886)
Photographer: Johannes Ganz
Source: ETH-Bibliothek, via Wikipedia
Public domain

bijection with a strict subset of itself. He also proposed an axiomatization of the integers which would be simplified shortly after by PEANO. This was the subject of his 1888 booklet *Was sind und was sollen die Zahlen?* ("What are numbers, and of what use are they?"), the preface of which starts with the very modern-sounding sentence:

> *In the sciences, what can be proved should not be believed without proof.*

His encounter in 1855 with DIRICHLET was pivotal for DEDEKIND's mathematical works. He published DIRICHLET's *Vorlesungen über Zahlentheorie*, (Lectures on number theory) which, in comparison with GAUSS's *Disquisitiones*, contained new results due to DIRICHLET himself, such as the arithmetic progression theorem. In the course of the successive editions of this book, DEDEKIND added 11 supplements, leading to his own notion of an *ideal*.

In order to restore uniqueness of decomposition into prime numbers, KUMMER and KRONECKER had defined what it means to be divisible by an "ideal prime factor" but, as explained by EDWARDS (1983), those "ideal numbers" themselves lacked a definition. DEDEKIND strived for a definition of an ideal which would be "delivered of all obscurity and the admission of ideal numbers". Such a definition is the subject of his 11th supplement (222 pages!) to the 4th edition (1894) of DIRICHLET's *Zahlentheorie*, where the modern definition is plainly stated, together with the unique decomposition of a (non-zero) ideal as a product of (non-zero) prime ideals.

As it seems, and as NOETHER would later say, *Es steht alles schon bei Dedekind*. ("Everything is already in Dedekind.") The idea of an "abstract algebra" was still missing, however, and the abstract notions of fields and rings would have to wait for the works of STEINITZ and NOETHER.

Theorem (9.6.8). — *Let A be an integral domain and let S be the set of its maximal ideals. Assume that every maximal ideal of A is invertible.*

(i) *The ring A is a Dedekind ring.*

(ii) *For every fractional ideal I of A, there exists a unique element $(a_M)_{M \in S}$ in $\mathbf{Z}^{(S)}$ such that $I = \prod_{M \in S} M^{a_M}$.*

(iii) *If $I = \prod_{M \in S} M^{a_M}$ and $J = \prod_{M \in S} M^{b_M}$, then $I \subset J$ is equivalent to the inequalities $b_M \leq a_M$ for every $M \in S$.*

Recall that we write M^n for $(M^{-1})^{-n}$ when $n < 0$.

This result serves as a partial replacement, in Dedekind rings, to the possible non-uniqueness of decomposition in irreducible elements.

Proof. — We first prove that for every non-zero ideal I of A, there exists a family $(a_M) \in \mathbf{N}^{(S)}$ such that $I = \prod_{M \in S} M^{a_M}$. This will also prove that I is invertible, hence A is a Dedekind ring.

Let us argue by contradiction; since A is noetherian, there exists a maximal counterexample I. Since A itself can be written $A = \prod_{M \in S} M^0$, one has $I \neq A$, hence there exists a maximal ideal P of A such that $I \subset P$. Set $J = IP^{-1}$; one has $J \subset A$ and $I = IP^{-1} \cdot P = PJ$. In particular, $I \neq J$, since otherwise, Nakayama's lemma (corollary 6.1.4) implies that $I = 0$. Consequently, there exists a family $(b_M)_{M \in S} \in \mathbf{N}^{(S)}$ such that $J = \prod_{M \in S} M^{b_M}$. Set $a_M = b_M$ for $M \neq P$ and $a_P = b_P + 1$; then $(a_M)_{M \in S} \in \mathbf{N}^{(S)}$ and $I = PJ = \prod_{M \in S} M^{a_M}$, a contradiction. In particular, the result holds for every non-zero ideal of A.

Let then I be an arbitrary fractional ideal. There exists an element $a \in A - \{0\}$ such that $aI \subset A$. Let $(b_M) \in \mathbf{N}^{(S)}$ be such that $aI = \prod_{M \in S} M^{b_M}$. Let also $(a_M) \in \mathbf{N}^{(S)}$ be such that $aA = \prod_{M \in S} M^{a_M}$. Then the ideal $J = \prod_{M \in S} M^{b_M - a_M}$ satisfies $aJ = aI$, hence $J = I$.

The uniqueness of a family $(a_M)_{M \in S} \in \mathbf{Z}^{(S)}$ such that $I = \prod_{M \in S} M^{a_M}$ will follow from the final assertion.

Let thus $I = \prod_{M \in S} M^{a_M}$ and $J = \prod_{M \in S} M^{b_M}$ be two fractional ideals. If $a_M \leq b_M$ for all $M \in S$, then $J = I \cdot \prod_{M \in S} M^{b_M - a_M}$, hence $J \subset I$. Conversely, assume that $J \subset I$ and let us prove that $a_M \leq b_M$ for all M. Multiplying both sides of this inclusion by $\prod_{M \in S} M^{-\inf(a_M, b_M)}$, we may assume that $\inf(a_M, b_M) = 0$ for every $M \in S$; we now need to prove that $a_M = 0$ for every M. Assume that there exists a prime ideal $P \in S$ with $a_P > 0$. Then $I = \prod_{M \in S} M^{a_M} \subset P$. On the other hand, $b_P = 0$, hence $J = \prod_{M \in S} M^{b_M} \not\subset P$ (choose an element $x_M \in M - P$ for every $M \in S$ such that $a_M > 0$; their product belongs to $J - P$). This contradicts the assumption that $J \subset I$ and concludes the proof of the theorem. □

Theorem (9.6.9). — *Let A be an integral domain. The following conditions are equivalent:*

(i) *The ring A is a Dedekind ring;*

(ii) *The ring A is noetherian and for every maximal ideal M of A, the ring A_M is principal;*

(iii) *The ring A is noetherian, integrally closed and $\dim(A) \leq 1$.*

Proof. — (i)⇒(ii). Assume that A is a Dedekind ring. We have already seen that every fractional ideal, a fortiori, every ideal of A is finitely generated; in particular, A is noetherian.

Let P be a maximal ideal of A. The local ring A_P is a Dedekind ring; by proposition 9.6.7, it is a principal ideal domain.

(ii)⇒(iii). By assumption, the ring A is noetherian and A_P is a principal ideal domain, for every maximal ideal P of A. In particular, $\dim(A_P) \leq 1$, that is, $ht_A(P) \leq 1$. Since P is arbitrary, this implies $\dim(A) \leq 1$.

Let us prove that A is integrally closed. Let $x \in K$ be integral over A. Let I be the set of $a \in A$ such that $ax \in A$; this is an ideal of A. Let P be a maximal ideal of A. The element x of K is integral over A_P, hence belongs to A_P because a principal ideal domain is integrally closed. This proves that there exists an $a \in A - P$ such that $ax \in A$; in other words, $I \not\subset P$. Since P is arbitrary, it follows from Krull's theorem (theorem 2.1.3) that $I = A$. In particular, $x \in A$.

(iii)⇒(i) Let A be an integral domain which is noetherian, integrally closed and such that $\dim(A) \leq 1$. If $\dim(A) = 0$, then A is a field, hence a Dedekind ring. We now assume that A is not a field; then $\dim(A) = 1$ and every non-zero prime ideal of A is maximal.

We first prove that every maximal ideal of A is invertible. Let M be a maximal ideal of A; in particular, $M \neq 0$. Then M^{-1} is a fractional ideal of A; let us prove that $M \cdot M^{-1} = A$.

One has $M \subset A$, hence $A \subset M^{-1}$. We first prove that $M^{-1} \neq A$. Let a be a non-zero element of M. The ring A/aA has dimension 0 and is noetherian. By Akizuki's theorem (theorem 6.4.14), it has finite length. Then there exists a finite sequence $(M_1, \dots, M_n)$ of maximal ideals of A such that $M_1 \dots M_n \subset aA$, and we may choose this sequence so that n is minimal. Then $M_1 \dots M_n \subset aA \subset M$. Since M is a prime ideal, there exists an $i \in \{1, \dots, n\}$ such that $M_i \subset M$ (otherwise, choose elements $a_i \in M_i - M$; the product $a_1 \dots a_n$ does not belong to M). Since M_i is maximal, this implies $M = M_i$. Up to renumbering the M_i, we thus assume that $M = M_1$. By the minimality assumption on n, one has $M_2 \dots M_n \not\subset aA$. Let $b \in M_2 \dots M_n - aA$. One has $bM \in MM_2 \dots M_n \subset aA$, hence $(ba^{-1})M \subset A$. By the definition of M^{-1}, this implies $ba^{-1} \in M^{-1}$. Since $ba^{-1} \notin A$, we thus have shown that $M^{-1} \neq A$.

From the inclusion $A \subset M^{-1}$, we deduce that $M \subset M \cdot M^{-1}$. Since M is a maximal ideal of A, it follows that either $M \cdot M^{-1} = A$, or $M \cdot M^{-1} = M$. Arguing by contradiction, let us assume that $M \cdot M^{-1} = M$.

Let $x \in M^{-1}$; one has $xM \subset M$. Since A is noetherian, M is a finitely generated A-module and theorem 4.1.5 implies that x is integral over A. Since A is integrally closed, this shows that $x \in A$. Consequently, $M^{-1} \subset A$, hence $M^{-1} = A$, a contradiction.

By theorem 9.6.8, the ring A is then a Dedekind ring. □

The following corollary will be a plethoric source of examples of Dedekind rings. We refer the reader to exercise 9.7.19 for a stronger result that does not require the separable hypothesis (the Krull–Akizuki theorem).

Corollary (9.6.10). — *Let A be a Dedekind ring and let K be its field of fractions. Let L be a separable extension of K and let B be the integral closure of A in L. Then B is a finitely generated A-module and a Dedekind ring.*

Proof. — By theorem 9.6.9, the ring A is noetherian, integrally closed and its dimension is at most 1. By proposition 9.2.1, B is a finitely generated A-module and a noetherian ring. Since B is integral over A, one has dim(B) = dim(A) ≤ 1 (corollary 9.3.10). Finally, B is integrally closed in its field of fractions, by construction. Theorem 9.6.9 then implies that B is a Dedekind ring. □

Example (9.6.11). — Let K be a number field, that is, a finite extension of **Q**, and let A be the integral closure of **Z** in K: it is called the *ring of integers of K*. Many exercises of this book consider particular cases of this situation.

Let $n = [K : Q]$. By corollary 9.6.10, A is a finitely generated **Z**-module and a Dedekind ring. Since $A \subset K$, it is torsion free, hence is a free **Z**-module. For every $\alpha \in K$, there exists an integer $c \geq 1$ such that $c\alpha \in A$, for example the leading coefficient of the minimal polynomial of α. This implies that A contains a basis of K as a **Q**-vector space, hence $A \simeq \mathbf{Z}^n$.

The study of these rings, in particular the discovery that they are not necessarily unique factorization domains, has been at the heart of the development of algebraic number theory since the eighteenth century. By corollary 9.6.10, A is a Dedekind ring, and an important result of Minkowski asserts that its class group $\mathscr{C}(A)$ is finite. However, many mysteries remain about these groups.

Theorem (9.6.12) (Minkowski). — *Let K be a finite extension of **Q** and let A be its ring of integers of K. The class group $\mathscr{C}(A)$ of A is finite.*

The proof that we give is due to Hurwitz; it follows the book of IRELAND & ROSEN (1990).

Proof. — Let $n = [K : Q]$. Let M be the integer provided by lemma 9.6.13 and let $\mathscr{J}$ be the set of ideals of A that contain M!. The elements of $\mathscr{J}$ are in bijection with the ideals of the quotient ring A/M!A. Since A is isomorphic to $\mathbf{Z}^n$ as an abelian group, this quotient ring is finite and the set $\mathscr{J}$ is finite. To prove that $\mathscr{C}(A)$ is finite, we will now prove that every non-zero ideal of A has the same class as an element of $\mathscr{J}$.

Let I be a non-zero ideal of A. Choose a non-zero element $a \in I$ such that $\left|N_{K/Q}(a)\right|$ is minimal (recall from corollary 4.7.6 that it is an integer). Let $b \in I$. By the following lemma there exists an integer M (depending on K only), an integer $u \in \{1, \ldots, M\}$ and an element $c \in A$ such that $\left|N_{K/Q}(ub - ac)\right| < \left|N_{K/Q}(a)\right|$. Since $ub - ac \in A$, the definition of a implies that $ub - ac = 0$. In particular $ub \in aA$, hence M!I $\subset aA$, so that J = M!a^{-1}I is an ideal of A. By construction, I and J have the same class in $\mathscr{C}(A)$. Since $a \in I$, one has M! $\in$ J, hence J $\in \mathscr{J}$. This concludes the proof. □

Lemma (9.6.13) (Hurwitz). — *There exists an integer M such that the following property holds: for every $a, b \in A$ with $a \neq 0$, there exists $u \in \{1, \ldots, M\}$ and $c \in A$ such that $\left|N_{K/Q}(ub - ac)\right| \leq \left|N_{K/Q}(a)\right|$.*

Proof. — Let $n = [K : Q]$ and let $\alpha_1, \ldots, \alpha_n$ be a basis of A as a **Z**-module; it is a basis of K as a **Q**-vector space. Let $\sigma_1, \ldots, \sigma_n \colon K \to \mathbf{C}$ be the n distinct embeddings of K into **C**. Set

$$C = \prod_{j=1}^{n} \left(\sum_{i=1}^{n} |\sigma_j(\alpha_i)| \right)$$

and let $m = \lfloor C^{1/n} \rfloor + 1$.

For $k = (k_1, \ldots, k_n) \in \{1, \ldots, m\}^n$, let P_k be the cube $[(k_1 - 1)/m, k_1/m] \times \cdots \times [(k_n - 1)/m, k_n/m]$ of side $1/m$ in $\mathbf{R}^n$. These m^n cubes cover the unit cube $P = [0,1]^n$. For $x \in \mathbf{R}^n$, we write $\lfloor x \rfloor = (\lfloor x_1 \rfloor, \ldots, \lfloor x_n \rfloor) \in \mathbf{Z}^n$ and $\{x\} = x - \lfloor x \rfloor \in P$.

Let $x = (x_1, \ldots, x_n) \in \mathbf{Q}^n$ be such that $b/a = \sum_{j=1}^{n} x_j \alpha_j$. By the pigeon-hole principle, there exists a $u \in \{1, \ldots, m\}^n$ such that P_u contains at least two terms of the sequence $(0, \{x\}, \ldots, \{m^n x\})$ of length $m^n + 1$; let $s < t$ be their indices and let $u = t - s$; observe that $1 \le u \le m^n = M$. Let $y = \lfloor tx \rfloor - \lfloor sx \rfloor$ and $z = \{tx\} - \{sx\}$; by construction, one has $y \in \mathbf{Z}^n$ and the coordinates of z satisfy $|z_j| \le 1/m$ for $j \in \{1, \ldots, n\}$. Moreover $ux = tx - sx = y + z$.

Set $c = y_1 \alpha_1 + \cdots + y_n \alpha_n$ and $d = z_1 \alpha_1 + \cdots + z_n \alpha_n$; one has $c \in A$, $r \in K$, and $ub = ac + ar$.

Moreover,

$$\left| N_{K/Q}(r) \right| = \prod_{j=1}^{n} \left(\sum_{i=1}^{n} z_i \sigma_j(\alpha_i) \right)$$

$$\le \sup(|z_1|, \ldots, |z_n|)^n \prod_{j=1}^{n} \left(\sum_{i=1}^{n} |\sigma_j(\alpha_i)| \right)$$

$$\le m^{-n} C < 1,$$

so that

$$\left| N_{K/Q}(ub - ac) \right| = \left| N_{K/Q}(ar) \right| = \left| N_{K/Q}(a) \right| \left| N_{K/Q}(r) \right| < \left| N_{K/Q}(a) \right|.$$

This concludes the proof of the lemma.   □

9.7. Exercises

Exercise (9.7.1). — Let k be a field and let A be a finitely generated k-algebra. Assume that A is a field.

a) Prove that there are an integer $n \ge 0$ and elements $x_1, \ldots, x_n \in A$ which are algebraically independent such that A is algebraic over $k[x_1, \ldots, x_n]$.

In the rest of the exercise, we will prove that $n = 0$: A is algebraic over k (this furnishes a rather direct proof of theorem 9.1.2). We argue by contradiction, assuming that $n \geq 1$.

b) Prove that there exists a polynomial $f \in k[x_1, \ldots, x_n] - \{0\}$ such that A is integral over $k[x_1, \ldots, x_n][1/f]$. Conclude that $k[x_1, \ldots, x_n][1/f]$ is a field.

c) Prove that f belongs to every maximal ideal of $k[x_1, \ldots, x_n]$.

d) Observing that $1 + f$ is a unit of $k[x_1, \ldots, x_n]$, deduce from this that $f \in k$ and derive a contradiction.

Exercise (9.7.2). — Let A be a noetherian integral domain which is not a field and let E be its field of fractions.

Prove the equivalence of the following assertions:

(i) The A-algebra E is finitely generated;

(ii) There exists a non-zero element of A which belongs to every non-zero prime ideal;

(iii) The ring A has only finitely many non-zero prime ideals of height 1;

(iv) dim(A) = 1 and the ring A has only finitely many maximal ideals.

(For (iii)$\Rightarrow$(iv), *use Krull's Hauptidealsatz. For* (ii)$\Rightarrow$(i), *take an element* $a \in A$ *and consider a primary decomposition of the ideal* aA.)

Exercise (9.7.3). — Let $j: A \to B$ be an injective morphism of integral domains; we assume that B is finitely generated over A. The goal of the exercise is to prove that for every non-zero $b \in B$, there exists an $a \in A$ such that for every algebraically closed field K and every morphism $f: A \to K$ such that $f(a) \neq 0$, there exists a morphism $g: B \to K$ such that $f = g \circ j$ and $g(b) \neq 0$.

a) Treat the case where $B = A[x]$, for some $x \in B$ which is transcendental over A.

b) Treat the case where $B = A[x]$, for some $x \in B$ which is algebraic over A.

c) Treat the general case.

d) Let V be a non-empty open subset of Spec(B); prove that its image $j^*(V)$ in Spec(A) contains a non-empty open subset.

Exercise (9.7.4). — Recall that a Jacobson ring is a commutative ring in which every prime ideal is the intersection of the maximal ideals that contain it (see p. 9.1).

a) Prove that $\mathbf{Z}$ is a Jacobson ring. What principal ideal domains are Jacobson rings?

b) Let A be a Jacobson ring and let I be an ideal of A. Prove that A/I is a Jacobson ring.

Let A be a Jacobson ring and let B be an A-algebra.

c) If B is integral over A, then B is a Jacobson ring.

d) Assume that B is finitely generated over A. Prove that for every maximal ideal M of B, the intersection M ∩ A is a maximal ideal of A, and the residue field B/M is a finite extension of A/(M ∩ A). Prove that B is a Jacobson ring.

e) Let M be a maximal ideal of $Z[T_1, \ldots, T_n]$. Prove that the residue field $Z[T_1, \ldots, T_n]/M$ is a finite field.

f) Let K be a field and let A be a K-algebra. Assume that there exist a set I such that $\mathrm{Card}(I) < \mathrm{Card}(K)$, and a family $(a_i)_{i \in I}$ such that $A = K[(a_i)]$. Prove that for every maximal ideal M of A, the residue field A/M is an algebraic extension of K, and that A is a Jacobson ring. (*Follow the arguments in the proof of theorem 2.3.1.*)

Exercise (9.7.5). — Let k be an infinite field and let A be a k-algebra.

a) Let $P \in k[T_1, \ldots, T_n]$ be a non-zero homogeneous polynomial. Show that there exists a $c \in k^n$ such that $P(c) \neq 0$ and $c_n = 1$.

b) Let $a_1, \ldots, a_n \in A$ and let $P \in k[T_1, \ldots, T_n]$ be a non-zero polynomial such that $P(a_1, \ldots, a_n) = 0$. Let $d = \deg(P)$ and let P_d be the homogeneous component of degree d of P. Let $c \in k^n$ be such that $P_d(c) \neq 0$ and $c_n = 1$; for $m \in \{1, \ldots, n-1\}$, let $b_m = a_m + c_m a_n$. Prove that $a_1, \ldots, a_n$ are integral over $k[b_1, \ldots, b_{n-1}]$.

c) Let $a_1, \ldots, a_N \in A$ be elements such that $A = k[a_1, \ldots, a_N]$. Prove that there exist an integer $n \geq 0$ and algebraically independent elements $b_1, \ldots, b_n$ which are k-linear combinations of $a_1, \ldots, a_N$ such that A is integral over its subring $k[b_1, \ldots, b_n]$. (*A version of Noether's normalization lemma.*)

Exercise (9.7.6). — Let k be a field and let E, F be two finitely generated field extensions of k. Let $R = E \otimes_k F$. The goal of the exercise is to prove that $\dim(R) = \inf(\mathrm{tr}\,\deg_k(E), \mathrm{tr}\,\deg_k(F))$, a formula due to Grothendieck.

a) Let $n = \mathrm{tr}\,\deg_k(E)$ and let $(a_1, \ldots, a_n)$ be a transcendence basis of E; let $E_1 = k(a_1, \ldots, a_n)$ and $R_1 = E_1 \otimes_k F$. Prove that R_1 is a subring of R and that R is integral over R_1. Conclude that $\dim(R) = \dim(R_1)$.

b) Show that $A_1 = k[a_1, \ldots, a_n] \otimes_k F$ is a subring of R of which R_1 is a fraction ring. Prove that A_1 is isomorphic to $F[T_1, \ldots, T_n]$ and that $\dim(R) \leq n$.

c) Assume, moreover, that $\mathrm{tr}\,\deg_k(F) \geq n$. Prove that there exists a surjective morphism of F-algebras $\psi \colon R_1 \to F$.

d) Using the fact that $\mathrm{Ker}(\psi) \cap A_1$ is a maximal ideal of A_1, prove that $\dim(R_1) \geq n$.

Exercise (9.7.7). — Let A be commutative ring.

a) Assume that there exists a prime ideal P of A which is not maximal, and let M be a maximal ideal containing P. Prove that multiplication by an element of M induces an endomorphism of A/P which is injective but not surjective.

For the rest of the exercise, we assume that $\dim(A) = 0$. Let M be a finitely generated A-module and let f be an *injective* endomorphism of M.

We endow M with the structure of an $A[X]$-module, where $P \cdot m = P(f)(m)$, for every $P \in A[X]$ and every $m \in M$; let $I \subset A[X]$ be its annihilator and let $B = A[X]/I$.

b) Using the Cayley–Hamilton theorem, prove that B is an integral A-algebra.

c) Prove that $\dim(B) = 0$.

d) Let x be the image of X in B and let J be its annihilator. Assume that x is not invertible. Prove that $J \neq 0$ and deduce a contradiction. (*Observe that x is nilpotent in the fraction ring B_P, for every prime ideal of B such that $x \in P$.*)

e) Conclude that f is surjective. (A theorem of VASCONCELOS (1970).)

Exercise (9.7.8). — Let A be a commutative noetherian ring. For $a \in A$, let J_a be the ideal of A generated by a and all elements $x \in A$ such that ax is nilpotent.

a) If a is invertible or nilpotent, prove that $J_a = A$.

b) Let P be a minimal prime ideal of A. Prove that for every $a \in A$, one has $J_a \not\subset P$. (*Assuming $a \in P$, deduce from the prime avoidance lemma, lemma 2.2.12, that there exists a $b \in A - P$ such that ab is nilpotent.*)

c) Let P be a minimal prime ideal of A and let Q be a prime ideal of A such that $P \subsetneq Q$. Prove that $J_a \subset Q$ for every $a \in Q - P$.

d) Let n be an integer. Prove that $\dim(A) \leq n$ if and only if $\dim(A/J_a) \leq n - 1$ for every $a \in A$. (A theorem of COQUAND ET AL (2005).)

Exercise (9.7.9). — Let A be a ring. For every $a \in A$, let S_a be the set of elements of A of the form $a^n(1 + ab)$, for $n \in \mathbf{N}$ and $b \in A$.

a) Let $a \in A$. Prove that S_a is a multiplicative subset of A. Prove that S_a contains 0 if and only if a is invertible or a is nilpotent.

b) Let $a \in A$. Prove that $M \cap S_a \neq \emptyset$ for every maximal ideal M of A. Conclude that $\dim(S_a^{-1}A) + 1 \leq \dim(A)$.

c) Let M be a maximal ideal of A and let P be a prime ideal of A such that $P \subset M$. Prove that $P \cap S_a = \emptyset$ for every $a \in M - P$. Conclude that there exists an $a \in A$ such that $\dim(S_a^{-1}A) + 1 = \dim(A)$.

d) Let $n \in \mathbf{N}$. Prove that $\dim(A) \leq n$ if and only if, for every $a_0, \ldots, a_n \in A$, there exist $b_0, \ldots, b_n \in A$ and $m_0, \ldots, m_n \in \mathbf{N}$ such that

$$a_0^{m_0}(a_0 b_0 + a_1^{m_1}(a_1 b_1 + \cdots + a_n^{m_n}(1 + a_n b_n))) = 0.$$

(*This characterization of Krull's dimension is due to COQUAND ET AL (2005).*)

e) Let K be a field. Using the previous characterization of Krull's dimension, reprove corollary 9.3.12 that $\dim(K[T_1, \ldots, T_n]) = n$. (See COQUAND & LOMBARDI (2005).)

Exercise (9.7.10). — Let A be a commutative noetherian ring. For $a \in A$, the ideal J_a of A is defined as in exercise 9.7.8. For every family $(a_1, \ldots, a_m)$ in A, we recall that $V(a_1, \ldots, a_m)$ is the closed subset of Spec(A) consisting of all prime ideals P which contain $a_1, \ldots, a_m$.

a) Let $a, b_1, \ldots, b_m, c \in A$ be such that $V(b_1, \ldots, b_m) \subset V(a)$ in $\mathrm{Spec}(A/J_c)$. Prove that there exists a $b \in A$ such that $V(a + bc) \subset V(a)$ in $\mathrm{Spec}(A/J_c)$. Conclude that $V(b_1, \ldots, b_m, a + bc) \subset V(a)$ in $\mathrm{Spec}(a)$.

b) Let m be an integer such that $m \geq \dim(A) + 2$ and let $a_1, \ldots, a_m \in A$. Prove by induction on $\dim(A)$ that there exists $b_2, \ldots, b_m \in A$ such that $V(a_1, \ldots, a_m) = V(a_2 + a_1 b_2, \ldots, a_m + a_1 b_m)$.

c) Let I be an ideal of A. Prove that there exist elements $(a_1, \ldots, a_n)$, with $n \leq \dim(A) + 1$, such that $V(I) = V(a_1, \ldots, a_n)$. (*A theorem of Kronecker; this proof is due to* COQUAND *(2004).*)

Exercise (9.7.11). — Let A be a ring and let $B = A[X_1, \ldots, X_n]$.

a) Prove that $\dim(B) \geq \dim(A) + n$.

b) Let P be a prime ideal of A and let $(Q_0, \ldots, Q_m)$ be a chain of prime ideals of B such that $Q_j \cap A = P$ for every j. Prove that $m \geq n$. (*Passing to the quotient by P and Q_0, reduce to the case where P = 0; by localization, then reduce to the case where A is a field.*)

c) Prove that $\dim(B) \leq \dim(A) + n(1 + \dim(A))$.

In the case $n = 1$, *i.e.,* $B = A[X]$, *this exercise says that* $\dim(A) + 1 \leq \dim(A[X]) \leq 2\dim(A) = 1$, *and* SEIDENBERG *(1954) has shown that all possibilities actually appear. When A is noetherian, exercise 9.7.17 shows that one has* $\dim(B) = \dim(A) + n$.

Exercise (9.7.12). — Let A be a noetherian ring and let I be an ideal of A. Let B be the subalgebra of $A[T]$ consisting of polynomials $P = \sum a_n T^n$ such that $a_n \in I^n$ for all n.

a) Using exercise 6.7.21, prove that B is a noetherian ring.

b) Assume that there exists a prime ideal P of A containing I such that $\dim(A/P) = \dim(A)$; prove that $\dim(B) = \dim(A) + 1$.

c) Otherwise, prove that $\dim(B) = \dim(A)$.

Exercise (9.7.13). — Let X be a topological space and let $A = \mathscr{C}(X; \mathbf{R})$ be the ring of real-valued continuous functions on X.

a) Assume that every function $f \in A$ is locally constant. Prove that A is a von Neumann ring and that $\dim(A) = 0$.

Let $\rho \in A$ be a continuous function which is not locally constant in any neighborhood of a point $\xi \in X$; we assume $\rho(\xi) = 0$.

b) Let P be the set of functions $f \in A$ such that

$$\lim_{x \to \xi} f(x)/\rho(x)^t = 0$$

for every $t \in \mathbf{N}$. Prove that P is an ideal of A, contained in the maximal ideal M_ξ of functions vanishing at ξ. Prove that P is a radical ideal and that $P \neq M_\xi$.

c) Let $\mathfrak{U}$ be an ultrafilter (see exercise 2.8.11) which converges to ξ, in the sense that all neighborhoods of ξ belong to $\mathfrak{U}$. For every $c \in \mathbf{R}_+$, let P_c be the set of functions $f \in A$ such that

$$\lim_{\mathfrak{U}} f(x) \exp(t\rho(x)^{-c}) = 0$$

for all $t \in \mathbf{R}$.

d) Prove that P_c is a prime ideal of A.

e) Let $c, c' \in \mathbf{R}_+$ be such that $c < c'$. Prove that $P_{c'} \subsetneq P_c$.

f) Prove that $\dim(A) = +\infty$.

Exercise (9.7.14). — Let K be a field and let $A = K[T_1, T_2, \dots]$ be the ring of polynomials in (countably) infinitely many indeterminates.

a) Prove that A is not noetherian and that $\dim(A)$ is infinite.

b) Let $(m_n)_{n \geq 1}$ be a strictly increasing sequence of strictly positive integers such that the sequence $(m_{n+1} - m_n)$ is unbounded. For every $n \geq 1$, let P_n be the ideal of A generated by the indeterminates T_i, for $m_n \leq i < m_{n+1}$. Prove that P_n is a prime ideal of A.

c) Let S be the intersection of the multiplicative subsets $S_n = A - P_n$, for all $n \geq 1$. Prove that $\dim(S^{-1}A)$ is infinite.

d) Prove that $S^{-1}A$ is a noetherian ring.

Exercise (9.7.15). — Let A be a local noetherian ring and let M be its maximal ideal, and let $K = A/M$ be its residue field .

a) Explain why the A-module structure of M/M^2 endows it with the structure of a K-vector space.

b) Let $a_1, \dots, a_n$ be elements of M which generate M/M^2. Prove that $M = (a_1, \dots, a_n)$. Deduce from this that $\dim(A) \leq \dim_K(M/M^2)$.

In the rest of the exercise, assume that $\dim(A) = \dim_K(M/M^2)$ (one says that A is a *regular ring*). The goal is to prove that A is an integral domain. We argue by induction on $\dim(A)$.

c) Treat the case $\dim(A) = 0$.

d) Assume that $\dim(A) > 0$. Let $P_1, \dots, P_n$ be the minimal prime ideals of A. Prove that there exists an $a \in M$ such that $a \notin M^2 \cup \bigcup_{i=1}^n P_i$.

e) Let $B = A/(a)$ and let $N = MB$ be its maximal ideal. Using the induction hypothesis, prove that (a) is a prime ideal of A.

f) Let $i \in \{1, \dots, n\}$ be such that $P_i \subset (a)$. Using Nakayama's lemma, prove that $P_i = (0)$. Conclude.

Exercise (9.7.16). — Let A be a commutative ring. Say that a polynomial $f \in A[X]$ in one indeterminate is primitive if its coefficients generate the unit ideal.

a) Prove that primitive polynomials are not zero divisors.

b) Prove that the set S of primitive polynomials is a multiplicative subset of A[X]. The ring of fractions $S^{-1} A[X]$ is denoted by A(X). Observe that this generalizes the construction of the field of fractions if A is field. (This construction is due to Nagata.)

c) Prove that the maximal ideals of A(X) are of the form $P = S^{-1} M[X]$, where M is a maximal ideal of A; identify its residue field with K(X), where $K = A/M$.

d) Assume that A is noetherian. With the notation of the preceding question, prove that $ht_{A(X)}(P) = ht_A(M)$. In particular, prove that $\dim(A(X)) = \dim(A)$.

Exercise (9.7.17). — Let A be a noetherian ring and let $B = A[X]$.

a) Let P be a prime ideal of A and let Q and Q′ be two prime ideals of B such that $Q \subsetneq Q'$ and $P = A \cap A = Q' \cap B$. Prove that $Q = PB$. (*Argue as in question b*) *of exercise 9.7.11.*)

b) Let I be an ideal of A and let P be a minimal prime ideal containing I. Prove that PB is a minimal prime ideal of B containing IB.

c) Let P be a prime ideal of A. Prove that $ht_A(P) = ht_B(PB)$.

d) Prove that $\dim(B) = \dim(A) + 1$.

e) More generally, prove that $\dim(A[X_1, \dots, X_n]) = \dim(A) + n$.

Exercise (9.7.18). — Let A be a noetherian integral domain, let K be its fraction field and let M be a finitely generated A-module. Assume that $\dim(A) = 1$.

a) Let $S = A - \{0\}$. Prove that $S^{-1}M$ is a finite-dimensional K-vector space. Let r be its dimension. Prove that r is the maximal number of A-linearly independent subsets of M.

b) Prove the equivalence of the following assertions: (i) Every element of M is torsion; (ii) $\ell_A(M)$ is finite; (iii) $r = 0$.

c) From now on, we assume that M is torsion-free. Show that M contains a free submodule L of rank r such that $\ell_A(M/L)$ is finite.

d) Prove that for every integer $n \geq 1$, one has

$$\ell_A(M/a^n M) \leq \ell_A(L/a^n L) + \ell_A(M/L).$$

Also prove that
$$\ell_A(M/a^n M) = n\, \ell_A(M/aM).$$

e) Prove that $\ell_A(M/aM) \leq r\, \ell_A(A/aA)$.

f) Generalize the previous inequality to any torsion-free A-module. Give an example where this inequality is strict.

Exercise (9.7.19). — Let A be a integral domain, let K be its field of fractions, let L be finite extension of K and let B be a subring of L containing A.

a) Let I be a non-zero ideal of B. Prove that $I \cap A \neq 0$.

Now assume that A is noetherian and $\dim(A) = 1$.

b) Let I be an ideal of B and let a be a non-zero element of $I \cap A$. Prove that B/aB and I/aB are B-modules of finite length. (*Use exercise 9.7.18.*)

c) Prove that B is noetherian and that $\dim(B) \leq 1$. (*The Krull–Akizuki theorem.*)

Exercise (9.7.20). — Let A be a noetherian integral domain such that every non-zero prime ideal of A is maximal.

a) Let $a \in A - \{0\}$. Prove that the A-module $A/(a)$ has finite length.

b) Let $a, b \in A - \{0\}$. Prove that $\ell_A(A/(ab)) = \ell_A(A/(a)) + \ell_A(A/(b))$.

c) Prove that there exists a unique morphism of groups $\omega_A \colon K^\times \to \mathbf{Z}$ such that $\omega_A(a) = \ell_A(A/(a))$ for every $a \in A - \{0\}$.

d) Assume that A is a principal ideal domain. Compute $\omega_A(a)$ in terms of the decomposition of a as a product of irreducible elements.

Exercise (9.7.21). — Let A be a ring and let $\varphi \colon M \to N$ be a morphism of A-modules. One says that φ is admissible if $\mathrm{Ker}(\varphi)$ and $\mathrm{Coker}(\varphi)$ have finite length; one then defines $e_A(\varphi) = \ell_A(\mathrm{Coker}(\varphi)) - \ell_A(\mathrm{Ker}(\varphi))$.

a) If M and N have finite length, then φ is admissible and $e_A(\varphi) = \ell_A(N) - \ell_A(M)$.

b) Let $\varphi \colon M \to N$ and $\psi \colon N \to P$ be morphisms of A-modules. If two among $\varphi, \psi, \psi \circ \varphi$ are admissible, then so is the third and

$$e_A(\psi \circ \varphi) = e_A(\psi) + e_A(\varphi).$$

c) Assume that A is an integral domain and that every non-zero prime ideal of A is maximal. Let M be a free and finitely generated A-module and let φ be an injective endomorphism of M. Show that φ is admissible and that $\ell_A(\mathrm{Coker}(\varphi)) = \ell_A(A/(\det(\varphi)))$. (*First treat the case where the matrix of φ in some basis of M is an elementary matrix.*)

d) In the case where A is a principal ideal domain, deduce the result of the previous question from theorem 5.4.3.

Exercise (9.7.22). — Let A be an integral domain and let K be its field of fractions.

a) Assume that for every $a \in A - \{0\}$ and every prime ideal P which is associated with $A/(a)$, the fraction ring A_P is a principal ideal domain. Prove that A is integrally closed.

Conversely, assume that A is noetherian and integrally closed. Let $a \in A - \{0\}$ and let P be a prime ideal of A which is associated with $A/(a)$.

b) Prove that there exists a $b \in A$ such that $P = \{x \in A \,;\, xb \in (a)\}$.

c) Prove that the element a/b of K belongs to A_P and that $PA_P = (a/b)A_P$. (*If $(b/a)PA_P \neq A_P$, prove that b/a is integral over A.*)

d) Prove that A_P is a principal ideal domain and that every prime ideal associated with $A/(a)$ is minimal.

Exercise (9.7.23). — Let k be a field and let $A = k[X, Y]$ be the ring of polynomials in two indeterminates.

 a) Let $I = (X, Y)$. Prove that $I^{-1} = A$. Conclude that I is not invertible.

 b) Prove that a fractional ideal of A is invertible if and only if it is principal.

Exercise (9.7.24). — Let A be a Dedekind ring.

 a) Assume that $\mathrm{Spec}(A)$ is finite. Prove that A is a principal ideal domain.

 b) Let I be an ideal of A and let $a \in I - \{0\}$. Prove that there exists a $b \in I$ such that $I = aA + bA$. (*Start by proving that the ring A/I is artinian.*)

Exercise (9.7.25). — Let A be a principal ideal domain which is a local ring. Let p be a generator of its maximal ideal and let K be its field of fractions. Let $P \in A[T]$ be an Eisenstein polynomial (see exercise 2.8.28 where it is proved that P is irreducible in $K[T]$), that is, $P = T^n + a_{n-1}T^{n-1} + \cdots + a_0$, where $a_0, \dots, a_{n-1} \in (p)$ and $a_n \notin (p^2)$.

Let $L = K[T]/(P)$, let ω be the class of T in L, and let $B = A[\omega]$.

 a) Prove that the ideal (ω) of B is maximal.

 b) Prove that (ω) is the unique maximal ideal of B. (*Use Nakayama's lemma to prove that every maximal ideal of B contains p.*)

 c) Prove that for every $b \in B - \{0\}$, there exists a unit $u \in B^\times$ and an integer $n \in \mathbf{N}$ such that $b = u\omega^n$.

 d) Prove that B is a principal ideal domain and is the integral closure of A in L.

Exercise (9.7.26). — Let A be a local ring, let K be its field of fractions, let M be its maximal ideal, and let $k = A/M$ be its residue field. Let $P \in A[T]$ be a monic irreducible polynomial whose image $\overline{P} \in k[T]$ is separable; let $\overline{P} = \prod_{j=1}^m \overline{P}_j$ be the decomposition as a product of monic irreducible polynomials; for every j, fix a monic polynomial $P_j \in A[T]$ with reduction $\overline{P}_j$ modulo M. Let $L = K[T]/(P)$ and let $B = A[T]/(P)$.

 a) Let N be a maximal ideal of B. Prove that there exists a unique element $j \in \{1, \dots, m\}$ such that $N = MB + (P_j)$.

 b) Assume that A is integrally closed in its field of fractions. Prove that B is the integral closure of A in L. (*Using the separability of the polynomial $\overline{P}$, prove that the determinant of the $n \times n$ matrix $(\mathrm{Tr}_{L/K}(\omega^{i+j}))_{0 \le i, j < n}$, the discriminant of P, is a unit of A.*)

 c) Assume that A is a principal ideal domain. Prove that B is a principal ideal domain. (*Deduce from theorem 9.6.9 that for every j, the local ring B_{N_j} is a principal ideal domain.*)

Exercise (9.7.27). — Let k be a field and let $A = k[X^2, X^3]$ be the subalgebra of $k[X]$ generated by X^2 and X^3.

 a) Prove that A is a noetherian integral domain and that $\dim(A) = 1$.

 b) Prove that A is not integrally closed. (*Prove that X is integral over A.*)

c) Prove that the ideal (X^2, X^3) of A is not invertible.

Exercise (9.7.28). — Let $A = R[X, Y]/(X^2 + Y^2 - 1)$.

a) Prove that A is a Dedekind ring.

b) Let M be a maximal ideal of A. Prove that $A/M \simeq R$ or $A/M \simeq C$. Give examples of both cases.

c) Let M be a maximal ideal of A such that $A/M \simeq C$. Prove that M is principal.

d) Let M be a maximal ideal of A such that $A/M \simeq R$; prove that there exists a unique pair $(a, b) \in R^2$ such that $a^2 + b^2 = 1$ and $A = (X - a, Y - b)$. Prove that M is not principal but that M^2 is principal. Prove also that all of these maximal ideals have the same class in the class group $\mathscr{C}(A)$ of A.

e) What is the class group of A?

Exercise (9.7.29). — Let *d* be an integer such that $d \geq 2$ and *d* is square free, let $K = Q(i\sqrt{d})$ and let A be the integral closure of Z in K; it is a Dedekind ring.

a) Prove that $A = Z[(-1 + i\sqrt{d})/2]$ is $d \equiv -1 \pmod 4$, and that $A = Z[i\sqrt{d}]$ otherwise.

b) Let I be a non-zero ideal of A. Let (e, f) be a basis of I as a Z-module. Applying exercise 5.7.11 to the quadratic form $(x, y) \mapsto N_{K/Q}(xe + yf)$, prove that there exists a non-zero element $a \in I$ such that $N_{K/Q}(a) \leq \text{Card}(A/I)\sqrt{\delta/3}$, where $\delta = d$ if $d \equiv -1 \pmod 4$, and $\delta = 4d$ otherwise.

c) (*continued*) Let $J = (a) : I$. Prove that J is non-zero ideal of A such that $IJ = (a)$ and $\text{Card}(A/J) \leq \sqrt{\delta/3}$.

d) Prove that $\mathscr{C}(A)$ is generated by the maximal ideals P of A such that $\text{Card}(A/P) \leq \sqrt{\delta/3}$. *This bound is better than the one implicitly given in the proof of theorem 9.6.12; it is also slightly better than the general Minkowski bound.*

Exercise (9.7.30). — Let $A = Z[i\sqrt{5}]$ (see exercise 2.8.25).

a) Prove that A is a Dedekind ring.

b) Prove that $M = (2, 1 + i\sqrt{5})$ is the unique maximal ideal of A containing 2. Prove that $M^2 = (2)$ and that M is not principal.

c) Prove that the ideals $P = (3, 1 + i\sqrt{5})$ and $P' = (3, 1 - i\sqrt{5})$ are the unique maximal ideals of A containing 3. Prove that $P \cdot P' = (3)$ and that neither P nor P' is principal.

d) Prove that the classes of P, P', M in the class group $\mathscr{C}(A)$ are equal.

e) Using exercise 9.7.29, prove that $\mathscr{C}(A) \simeq Z/2Z$.

Exercise (9.7.31). — Let *k* be a finite field, let $K = k(T)$ be the field of rational functions in one indeterminate and let L be a finite separable extension of K. Let $A = k[T]$ and let B be the integral closure of A in L.

a) Show that B is a Dedekind ring and a free A-module of rank n, where $n = [L : K]$.

b) Let $a \in k(T)$; prove that there exists a unique pair (b, c) where $b \in k[T]$ and $c \in k(T)$ are such that $a = b + c$ and $\deg(c) < 0$. (The degree $\deg(a)$ of an element $a \in k(T)$ is defined as $\deg(f) - \deg(g)$, where $f, g \in k[T]$ are polynomials such that $g \neq 0$ and $a = f/g$.)

c) Let $(e_1, \ldots, e_n)$ be a basis of L as an K-vector space. Prove that there exists an integer M such that $\deg(N_{L/K}(\sum a_i e_i)) \leq M + n \sup_i \deg(a_i)$ for every $(a_1, \ldots, a_n) \in K^n$.

d) Prove that there exists an integer M such that the following property holds: for every $a, b \in B$ with $a \neq 0$, there exists a polynomial $f \in A$ such that $\deg(f) \leq M$ and $g \in B$ such that $\deg(N_{L/K}(fb - ac)) \leq \deg(N_{L/K}(a))$.

e) Prove that the class group $\mathscr{C}(B)$ is finite. (*Adapt the proof of theorem 9.6.12, using the preceding question to replace Hurwitz's lemma.*)

Exercise (9.7.32). — Let A be a Dedekind ring. Prove that every divisible A-module is injective.

(*Conversely, it follows from exercise 7.8.24 that if A is a ring such that every divisible right A-module is injective, then A is right hereditary. If A is an integral domain, it is then a Dedekind ring.*)

Exercise (9.7.33). — Let A be a Dedekind ring and let M be a finitely generated A-module.

a) If M is torsion-free, then M is projective.

b) Prove that the torsion submodule T(M) of M has a direct summand.

c) Let I, J be fractional ideals of A and let $f : J \to I$ be a morphism of A-modules. Prove that there exists a unique element $a \in A$ such that $f(x) = ax$ for every $x \in J$. Prove that $a \in IJ^{-1}$.

d) Let $I_1, \ldots, I_m$ and $J_1, \ldots, J_n$ be fractional ideals of A, let $M = I_1 \oplus \cdots \oplus I_m$ and $N = J_1 \oplus \cdots \oplus J_n$, and let $f : N \to M$ be an isomorphism of A-modules. Prove that $m = n$ and $I_1 \ldots I_m \simeq J_1 \ldots J_n$. (*Represent f by a matrix $U = (a_{ij}) \in \mathrm{Mat}_{m,n}(K)$, where $a_{ij} \in I_i J_j^{-1}$. Prove that $I = (\det(U))J$.*)

e) Let I, J be fractional ideals of A. Prove that there exist non-zero elements $a \in I$ and $b \in J$ such that $(I \oplus J)/(a, b)A$ is torsion-free. Conclude that $I \oplus J \simeq A \oplus IJ$.

f) Let M be a finitely generated A-module. Assume that M is torsion free and non-zero. Prove that there exists an integer $n \geq 1$ and a fractional ideal I such that $M \simeq A^{n-1} \oplus I$.

Appendix

This appendix consists of three different sections.

The first one, called algebra, collects a few definitions and facts which are essentially prerequisites for the reading of this book. Their role is also to establish a few conventions on which there is no unanimity, such as the precise meaning of a "positive" number or an "increasing" sequence.

The second section is devoted to results of set theory which are used at a few places of the book to manage infinite cardinals. The Cantor–Bernstein theorem will imply that the dimension of a vector space, or the transcendence degree of a field extensions, are well defined. On the other hand, Zorn's lemma asserts the existence of maximal elements in adequate ("inductive") ordered sets; it is used to prove the existence of bases of a vector space, of maximal ideals, of minimal prime ideals, characterize injective modules, etc. The proof of Kaplansky's theorem, however, relies on a transfinite inductive construction, and I couldn't "zornify" it fully. For this reason, rather than giving a direct (but maybe unnatural proof) of Zorn's lemma, I chose to start from the inductive principle for well-ordered sets, and Hartogs's lemma. To make this appendix reasonably complete, I also give a proof of the theorems of Cantor and Zermelo.

Finally, I give a brief introduction to category theory. Mathematical practice progressively showed the importance of considering not only mathematical objects but morphisms between them, and a category is the abstraction of this idea. It then appeared that categories can themselves be treated as mathematical objects, their morphisms are called functors, and that this point of view is extremly fruitful in many fields of mathematics. While categories are only used as a language in the first chapters of this book, they really are a useful tool in the chapters devoted to homological algebra and tensor products.

© Springer Nature Switzerland AG 2021
A. Chambert-Loir, *(Mostly) Commutative Algebra*, Universitext,
https://doi.org/10.1007/978-3-030-61595-6

A.1. Algebra

A.1.1. Numbers

A.1.1.1. — **N**, **Z**, **Q**, **R** and **C** respectively denote the sets of positive integers $(0, 1, 2, \dots)$, integers $(0, 1, 2, \dots, -1, -2, -3, \dots)$, rational numbers (fractions of integers), real numbers and complex numbers.

A.1.1.2. — Between real numbers, the ordering relation $\leq$ is pronounced as "less than", or sometimes "less than or equal to" if we feel it necessary to insist on the possibility of equality. The expression "$x < y$" reads "x is strictly smaller than y". Similarly for the opposite ordering relation $\geq$ ("greater than", or "greater than or equal to"), and the relation $>$ ("strictly greater than").

In particular, a real number x is said to be positive if $x \geq 0$, strictly positive if $x > 0$, negative if $x \leq 0$, and strictly negative if $x < 0$.

A.1.1.3. Integers — An integer n is said to be a prime number if it is strictly greater than 1 and if its only positive divisors are 1 and itself. The first prime numbers are $2, 3, 5, 7, 11, \dots$

A.1.1.4. Binomial coefficients — Let n be a positive integer. One sets $n!$ ("factorial n") to be the product

$$n! = 1 \cdot 2 \dots n.$$

One has $0! = 1$.

Let m, n be positive integers such that $m \leq n$. The binomial coefficient $\binom{n}{m}$ is defined by

$$\binom{n}{m} = \frac{n!}{m!(n-m)!}.$$

It is always an integer. This follows by induction from the equalities $\binom{n}{0} = 1$ and from the recurrence relation

$$\binom{n+1}{m+1} = \binom{n}{m} + \binom{n}{m+1},$$

whenever $0 \leq m \leq n$.

A.1.2. Sets, subsets, maps

A.1.2.1. — As usual, I write $x \in S$ to mean that x is an element of the set S.

If S and T are sets, then $S - T$ is the set of elements $x \in S$ such that $x \notin T$. If the set S is clear from the context and T is a subset of S, then this set $S - T$, the complementary subset to T in S, is also denoted by $\complement T$.

If S and T are two sets, I write $S \subset T$ or $T \supset S$ to mean the equivalent formulations S is a subset of T, S is contained in T, or T contains S, that is: every element of S belongs to T. The symbols $S \subsetneq T$ mean that $S \subset T$ but $S \neq T$.

The set of all subsets of a set S is denoted by $\mathfrak{P}(S)$.

A.1.2.2. — A map $f: S \to T$ determines, for every element $s \in T$, an element $f(s) \in T$; the element $f(s)$ of T is called the image of s by f. The map f has a graph Γ_f which is the subset of $S \times T$ consisting of all pairs of the form $(s, f(s))$, for $s \in S$.

The identity map, $id_S: S \to S$, is such that every element of S is its own image.

A.1.2.3. — If A is a subset of S, one writes $f(A)$ for the set of all elements $f(a)$, for $a \in A$.

If B is a subset of T, one writes $f^{-1}(T)$ for the set of all elements $s \in S$ such that $f(s) \in T$.

A.1.2.4. — One says that the map f is *injective* if distinct elements of S have distinct images. One says that the map f is *surjective* if every element of T is the image of some element of S.

One says that the map f is *bijective* if it is both injective and surjective. Then there exists a unique map $g: T \to S$ such that $g \circ f = id_S$ and $f \circ g = id_T$; the map g is called the inverse of f and is often denoted by f^{-1}.

Two sets S and T are said to be *equipotent* if there exists a bijection from S to T.

A.1.3. Equivalence relations

A.1.3.1. — A *binary relation* R on a set X can be described by its graph Γ_R, which is the set of all pairs $(x, y) \in X^2$ such that x R y.

One says that the relation R is *reflexive* if x R x for all $x \in X$.

One says that it is *symmetric* if for all $x, y \in X$ such that x R y, one has y R x.

One says that it is *transitive* if for all $x, y, z \in X$ such that x R y and y R z, one has x R z.

One says that the relation R is an *equivalence relation* if it is reflexive, symmetric and transitive.

A.1.3.2. — Let R be an equivalence relation on a set X. For $x \in X$, the equivalence class R_x of x modulo R is the set of all $y \in X$ such that x R y.

Two equivalence classes are either disjoint or equal. Indeed, let $x, y \in X$ be such that $R_x \cap R_y \neq \emptyset$, and let $z \in R_x \cap R_y$; let us show that $R_x = R_y$. Let $u \in R_x$; by assumption, one has x R u. Moreover, x R z and y R z; by symmetry, one has z R x; by transitivity, one has y R x; by transtivity again, one has y R u. This proves that $R_x \subset R_y$, and the other inclusion holds by symmetry.

It follows that $R_x = R_y$ if and only if x R y.

Let X/R be the set of all equivalence classes modulo R; it is a subset of $\mathfrak{P}(X)$. Let $c: X \to X/R$ be the map $x \mapsto R_x$. By construction, one has $c(x) = c(y)$ if and only if x R y.

A.1.4. Ordered sets

A.1.4.1. — Let us recall that a (strict) order $\prec$ on a set S is a transitive binary relation for which the assertions $x \prec y$ and $y \prec x$ are incompatible.

An *ordered set* is a set together with an order on it. In an ordered set, one writes $x \leq y$ as a shorthand for $x \prec y$ or $x = y$. One also defines $x > y$ and $x \geq y$ as synonyms for $y \prec x$ and $y \leq x$.

An *initial segment* of an ordered set S is a subset I such that for every $x \in$ I and any $y \in$ S, if $y \prec x$, then $y \in$ I.

An ordered set is sometimes called a "partially ordered set". Indeed, one says that two elements x, y of S are comparable if one has either $x \prec y$, or $y \prec x$, or $x = y$; if any two elements of S are comparable, then one says that the order $\prec$ is total, or that S is *totally ordererd*.

A.1.4.2. — Let S be an ordered set and $(a_n)_{n \in \mathbf{N}}$ be a sequence of elements of S. One says that it is *increasing* if $a_{n+1} \geq a_n$ for every $n \in \mathbf{N}$, and that it is *decreasing* if $a_{n+1} \leq a_n$ for every $n \in \mathbf{N}$.

This would probably be called "non-decreasing" and "non-increasing" in the usual US terminology. However, this terminology is terribly misleading, since the two sentences "the sequence is non-decreasing" and "the sequence is not decreasing" have totally different meanings: while the first one means that $a_{n+1} \geq a_n$ for every $n \in \mathbf{N}$, the second one holds if and only if there exists an integer $n \in \mathbf{N}$ such that $a_{n+1} \geq a_n$ is false.

If one wants to insist that $a_{n+1} > a_n$ for every n, one says that it is *strictly increasing*. One defines similarly *strictly decreasing* sequences.

Finally, a sequence (a_n) is *stationary* if there exists an integer $m \in \mathbf{N}$ such that $a_n = a_m$ for every integer $n \geq m$.

A.1.4.3. — Let A be a subset of an ordered set S.

An *upper bound* of A is an element $u \in$ S such that $a \leq u$ for every $a \in$ A; if A has an upper bound, one says that A is bounded above. A *maximal element* of A is an element a of A such that there is no $x \in$ A such that $x > a$, which is an upper bound of A. Observe that A may have an upper bound but no maximal element, and that a maximal element of A may not be an upper bound of A, unless A is totally ordered. A *largest element* of A is an element a of A such that $x \leq a$ for every $x \in$ A.

A *lower bound* of A is an element $l \in$ S such that $l \leq a$ for every $a \in$ A; if A has a lower bound, one says that it is bounded below. A minimal element of A is an element a of A such that there is no $x \in$ A such that $x \prec a$. A *smallest element* of A is an element a of A such that $a \leq x$ for every $x \in$ A.

A *least upper bound* is a minimal element of the set of all upper bounds for A. A *largest lower bound* is a maximal element of the set of all lower bounds for A.

A.1.4.4. — Let S be an ordered set. A subset A of S which is totally ordered is also called a *chain* in S; its *length* is defined as one less than its cardinality: $\ell(A) = \mathrm{Card}(A) - 1$.

Finite chains are particularly easy to describe. Indeed, let A be a non-empty finite chain in S and let $n = \text{Card}(A) - 1$. One proves by induction on n that A has a largest element, a_n. Then, again by induction, the elements of A can be enumerated as a strictly increasing sequence $(a_0, \ldots, a_n)$.

Chains can be ordered by inclusion. Saying that a finite chain $A = \{a_0, \ldots, a_n\}$ as above is maximal means that there is no element $s \in S$ such that $s < a_0$, of $a_{m-1} < s < a_m$ for some $j \in \{1, \ldots, n\}$, or $a_n < s$.

A.1.5. Monoids, groups

A.1.5.1. — A binary law on a set X is a map $X^2 \to X$; we often use an "infix" notation for such laws, choosing a symbol such as $*, +, \cdot, \times, \ldots$ and denoting by $x * y, x + y, x \cdot y, x \times y, \ldots$ the image of the pair (x, y).

One says that the law $*$ is *commutative* if $x * y = y * x$ for all $x, y \in X$.

One says that it is *associative* if $(x * y) * z = x * (y * z)$ for all $x, y, z \in X$. In that case, parentheses are not necessary to make sense of the potentially ambiguous formula $x * y * z$.

One says that an element $e \in X$ is a left neutral element if $e * x = x$ for all $x \in X$, and that it is a right neutral element if $x * e = x$ for all $x \in X$.

If an associative law has a left neutral element e and a right neutral element e', then $e = e * e' = e'$. In this case, we say that e is a *neutral element*, and it is then the only (left or right) neutral element.

A *monoid* is a set endowed with an associative binary law which admits a neutral element.

A.1.5.2. — Let S be a monoid with law $*$ and neutral element e.

One says that an element x is left invertible if there exists a $y \in S$ (a left inverse of x) such that $y * x = e$; one says that it is right invertible if there exists a $z \in S$ (a right inverse of x) such that $x * z = e$.

Assume that x is both left and right invertible, let y be a left inverse of x and let z be a right inverse of x. Then one has $y = y * e = y * x * z = e * z = z$. This implies that y is the only left inverse of x and that it is the only right inverse of x; this element is called the *inverse* of x.

A *group* is a monoid in which every element is invertible.

The set of invertible elements of a monoid is a group with respect to the induced law.

A.1.6. — Let X be a set. With composition of maps, the set $\mathfrak{F}_X$ of all maps from X to X is a monoid, with neutral element the identity map id_X.

In this monoid $\mathfrak{F}_X$, an element $f : X \to X$ is left invertible if and only if it is injective; it is right invertible if and only if it is surjective; it is invertible if and only if it is bijective. Its inverse is denoted by f^{-1}.

The group of invertible elements of $\mathfrak{F}_X$ is called the *permutation group* of X, and is denoted by $\mathfrak{S}_X$. When $X = \{1, \ldots, n\}$, we rather use the notation $\mathfrak{S}_n$.

A.2. Set Theory

Theorem (A.2.1) (Cantor–Bernstein theorem). — *Let* A *and* B *be sets. Assume that there exists an injection f from* A *into* B, *as well as an injection g from* B *into* A. *Then, the sets* A *and* B *are equipotent.*

Proof. — Let $f': f(A) \to A$ be the map such that for any $b \in f(A)$, $f'(b)$ is the unique preimage of b by f, so that $f \circ f' = \mathrm{id}_{f(A)}$ and $f' \circ f = \mathrm{id}_A$. Define similarly $g': g(B) \to B$. The idea of the proof consists in iterating successively f' and g' as much as possible. Let A_e be the set of elements of A of the form $gfgf \dots f(b)$, where $b \in B - f(A)$; let A_o be the set of elements of A of the form $gfgfgf \dots fg(a)$, where $a \in A - g(B)$. The set A_e consists of all elements of A where one can apply f' and g' an even number of times, so that one ends with g' but cannot apply f' again; the set A_o consists of all elements of A where one can apply f' and g' an odd number of times, ending with f' but cannot apply g' again. Let A_∞ be the complementary subset of $A_o \cup A_e$ in A. Define analogously B_e, B_o and B_∞.

By construction, the map f induces bijections from A_e to B_o and from A_∞ to B_∞. The map g induces a bijection from B_e to A_o.

Let $h: A \to B$ be the map that coincides with f on $A_e \cup A_\infty$ and with g' on A_o. It is a bijection. This concludes the proof of the theorem.   □

A.2.2. Cardinals — Set theory defines the *cardinal* Card(A) of an arbitrary set in such a way that Card(A) = Card(B) if and only if A and B are equipotent. (The precise construction often uses von Neumann ordinals but is irrelevant in this book.)

The Cantor–Bernstein theorem implies that one defines an ordering relation on cardinals by writing Card(A) ≤ Card(B) if there exists an injection from A to B.

Cantor's theorem (corollary A.2.17 below) will imply that this ordering relation is total.

A.2.3. Well-orderings — Let S be an ordered set. One says that S is *well-ordered* (or that the given order on S is a *well-ordering*) if every non-empty subset of S possesses a smallest element.

Let S be a non-empty well-ordered set. Applying the definition to S itself, we see that S has a smallest element. In particular, S has a lower bound.

Applying the definition to a pair (x, y) of elements of S, we obtain that S is totally ordered: the smallest element of $\{x, y\}$ is smaller than the other.

The set of all natural integers with the usual order is well-ordered. In contrast, the set of all real numbers which are positive or zero is not well-ordered: the set of all real numbers $x > 1$ has no smallest element.

A subset of a well-ordered set is itself well-ordered.

Properties depending on an element of a well-ordered set can be proved by induction, in a similar way to the induction principle for properties depending on an integer.

Lemma (A.2.4) (Transfinite induction principle). — *Let* S *be a well-ordered set and let* A *be a subset of* S. *Assume that for every* $a \in$ S *such that* A *contains* $\{x \in S; x < a\}$, *one has* $a \in$ A *("induction hypothesis"). Then* A = S.

Proof. — Arguing by contradiction, assume that A ≠ S and let B = S—A. Then B is a non-empty subset of S; let a be the smallest element of B. By assumption, A contains every element $x \in$ S such that $x < a$; by the induction hypothesis, one has $a \in$ A, but this contradicts the definition of a. Consequently, A = S.□

A.2.5. — The induction principle can also be reformulated by considering the nature of the elements of a well-ordered set S. If S has a largest element, we set $S_* = S - \{\sup(S)\}$; otherwise, we set $S_* = S$.

Let $a \in S_*$; since a is not the largest element of S, the set $\{x \in S; x > a\}$ is non-empty; its smallest element is called the *successor* of a; let us denote it by $\sigma(a)$. One has $a < \sigma(a)$.

Let $a, b \in S_*$ be such that $a < b$. Then $\sigma(a) \le b$, by definition of $\sigma(a)$, hence $\sigma(a) < \sigma(b)$. This proves that the map $\sigma \colon S_* \to S$ is strictly increasing; in particular, it is injective.

An element $a \in$ S is a successor if and only if $\{x \in S; x < a\}$ has a largest element a'; then $a = \sigma(a')$. Otherwise, one says that a is a *limit*.

A.2.6. Transfinite constructions — It is also possible to make transfinite constructions indexed by a well-ordered set S. For $a \in$ S, let us write $S_a = \{x \in S; x < a\}$, so that a is the smallest element of $S - S_a$.

Proposition (A.2.7) (Transfinite construction principle). — *Let* X *be a set. Let* S *be a well-ordered set and let* $\mathscr{D}_X$ *be the set of pairs* (a, f), *where* $a \in$ S *and* $f \colon S_a \to X$ *is a function. For every* $\Phi \in \mathscr{D}_X$, *there exists a unique function* F: S → X *such that* $F(a) = \Phi(a, F|_{S_a})$ *for every* $a \in$ S.

Proof. — Let F, F' be two such functions. Let us prove that F = F' by transfinite induction: let A be the set of $a \in$ S such that $F(a) = F'(a)$ and let us prove that A satisfies the transfinite induction hypothesis. Let $a \in$ A be such that $S_a \subset$ A; by definition, one has $F(x) = F'(x)$ for every $x \in S_a$, so that $F|_{S_a} = F'|_{S_a}$; by definition of F and F', one has $F(a) = \Phi(F|_{S_a}, a) = \Phi(F'|_{S_a}, a) = F'(a)$; this proves that $a \in$ A.

The existence of F is proved by transfinite induction as well. Let S* be the well-ordered set obtained by adding to S a largest element ω and let A be the set of $a \in$ S* such that there exists a function F: $S_a \to$ X satisfying $F(b) = \Phi(F|_{S_b}, b)$ for every $b \in S_a$, and let us prove that A satisfies the transfinite induction hypothesis. Let $a \in$ S* be such that $S_a \subset$ A. We distinguish two cases:

– If a is a successor, let $b \in S_a$ be such that $a = \sigma(b)$ and let F: $S_b \to$ X be a function such that $F(x) = \Phi(F|_{S_x}, x)$ for every $x \in S_b$. Then $S_a = S_b \cup \{b\}$; let F': $S_a \to$ X be defined by $F'(x) = F(x)$ if $x \in S_b$ and $F'(b) = \Phi(F|_{S_b}, b)$. Then F' satisfies the required hypothesis, hence $a \in$ A.

– Otherwise, a is a limit; for every $b \in S_a$, there exists a function $F_b \colon S_b \to$ X satisfying the relation $F(x) = \Phi(F|_{S_x}, x)$ for every $x \in S_b$. By uniqueness, if

$b \prec c$, then F_b and F_c coincide on S_b. Since a is a limit, the union of all S_b, for $b \in S_a$, is equal to S_a, so that there exists a unique function $F : S_a \to X$ which coincides with F_b on S_b. This function F satisfies the desired relation $F(x) = \Phi(F|_{S_x}, x)$ for all $x \in S_a$ since for such x, there exists an element $b \in S_a$ such that $x \in S_b$, and $F(x) = F_b(x) = \Phi(F_b|_{S_x}, x) = \Phi(F|_{S_x}, x)$.

By transfinite induction, one has $\omega \in A$. Since $S_\omega = S$, this proves that there exists a function $F : S \to X$ satisfying the relation $F(a) = \Phi(F|_{S_a}, a)$ for every $a \in S$.   □

Corollary (A.2.8). — *Let* S, T *be two well-ordered sets. One and only one of the following assertions holds:*

   (i) *There exists an increasing bijection* $f : S \to T$;

   (ii) *There exists* $a \in S$ *and an increasing bijection* $f : S_a \to T$;

   (iii) *There exists* $b \in T$ *and an increasing bijection* $f : S \to T_b$.

Moreover, in each case, the indicated bijection (and the element a, *resp.* b) *is uniquely determined.*

This is a very strong rigidity property of well-ordered sets.

Proof. —   a) Let us add to T an element ω and endow the set $T^* = T \cup \{\omega\}$ with the order such that $\omega \geq b$ for all $b \in T$, so that ω is the largest element of T^*. Then T^* is well-ordered. By transfinite induction, there exists a unique map $f : S \to T^*$ such that $f(a) = \inf((T^* - f(S_a)) \cup \{\omega\})$ for every $a \in S$. Explicitly, one has $f(a) = \inf(T - f(S_a))$ if $f(S_a) \neq T$, and $f(a) = \omega$ if $f(S_a) \supset T$.

It follows from this definition that if $\omega \in f(S)$, then $f(S) = T^*$. Let moreover $x, y \in S$ such that $x < y$ and $f(x) = f(y)$; by the definition of f, one has $f(y) = \omega$. Consequently, the map f induces a bijective increasing map from $f^{-1}(T)$ to $f(f^{-1}(T))$.

We will show that the three cases of the corollary correspond to the three mutually incompatible possibilities: $f(S) = T$, $\omega \in f(S)$ and $f(S) \subsetneq T$.

First assume that $f(S) = T$. Then f is a bijection.

Assume that $\omega \in f(S)$ and let $a = \inf(f^{-1}(\omega))$. Then $f(S_a) \subset T$ and $f(a) = \omega = \inf(T^* - f(S_a))$, so that $f(S_a) = T$. Consequently, f induces an increasing bijection from S_a to T.

Assume finally that $f(S) \subsetneq T$ and let $b = \inf(T - f(S))$, so that $T_b \subset f(S)$ and $b \notin f(S)$. Conversely, if $x \in S$, then $f(x) = \inf(T - f(S_x)) \leq \inf(T - f(S))$, since $S_x \subset S$. This implies that $f(S) = T_b$, and f is an increasing bijection from S to T_b.

b) To establish the uniqueness assertion, we first prove that if $f : S \to T$ is a strictly increasing map, then $f(x) \geq x$ for all $x \in S$. By transfinite induction, it suffices to prove that for any $a \in S$ such that $f(x) \geq x$ for all $x < a$, then $f(a) \geq a$. If this does not hold, that is, if $f(a) < a$, then one has $f(f(a)) < f(a)$ because f is strictly increasing, which contradicts the induction hypothesis.

Let then $f, g : S \to T$ be two increasing bijective maps and let us prove that $f = g$. Applying the initial result for $g^{-1} \circ f$, we have $g^{-1} \circ f(x) \geq x$ for all $x \in S$, hence $f(x) \geq g(x)$ because g is increasing. By symmetry, one has $g(x) \geq f(x)$ for all $x \in S$, hence $f = g$.

Let $a \in S$ and let $f : S \to S_a$ be an increasing bijection. Applying the initial result, one has $a \leq f(a)$; by assumption, $f(a) \in S_a$, hence $f(a) < a$, a contradiction.

This implies that case (i) of the corollary is incompatible with the cases (ii) or (iii), the uniqueness of a bijection f as indicated in each case, as well as the fact that the elements a (in (ii)) or b (in (iii)) are well defined. Cases (ii) and (iii) are themselves incompatible: if $f : S_a \to T$ and $g : S \to T_b$ are increasing bijections, let $a' \in S$ be such that $f(a') = b$; then f induces an increasing bijection from $S_{a'}$ to T_b, hence $g^{-1} \circ f$ induces an increasing bijection from $S_{a'}$ to S, a contradiction. $\qquad\square$

Proposition (A.2.9) (Hartogs). — *Let X be a set. There exists a well-ordered set S such that no map $f : S \to X$ is injective.*

Proof. — Let $\mathscr{S}$ be the set of well-orderings on a subset A of X. Let us define two relations $\simeq$ and $\leq$ on $\mathscr{S}$. First, we say that $(A, \leq_A) \simeq (B, \leq_B)$ if there exists an increasing bijection from A to B; this is an equivalence relation. We then say that that $(A, \leq_A) \leq (B, \leq_B)$ if there exists an increasing bijection from A to B, or an increasing bijection from A to B_b, for some $b \in B$. The relation $\leq$ is compatible with the equivalence relation $\simeq$ and corollary A.2.8 implies that on the quotient set $S = \mathscr{S}/\simeq$, the relation $\leq$ induces a total order.

In fact, S is even well-ordered. Indeed, let $\mathscr{A}$ be a non-empty subset of S and let $\mathscr{A}'$ be its preimage in $\mathscr{S}$. It suffices to find an element $(B, \leq_B) \in \mathscr{A}'$ such that $(B, \leq_B) \leq (C, \leq_C)$ for every $(C, \leq_C) \in \mathscr{A}'$.

Fix an element $(A, \leq_A)$ in $\mathscr{A}'$. Let then $\mathscr{B}'$ be the set of all elements $(B, \leq_B)$ in $\mathscr{A}'$ such that $(B, \leq_B) \prec (A, \leq_A)$. If $\mathscr{B}'$ is empty, then the class of $(A, \leq_A)$ modulo $\simeq$ is the smallest element of $\mathscr{A}$. Otherwise, for every $B \in \mathscr{B}'$, let $\varphi(B)$ be the unique element of A such that there exists an increasing bijection f_B from B to $A_{\varphi(B)}$; let then a be the smallest element of A which is of this form; fix $B \in \mathscr{B}'$ such that $a = \varphi(B)$.

Let $(C, \leq_C) \in \mathscr{A}'$. If $(C, \leq_C) \notin \mathscr{A}'$, then $(A, \leq_A) \leq (C, \leq_C)$, hence $(B, \leq_B) \leq (C, \leq_C)$. Otherwise, one has $(C, \leq_C) \in \mathscr{A}'$ and $\varphi(C) \geq \varphi(B)$. Let then $f_B : B \to A_{\varphi(B)}$ and $f_C : C \to A_{\varphi(C)}$ be increasing bijections. If $\varphi(B) = \varphi(C)$, then $(B, \leq_B) \simeq (C, \leq_C)$; in particular, $(B, \leq_B) \leq (C, \leq_C)$. Otherwise, one has $\varphi(B) < \varphi(C)$, so that there exists a $c \in C$ such that $f_C(c) = \varphi(B)$; then f_C induces an increasing bijection from C_c to $A_{\varphi(B)}$, so that $C_c \simeq B$ and $(B, \leq_B) \prec (C, \leq_C)$. This proves that the class of $(B, \leq_B)$ is the smallest element of $\mathscr{A}$, and concludes the proof that S is well-ordered.

To conclude the proof of the proposition, it remains to establish that there does not exist an injection from S to X. Let us argue by contradiction and consider such a map $h : S \to X$. Let $A = h(S)$. Transferring the ordering of S to A by h, one gets a well-ordering $\leq_A$ on A, so that $(A, \leq_A)$ is an element of $\mathscr{S}$; let α be its class in S. For every $x \in S$, $A_{f(x)}$ is a well-ordered subset of A; let $g(x)$ be its class in S; one has $g(x) < \alpha$ by corollary A.2.8. The map

$g \colon S \to S$ is strictly increasing, so that $g(x) \leq x$ for every $x \in S$. Taking $x = \alpha$, we get the desired contradiction $\alpha \leq g(\alpha) < \alpha$. □

A.2.10. — From now on, we will make use of the *axiom of choice*: the product $\prod_{i \in I} S_i$ of any family $(S_i)_{i \in I}$ of non-empty sets is non-empty.

It has various alternatives formulations:

– *For every set S, there exists a map* $f \colon \mathscr{P}(S) - \{\emptyset\} \to S$ *such that* $f(A) \in A$ *if* $A \neq \emptyset$ *("choice function")*. Let $I = \mathscr{P}(S) - \{\emptyset\}$; for $i \in I$, let S_i be the corresponding non-empty subset. An element of $\prod S_i$ is precisely a choice function.

– *If* $f \colon S \to T$ *is a surjective map, there exists a map* $g \colon T \to S$ *such that* $f \circ g = \mathrm{id}_T$. Let $I = T$ and for $t \in T$, let $S_t = f^{-1}(t)$; by assumption, $(S_t)_{t \in T}$ is a family of non-empty sets. An element of $\prod_{t \in T} S_t$ is precisely a function $g \colon T \to S$ such that $g(t) \in f^{-1}(t)$ for every $t \in T$, that is, $f \circ f = \mathrm{id}_T$.

Theorem (A.2.11) (Zorn's lemma). — *Let S be an ordered set. Assume that every well-ordered subset of S has an upper bound. Then S has a maximal element.*

Proof. — Let us argue by contradiction, assuming that S doesn't have a maximal element. Let S^* be the ordered set obtained by adjoining to S a largest element ω.

Let A be a well-ordered subset of S. Let a be an upper bound of A; since a is not a maximal element of S, there exists an element $a' \in S$ such that $a < a'$. Using the axiom of choice, there exists a function f that assigns, to any well-ordered subset A of S, an element $f(A) \in S$ such that $x < f(A)$ for every $x \in A$. We extend f to $\mathfrak{P}(S^*)$ by setting $f(A) = \omega$ if A is not a well-ordered subset of S.

Let T be a well-ordered set that admits no injection to S. By the transfinite construction principle, there exists a unique map $h \colon T \to S^*$ such that $h(t) = f(T_t)$ for every $t \in T$. Let us prove that h is strictly increasing and that $h(T) \subset S$. By the transfinite induction principle, it suffices to prove that if $h|_{T_t}$ is strictly increasing and its image is contained in S, then $h(t) \in S$ and $h(x) < h(t)$ for every $x \in T_t$. Then $h(T_t)$ is a well-ordered subset of S, hence $h(t) = f(h(T_t)) \in S$ and satisfies $h(t) > h(x)$ for every $x \in T_t$, hence the claim.

In particular, h is an injective map from T to S, and this contradicts the choice of T. □

A.2.12. — Classical applications of Zorn's lemma (theorem A.2.11) happen in situations where the ordered set S satisfies stronger assumptions.

One says that an ordered set S is *inductive* if every totally ordered subset of S has an upper bound. Applying the definition to the empty subset, we observe that an inductive set is not empty. Moreover, if S is an inductive set, then for every $a \in S$, the set $\{x \in S; a \leq x\}$ is also inductive. The following result is then an immediate corollary of Zorn's lemma.

Corollary (A.2.13). — *Every inductive set has a maximal element. More precisely, if S is an inductive set and a is an element of S, then S has a maximal element b such that $a \leq b$.*

A.2.14. — Let X be a set and let S be a subset of $\mathfrak{P}(X)$, ordered by inclusion. One says that S is of *finite character* if it contains the empty set and if a subset A of X belongs to S if and only if every finite subset of A belongs to S. Two extremly important examples appear in the text: the set of free subsets of a vector space (theorem 3.7.3), and the set of algebraically free subsets of a field extension (theorem 4.8.4).

Lemma (A.2.15). — *Let X be a set and let S be a subset of $\mathfrak{P}(X)$, ordered by inclusion. If S is of finite character, then S is inductive.*

Proof. — Let T be a totally ordered subset of S; let us prove that T has an upper bound in S. If T is empty, then $\emptyset$ is an upper bound for T, and $\emptyset \in S$, by definition. Otherwise, let V be the union of all elements of T. By definition, one has $A \subset V$ for every $A \in T$, hence V is an upper bound for T. To conclude the proof of the lemma, it suffices to prove that V belongs to S.

By the definition of a set of finite character, in order to prove that $V \in S$, it suffices to prove that every finite subset A of V belongs to S.

In fact, let us prove by induction on the cardinality of A that there exists a subset $B \in T$ such that $A \subset B$. This holds if $A = \emptyset$, since $\emptyset \in S$, by assumption. Otherwise, let $a \in A$ and let $A' = A - \{a\}$; by induction, there exists a subset $B' \in T$ such that $A' \subset B'$. Since $a \in A \subset V$, there exists a set $B'' \in T$ such that $a \in B''$. Since T is totally ordered, one has either $B' \subset B''$, or $B'' \subset B'$; since $A = A' \cup \{a\}$, we get $A \subset B''$ or $A \subset B'$, which proves the claim by induction.

Since A is finite, the definition of a set of finite character implies that A belongs to S.

Applying once more the definition of a set of finite character, this proves that $V \in S$. $\qquad\square$

Corollary (A.2.16) (Tukey). — *Let X be a set and let S be a subset of $\mathfrak{P}(X)$ which is of finite character. Then, for every $A \in S$, there exists a maximal element M of S such that $A \subset M$.*

Proof. — By the preceding lemma, the set S is inductive. The result thus follows from Zorn's lemma (corollary A.2.13). $\qquad\square$

Corollary (A.2.17) (Cantor). — *Let X, Y be sets. At least one of the two possibilities hold:*

(i) *There exists an injective map $f : X \to Y$;*
(ii) *There exists an injective map $f : Y \to X$.*

Proof. — Let $\mathscr{C}$ be the set of all pairs (A, f), where A is a subset of X and $f : A \to Y$ is an injective map. We order $\mathscr{C}$ as follows: $(A, f) \le (B, g)$ if $A \subset B$ and $g|_A = f$. Let us prove that the set $\mathscr{C}$ is inductive.

Let $((A_i, f_i))_{i \in I}$ be a totally ordered family of elements of $\mathscr{C}$; let $A = \bigcup_i A_i$. I claim that there exists a unique map $f : A \to Y$ such that $f(x) = f_i(x)$ for every $i \in I$ and every $x \in A_i$. Indeed, let $i, j \in I$ be such that $x \in A_i$ and $x \in A_j$; if $(A_i, f_i) \le (A_j, f_j)$, then $A_i \subset A_j$ and $f_i = f_j|_{A_i}$, so that $f_i(x) = f_j(x)$; by symmetry, the same equality holds if $(A_j, f_j) \le (A_i, f_i)$.

Let $x, y \in A$ be such that $f(x) = f(y)$. Let $i, j \in I$ be such that $x \in A_i$ and $y \in A_j$. If $(A_i, f_i) \leq (A_j, f_j)$, then $x, y \in A_j$, hence $f(x) = f_j(x)$ and $f(x) = f_j(y)$, hence $f_j(x) = f_j(y)$; since f_j is injective, this implies $x = y$. By symmetry, the same property holds if $(A_j, f_j) \leq (A_i, f_i)$. This proves that f is injective.

Consequently, (A, f) is an element of $\mathscr{C}$ and one has $(A_i, f_i) \leq (A, f)$ for every $i \in I$. This proves that the ordered set $\mathscr{C}$ is inductive.

By Zorn's lemma (corollary A.2.13), the set $\mathscr{C}$ admits a maximal element, say (A, f). If $A = X$, then f is an injective map from X to Y. Assume that $A \neq X$; I then claim that f is surjective. Otherwise, there exists $a \in X - A$ and $b \in Y - f(A)$; the map $g : A \cup \{a\} \to Y$ such that $g|_A = f$ and $g(a) = b$ is injective, and one has $(A, f) \leq (A \cup \{a\}, g)$, contradicting the hypothesis that (A, f) is a maximal element of $\mathscr{C}$. The map $f : A \to Y$ is thus bijective; its inverse $f^{-1} : Y \to A$ is an injective map from Y to X. This concludes the proof of Cantor's theorem. □

Corollary (A.2.18) (Zermelo, 1904). — *Every set can be well-ordered.*

Proof. — Let X be a set. Let S be a set such that there is no injection from S to X. By Cantor's theorem (corollary A.2.17), there exists an injection f from X to S. Then there exists a unique ordering relation on X such that f is increasing; this ordering is a well-ordering on X, establishing the corollary. □

A.3. Categories

Category theory provides a very useful and common vocabulary to describe certain algebraic structures (the so-called categories) and the way by which one can connect them (functors).

A.3.1. Categories — A *category* C consists of the following data:

  – A collection ob C of *objects*;

  – For any two objects M, N, a set $C(M, N)$ called the *morphisms* from M to N;

  – For any three objects M, N, P, a *composition map* $C(M, N) \times C(N, P)$, $(f, g) \mapsto g \circ f$,

so that the following axioms are satisfied:

  (i) For any object M, there is a distinguished morphism $\mathrm{id}_M \in C(M, M)$, called the identity;

  (ii) $\mathrm{id}_N \circ f = f$ for any $f \in C(M, N)$;

  (iii) $g \circ \mathrm{id}_N = g$ for any $g \in C(N, P)$;

  (iv) For any four objects M, N, P, Q, and any three morphisms $f \in C(M, N)$, $g \in C(N, P)$, $h \in C(P, Q)$, the two morphisms $h \circ (g \circ f)$ and $(h \circ g) \circ f$ in $C(M, Q)$ are equal (associativity of composition).

A common notation for $C(M, N)$ is also $\mathrm{Hom}_C(M, N)$. Finally, instead of $f \in C(M, N)$, one often writes $f : M \to N$.

Let $f : M \to N$ be a morphism in a category C. One says that f is left-invertible, resp. right-invertible, resp. invertible, if there exists a morphism $g : N \to M$ such that $g \circ f = \mathrm{id}_M$, resp. $f \circ g = \mathrm{id}_N$, resp. $g \circ f = \mathrm{id}_M$ and $f \circ g = \mathrm{id}_N$.

One proves in the same way as for rings (see p. 9) or for monoids (see p. 439) that if f is both left- and right-invertible, then it has a unique left-inverse and a unique right-inverse, and both are equal, so that f is invertible. Indeed, if g is a left inverse of f and h is a right inverse of f, then $g = g \circ (f \circ h) = (g \circ f) \circ h = h$.

An invertible morphism is also called an isomorphism.

Let us now give examples of categories. It will become clear that all of the basic algebraic structures fall within the categorical framework.

Examples (A.3.2). — *a)* The category $\boldsymbol{Set}$ of sets has for objects the sets, and for morphisms the usual maps between sets.

b) The category $\boldsymbol{Gr}$ of groups has for objects the groups and for morphisms the morphisms of groups. The category $\boldsymbol{AbGr}$ of abelian groups has for objects the Abelian groups and for morphisms the morphisms of groups. Observe that objects of $\boldsymbol{AbGr}$ are objects of $\boldsymbol{Gr}$, and that morphisms in $\boldsymbol{AbGr}$ coincide with those in $\boldsymbol{Gr}$; one says that $\boldsymbol{AbGr}$ is a full subcategory of $\boldsymbol{Gr}$.

c) The category $\boldsymbol{Ring}$ of rings has for objects the rings and for morphisms the morphisms of rings.

d) Similarly, there is the category $\boldsymbol{Field}$ of fields and, if k is a field, the category $\boldsymbol{Ev}_k$ of k-vector spaces. More generally, for any ring A, there is a category $\boldsymbol{Mod}_A$ of right A-modules, and a category $\boldsymbol{Mod}_A$ of left A-modules.

Example (A.3.3). — Let C be a category; its *opposite category* C° has the same objects as C, but the morphisms of C° are defined by $C^\circ(M, N) = C(N, M)$ and composed in the opposite direction.

This resembles the definition of an opposite group. However, a category is usually different from its opposite category.

Remark (A.3.4). — Since there is no set containing all sets, nor a set containing all vector spaces, the word *collection* in the above definition cannot be replaced by the word *set* (in the sense of the Zermelo–Fraenkel theory of sets). In fact, a proper treatment of categories involves set-theoretic issues. There are at least three ways to resolve them:

– The easiest one is to treat a category as a formula (in the sense of first order logic). For example $\boldsymbol{Ring}$ is a formula φ_{Ring} with one free variable A that expresses that A is a ring. This requires us to encode a ring A and all its laws as a tuple: for example, one may consider a ring to be a tuple (A, S, P) where A is the ring, S is the graph of the addition law and P is the graph of the multiplication law. The formula $\varphi_{Ring}(x)$ then checks that x is a triplet of the form (A, S, P), where $S \subset A^3$ and $P \subset A^3$, that S is the graph of a map

$A \times A \to A$ which is associative, commutative, has a neutral element, and for which every element has an opposite, etc.

Within such a framework, one can also consider functors (defined below), but only those which can be defined by a formula.

This treatment would be sufficient at the level of this book.

– One can also use another theory of sets, such as the one of Bernays–Gödel–von Neumann, which allows for two kinds of collections: sets and classes. Sets obey the classical formalism of sets, but classes are more general, so that one can consider the class of all sets (but not the class of all classes). Functors are defined as classes.

This is a very convenient possibility at the level of this book. However, at a more advanced development of algebra, one is led to consider the category of categories, or categories of functors. Then, this approach also becomes insufficient.

– Within the classical theory of sets, Grothendieck introduced *universes*, which are very large sets, so large that every usual construction of sets does not leave a given universe. One needs to add the axiom that there is a universe, or, more generally, that any set belongs to some universe. This axiom is equivalent to the existence of *inaccessible cardinals*, an axiom which is well studied and often used in advanced set theory.

In this book, categories mostly provide a *language* to state algebraic results of a quite formal nature.

Definition (A.3.5). — Let C be a category, let M, N be objects of C and let $f \in C(M, N)$.

One says that f is a *monomorphism* if for any object P of C and any morphisms $g_1, g_2 \in C(N, P)$ such that $g_1 \circ f = g_2 \circ f$, one has $g_1 = g_2$.

One says that f is an *epimorphism* if for any object L of C and any morphisms $g_1, g_2 \in C(P, M)$ such that $f \circ g_1 = f \circ g_2$, one has $g_1 = g_2$.

Example (A.3.6). — Monomorphisms and epimorphisms in *Set* or in categories of modules are respectively injections and surjections (see exercise 7.8.20).

A.3.7. Functors — Functors are to categories what maps are to sets.

Let C and D be two categories.

A *functor* F from C to D, also called a *covariant functor*, consists of the following data:

– an object F(M) of D for any object M of C;

– a morphism $F(f) \in D(F(M), F(N))$ for any objets M, N of C and any morphism $f \in C(M, N)$,

subject to the two following requirements:

(i) For any object M of C, $F(\mathrm{id}_M) = \mathrm{id}_{F(M)}$;

(ii) For any objects M, N, P of C and any morphisms $f \in C(M, N)$ and $g \in C(N, P)$, one has
$$F(g \circ f) = F(g) \circ F(f).$$

A *contravariant functor* F from C to D is a functor from C^o to D. Explicitly, it consists of the following data

- an object F(M) of D for any object M of C;
- a morphism $F(f) \in D(F(N), F(M))$ for any objects M, N of C and any morphism $f \in C(M, N)$,

subject to the two following requirements:

(i) For any object M of C, $F(\mathrm{id}_M) = \mathrm{id}_{F(M)}$;

(ii) For any objects M, N, P of C and any morphisms $f \in C(M, N)$ and $g \in C(N, P)$, one has
$$F(g \circ f) = F(f) \circ F(g).$$

One says that such a functor F is *faithful*, resp. *full*, resp. *fully faithful* if for any objects M, N of C, the map $f \mapsto F(f)$ from $C(M, N)$ to $C(F(M), F(N))$ is injective, resp. surjective, resp. bijective. A similar definition applies for contravariant functors.

Example (A.3.8) (Forgetful functors). —Many algebraic structures are defined by enriching other structures. Often, forgetting this enrichment gives rise to a functor, called a forgetful functor.

For example, a group is already a set, and a morphism of groups is a map. There is thus a functor that associates to every group its underlying set, thus forgetting the group structure. One gets a forgetful functor from Gr to Set. It is faithful, because a group morphism is determined by the map between the underlying sets. It is however not full because there are maps between two (non-trivial) groups which are not morphism of groups.

Example (A.3.9). — Let C be a category and let P be an object of C.

One defines a functor F from the category C to the category of sets as follows:

- For any object M of C, one sets $F(M) = C(P, M)$;
- For any morphism $f : M \to N$ in C, $F(f)$ is the map $u \mapsto f \circ u$ from $C(P, M)$ to $C(P, N)$.

This functor is often denoted $\mathrm{Hom}_C(P, \bullet)$. Such a functor is also called a *representable functor*.

One can also define a contravariant functor G, denoted $\mathrm{Hom}_C(\bullet, P)$, as follows:

- For any object M of C, one sets $F(M) = C(M, P)$;
- For any morphism $f : M \to N$ in C, $F(f)$ is the map $u \mapsto u \circ f$ from $C(N, P)$ to $C(M, P)$.

In other words, G is the functor $\mathrm{Hom}_{C^o}(P, \bullet)$. One says that it is *corepresentable*.

Let F and G be two functors from a category C to a category D. A morphism of functors α from F to G consists of the datum, for every object M of C, of a morphism $\alpha_M\colon F(M) \to G(M)$ such that the following condition holds: For any morphism $f\colon M \to N$ in C, one has $\alpha_N \circ F(f) = G(f) \circ \alpha_M$.

Morphisms of functors can be composed, and for any functor F, one has an identity morsism from F to itself. Consequently, functors from C to D collectively form a category, denoted by $F(C, D)$.

A.3.10. Universal properties and representable functors — This book contains many *universal properties*: the free module on a given basis, quotient ring, quotient module, direct sum and product of modules, localization, algebra of polynomials on a given set of indeterminates. They are all of the following form: "in such algebraic situation, there exists an object and a morphism satisfying such property and such that any other morphism which satisfies this property factors through it".

The prototype of a universal property is the following.

Definition (A.3.11). — Let C be a category.

One says that an object I of C is an *initial object* if for every object M of C, the set $C(I, M)$ has exactly one element.

One says that an object T of C is a *terminal object* if for every object M of C, the set $C(M, T)$ has exactly one element.

Observe that an initial object of a category is a terminal object of the opposite category, and vice versa.

Examples (A.3.12). —   *a*) In the category *Set* of sets, the empty set is the only initial object, and any set of cardinality one is a terminal object.

b) In the category *Ring* of rings, the ring $\mathbf{Z}$ is an initial object (for any ring A, there is exactly one morphism from $\mathbf{Z}$ to A, see Example 1.3.4). Moreover, the ring 0 is a terminal object.

c) In the category of modules over a ring A, the null module is both an initial and a terminal object.

The property for an object I to be an initial object can be rephrased as a property of the representable functor Hom $C(I, \bullet)$, namely that this functor coincides with (or, rather, is isomorphic to) the functor F that sends any object of C to a fixed set with one element.

This allows us to rephrase the definition of an initial object as follows: an object I is an initial object if it *represents* the functor F defined above.

Definition (A.3.13). — Let C be a category and let F be a functor from C to the category *Set* of sets. Let P be an object of C. One says that P represents the functor F is the functor $\mathrm{Hom}_C(P, \bullet)$ is isomorphic to F.

Similarly, if G is a contravariant functor from C to *Set*, one says that an object P of C co-represents G if the functors $\mathrm{Hom}_C(\bullet, P)$ and G are isomorphic.

Objects that represent a given functor are unique up to an isomorphism:

Proposition (A.3.14) (Yoneda's lemma). — *Let C be a category, let A and B be two objects of C.*

For any morphism of functors φ from $\mathrm{Hom}_C(A, \bullet)$ to $\mathrm{Hom}_C(B, \bullet)$, there is a unique morphism $f : B \to A$ such that $\alpha_M(u) = u \circ f$ for any object M of C and any morphism $u \in C(A, M)$.

In particular, φ is an isomorphism if and only if f is an isomorphism.

Proof. — Let us write $F = \mathrm{Hom}_C(A, \bullet)$ and $G = \mathrm{Hom}_C(B, \bullet)$. Recall the definition of a morphism of functors: for every object M of C, one has a map $\varphi_M : F(M) \to G(M)$ such that $G(u) \circ \varphi_M = \varphi_N \circ F(u)$ for every two objects M, N of C and every morphism $u : M \to N$. In the present case, this means that for every object M of C, φ_M is a map from $C(A, M)$ to $C(B, M)$ and that $u \circ \varphi_M(f) = \varphi_N(u \circ f)$ for every $f \in C(A, M)$ and every $u \in (M, N)$. In particular, taking M = A and $f = \mathrm{id}_A$, one obtains $\varphi_N(u) = u \circ \varphi_A(\mathrm{id}_A)$. This proves the existence of a morphism f as required by the lemma, namely $f = \varphi_A(\mathrm{id}_A)$. The uniqueness of f also follows from this formula, applied to N = A and $u = \mathrm{id}_A$: if $f' : N \to M$ is any morphism possessing the required property, one has $f = \varphi_A(\mathrm{id}_A) = \mathrm{id}_A \circ f' = f'$. ☐

A.3.15. Adjunction — Let C and D be two categories, let F be a functor from C to D and G be a functor from D to C.

An *adjunction* for the pair (F, G) is the datum, for any objects M of C and N of D, of a *bijection*

$$\Phi_{M,N} : D(F(M), N) \overset{\sim}{\to} C(M, G(N))$$

such that the following holds: for any objects M, M' of C, any morphism $f \in C(M, M')$, any objects N, N' of D, any morphism $g \in D(M, M')$, and any morphism $u \in D(F(M'), N)$,

$$g \circ \Phi_{M',N}(u) \circ F(f) = \Phi_{M,N'}(G(g) \circ u \circ f).$$

If there exists an adjunction for the pair (F, G), one says that F is a left adjoint to G, or that G is a right adjoint to F.

Section 7.6 of the book offers a first study of adjoint morphisms in the framework of categories of modules. The reader is invited to try to generalize what is explained there to other categories.

Remark (A.3.16). — Assume that F and G are adjoint functors. Then for any object M of C, F(M) represents the functor $\mathrm{Hom}_C(M, G(\bullet))$ from D to *Set*.

Conversely, let G be a functor from D to D. One can prove that if the functor $\mathrm{Hom}_C(M, G(\bullet))$ is representable for any object M of C, then G has a left adjoint. The proof consists in choosing, for any object M of C, an object F(M) which represents the given functor. Moreover, any morphism $f : M \to M'$ in C gives rise to a morphism from the functor $\mathrm{Hom}_C(M', G(\bullet))$ to the functor $\mathrm{Hom}_C(M, G(\bullet))$, hence, by the contravariant version of Yoneda's lemma, to a morphism $F(f) : F(M) \to F(M')$.

Credits

Page 10
  – Portrait of William Rowan Hamilton. Etching by John Kirkwood, after Charles Grey. Source: Wellcome Library no. 3972, via Wikipedia. Copyright: CC-BY-4.0.
  – Photograph of Broom Bridge plaque, source: Wikipedia (cropped), uploaded by: *Cone83*, CC BY-SA 4.0

Page 66
  – Portrait of David Hilbert, source: Wikipedia, unknown photographer, public domain
  – Photograph of David Hilbert grave, source: Wikipedia, uploaded by: *Kassandro*, CC BY-SA 3.0

Page 84
  – Portrait of Carl Friedrich Gauss, source: Wikipedia, artist: Siegfried Detlev Bendixen (1828), public domain
  – Postage stamp of the Federal Republic of Germany, issued 1977; copyright: unapplicable, source: `www.mathematicalstamps.eu` (cropped)

Page 142
  – Portrait of Arthur Cayley (circa 1860), source: Wikipedia, photographer: Herbert Beraud, public domain
  – Representation of the Cayley surface made by the author with the help of the software SURFER

Page 178
  – Portrait of Évariste Galois (circa 1826), source: Wikipedia, unknown artist, public domain
  – French stamp (1984), engraver: Jacques Combet, ©Musée de La Poste, Paris / La Poste 2020, source: `wikitimbres.fr`

Page 220
  – Portrait of Ferdinand Frobenius (circa 1886); photographer: Carl Günther, source: ETH-Zürich, Bibliothek Bildarchiv (doi: 10.3932/ethz-a-000046508), public domain

© Springer Nature Switzerland AG 2021
A. Chambert-Loir, *(Mostly) Commutative Algebra*, Universitext,
https://doi.org/10.1007/978-3-030-61595-6

– Character table, from G. Frobenius (1899), "Über die Composition der Charaktere einer Gruppe", *Sitzungsberichte der Königlich Preußischen Akademie der Wissenschaften zu Berlin*, p. 330–339. In: *Gesammelte Abhandlungen*, Band III, edited by J-P. Serre, 1968, Springer-Verlag, p. 128

Page 256

– Portrait of Emil Artin (1920s), photographer: Natascha Artin-Brunswick, by courtesy of Tom Artin
– Braids drawing, from E. Artin (1947), "Theory of braids", *Annals of Mathematics* **48** (1), p. 101–126, by courtesy of the Department of Mathematics at Princeton University

Page 272

– Portrait of Emmy Noether (before 1910), source: Wikipedia, unknown photographer, public domain
– Copy of E. Noether, "Ableitung der Elementarteilertheorie aus der Gruppentheorie", Nachrichten der 27 Januar 1925, *Jahresbericht Deutschen Math. Verein.* (2. Abteilung) 34 (1926), p. 104. Source: Göttinger Digitalisierungszentrum

Page 378

– Portrait of Alexander Grothendieck (1950), photographer: Paulo Ribenboim (digitized by Philippe Douroux), courtesy of Paulo Ribenboim
– Photograph of Grothendieck's seminar (1967), unknown photographer, unknown copyright

Page 408

– Portrait of Wolfgang Krull (1969), photographer: Paul Halmos. Source: *Who's That Mathematician? Paul R. Halmos Collection*, Mathematical Association of America. Digitization: Archives of American Mathematics, Dolph Briscoe Center for American History, University of Texas, Austin, under the direction of Archivist Carol Mead. Unknown copyright
– Cover of Krull's *Idealtheorie* (1935), Springer-Verlag

Page 418

– Portrait of Richard Dedekind (before 188§), photographer: Johannes Ganz, source: ETH-Zürich, Bibliothek Bildarchiv (doi:10.3932/ethz-a-000046496); public domain
– Postage stamp of the German Democratic Republic, issued in 1981 on the occasion of Dedekind's 150th birthday; engraver: Gerhard Stauf; Michel catalogue: 2605; Source: *colnect.com* website; copyright, unknown.

References

M. Aigner & G. M. Ziegler (2014), *Proofs from The Book*, Springer-Verlag, Berlin, fifth edition. URL `http://dx.doi.org/10.1007/978-3-662-44205-0`, including illustrations by Karl H. Hofmann.

E. Artin (1998), *Galois theory*, Dover Publications Inc., second edition. Edited and with a supplemental chapter by Arthur N. Milgram.

M. F. Atiyah & I. G. Macdonald (1969), *Introduction to Commutative Algebra*, Addison–Wesley.

H. Bass (1960), "Finitistic dimension and a homological generalization of semi-primary rings". *Transactions of the American Mathematical Society*, **95** (3), pp. 466–488.

N. Bourbaki (1989a), *Elements of Mathematics. Algebra. Chapters 1-3*, Berlin etc.: Springer-Verlag. Transl. from the French. 2nd Printing.

N. Bourbaki (1989b), *Elements of Mathematics. Commutative Algebra. Chapters 1-7*, Berlin etc.: Springer-Verlag. Transl. from the French. 2nd Printing.

N. Bourbaki (1999), *Elements of the History of Mathematics*, Berlin: Springer. Transl. from the French by John Meldrum. 2nd Printing.

N. Bourbaki (2003), *Elements of Mathematics. Algebra. Chapters 4-7*, Berlin: Springer. Transl. from the French by P. M. Cohn and J. Howie. Reprint of the 1990 English translation.

N. Bourbaki (2012), *Éléments de mathématique. Algèbre. Chapitre 8. Modules et anneaux semi-simples*, Springer, Berlin. Second revised edition of the 1958 edition.

T. Coquand (2004), "Sur un théorème de Kronecker concernant les variétés algébriques". *C. R. Acad. Sci. Paris Sér. I Math.*, **338** (4), pp. 291–294.

T. Coquand & H. Lombardi (2005), "A Short Proof for the Krull Dimension of a Polynomial Ring". *The American Mathematical Monthly*, **112** (9), p. 826.

T. Coquand, H. Lombardi & M.-F. Roy (2005), "An elementary characterization of Krull dimension". *From Sets and Types to Topology and Analysis*, Oxford Logic Guides **48**, pp. 239–244, Oxford Univ. Press, Oxford.

L. Corry (2004), *Modern algebra and the rise of mathematical structures*, Birkhäuser Verlag, Basel, second edition.

© Springer Nature Switzerland AG 2021
A. Chambert-Loir, *(Mostly) Commutative Algebra*, Universitext,
https://doi.org/10.1007/978-3-030-61595-6

B. Deschamps (2001), "À propos d'un théorème de Frobenius". *Annales mathématiques Blaise Pascal*, **8** (2), pp. 61–66.

H.-D. Ebbinghaus, H. Hermes, F. Hirzebruch, M. Koecher, K. Mainzer, J. Neukirch, A. Prestel & R. Remmert (1991), *Numbers*, Graduate Texts in Mathematics **123**, Springer New York, New York, NY.

H. M. Edwards (1983), "Dedekind's invention of ideals". *Bull. London Math. Soc.*, **15** (1), pp. 8–17.

C. Ehrhardt (2011), "Évariste Galois and the social time of mathematics". *Revue d'histoire des mathématiques*, **17** (2), pp. 175–210.

D. Eisenbud (1995), *Commutative Algebra with a View towards Algebraic Geometry*, Graduate Texts in Mathematics **150**, Springer-Verlag.

B. Fine & G. Rosenberger (1997), *The Fundamental Theorem of Algebra*, Undergraduate Texts in Mathematics, Springer New York, New York, NY.

A. Geraschenko (2009), "What do epimorphisms of (commutative) rings look like?" https://mathoverflow.net/questions/109/what-do-epimorphisms-of-commutative-rings-look-like. Accessed: 2020-07-31.

J. Gray (2018), *A History of Abstract Algebra: From Algebraic Equations to Modern Algebra*, Springer Undergraduate Mathematics Series, Springer International Publishing, Cham.

D. Grinberg (2016), "A constructive proof of Orzech's theorem". URL http://www.cip.ifi.lmu.de/~grinberg/algebra/orzech.pdf, accessed: 2020-07-31.

A. Grothendieck (1960), "Technique de descente et théorèmes d'existence en géométrie algébrique. I. Généralités. Descente par morphismes fidèlement plats". *Séminaire Bourbaki : années 1958/59 - 1959/60, exposés 169-204*, Séminaire Bourbaki **5**, pp. 299–327, Société mathématique de France. URL http://www.numdam.org/item/SB_1958-1960__5__299_0, talk:190.

D. W. Henderson (1965), "A Short Proof of Wedderburn's Theorem". *The American Mathematical Monthly*, **72** (4), p. 385.

D. Hilbert (1890), "Über die Theorie der algebraischen Formen". *Mathematische Annalen*, **36** (4), pp. 473–534.

K. Ireland & M. Rosen (1990), *A Classical Introduction to Modern Number Theory*, Graduate Texts in Mathematics **84**, Springer New York, New York, NY.

N. Jacobson (1985), *Basic Algebra. I*, W. H. Freeman and Company, New York, second edition.

C. W. Kohls (1958), "Prime ideals in rings of continuous functions". *Illinois Journal of Mathematics*, **2** (4A), pp. 505–536.

A. I. Malcev (1937), "On the immersion of an algebraic ring into a field". *Math. Ann.*, **113** (1), pp. 686–691.

H. Matsumura (1986), *Commutative Ring Theory*, Cambridge Studies in Advanced Mathematics, Cambridge Univ. Press.

P. Mazet (1967), "Caractérisation des épimorphismes par relations et générateurs". *Séminaire Samuel. Algèbre commutative*, **2**, pp. 1–8.

W. K. Nicholson (1993), "A short proof of the Wedderburn–Artin theorem". *New Zealand Journal of Mathematics*, **22** (1), pp. 83–86.

E. Noether (1921), "Idealtheorie in Ringbereichen". *Mathematische Annalen*, **83** (1), pp. 24–66.

M. Orzech (1971), "Onto Endomorphisms are Isomorphisms". *The American Mathematical Monthly*, **78** (4), pp. 357–362.

R. S. Palais (1968), "The classification of real division algebras". *The American Mathematical Monthly*, **75**, pp. 366–368.

H. Perdry (2004), "An Elementary Proof of Krull's Intersection Theorem". *The American Mathematical Monthly*, **111** (4), p. 356.

M. A. Rieffel (1965), "A General Wedderburn Theorem". *Proceedings of the National Academy of Sciences*, **54** (6), pp. 1513–1513.

A. Seidenberg (1954), "On the dimension theory of rings. II." *Pacific Journal of Mathematics*, **4** (4), pp. 603–614.

A. Seidenberg (1974), "What is Noetherian?" *Rendiconti del Seminario Matematico e Fisico di Milano*, **44** (1), pp. 55–61.

E. S. Selmer (1956), "On the irreducibility of certain trinomials". *Math. Scand.*, **4** (2), pp. 287–302.

H. Tverberg (1964), "A remark on Ehrenfeucht's criterion for irreducibility of polynomials". *Commentationes Mathematicae*, **8** (2).

W. V. Vasconcelos (1970), "Injective endormorphisms of finitely generated modules". *Proceedings of the American Mathematical Society*, **25** (4), pp. 900–901.

B. L. van der Waerden (1930), *Moderne Algebra. Unter Benutzung von Vorlesungen von E. Artin und E. Noether*, **23-24**, Springer, Berlin.

B. L. van der Waerden (1976), "Hamilton's Discovery of Quaternions". *Mathematics Magazine*, **49** (5), pp. 227–234.

Index

© Springer Nature Switzerland AG 2021   459
A. Chambert-Loir, *(Mostly) Commutative Algebra*, Universitext,
https://doi.org/10.1007/978-3-030-61595-6

Printed in the United States
by Baker & Taylor Publisher Services